Photochemistry

Volume 34

A Specialist Periodical Report

Photochemistry

Volume 34

A Review of the Literature Published between
July 2001 and June 2002

Senior Reporter
I. Dunkin, *University of Strathclyde, Glasgow, UK*

Reporters
N.S. Allen, *Manchester Metropolitan University, UK*
W.M. Horspool, *University of Dundee, UK*
A. Gilbert, *Department of Chemistry, University of Reading, UK*

RS•C
advancing the chemical sciences

NEW FROM 2003

If you buy this title on standing order, you will be given FREE access to the chapters online. Please contact sales@rsc.org with proof of purchase to arrange access to be set up.

Thank you.

ISBN 0-85404-440-X
ISSN 0556-3860

A catalogue record for this book is available from British Library

Published by The Royal Society of Chemistry,
Thomas Graham House, Science Park, Milton Road, Cambridge CB4 0WF, UK

Registered Charity Number 207890

For further information see our web site at www.rsc.org

Typeset by Vision Typesetting, Manchester, UK
Printed and bound by Athenaeum Press Ltd, Gateshead, Tyne & Wear

Contents

Introduction and Review of the Year

BY IAN R. DUNKIN

The first things to note in this latest volume of *Photochemistry* are several changes amongst the reporters. After many successful years as the Senior Reporter, Andrew Gilbert has finally passed on this role to me, though he is continuing to contribute individual chapters. I am sure I speak for all regular users of these volumes in thanking him for maintaining the series at such a high level. Secondly, we must also say goodbye and thank you to Albert Pratt and Alan Cox, both of whom have decided to retire as reporters, also after many years of devoted service. Finally, we are missing Anthony Harriman's chapter on photophysical processes this year, owing to pressing personal reasons. The overall result is a rather slimmer volume than normal, but we shall be welcoming new contributors for volume 35, and it is hoped that any gaps in the coverage this year will be remedied at that time.

In addition to changes amongst the authors, we are looking forward to a new era in the publication of Specialist Periodical Reports. These will soon become available electronically, and with a considerably faster time to press than has been possible hitherto. We all hope that this development will assist our readers in keeping more up-to-date than ever with the photochemical literature.

A selection of photochemical highlights of the period reviewed in this volume will be at least as subjective this year as under the previous editor. On reading the various chapters, I was first of all struck by the number of high-yielding synthetic reactions which have been reported. So many photochemical reactions seem to have disappointingly poor chemical yields or quantum yields, or give rather too many by-products; so that it is pleasing to see a selection of efficient photoreactions which can compete with the best thermal processes in the design of syntheses. I was also struck by the growing number of solid-state photoreactions reported. These often profit from the constraints imposed by the crystal environment in providing highly selective reactions, again with synthetic potential. To these, we can add studies of photochemistry in zeolites, in supercritical media or in ionic liquids, all of which promise to deliver benefits from carrying out the reactions in media with very specific properties. So far, reports of photochemistry in supercritical fluids or ionic liquids have not been very numerous, but one senses here an area which is likely to grow significantly in the coming years.

My review this year will therefore concentrate on these themes, and I offer my apologies to those who would prefer to read about developments in, for example,

Photochemistry, Volume 34

mechanistic studies, environmental photochemistry or new polymerization techniques. No doubt these themes will be picked up in future volumes.

A good example of a high-yielding photoreaction of synthetic utility has been reported by Kadota and Ogasawara (Chapters 1 and 6), who have studied the photochemical decarbonylation of oxabicyclo[3.2.1]octanone derivatives as a key step in a convenient synthesis in reasonable yields of pentose and hexose sugars. Ketone (1), for example, loses CO to give (2), which can be hydrolysed to D-ribose (3). Nicolaou *et al.* (Chapters 1 and 5) have studied very efficient ring-closure reactions based on Norrish Type II pathways, leading to the final products via intermediate photoenols. This approach is exemplified by the conversion of (4) into (6) via (5) in 95% yield, and has been extended to the synthesis of some natural products of the hamigeran family, *e.g.* (7), which were also obtained in >90% yield.

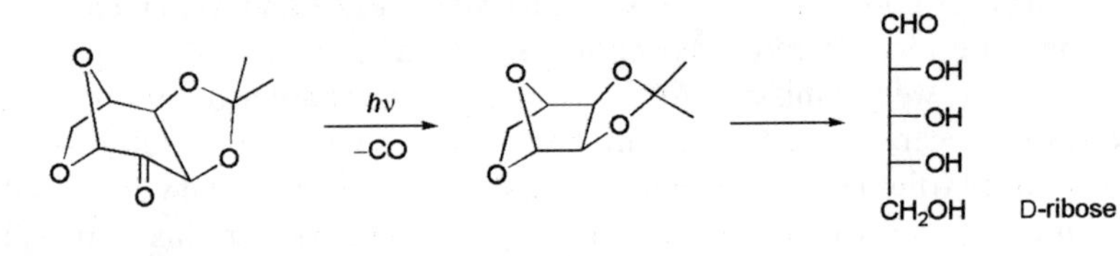

(1) (2) (3)

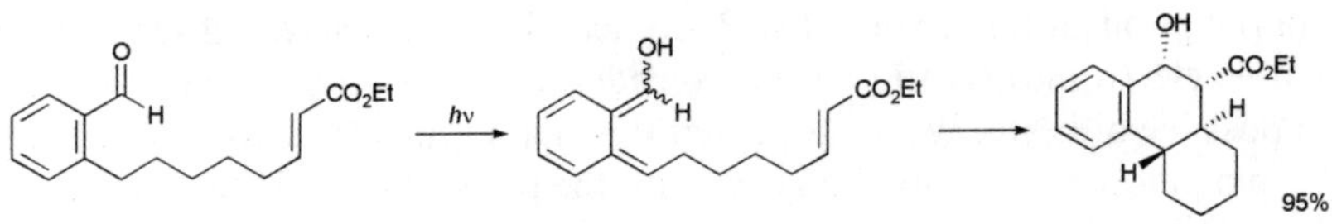

(4) (5) (6)

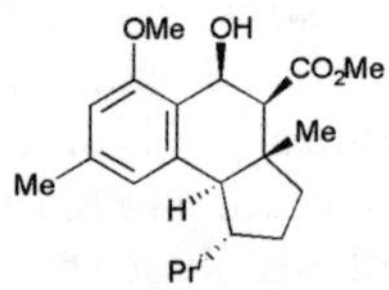

(7)

Lu *et al.* (Chapter 4) have exploited a sequential photoisomerization-aziridine ring opening of pyridine salts based on earlier studies (Ling *et al.*) to provide a successful synthetic approach to aminocyclopentendiols, *e.g.* (8). This reaction also furnished a key intermediate in a concise synthesis of (−)-allosamizoline (9). A photocatalysed procedure has been developed for the synthesis of 2-substituted pyridines (Heller *et al.*, Chapter 4). The pyridines are formed in a (2+2+2) cycloaddition of two molecules of acetylene with one of a nitrile, which is induced by 350-500 nm radiation in the presence of a precatalyst such as η^5-cyclopentadienyl-η^4-cyclooct-1,5-diene-cobalt. This reaction proceeds at room temperature and under atmospheric pressure – very mild conditions compared with those required for thermally initiated variants of the process. Efficient photoaddition of *E*-1,2-dichloroethene to the enone (10) gave the

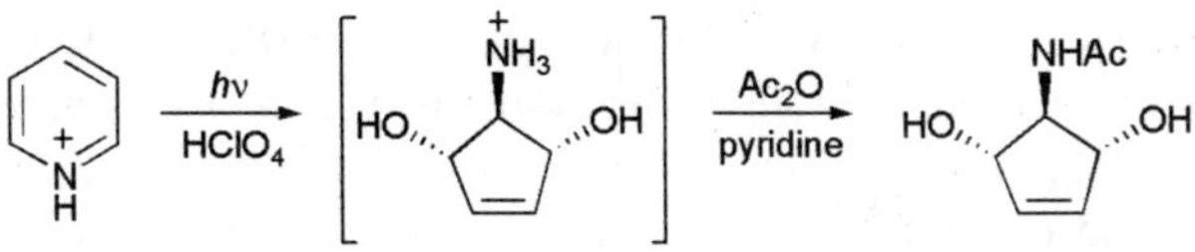

(8)

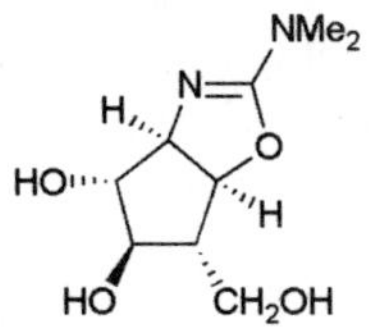

(9)

stereoisomeric adducts (11) in 95% yield (Mehta and Sreenivas, Chapter 2). The latter were key intermediates in a new synthesis of the sesquiterpene, sterpurene.

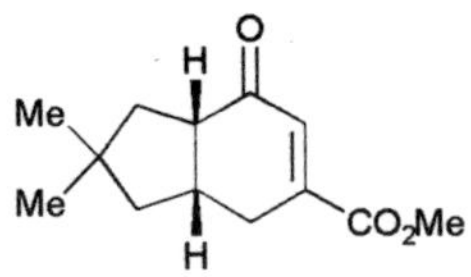

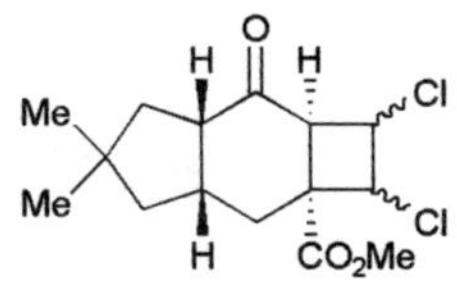

(10) (11)

The addition of singlet oxygen, generated by tetraphenyl porphyrin sensitization, to cycloheptatriene (12) gave a 90% yield of endoperoxide (13), which, on treatment with triethylamine, afforded a new isomer of stipatatic acid (14) (Dastan *et al.*, Chapter 5). A new synthetic method has been reported for the formation of carbonyl compounds from corresponding alcohols (Itoh *et al.*, Chapter 5). The process involves irradiation of di-*iso*-propyl ether solutions of the alcohol in the presence of iodine; *e.g.* cinnamaldehyde was obtained in 99% yield from cinnamyl alcohol. A building block for Manumycin antibiotics (Adam *et al.*, Chapter 5) has been prepared in a sequence of photo-oxygenation, reduction and Weitz-Scheffer epoxidation. The first step is the reaction of the readily accessible anilide (15) with singlet oxygen to give (16), which is then converted into the required hydroxy-epoxy-enone (17).

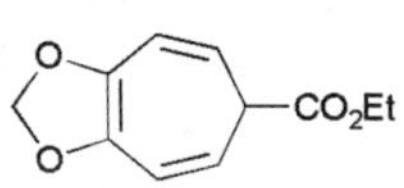

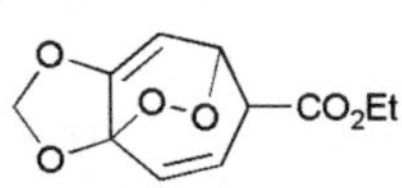

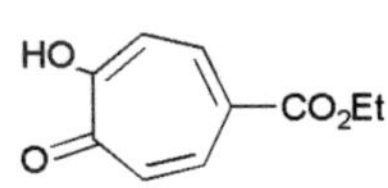

(12) (13) (14)

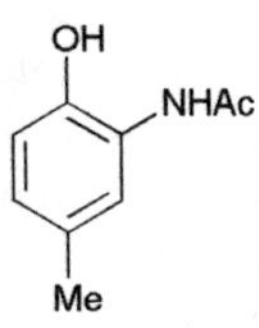

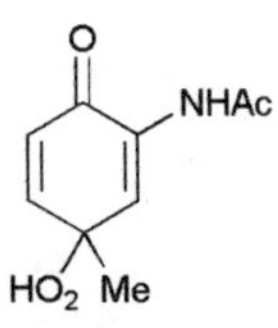

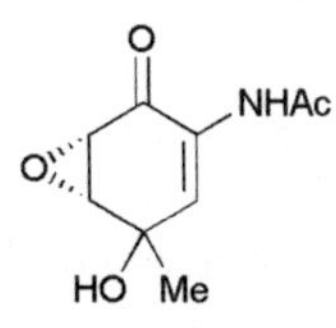

(15) (16) (17)

An interesting paper has appeared on the subject of the orthogonality of photolabile protecting groups, *i.e.*, the possibility of selectively removing two or more such groups in any chosen chronological order (Bochet, Chapter 6). For this approach to succeed, the two groups must behave as independent chromophores with well separated absorptions. The mixed diester (18) showed how this might be achieved reasonably successfully: 254 nm irradiation for 5 minutes in MeCN selectively cleaved the dimethoxybenzoin ester to give (19), which was isolated as its methyl ester in 70% yield; while 420 nm irradiation for 24 hours in MeCN selectively cleaved the nitrobenzyl derivative, giving (20), also isolated as its methyl ester in 70% yield.

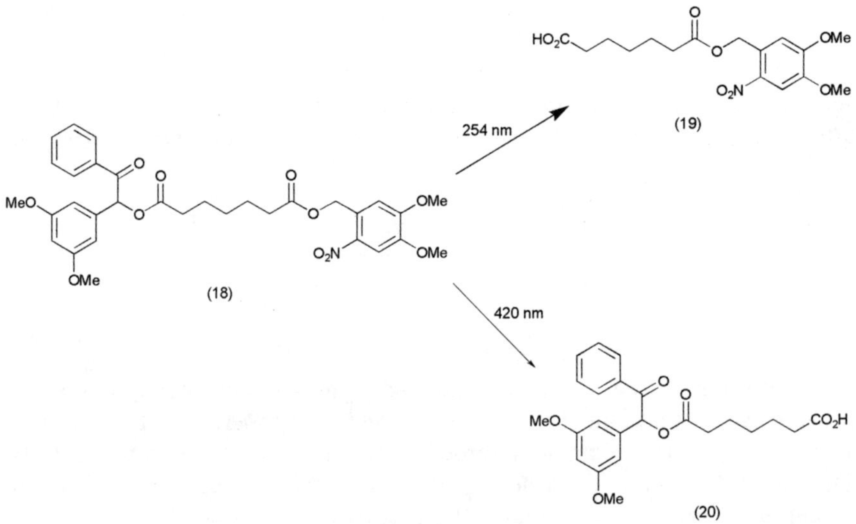

Although *o*-alkylbenzaldehydes give complex mixtures of irradiation products, solid-phase irradiation is much more specific (Moorthy *et al.*, Chapter 1), yielding cyclobutenols, *e.g.* (22) from (21). Koshima *et al.* (Chapter 1) have studied the enantiospecific cyclization of a diisopropyl 4-carboxybenzophenone in the solid phase, controlled by the presence of (*S*)-phenylethylamine. The product, the (*R*)-(+)-cyclobutenol (23), was formed in a Norrish Type II H-abstraction process. Kang and Scheffer (Chapter 1) have shown that the ketone (24) in the solid state forms diastereomeric oxetanes (25) on irradiation, whereas in solution the usual Norrish Type I reactvity is observed.

Hasegawa *et al.* (Chapter 2) reported a highly efficient dimerization of the enone (26). Pyrex-filtered UV irradiation for one hour resulted in quantitative conversion into the dimer (27). The dopamine (28) undergoes a 2+2 cycloaddition in which the enone carbon-carbon double bond in one molecule adds to a benzene ring in another (Ito *et al.*, Chapter 2). This seems to be the result of the way the molecules line up in the crystal. Scheffer and Wang (Chapter 2) reported Norrish Type II reactivity of the amide (29) in crystals prepared with the amines (*R*)-(+)-1-phenylethylamine and L-(−)-prolinamine. Irradiation afforded single products, which were converted to the ester (30) with ee >99%.

The quantum yield for photocyclization of (31) in the crystal has been measured as unity (Shibata *et al.*, Chapters 3 and 4). This high value is attributed to

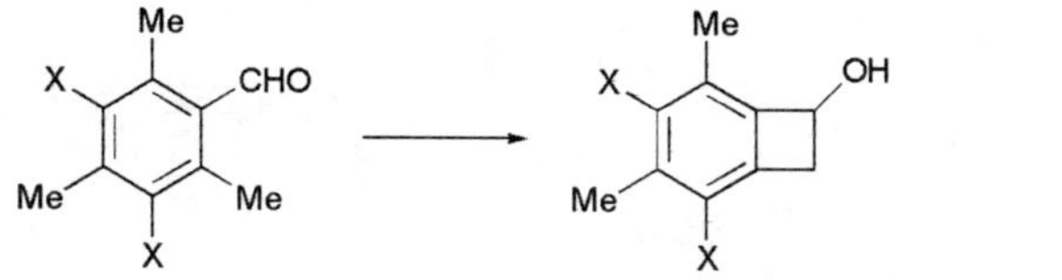

(21) (22) X = H, Br, CN, CHO

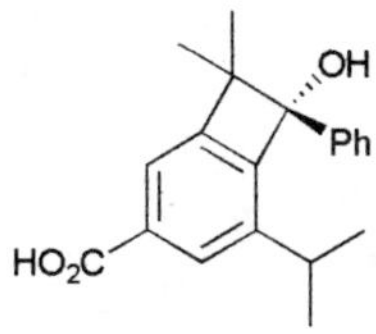

(23)

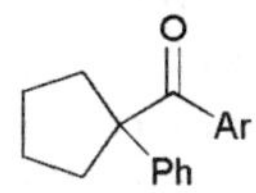

(24)

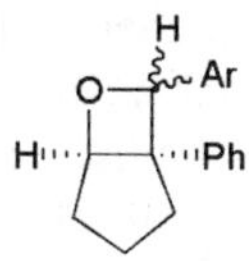

(25)

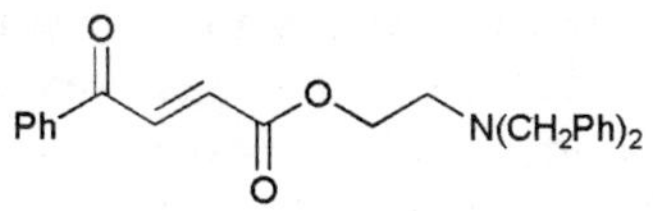

(26)

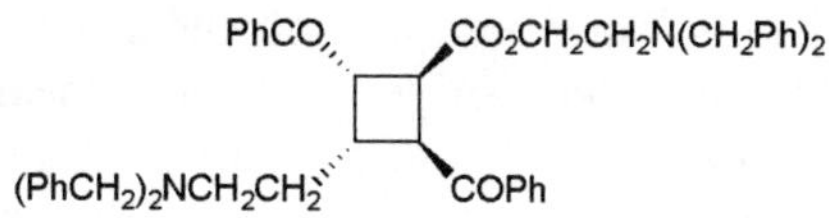

(27)

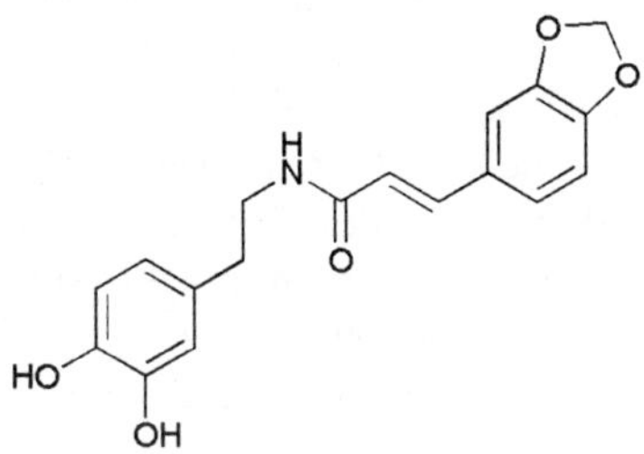

(28)

the conformational restraint within the crystal, such that only the photoactive anti-parallel conformers are packed in the crystal, whereas more than one conformer is present in solution. Some analogues of (31) showed similar behaviour. The stilbene derivative (32) has been co-crystallized with 1,8-naphthalenedicarboxylic acid, which acts as template, with H-bonding to the pyridine nitrogens. The C=C bonds of the two stilbene molecules thus lie in close proximity, and 300 nm irradiation of the crystals results in 100% stereospecific conversion into the corresponding cyclobutane (Papaefstathiou *et al.*, Chapter 3).

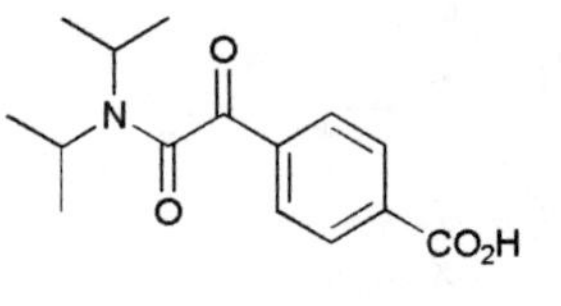

(29)

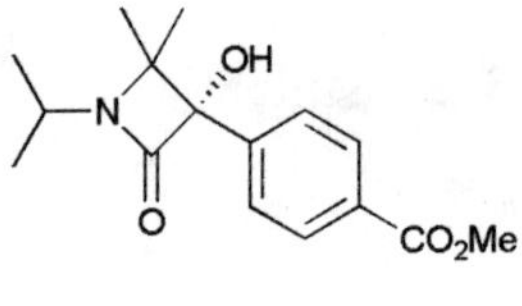

(30)

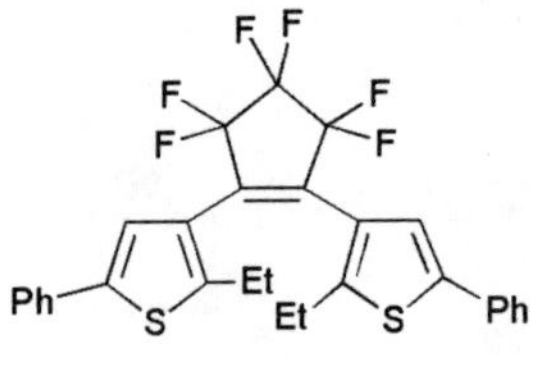

(31)

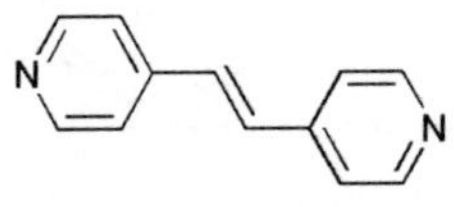

(32)

Besides solid-state photochemical reactions, there have been a significant number of studies of photochemistry in zeolites and clays, though most of these have concentrated on mechanistic investigations rather than synthetic goals. Turro *et al.* (Chapter 1), for example, have studied by EPR the persistent radicals formed in zeolites on photolysis of a series of dibenzyl ketones. Two type of Ti-β zeolites have been synthesized which display photocatalytic activity for reduction of CO_2 with water at 323 K, producing methane and methanol (Ikeuee *et al.*, Chapter 5).

The investigation of photochemistry in supercritical media is an interesting field, which has not yet been well researched, probably owing to the need for specialized high-pressure reactors. Tankut and Pacut (Chapter 1) have examined the Norrish Type I behaviour of the ketone (33) in supercritical CO_2, and an enhanced cage effect was detected near the supercritical pressure. In a continuation of mechanistic investigations on the photo-extrusion of N_2 from diazabicylo[2.2.1]hept-2-ene (DBH), it was found that the stereoselectivity of the formation of (35) from the stereolabelled DBH derivative (34) in supercritical media (sc-CO_2, sc-C_2H_6) decreased by a factor of about 2.3 when the pressure was increased up to 200bar. According to the authors, this observation is best accounted for in terms of the pressure-dependency of the viscosity of the reaction medium, and the observations imply a stepwise denitrogenation (Adam *et al.*, Chapter 6).

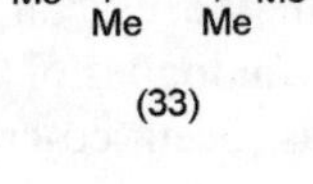

(33)

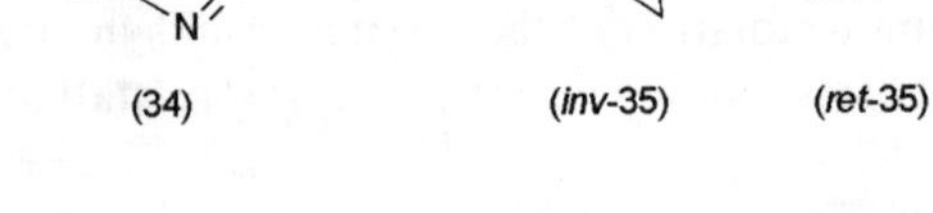

(34) (*inv*-35) (*ret*-35)

Ionic liquids (also known as room-temperature molten salts) have received much attention recently, because of their potential uses as 'clean' solvents as well as some more specific advantages that they have as reaction media. Photochemical studies exploiting these materials are still rare, however. An example of the likely influence of such media on photoreactions has been reported by Ozawa and Hamaguchi (Chapter 3), who have discovered that the photoisomerization of *trans*-stilbene proceeds with a greater rate than expected in the ionic liquid 1-butyl-3-methylimidazolium hexafluorophosphate.

Finally, some interesting results in polymer chemistry, depending on molecular organization, have been published. For example, photopolymerization of acrylamide has been correlated with monomer organization in various phases of a liquid crystal template (Lester and Guymon, Chapter 7). The polymerization rate is enhanced in the more ordered phases, owing to orientation of the monomer molecules. The propagation rate of MMA has also been measured in miniemulsions (Jung *et al.*, Chapter 7): in this case the molecular weight exhibits spikes or peaks due to pulses in the initiation process, where compartmentalization takes place. Liquid crystalline poly(ester imides) with cinnamoyl groups form Grandjean textures with a blue iridescence on annealing, and this is frozen in after photocuring (Sapich *et al.*, Chapter 7).

The last few paragraphs show how greater selectivity can be gained by carrying out photochemical reactions in solid phases or in other ordered or special environments. Studies in these fields are likely to mushroom as the technologies become better known and more widely available.

1
Photolysis of Carbonyl Compounds

BY WILLIAM M. HORSPOOL

1 Norrish Type I Reactions

Formaldehyde undergoes photochemical decomposition in the 269 to 339 nm range in the gas phase. There are various dissociation paths for this molecule, affording hydrogen atoms and CHO radicals, CO and hydrogen atoms and hydrogen atoms and CO. The quantum yields for the processes were measured.[1] The photochemical decomposition by Norrish Type I reactivity of propionaldehyde has been studied in the 280-330 nm range.[2] Again the formation of CHO radicals was detected.

The multi-photon ionization processes arising within propanone in the irradiation range of 243-263 nm have been studied. The ionization processes that were detected arise from within the S_1 and T_1 states.[3] Photodissociation (243 nm) of propanone, ethanal and ethanoic acid brings about release of hydrogen atoms. These were detected using two-photon absorption and induced fluorescence.[4] Studies of propanone decomposition in air have been used to assess possible dissociation processes in the troposphere.[5]

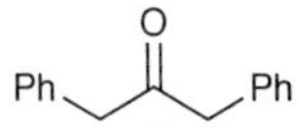

(1)

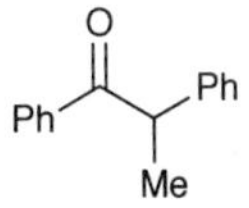

(2)

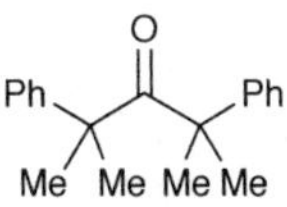

(3)

The stimulated nuclear polarization spectra from irradiation of the ketones (1) and (2) has been reported.[6] A study of the Norrish Type I behaviour of the ketone (3) in supercritical CO_2 has been reported, and an enhanced cage effect has been detected near the supercritical pressure.[7] Turro and his co-workers[8] have carried out a detailed EPR study of the persistent radicals formed on photolysis of the dibenzyl ketones (4) in zeolites. Some aspects of supramolecular chemistry have been reviewed.[9] A short review has highlighted some of the research carried out in zeolites, focusing particularly on the exploitation of triplet-triplet energy transfer.[10] A CIDNP study of the photochemical Norrish type I processes brought about by irradiation at 308 nm in of the two ketones (5) and (6) has been reported.[11]

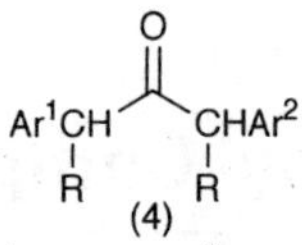

(4)

Ar¹	Ar²	R
Ph	Ph	H
C_6D_5	C_6D_5	H
o-MeC_6H_4	Ph	H
o-MeC_6D_4	Ph	H
o-MeC_6H_4	C_6D_5	H
o-MeC_6H_4	*o*-MeC_6H_4	H
Ph	Ph	Me
o-MeC_6H_4	Ph	Me
C_6D_5	C_6D_5	Me
Ph	Ph	Et
Ph	Ph	C_5H_{11}

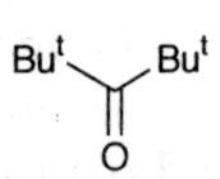

(5)

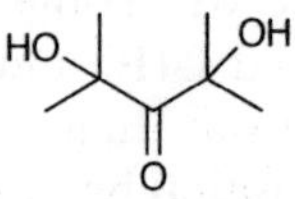

(6)

The study of some benzylbenzoin benzyl ethers has shown that they undergo Norrish Type I fission, affording benzoyl and benzyloxybenzyl radicals.[12] The intermediates were characterized by laser flash photolysis. Previously the Norrish Type I fission reactions of ketones related to (7) had been reported; further work has shown that irradiation (305 nm in water-methanol) of (7) brings about its conversion into (8) in 94% yield.[13a] The reaction sequence was also demonstrated in oligonucleotides.[13b] Norrish Type I fission also occurs in systems like the cyclophane dione (9).[14] This brings about sequential decarbonylation to yield

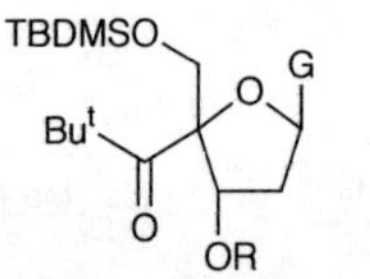

(7) R = $P(O)(OEt)_2$, G = guanine

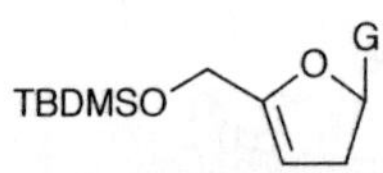

(8)

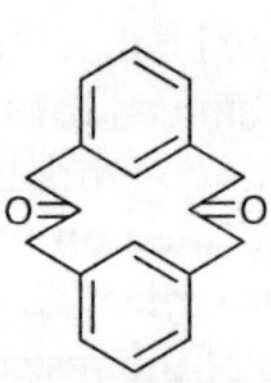

(9)

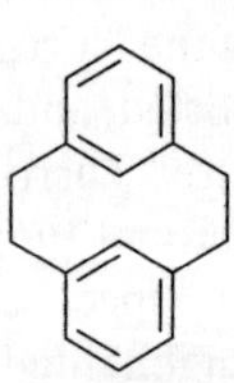

(10)
20 h 95%
4 h 9%

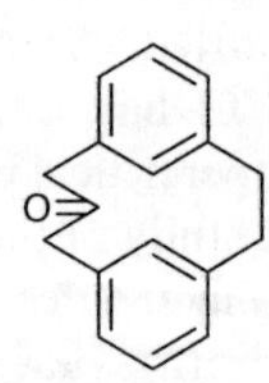

(11)
–
57%

the cyclophane (10) and the monoketone (11). Proof of the sequential nature of the reaction was demonstrated by the decarbonylation of (11) to yield (10). The time-dependency of the irradiations are shown below the structures. The α-fission of the ketone (12) affords the ring-opened ester (13) in 57% yield when the irradiations are carried out in methanol.[15] The reaction is a conventional process

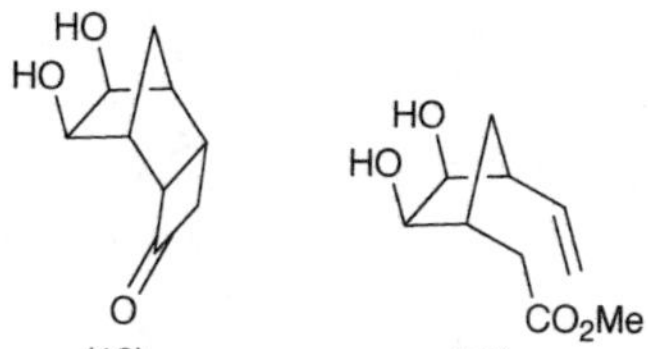

(12) (13)

and affords a ketene as a result of fission in the resultant 1,4-biradical produced by photochemical fission of the α-bond. Another facet of the Norrish Type I reaction is ring expansion of a cyclobutanone to a dihydrofuran. This process has been used by Lee-Ruff and co-workers[16] in the photochemical ring expansion of (±)-3-[2′-(benzoyloxy)ethyl]-2,2-dimethylcyclobutanone. This has been used as a route for the synthesis of 2′,3′-dideoxynucleosides based on the apiose family.

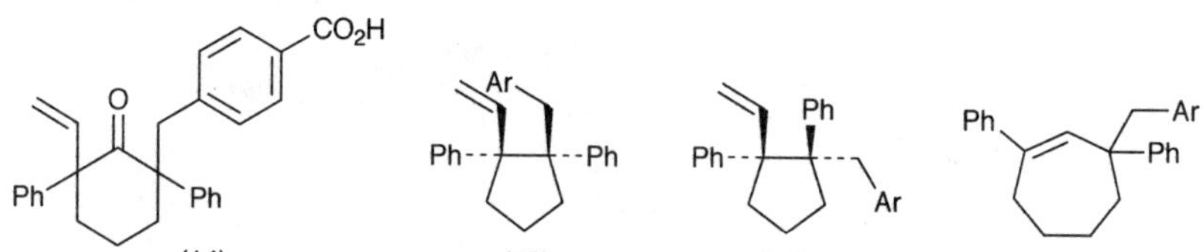

(14) (15) (16) (17)

Larger ring ketones undergo decarbonylation, as has been described following the irradiation of the cyclohexanone derivative (14) as a dilute solution in benzene with $\lambda > 300$ nm.[17] The resultant biradical produced by the decarbonylation undergoes ring closure to give a mixture of the isomeric cyclopentanes (15) and (16) as well as the ring expanded product (17). Interestingly the compound (14) is unreactive in the crystalline phase. The authors[17] reason that the failure to decarbonylate is a result of deactivation of the carbonyl excited state by interaction with the proximate benzyl group. Kadota and Ogasawara[18] have described the photochemical decarbonylation of cyclic ketones containing the bicyclo[3.2.1]octane skeleton (Scheme 1). This process, carried out in methanol with Pyrex filtered light, provides reasonable yields of the ring-contracted compounds shown. These products can be readily converted in high yield into the pentose and hexose sugars illustrated.

The irradiation of the esters (18) results in a Norrish Type I fission, rupture of the ester carbonyl-O bond, with the formation of the xanthenyl radical and the corresponding formyl radical.[19] The reactivity of these species was investigated, and some of the results obtained are shown in Scheme 2, where the principal process is shown to be cyclization of the unsaturated formyl radicals to yield a lactone or lactones. The yields obtained can be variable as indicated. Other products such as the alcohols (18) and formates (20) are also produced.

2 Norrish Type II Reactions

2.1 1,5-Hydrogen Transfer.

2.1 1,5-Hydrogen Transfer. – While solution phase photochemistry of *o*-alkyl-

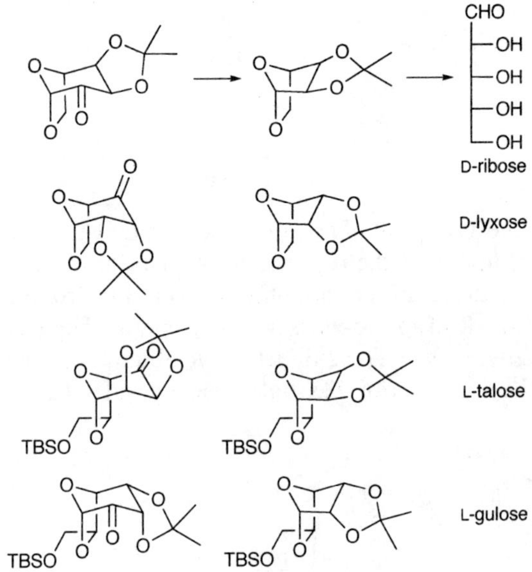

Scheme 1

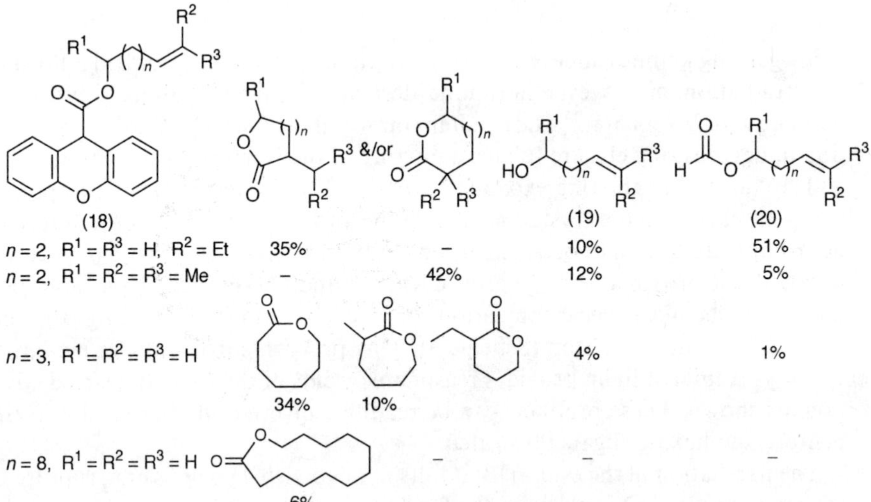

Scheme 2

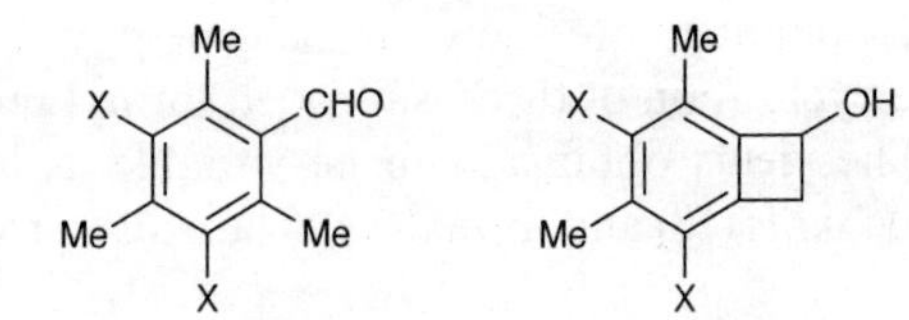

(21) X = H, Br, CN, CHO (22)

benzaldehydes affords a complex mixture of products, irradiation in the solid phase is much more specific. The aldehydes (21) are all photoreactive in the solid state and give the cyclobutenols (22).[20] Even the liquid aldehyde (23) (Scheme 3),

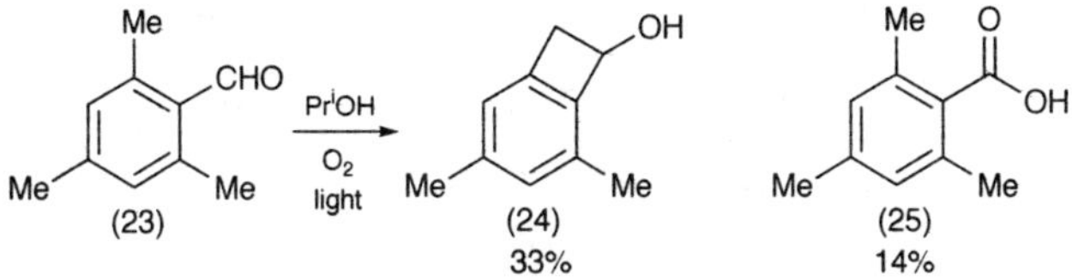

Scheme 3

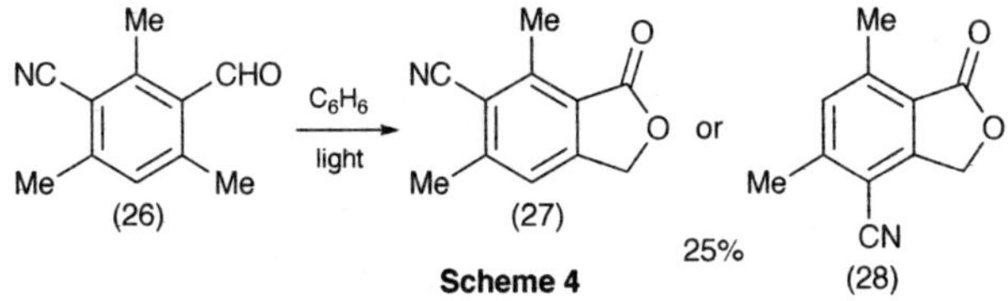

Scheme 4

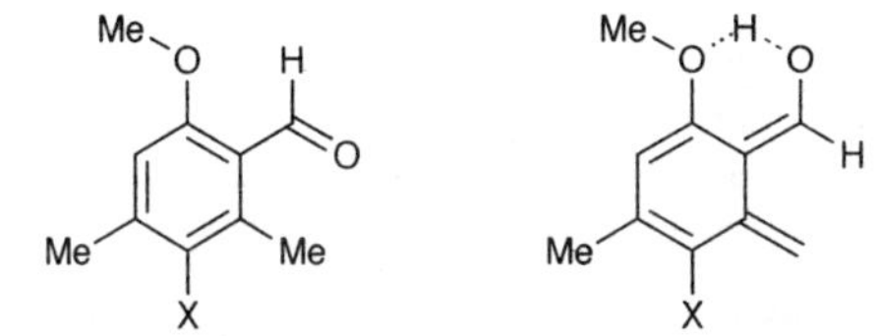

(29) X = Br, CN or CHO (30)

in a solid inclusion complex, is readily converted into the cyclobutenol (24), by a conventional γ-hydrogen abstraction and cyclization within the resultant biradical. The conditions used are aerobic, and oxidation of the aldehyde to the acid (25) occurs in competition with the cyclization. The irradiation of the cyano-substituted aldehyde (26) (Scheme 4) in benzene affords the lactones (27) and (28) in a total yield of 25%.[20] Interestingly, the related anisaldehyde derivatives (29) are all photochromic in the solid state. The reaction involves an intramolecular proton transfer with the formation of the photoenols (30). In the case of the derivative (29, X = CHO), the resultant enol is stated to be 'remarkably stable'.[21] Nicolaou *et al.*[22] have studied the scope of the reaction shown in Scheme 5. Irradiation of (31) follows the Norrish Type II path with the formation of a photoenol (32). This then undergoes intramolecular addition to afford the tricyclic product (33) in high yield. Several examples were reported, such as the cyclization of (34) to afford (35) and of (36) to yield (37). In all cases the yields of products obtained are > 90%. They[23] have extended the study to provide a path to some natural products of the hamigeran family. This was achieved using the cyclization of (38) into (39) as the key step. The photolysis of an adduct obtained from a thermal reaction of benzoquinone and a mixture of isopentafulvenes has been described.[24] The reaction observed on irradiation is a Norrish Type II hydrogen abstraction process followed by cyclization within the resultant biradical. Irradiation of 3β-formyloxy-5α-bromo-6β-hydroxy-21-acetyloxypregnan-20-one has been used as a key step in a synthesis of a pregnenolone derivative.[25]

Efficient decarboxylation of the keto acids (40) to the arylalkyl ketones (41) has been reported following their irradiation.[26] The deuteriated compounds shown demonstrate that there is an intramolecular hydrogen transfer as the key step in the process. The *m*- and *p*-isomers are unreactive.

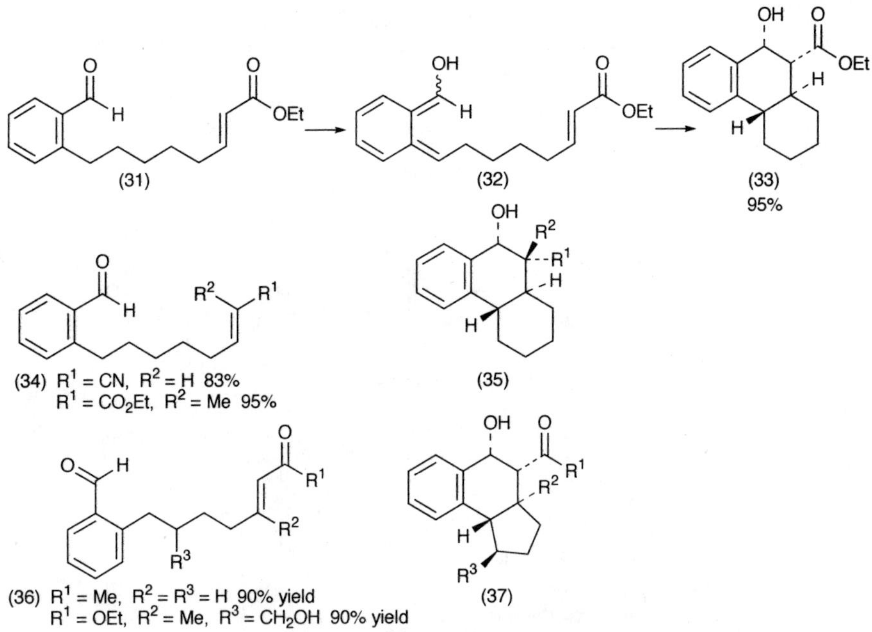

Scheme 5

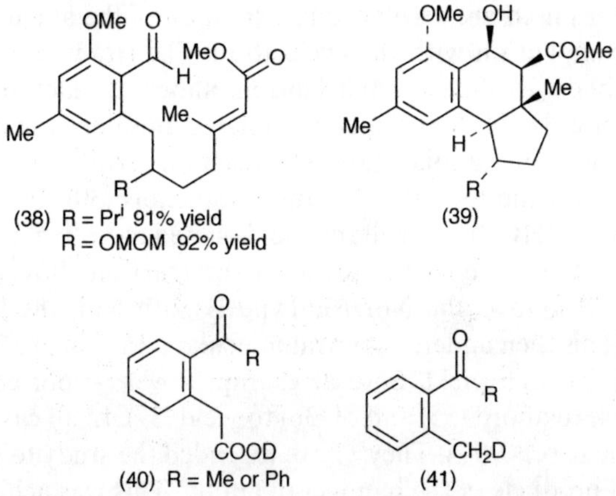

An account of the enantiospecific photocyclization of 2,5-diisopropyl-4′-carboxybenzophenone in the solid phase has been described.[27] The cyclization involves a Norrish Type II hydrogen abstraction and the outcome is controlled by the presence of (*S*)-phenylethylamine. The product obtained is the (*R*)(+)-cyclobutenol (42). The regioselectivity of the Norrish-Yang hydrogen abstraction process of the ketone (43) in the crystalline phase has been examined.[28] The hydrogen abstractions occur at both positions ‘a’ and ‘b’ in the cyclohexane ring. Abstraction from ‘a’ affords (44) while (45) arises from the biradical afforded by abstraction from ‘b’. The selectivity observed depends on the nature of substituents on the aryl ring.

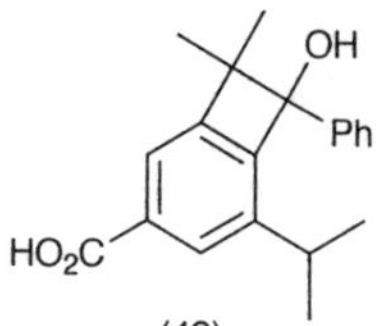

(42)

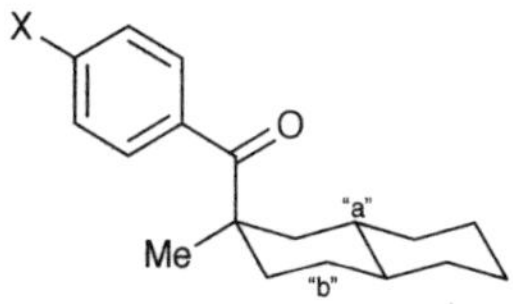

(43)

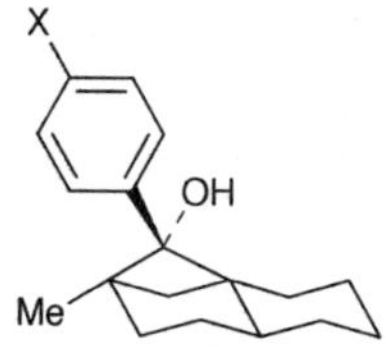

(44)

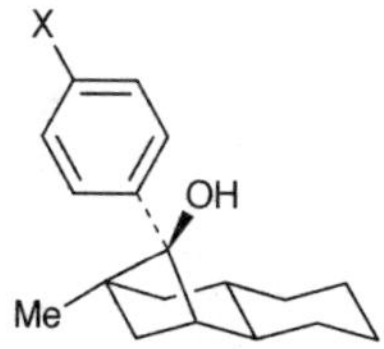

(45)

Others have examined the influence of a variety of media on the Norrish Type II process. For example, the Norrish Type II reactivity of the arylketones (11-mercapto-1-phenylundecanone, 1-[4-dodecylphenyl]-11-mercaptoundecanone, 1-[4-hexylphenyl]-11-mercaptoundecanone, 1-[4-(11-mercaptoundecyl)phenyl]hexanone and 1-[4[(11-mercaptoundecyl)-phenyl]-undecanone) has been studied with the ketones anchored as monolayers on gold nanoclusters.[29] The photochemical efficiency is reasonable, with the cleavage process giving alkenes in 75% yield. The photoreactivity of pentan-2-one when it is include in zeolites has been examined with respect to the influence of the alkali metal cation.[30] Changing the metal ion from Cs^+ to Li^+ brings about a decrease both in the Norrish Type II activity and in the photochemical reactivity. A study has highlighted the usefulness of irradiation in zeolites.[31] The control that these substances exercise on the enantio- and diastereo-selectivity in some reactions was assessed. A short review has highlighted examples of time-related chirality memory in some photochemical reactions involving Norrish-Yang cyclizations.[32] Norrish type II reactivity has been observed on photolysis at wavelengths > 300 nm of poly(4′-ethoxyacrylophenone).[33] The reactivity arises from the triplet state of the carbonyl function, but the quantum yield for the process is lower than that observed in solution. The results from a study of temperature dependent photochemical reactions in a microwave field have been reported.[34] Griesbeck and Heckroth[35] have carried out a detailed study of the photochemical reactivity of a series of ketoamines (46). These undergo a variety of reactions dependent to some extent on the nature of the substitution. The two principal reactions are either Norrish Type II hydrogen abstraction or Norrish Type I fission. The fate of the former reaction is the formation of a biradical that either undergoes cyclization to yield (47) or fission of the 1,4-biradical to afford (48). The Norrish Type I process affords the amines (49). Norrish Type II reactivity is also shown in the propenamide derivatives (50).[36] In this section, only the hydrogen abstraction process will be considered, and elsewhere (Part II, Chapter 2) the intramolecular cycloaddition reactions will be discussed. The three examples shown are induced by benzene sensitization and they all undergo Norrish type II hydrogen abstraction with the formation of the azetidines (51) in moderate to poor yields.

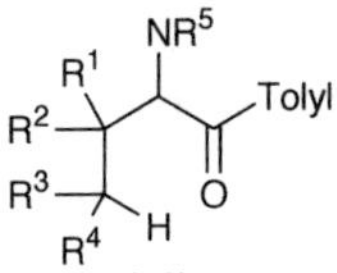
(46)
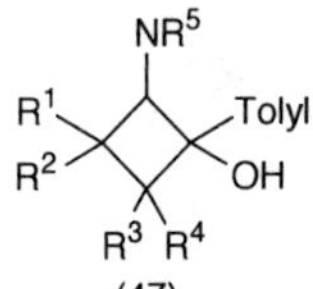
(47)
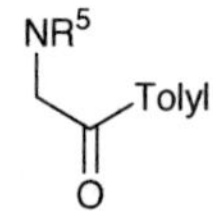
(48)
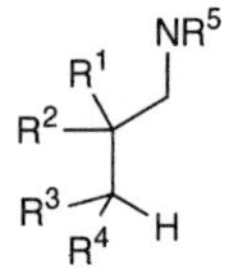
(49)

	R^1	R^2	R^3	R^4	R^5	cyclization	cleavage (type II)	cleavage (type I)
a	H	H	H	H	HAc	–	33	67
b	H	H	Me	H	HAc	33	40	~
c	H	H	Et	H	HAc	48	35	12
d	H	H	Me	Me	HAc	49	35	15
e	Me	H	H	H	HAc	74	8	18
f	Me	Me	H	H	HAc	50	30	20
g	Me	H	Me	H	HAc	–	>95	5
h	Me	H	H	H	HBz	65	13	22
i	Me	H	H	H	Succ	65	30	5
j	Me	H	H	H	HTfAc	–	20	40
k	H	H	SMe	H	HAc	–	65	–

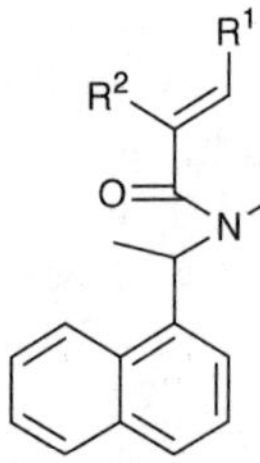
(50)
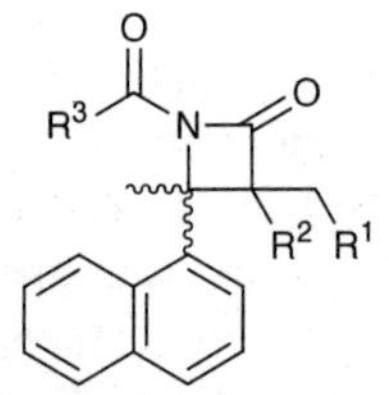
(51)

	R^1	R^2	R^3	yield (%)
a	H	H	Me	59
b	H	Me	Me	23
c	H	Me	Ph	46

2.2 Other Hydrogen-Transfer Processes. – Hydrogen abstraction can also be brought about at sites other than the γ-position. Wessig and his co-workers described an example of this in last year's account.[37] This work involved a new route to cyclopropanes. A further report of this type of reactivity has been made.[38] Succinimido and glutarimido substituted glycosans have been shown to undergo Norrish-Yang type cyclizations on irradiation at 254 nm.[39] δ-Hydrogen abstraction occurs on excitation of the diketones (52). Cyclization within the resultant biradical provides a convenient route to the oxazinone derivatives (53).[40] A further account of the intramolecular hydrogen abstraction processes within the cyclophanes (54), with a variety of linkers, has been published.[41] The irradiation brings about the conversion into the products (55) by a 1,6-hydrogen transfer. The yields are variable and these are shown below the structures. Park and his co-workers have reported a further account of such cyclizations using excitation at 350 nm in benzene.[42] These results are shown in Scheme 6. As can be seen, excitation results in δ-hydrogen abstraction from the side chains, and the resultant 1,5-biradicals undergo ring closure to yield the diols. These products are readily dehydrated to afford the difuran derivatives in 40% overall yield. These compounds were used to synthesize novel cyclophanes. A mechanistic

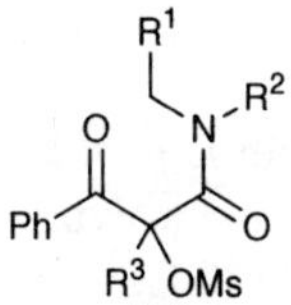

(52)

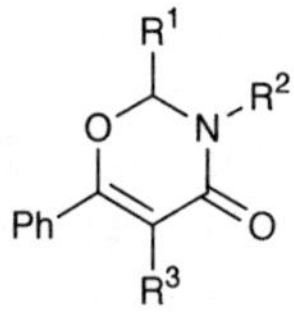

(53)

	R^1	R^2	R^3	yield (%)
a	Me	Me	H	44
b	CH_2-CH_2		H	58
c	$(CH_2)_3$		H	42
d	$(CH_2)_4$		H	57
e	$(CH_2)_5$		H	64
f	$CH_2OCH_2CH_2$		H	42
g	$CH(Me)-CH(Me)CH_2CH_2$		H	73
h	Pr^i	Pr^i	H	43
i	$(CH_2)_3$		Me	44

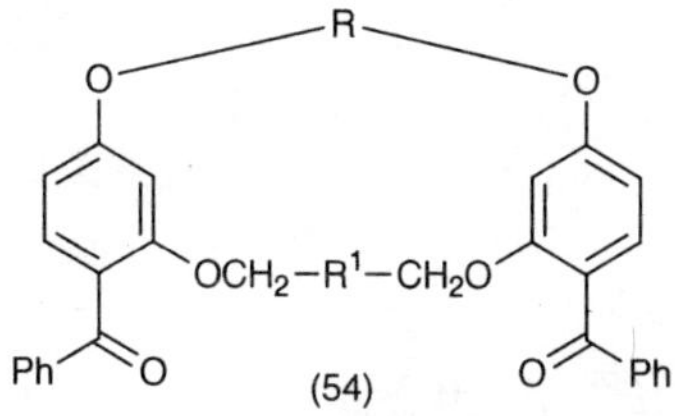

(54)

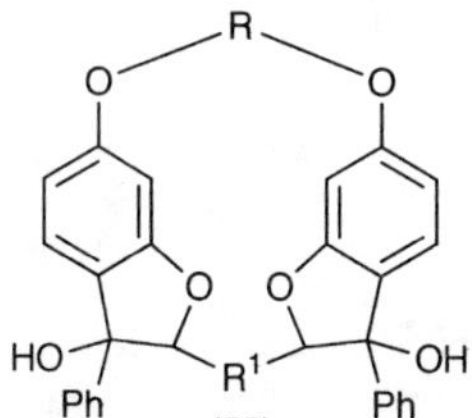

(55)

R	R^1	yield (%)
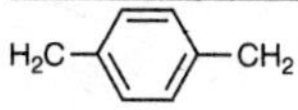	$(CH_2)_4$	50
	$(CH_2)_6$	40
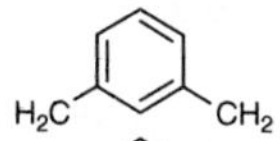	$(CH_2)_4$	55
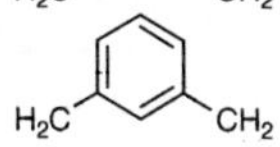	$(CH_2)_{10}$	58
$(CH_2)_6$	$(CH_2)_6$	32
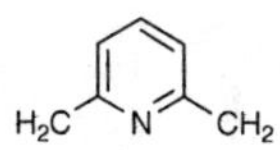	$(CH_2)_{10}$	20

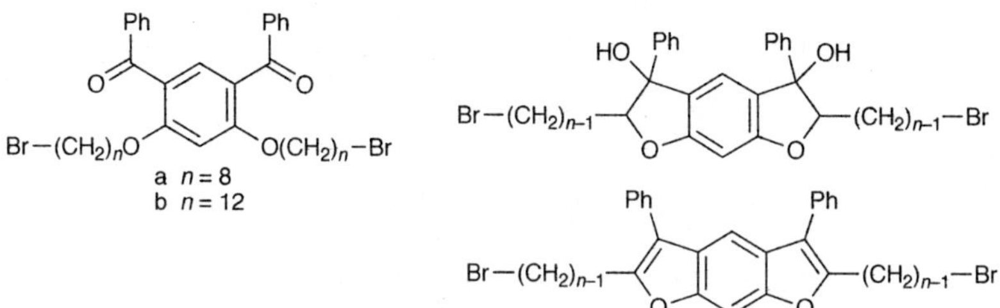

Scheme 6

study of the photochemical behaviour of a series of ring-substituted benzyl alkanoates has been reported.[43]

Bochet[44] has reviewed the area of photolabile protecting groups. Cano, Ladlow and Balasubramanian[45] have described a polymer-linked system for the protection of amino acids. The systems are illustrated by (56): its irradiation affords good yields of the free acid (57).

3 Oxetane Formation

The photochemical addition of aldehydes and ketones (58) to the alkenols (59) has been described.[46] The reactions show marked regio- and diastereo-selectivity

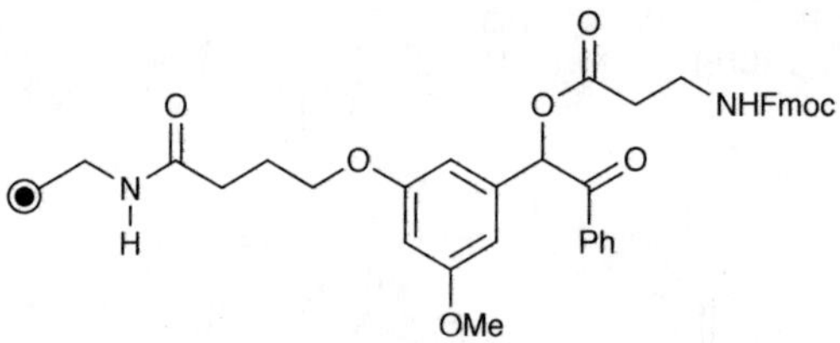

(56)

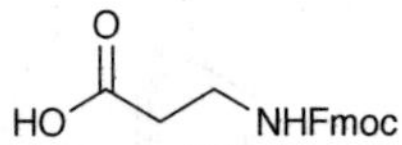

(57)

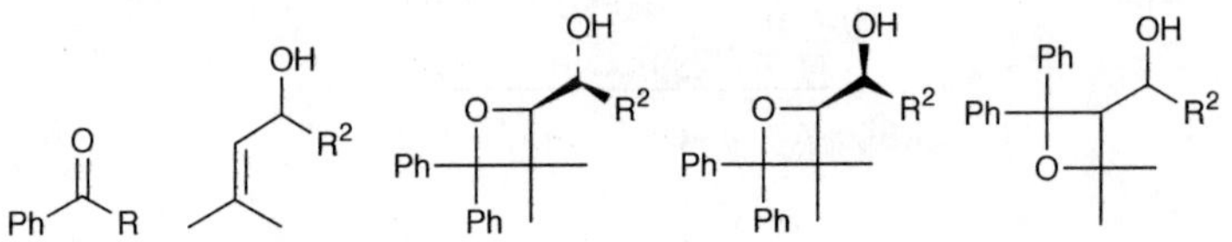

(58) (59)

R	R^2	regio ratio	diastereoselectivity *threo* : *erythro*
Me	Me	95 : 5	82 : 8
Me	Et	95 : 5	88 : 12
Me	Pr^i	95 : 5	92 : 8
Me	Bu^t	95 : 5	>95 : 5
H	Me	67 : 33	72 : 28
H	Et	65 : 35	81 : 19
H	Pr^i	60 : 40	91 : 9
H	Bu^t	59 : 41	>95 : 5

Scheme 7

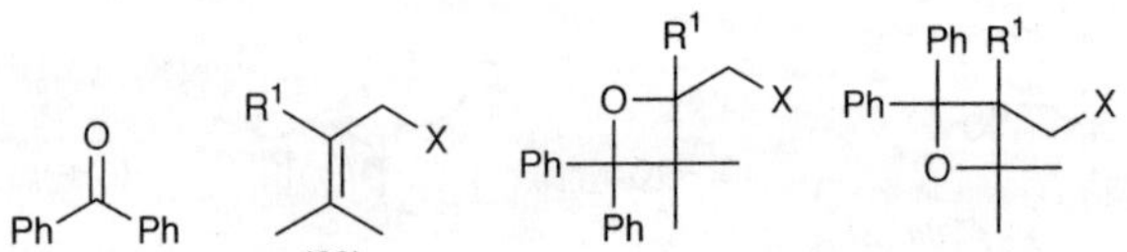

(60)

	R^1	X	ratio
a	H	OH	95 : 5
b	H	H	84 : 16
c	Me	OH	61 : 39
d	Me	OAc	55 : 45

Scheme 8

(Scheme 7). A companion study (Scheme 8) has examined the results from the addition of benzophenone to the derivatives (60). Here the influence of the presence and absence of the hydroxy function on the outcome of the reaction was established. This is seen to best effect with the alkenols (60c,d), where acetylation of the hydroxy group (60c) virtually eliminates the regioselectivity. Others[47] have demonstrated that pyrylium and thiapyrylium salts induce photochemical electron transfer from an oxetane to yield the resultant radical cation (61). This undergoes collapse to the radical cation of *trans*-stilbene.

Two oxetanes (62) and (63) are formed on photochemical addition of aryl aldehydes (*p*-CNC_6H_4, phenyl and 2-naphthyl) to the ketene acetals (64).[48] There is a marked regioselectivity, with a ratio of (62):(63) of 95:5. Within the major product (62) the ratios of *trans:cis* are as shown below the structure. A short review of the above additions has also been published.[49]

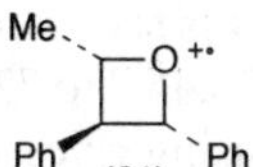
(61)

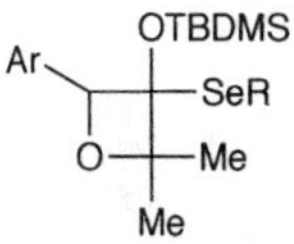
(62)
t/c
68:32
62:38
68:22
86:14

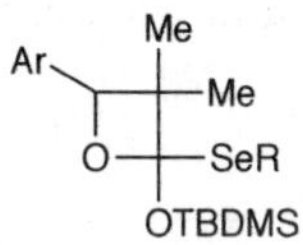
(63)

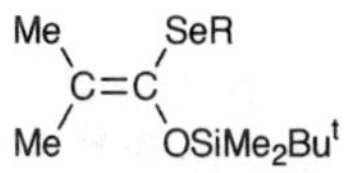
(64) R = Ph or Me

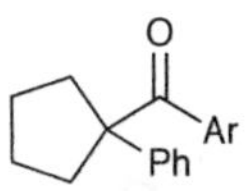
(65)

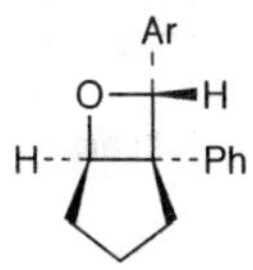
(66)

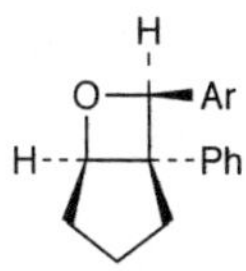
(67)

Kang and Scheffer[50] have studied the photochemical behaviour of the ketone (65) in the solid state. Irradiation brings about the formation of the two oxetanes (66) and (67). Interestingly this behaviour is markedly different from that of (65) in solution, when the usual Norrish Type I reactivity is observed. This is also thought to be the case in the crystal. Thus irradiation essentially affords an aldehyde trapped close to the phenylcyclopentene. Photochemical addition affords the two products (66) and (67). Adam and his co-workers[51] reported the photochemical addition of benzophenone to both *cis*- and *trans*-cyclooctene last year. Within this study they uncovered a remarkable temperature effect on the formation of the oxetane products. A further study has looked at this reaction again.[52] Full details have been reported of the control observed in the oxetane forming reaction between the isomers of the cyclooctenes (68) and benzophenone and quinone. The detailed analysis of the results suggests that the outcome is the result of a variety of factors. The authors suggest these to be the *syn* or *anti* approach of the components, conformation changes in the triplet biradicals and competition between cyclization and cleavage of these biradicals.

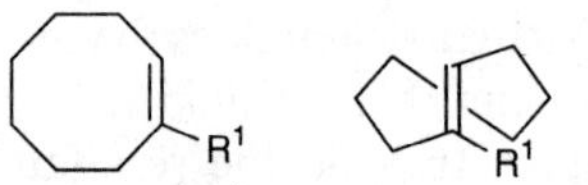

(68) R^1 = H or Me

Griesbeck *et al.*[53] have examined the influence of solvent viscosity on the oxetane formation observed between aldehydes and dihydrofuran. Several solvents of varying viscosity were used in the study. The results shown in Scheme 9 are only a few of those recorded. The three solvents chosen in the scheme range from the lowest to the highest viscosity. It can be seen that there is an influence on the *endo*:*exo* ratio. The greatest effect is seen with the addition of propionaldehyde, where the ratio changes from 45.3:45.7 to 72.6:27.4 going from the lowest to the highest viscosity. This is bettered using glycerol, when a ratio of 80.2:19.8 is observed. A further study has reported on the oxetane formation between aromatic ketones and thiophenes and selenophenes.[54]

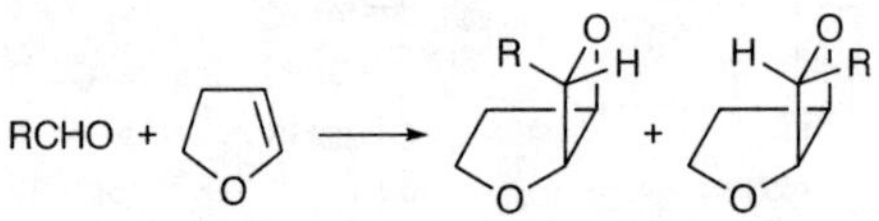

solvent	R	*endo*	*exo*
n-hexane	Ph	82	18
	Et	45.3	45.7
ethanol	Ph	86	14
	Et	50.4	49.6
1,4-butanediol	Ph	91	9
	Et	72.6	27.4

Scheme 9

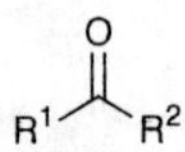

(69)

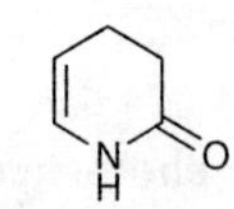

(70)

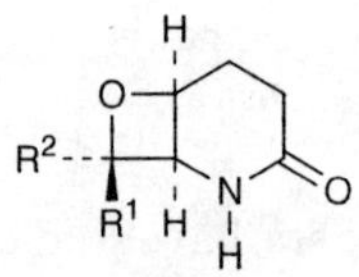

(71)

R^1	R^2	yield (%)	ratio of isomers
Ph	H	63	92 : 8
Ph	Ph	56	–
3-PivOPh	H	54	88 : 12
Ph	Me	51	90 : 10
Ph	CO_2Me	52	90 : 10

The photoaddition of aldehydes (69) to the dihydropyridone (70) affords the oxetane derivatives (71), and Bach and his co-workers[55] claim that this is a versatile building block and yields products with high regio- and diastereoselectivity. The enantioselectivity of the system was assessed using the addition of the (+)-aldehyde (72). This affords the (−)-adduct (73) with a 95% ee. A review has described photochemical processes such as oxetane forming reactions involving Lewis acid-catalysis.[56]

4 Miscellaneous Reactions

4.1 Decarboxylation. – A kinetic model for the photochemical decomposition of formic acid on a pilot-plant scale has been studied.[57] The photodissociation dynamics at 193 nm of ethanoic acid have been studied.[58,59] An FTIR examination of the photochemical decomposition of acetic acid on TiO_2 has been

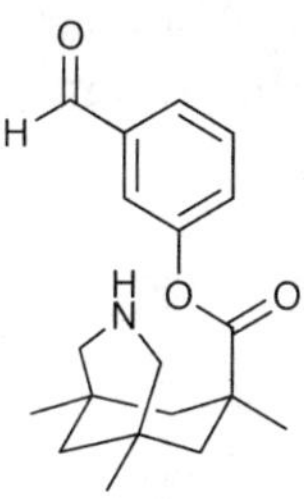

(72)
(+)

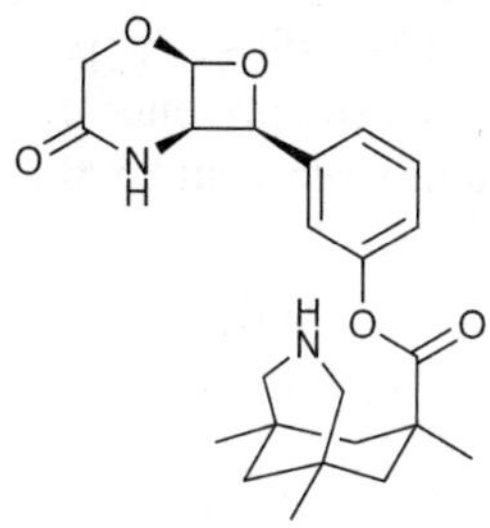

(73)
(–) >95% ee

reported.[60] Results have been obtained demonstrating that matrix isolation of carboxylic acids provides a good method for the study of the various photochemical paths that are open to such molecules.[61] The influence of UV irradiation in combination with ultrasound on the decomposition of trichloroacetic acid has been reported.[59,62]

Petrenko,[63] using a quantum mechanical simulation, has examined the possible paths for the formation of radicals following the irradiation of L-α-alanine in the crystalline state. The decarboxylation of 4-chloro-2-methylphenoxyacetic acid on TiO_2 in aqueous suspension affords 4-chloro-2-methylphenol as the principal product.[64] 2-Phenylpropionic acid undergoes photochemical decarboxylation on irradiation under a variety of conditions.[65] The present work has demonstrated the influence of irradiation in a variety of cation-exchanged zeolites. Some of the results obtained are shown in Scheme 10.

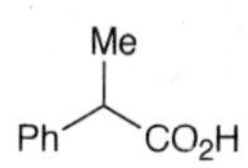

condition	Ph	Ph–Ph *meso*	*dl*	$PhCOCH_2$	*dl* : *meso* ratio
CH_3CN	16%	45	34	–	0.76
β-CN	–	66	34	–	0.52
LiY	3.0	2.0	95	–	48
NaY	5.0	1.0	97		97
RbY	43	5.0	52		10

Scheme 10

A review has highlighted the photochemical reactivity of some arylpropionic acids used as nonsteroidal anti-inflammatory materials.[66] The compounds discussed are benoxaprofen, carprofen, naproxen, ketoprofen, tiaprofenic acid (5-benzoyl-α-methylthiophene-2-acetic acid) and suprofen [(α-methyl-4-(2-thienyl-carbonyl)benzeneacetic acid)]. The photodegradation of triaprofenic acid and suprofen is enhanced when the acids are complexed within β-cyclodextrin; the transients observed are the same as those observed in the non-complexed state.[67] The anionic form of fenofibric acid undergoes facile photodecarboxylation to afford a carbanion.[68]

The triplet state is involved in the photochemical degradation of 2-hydroxy-4-trifluoromethylbenzoic acid.[69] The major photodegradation path is by nucleophilic attack on the trifluoromethyl group. The irradiation of 3,6-dichloro-2-methoxybenzoic acid in aqueous solution without the exclusion of air results in the formation of (74) and (75) as the two principal products.[70] The formation of the latter of these needs the presence of oxygen for its formation. The catalysed decomposition of cinnamic acid in the presence of TiO_2 has been studied.[71]

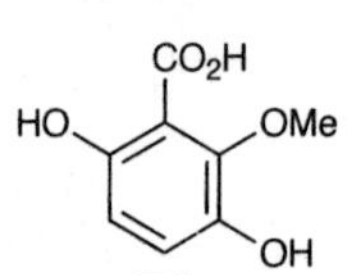

(74)

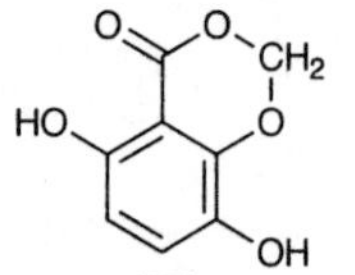

(75)

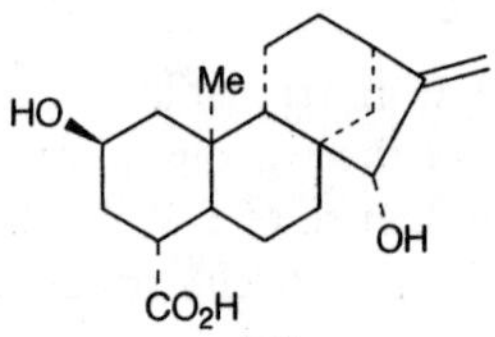

(76)

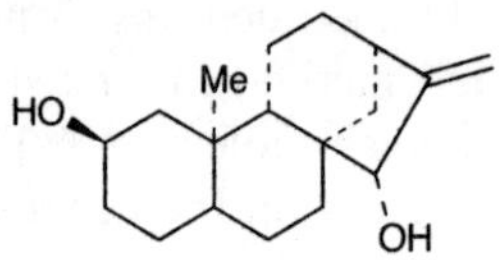

(77)

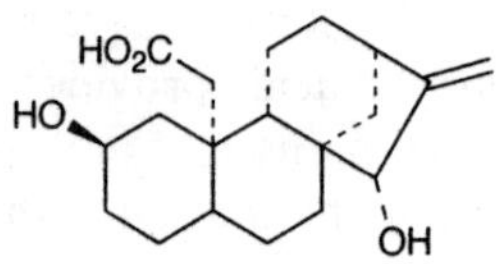

(78)

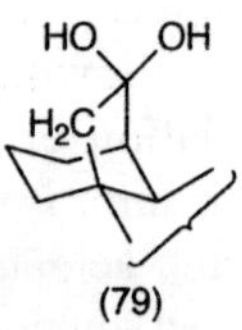

(79)

Interestingly in the pH range of 3-11 little degradation was detected. Irradiation at 254 nm in methanol of the nor-diterpene acid (76) affords the decarboxylated compound (77) in 25% yield.[72] A minor product was also isolated and was identified as the C-20 methyl substituted derivative (78, 14%). The authors suggest that remote functionalization brings about the formation of this product: the C-20 methyl and the carboxyl function are ideally disposed to undergo a hydrogen abstraction reaction. Bond closure affords an intermediate such as (79) that can undergo the transfer of the carboxyl group to afford (78).

4.2 Reactions of Haloketones. – Irradiation of acetyl bromide at 234 nm brings about C-Br fission.[73] This process is thought to be initiated by an *n*σ* transition localized on the CBr chromophore. The C-Br bond in bromoacetyl-2-naphthalene is photochemically labile, and irradiation brings about the formation of the corresponding naphthoylmethyl radical.[74]

4.3 Other Fission Processes. – The use of triphenylpyrylium tetrafluoroborate as a catalyst for the ring opening of α-epoxyketones in methanol has been described.[75] The reaction involves an electron-transfer process. The influence of electron donating groups on the outcome of the reaction was assessed. The keto epoxides (80) undergo photochemical conversion (λ > 280 nm, THF) to the β-hydroxyketones (81) in good to excellent yields when they are irradiated in the presence of 1,3-dimethyl-2-phenylbenzimidazoline and a carboxylic acid.[76] Several acids were tried but the most efficient was found to be ethanoic acid. Interesting observations have been made relating to the photochemistry of the oxirane (82) and the carbonate (83).[77] The irradiations were carried out in acetonitrile or hexane using 254 nm light. This converts both into the ketone (84). The *cis*-isomer of the oxirane is photochemically inert as is the dimethyl-substituted derivative of the carbonate. The authors suggest that this change in reactivity is associated with the energy required to obtain a conformation where the aromatic ring is orthogonal to the orbitals of the oxirane (or carbonate).

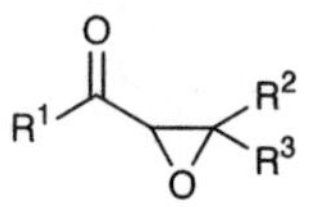
(80)

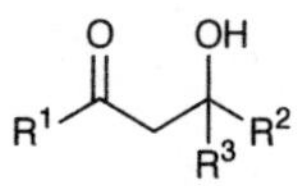
(81)

	R^1	R^2	R^3	yield (%)
trans	Ph	H	Ph	96
cis	Ph	H	Ph	94
	α-naphth	H	Ph	93
	Ph	H	α-naphth	90
trans	Ph	Me	Ph	66
cis	Ph	Ph	Me	58

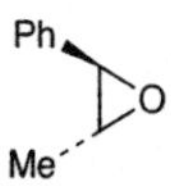
(82)

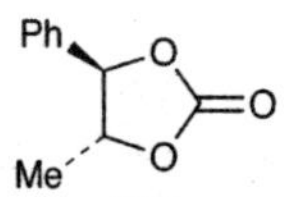
(83)

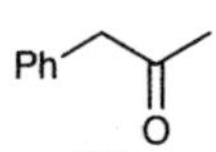
(84)

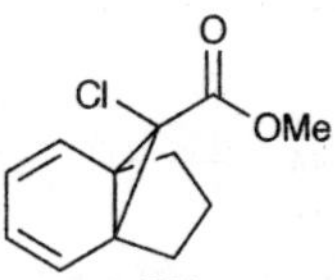
(85)

Irradiation of (85) results in the formation of carbomethoxychlorocarbene by extrusion; the by-product of the reaction is indane.[78] The thiazolethione derivatives (86) are photoreactive on irradiation at 350 nm in the presence of tributyltin hydride.[79] This brings about the formation of the aldehydes (87) in good yield. The photochemical step is fission of the O-N bond to yield an alkoxy radical. Ring opening within this generates the carbonyl group of the aldehyde. Photochemical cleavage of the lactone (88) can be brought about using I_2-HgO.[80] This treatment affords the lactone (89, 65%).

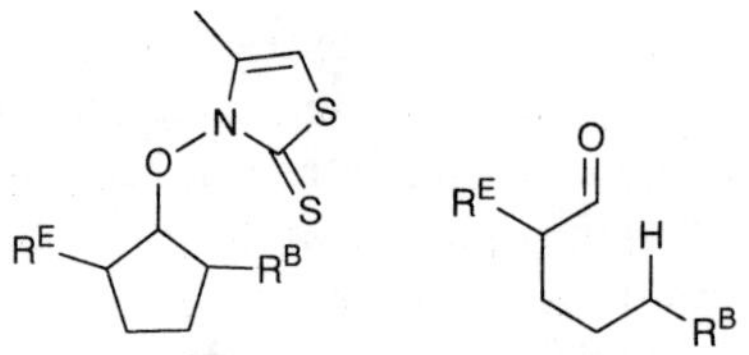

(86) (87)

R^E	R^B	yield (%)
H	Me	79
Ph	H	67
Ph	Me	76

Kako *et al.*[81] report that irradiation of aromatic ketones in benzene solution in the presence of the cyclic silanes (90) results in the formation of new products.

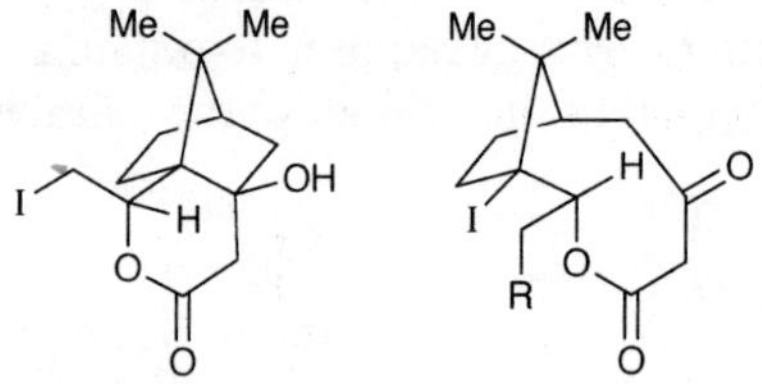

(88) (89)

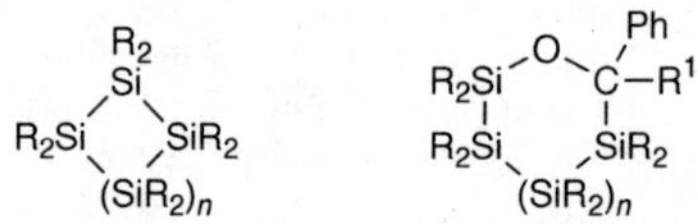

(90) (91)

$n = 3$, R = Me
$n = 2$, R = Et
$n = 1$, R = Pr^i

n	yield (%)	ketone	R^1
3	85	A	CF_3
2	20	A	CF_3
1	29	A	CF_3
2	57	B	Me
1	28	B	Me
2	12	C	Ph
1	7	C	Ph

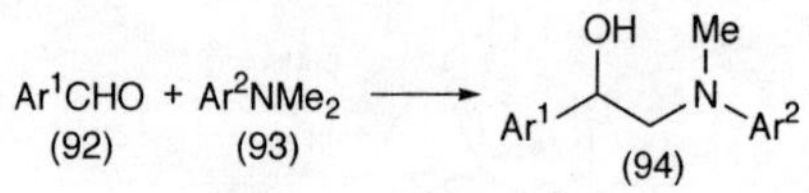

Ar^1	Ar^2	yield (%)
Ph	Ph	79
1-naphthyl	Ph	66
3-HO-4-MeOC_6H_3	Ph	71
2-styryl	Ph	73
Ph	2-pyridyl	47

Scheme 11

These have been identified as the silanes (91). The yields are variable, with the best yields obtained using 2,2,2-trifluoroacetophenone.

A study of hydrogen abstraction by laser flash-photolysed benzaldehydes has been carried out.[82] Irradiation of the aldehydes (92) in the presence of *N,N*-

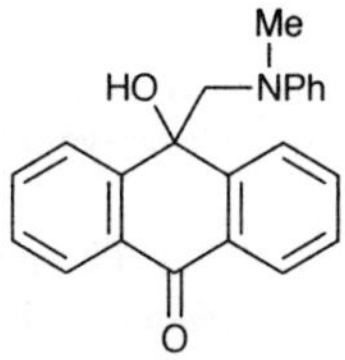

(95)

dimethylamino arenes (93) results in hydrogen abstraction from the *N*-methyl group. The resultant radical pair undergoes bonding to yield the products (94) in the yields shown in Scheme 11. Anthraquinone behaves in a like manner and yields adducts such as (95).[83]

References

1. G. D. Smith, L. T. Molina and M. J. Molina, *J. Phys. Chem. A*, 2002, **106**, 1233.
2. Y. Q. Chen and L. Zhu, *J. Phys. Chem. A*, 2001, **105**, 9689.
3. L. Zhang, H. Chen, J. Chen, H. Xu, J. Dai, S. Liu and X. Ma, *Huaxue Wuli Xuebao*, 2001, **14**, 135.
4. S. K. Shin, S. K. Kim, H. L. Kim and C. R. Park, *J. Photochem. Photobiol. A- Chem.*, 2001, **143**, 11.
5. P. Warneck, *Atmos. Environ.*, 2001, **35**, 5773.
6. N. V. Lebedeva, E. G. Bagryanskaya, V. R. Gorelik, I. V. Koptyug and R. Z. Sagdeev, *J. Phys. Chem. A*, 2001, **105**, 4640.
7. J. M. Tanko and R. Pacut, *J. Am. Chem. Soc.*, 2001, **123**, 5703.
8. N. J. Turro, X.-G. Lei, S. Jockusch, W. Li, Z. Liu, L. Abrams and M. F. Ottaviani, *J. Org. Chem.*, 2002, **67**, 2606.
9. N. J. Turro, *Proc. Nat. Acad. Sci. U. S. A.*, 2002, **99**, 4805.
10. S. Hashimoto, *Kokagaku*, 2000, **31**, 194.
11. G. Ananchenko and H. Fischer, *J. Chem. Soc., Perkin Trans. 2*, 2001, 1887.
12. H. Tomioka, H. N. Takase, Y. Maeyama, K. Hida, V. Lemee, J. P. Fouassier and D. Burget, *Res. Chem. Intermed.*, 2001, **27**, 47; *Chem. Abstr.*, 2001, **135**, 99702.
13. a. Korner, A. Bryant-Friedrich and B. Giese, *J. Org. Chem.*, 1999, **64**, 1559.
13. b. M. Spormann and B. Giese, *Synthesis*, 2001, 2156.
14. H. Isaji, M. Tasutake, H. Takemura, K. Sato, H. Tatemitsu, T. Inazu and T. Shinmyozu, *Eur. J. Org. Chem.*, 2001, 2487.
15. R. M. Kanada and K. Ogasawara, *Tetrahedron Lett.*, 2001, **42**, 7311.
16. E. Lee-Ruff and R. Margau, *Nucleosides, Nucleotides Nucleic Acids*, 2001, **20**, 185.
17. D. Ng, Z. Yang and M. A. Garcia-Garibay, *Tetrahedron Lett.*, 2001, **42**, 9113.
18. K. Kadota and K. Ogasawara, *Tetrahedron Lett.*, 2001, **42**, 8661.
19. C. Plessis and S. Derrer, *Tetrahedron Lett.*, 2001, **42**, 6519.
20. J. N. Moorthy, P. Mal, R. Natarajan and P. Venugopalan, *J. Org. Chem.*, 2001, **66**, 7013.
21. J. N. Moorthy, P. Mal, R. Natarajan and P. Venugopalan, *Org. Lett.*, 2001, **3**, 1579.
22. K. C. Nicolaou, D. Gray and J. Tae, *Angew. Chem. Int. Edn.*, 2001, **40**, 3675.
23. K. C. Nicolaou, D. Gray and J. Tae, *Angew. Chem. Int. Edn.*, 2001, **40**, 3679.
24. A. G. Griesbeck and D. Scheutzow, *J. Photochem. Photobiol. A: Chem.*, 2002, **147**, 109.
25. G. Burton, C. P. Lantos and A. S. Veleiro, PCT, Int. Appl. WO 02 22,646; *Chem.*

Abstr., 2002, **136**, 263299.

26. M. Sobczak and P. J. Wagner, *Org. Lett.*, 2002, **4**, 379.
27. H. Koshima, D. Matsushige and M. Miyauchi, *Cryst. Eng. Comm.*, 2001; *Chem. Abstr.*, 2002, **136**, 158658.
28. K. Vishnumurthy, E. Cheung, J. R. Scheffer and C. Scott, *Org. Lett.*, 2002, **4**, 1071.
29. A. J. Kell and M. S. Workentin, *Langmuir*, 2001, **17**, 7355; *Chem. Abstr.*, 2002, **136**, 45544.
30. H. Yamashita, S. Takeda, M. Nishimura, H. Bessho and M. Anpo, *Stud. Surf. Sci. Catal.*, 2001, **135** (Zeolites and Mesoporous Materials at the Dawn of the 21st Century), 2493.
31. J. Shailaja, J. Sivaguru, S. Uppili, A. Joy and V. Ramamurthy, *Microporous Mesoporous Mater.*, 2001, **48,** 319.
32. M. Abe and M. Nojima, *Kokagaku*, 2000, **31**, 28.
33. N. A. Weir and M. Delaney-Luu, *Eur. Polym. J.*, 2001, **37**, 1339.
34. P. Klan, J. Literak and S. Relich, *J. Photochem. Photobiol. A- Chem.*, 2001, **143**, 49.
35. A. G. Griesbeck and H. Heckroth, *J. Am. Chem. Soc.*, 2002, **124**, 396.
36. S. Kohmoto, Y. Miyaji, M. Tsuruoka, K. Kishikawa, M. Yamamoto and K. Yamada, *J. Chem. Soc., Perkin Trans. 1*, 2001, 2082.
37. P. Wessig and O. Muhling, *Angew. Chem. Int. Ed.*, 2001, **40**, 1064.
38. O. Muhling and P. Wessig, *Proc. ECSOC-3, Proc. ECSOC-4, 1999, 2000 [computer optical disk]* 1999-2000 (Pub. **2000**), 534; *Chem. Abstr.*, 2001, **135**, 33337.
39. S. Thierung, C. Sund, J. Thiem, A. Giesler and J. Kopf, *Carbohydr. Res.*, 2001, **336**, 271; *Chem. Abstr.*, 2002, **136**, 247786.
40. P. Wessig, J. Schwarz, U. Lindemann and M. C. Holthausen, *Synthesis*, 2001, 1258.
41. K. K. Park, I. K. Han and J. W. Park, *J. Org. Chem.*, 2001, **66**, 6800.
42. K. K. Park, H. Lim, S. H. Kim and D. H. Bae, *J. Chem. Soc., Perkin Trans. 1*, 2002, 310.
43. T. Goshima, Y. Itoh, H. Shirai and M. Kojima, *J. Photochem. Photobiol. A- Chem.*, 2001, **141**, 139.
44. C. G. Bochet, *J. Chem. Soc., Perkin Trans. 1*, 2002, 125.
45. M. Cano, M. Ladlow and S. Balasubramanian, *J. Org. Chem.*, 2002, **67**, 129.
46. W. Adam and V. R. Stegmann, *Synthesis*, 2001, 1203.
47. M. A. Miranda, M. A. Izquierdo and F. Galindo, *Org. Lett.*, 2001, **3**, 1965.
48. M. Abe, K. Tachibana, K. Fujimoto and M. Nojima, *Synthesis*, 2001, 1243.
49. M. Abe and M. Nojima, *Yuki Gosei, Kagaku Kyokaishi*, 2001, **59**, 855; *Chem. Abstr.*, 2001, **135**, 371657.
50. T. Kang and J. R. Scheffer, Org. *Lett.*, 2001, **3**, 3361.
51. W. Adam, V. R. Stegmann and S. Weinkotz, *J. Am. Chem. Soc.*, 2001, **123**, 2452.
52. W. Adam and V. R. Stegmann, *J. Am. Chem. Soc.*, 2002, **124**, 3600.
53. A. G. Griesbeck, S. Bondock and M. S. Gudipati, *Angew. Chem. Int. Edn.*, 2001, **40**, 4684.
54. F. Vargas and C. Rivas, *Int. J. Photoenergy*, 2000, **2**, 97.
55. T. Bach, H. Bergmann, H. Brummerhop, W. Lewis and K. Harms, *Chem.-Eur. J.*, 2001, **7**, 4512.
56. F. Kitagawa and N. Kitamura, *Kokagaku*, 2002, **32**, 153.
57. M. D. Labas, C. S. Zalazar, R. J. Brandi, C. A. Martin and A. E. Cassano, *Helv. Chim. Acta*, 2002, **85**, 82.
58. H. T. Kwon, S. K. Shin, S. K. Kim, H. L. Kim and C. R. Park, *J. Phys. Chem. A*, 2001, **105**, 6775.
59. V P. D. Naik, H. P. Upadhyaya, A. Kumar, A. V. Sapre and J. P. Mittal, *Chem.*

Phys. Lett., 2001, **340**, 116; *Chem. Abstr.*, 2001, **135**, 180517.
60. L.-F. Liao, C.-F. Lien and J.-L. Lin, *Phys. Chem. Chem. Phys.*, 2001, **3**, 3831; *Chem. Abstr.*, 2001, 603224.
61. R. Fausto and E. M. S. Macoas, *J. Mol. Struct.*, 2001, **563-564**, 27.
62. C. Wu, D. Wei, J. Fan and L. Wang, *Chemosphere*, 2001, **44**,1293; *Chem. Abstr.*, 2001, **135**, 261916.
63. T. L. Petrenko, *J. Phys. Chem. A*, 2002, **106**, 149.
64. A. Zertal, T. Sehili and P. Boule, *J. Photochem. Photobiol. A: Chem.*, 2001, **146**, 37.
65. A. Lalitha, K. Pitchumani and C. Srinivasan, *Tetrahedron*, 2001, **57**, 4455.
66. F. Bosca, M. L. Marin and M. A. Miranda, *Photochem. Photobiol.*, 2001, **74**, 637.
67. S. Monti, S. Encinas, A. Lahoz, G. Marconi, S. Sortino, J. Perez-Prieto and M. A. Miranda, *Helv. Chim. Acta*, 2001, **84**, 2452.
68. G. Cosa, S. Purchit, J. C. Scaiano, F. Bosca and M. A. Miranda, *Photochem. Photobiol.*, 2002, **75**, 193.
69. F. Bosca, M. C. Cuquerella, M. L. Marin and M. A. Miranda, *Photochem. Photobiol.*, 2001, **73**, 463.
70. J.-P. Aguer, F. Blachere, P. Boule, S. Garaudee and C. Chantal, *Int. J. Photoenergy*, 2000, **2**, 81; *Chem. Abstr.*, 2001, **135**, 199963.
71. S. M. Grimes, L. K. Mehta and H. C. Ngwang, *J. Environ. Sci. Health, Part A: Toxic/Hazard. Subst. Environ. Eng.*, 2001, **A36**, 599.
72. S. Buscemi, S. Rosselli, M. Bruno, N. Vivona and F. Piozzi, *Tetrahedron Lett.*, 2001, **42**, 8289.
73. K. W. Lee, Y.-J. Jee and K.-H. Jung, *J. Chem. Phys.*, 2002, **116**, 4490.
74. T. Suzuki, Y. Kaneko, K. Maeda, T. Arai, K. Akiyama and S. Tero-Kubota, *Riken Review*, 2002, **44** 121.
75. H. R. Memarian, A. Hesami, F. Nikpour and D. Dopp, *Indian J. Chem., Sect. B: Org. Chem. Incl. Med. Chem.*, 2001, **40B**, 662; *Chem. Abstr.*, 2001, **135**, 344100.
76. E. Hasegawa, N. Chiba, A. Nakajima, K. Suzuki, A. Yoneoka and K. *Synthesis*, 2001, 1248.
77. S. Linder, K. White, M. Palmer, B. Arney and R. White, *Tetrahedron Lett.*, 2002, **43**, 1169.
78. I. Likhotvorik, Z. D. Zhu, E. L. Tae, E. Tippmann, B. T. Hill and M. S. Platz, *J. Am. Chem. Soc.*, 2001, **123**, 6061.
79. J. Hartung, T. Gottwald and R. Kneuer, *Synlett*, 2000, 749.
80. N. S. Vostrikov, A. V. Abutkov, L. V. Sprirkhin, A. A. Fatykhov and M. S. Miftahov, *Russ. Chem. Bull.*, 2001, **50**, 654; *Chem. Abstr.*, 2001, **135**, 357708.
81. M. Kako, M. Gu, Y. Takenaka, K. Takagi, K. Kondo, Y. Nakadaira, M. Wakasa and H. Hayashi, *Chem. Lett.*, 2001, 1220.
82. J.-H. Mu, G.-Z. Li, X.-H. Tu, T.-X. Lu and K.-S. Zhao, *Chem. Phys. Lett.*, 2002, **354**, 186; *Chem. Abstr.*, 2002, 214174.
83. S. S. Kim, Y. J. Mah and A. R. Kim, *Tetrahedron Lett.*, 2001, **42**, 8315.

2

Enone Cycloadditions and Rearrangements: Photoreactions of Dienones and Quinones

BY WILLIAM M. HORSPOOL

1 Cycloaddition Reactions

1.1 Intermolecular Cycloaddition. – *1.1.1 Open-chain Systems.* A study of the α- and the γ-polymorphs of *o*-ethoxycinnamic acid has shown that the γ-polymorph is unreactive photochemically owing to the mode of intermolecular hydrogen bonding. This leads to a deviation from planarity that prevents dimerization.[1] The photodimerization of cinnamates has been studied by scanning tunnelling microscopy (STM). Surprisingly the results obtained are contradictory, with STM suggesting the failure of a reaction that actually does take place.[2] A review has given details of the use of atomic force microscopy applied to the photochemical reactivity of cinnamic acids in the solid phase.[3] Hasegawa *et al.*[4] have reported the highly efficient dimerization of the enone (1). This is irradiated for one hour in the crystalline phase through Pyrex and is converted quantitatively into (2). A study of the crystals formed from phthalic acid and *trans*-cinnamide (1:2) has shown that the double bonds lie in a criss-cross fashion.[5]

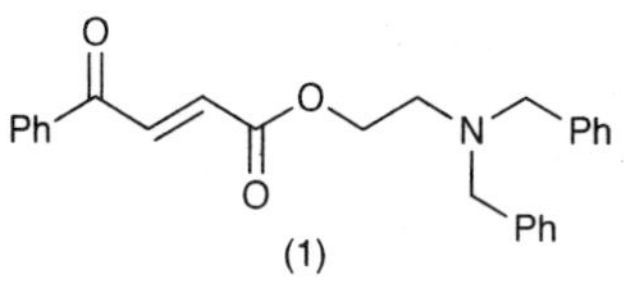

(1)

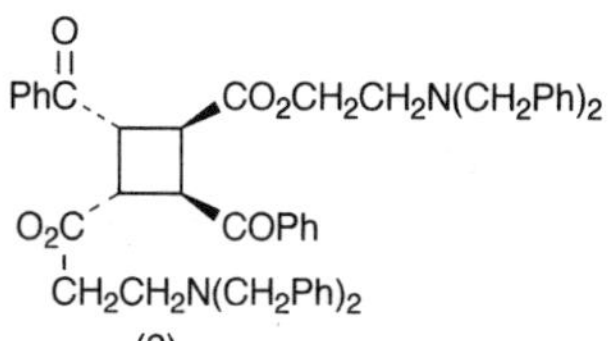

(2)

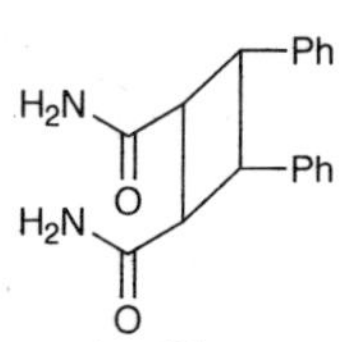

(3)

However, irradiation does afford the adduct (3), and the formation of this suggests that a conformational change occurs within the crystal during the irradiation. In the crystals of the dopamine (4), the molecules line up as shown.[6]

Photochemistry, Volume 34

When these are irradiated a (2+2)-cycloaddition affords (5), where addition of one double bond has taken place to a benzene ring in another. Either daylight or UV irradiation of (6) in ethanol solution brings about dimerization and the formation of the cyclobutane (7).[7] Irradiation of crystals of the cyclopentenone (8) results in dimerization, with the formation of (9) in a first order reaction.[8] Apparently there is molecular movement within the crystal and migration of one enone molecule towards another prior to dimerization.[8]

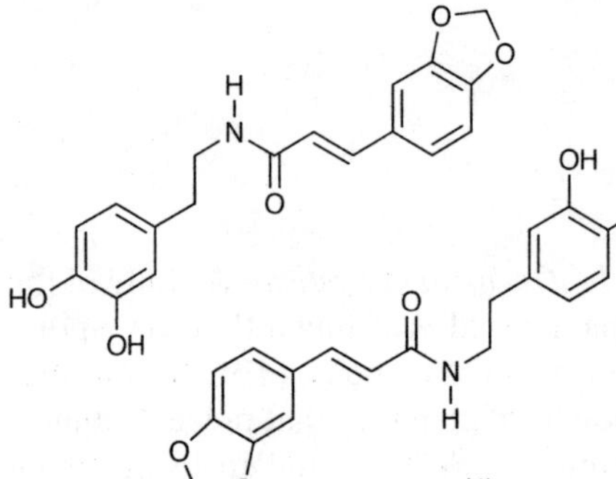

(4)

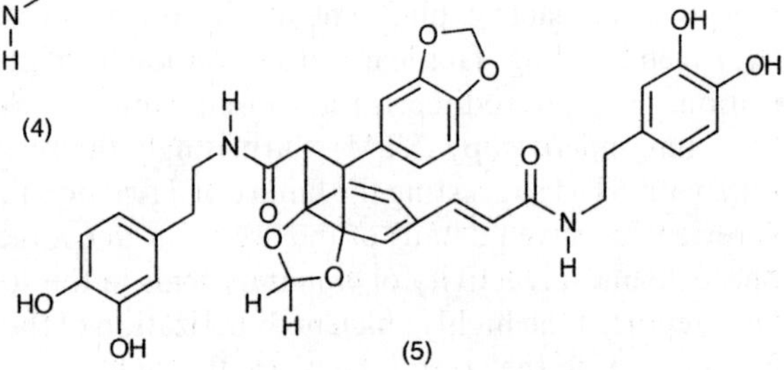

(5)

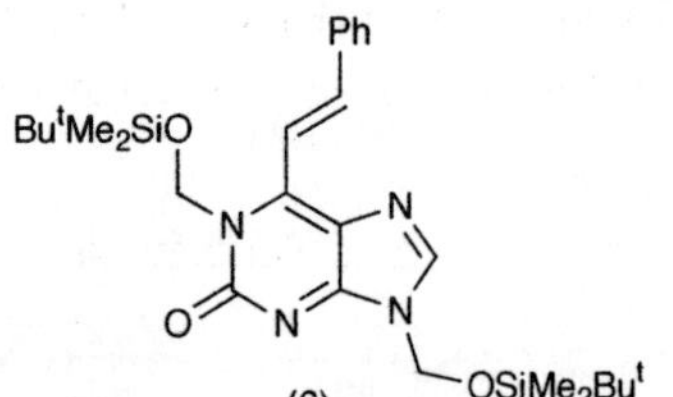

(6)

(7) R =

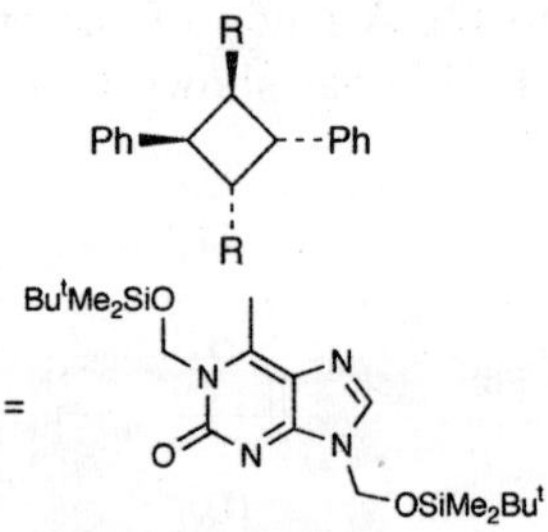

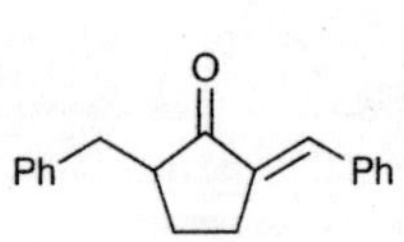

(8)

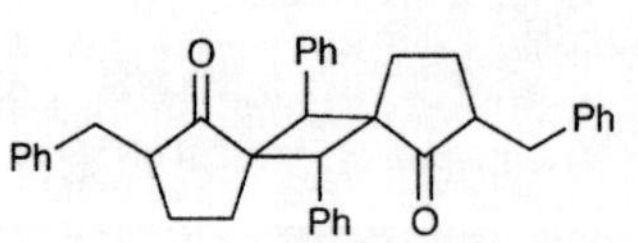

(9)

1.1.2 Additions to Cyclopentenones and Related Systems. The enone (10) undergoes (2+2)-photochemical cycloaddition of ethene to afford an intermediate that is the key in a new synthesis of racemic-kelsoene.[9] Studies by Alibés *et al.*[10] have shown that the photochemical addition of ethyne to the enones (11) affords (2+2)-adducts in varying yields. The enones (11 a-c) are reactive and yield the isomeric products (12) and (13) in the yields shown when the reactions are carried out in acetone or acetonitrile. It can be seen from these results that there is a small solvent dependence. However, the main effect is the preference for the formation of the *anti*-isomer (12). The bis lactone (14) is also reactive and irradiation affords the bis adduct (15). A theoretical study has examined the addition of ethylene to acrolein as a model for cycloadditions in enone systems.[11]

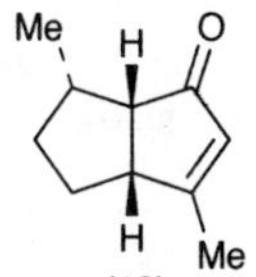

(10)

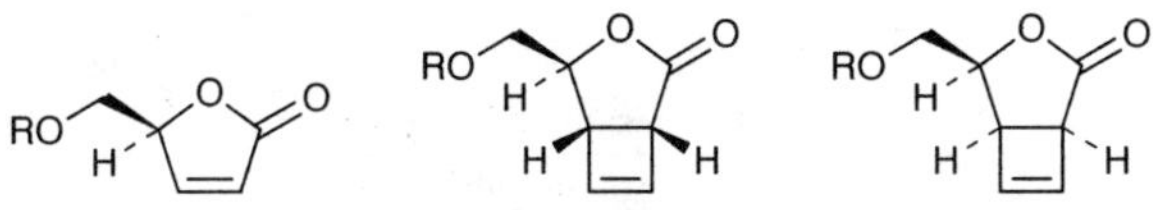

(11) (12) (13)

	R	in acetone yield (%)	ratio	in CH_3CN yield (%)	ratio
a	$COBu^t$	53	70 : 30	74	66 : 34
b	CO_2Mnt	51	66 : 34	57	59 : 41
c	COPh	26	68 : 32	25	66 : 34
d	TBDBS				

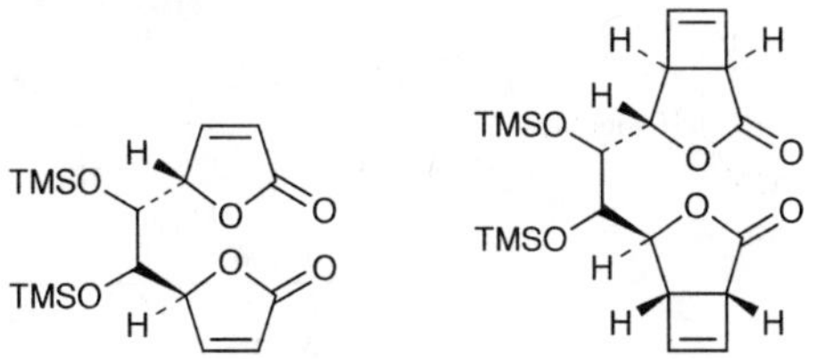

(14) (15)
56%

Albini and his co-workers[12] have demonstrated the use of alkylation of enones using 1,3-dioxolan-2-yl radicals. These radicals are produced from the corresponding dioxolane by hydrogen abstraction using photoexcited benzophenone. The radicals generated in this manner can be added to cyclic enones (Scheme 1) or to open chain systems (Scheme 2). Brule and Hoffmann[13] have studied the addition of ketyl radicals formed from cyclic ketones to the enone (16). The ketyl radicals are formed from the ketones (17) by an electron-transfer process, using either *N*,*N*-dimethylbenzylamine (A) or triethylamine (B) as the electron donating species. The varying yields of the adducts (18) depending on the use of A or B are shown. Interestingly the yields obtained with the benzyl amine are better than those using triethylamine. The reaction fails with cyclobutanone as the radical source. Nevertheless, the addition is also efficient with aryl ketones. The use of semiconductors to control the addition of amines to enones under photochemical excitation has shown that TiO_2 is most effective.[14] The results are shown below in scheme 3. The ratio of products is generally invariant at *ca.* 45:55 but the overall yield rises to 88 or 90% with TiO_2.

1.1.3 Additions to Cyclohexenones and Related System. The efficient photoaddition of *trans*-1,2-dichloroethene to the enone (19) affords the adduct (20) in

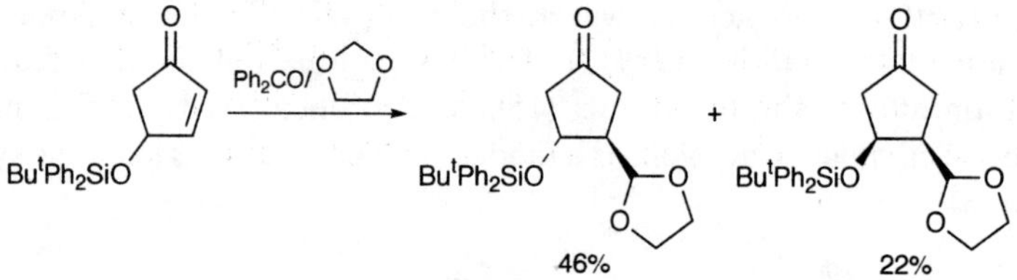

46% 22%

Scheme 1

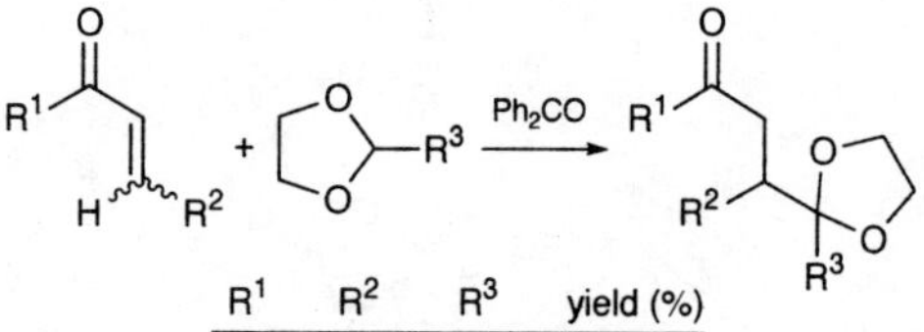

R^1	R^2	R^3	yield (%)
Me	H	menthyl	49
Et	H	Et	21
$-(CH_2)_2-$		menthyl	70
$-(CH_2)_3-$		*n*-hexyl	57

Scheme 2

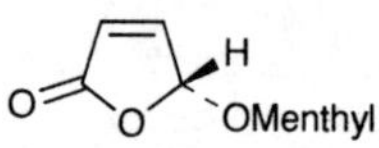

(16)

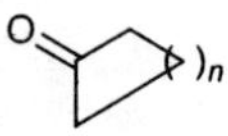

(17)

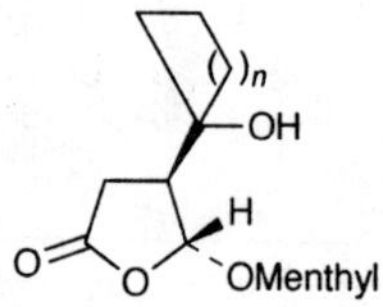

(18)

n	yield (%) (A)	(B)
2	42	17
3	69	38
4	44	29
5	37	35
9	37	12

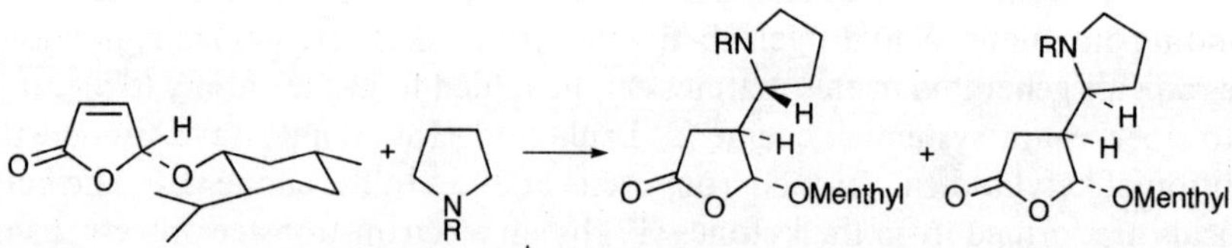

R = Me or Bu^t

Scheme 3

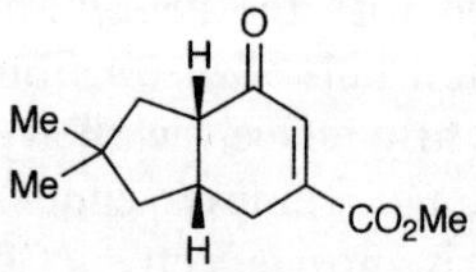

(19)

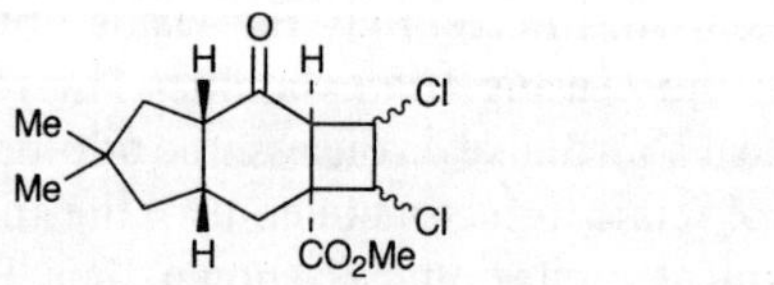

(20)

95% yield.[15] This compound was a key molecule in the development of a new route to the sesquiterpene, sterpurene. Interestingly the enone (21) does not undergo dimerization in complexes with host compounds such as (22).[16] Photochemical addition of 2,3-dimethylbut-2-ene to the enones (23) affords the adducts

(24) in > 75% yield.[17] The reaction is essentially a 1,3-addition of the alkene to the three carbon atoms of the cyanoalkene moiety. The enones (25) are photochemically unreactive in the solid state. Irradiation at 270 nm in solution, however, converts them into the dimers (26). The dimers (26) are conventional *cis,syn,cis*-(2+2)-adducts that are themselves photochemically reactive. Irradiation at 313 nm converts them into the homocubanes (27) and (28).[18]

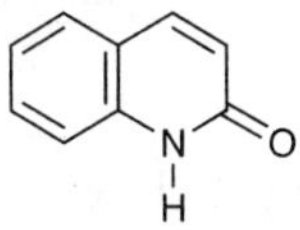

(21)

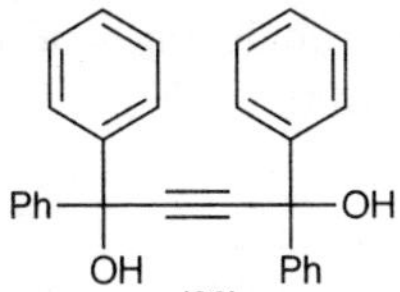

(22)

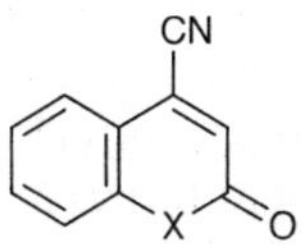

(23) X = O or S

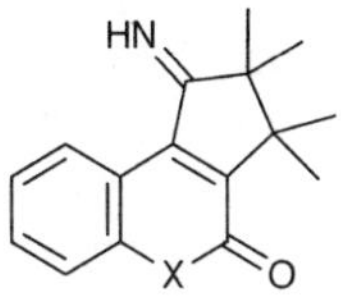

(24)

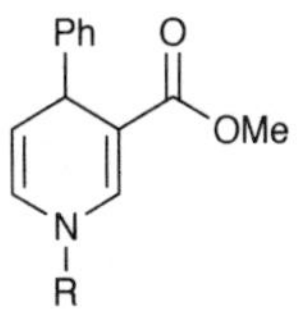

(25) R = CH_3 or $PhCH_2$

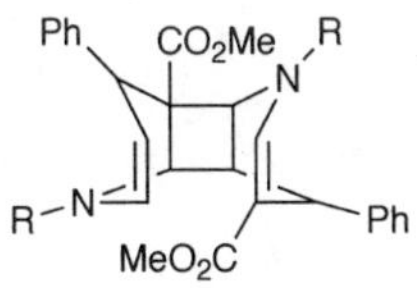

(26)

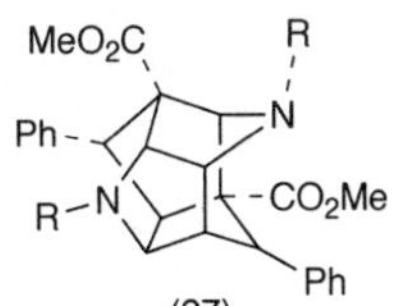

(27)

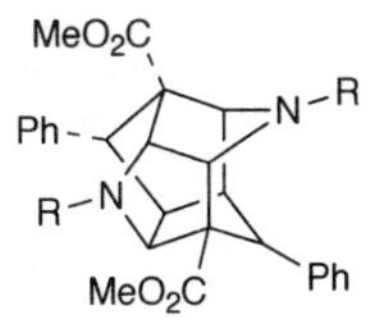

(28)

A Raman spectroscopic study of the photodimerization of 7-bromocoumarin in the crystalline state has been reported.[19] The dimerization results in the formation of a cyclobutane derivative. The photodimerization of 7-(8-trimethoxysilyloctyloxy)coumarin has been studied in self-assembled monolayers.[20] The photodegradation of 7-amino-4-methylcoumarin can be suppressed by forming complexes with β-cyclodextrin.[21] A detailed examination of the synthesis and reactivity of a series of 3-(*N*-arylformimidoyl)-4-hydroxycoumarins has been carried out.[22]

(2+2)-Photocycloaddition of the enone (29) to the pyrrolidones (30) affords adducts in good yield.[23] The adducts readily undergo methoxide-induced ring opening. The resultant compounds were used in model studies towards a synthesis of kainic acid. The quenching rate constants for the interaction of triplet excited 1-acetonaphthones by 2-morpholinoacrylonitrile and 2-piperidinoacrylonitrile have been measured.[24] Döpp and his co-workers[25] have described the photochemical addition of 2-morpholinoprop-2-enenitrile to perinaphthenone.

The irradiation of 3- and 4-fluorochalcone in the solid state brings about dimerization.[26] This results in the formation of a mixture of *anti-head-to-head*, *syn-head-to-tail* and *anti-head-to-tail* cyclobutane derivatives. Di-substitution, however, with 3,4- and 3,5-difluorochalcones influences the dimerization and a

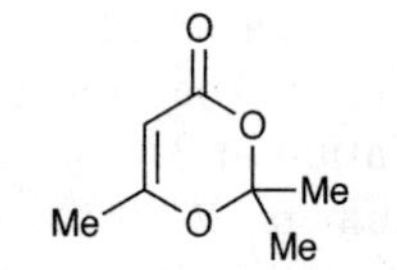
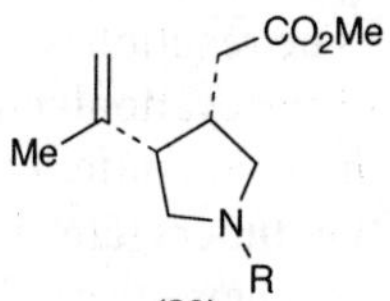

(29) R = Boc or $COCF_3$ (30)

stereoselective addition occurs affording the *syn-head-to-tail* dimer. The photochemical addition of tetrachloroethene to isocoumarins affords high yields of *cis*-fused cyclobutane adducts.[27]

1.2 Intramolecular Additions. – A further report of the photocycloaddition of vinylogous imides has demonstrated that (31) undergoes cycloaddition to yield (32, 87%) when irradiated in acetonitrile using a Pyrex filter.[28] The rearrangement of this product provides a path to the hetisine alkaloids. Asokan and his co-workers[29] have described a method whereby topochemical control can be exercised on the (2 + 2)-photocycloaddition reactions of cinnamoyl groups. This involves the synthesis of the alkenoylketenethioacetals (33). The irradiation of these in benzene solution with Pyrex-filtered light brings about the formation of the cycloadducts (34). The yields of these are moderate, as can be seen from the details below the structures. The stereochemistry of the additions was verified by X-ray crystallography.

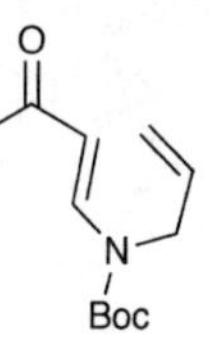
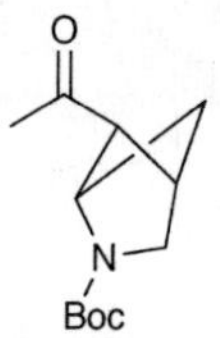

(31) (32)

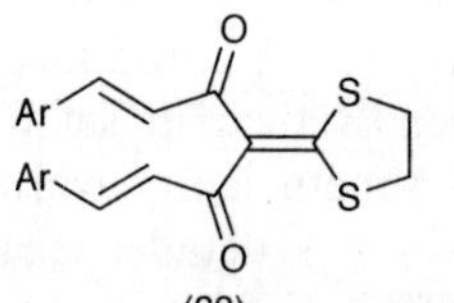
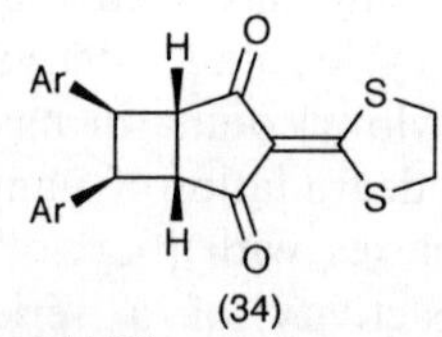

(33) (34)

Ar	yield (%)
Ph	50
p-$MeOC_6H_4$	55
p-ClC_6H_4	57
m-$MeOC_6H_4$	62
o-$MeOC_6H_4$	55
3,4-di$MeOC_6H_3$	70
thienyl	40

1.2.1 Intramolecular Additions to Cyclopentenones. Irradiation of the cyclopentenone derivative (35) brings about intramolecular (2 + 2)-cycloaddition with the formation of (36).[30] Subsequent thermal transformation by cleavage of the lactone system followed by a Cope rearrangement affords the cyclooctadiene derivative (37). Booker-Milburn and his co-workers[31] have described the intramolecular cycloaddition on irradiation through Pyrex of (38) in acetonitrile. The

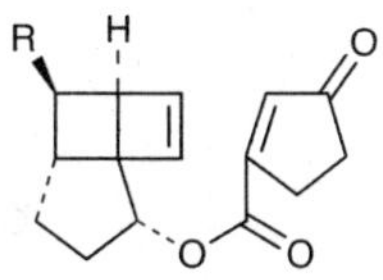
(35)

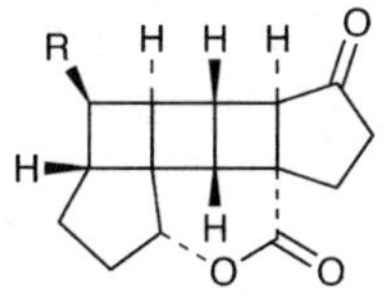
(36)

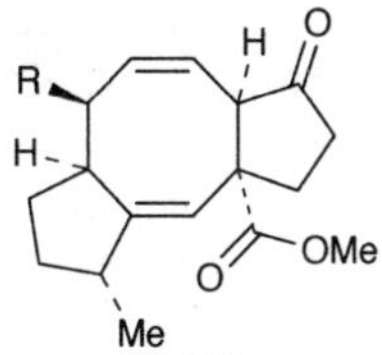
(37)

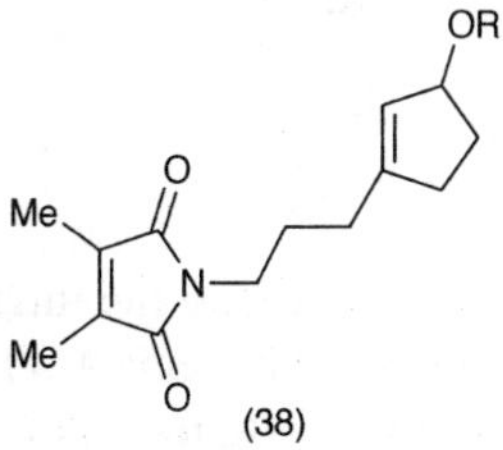
(38)

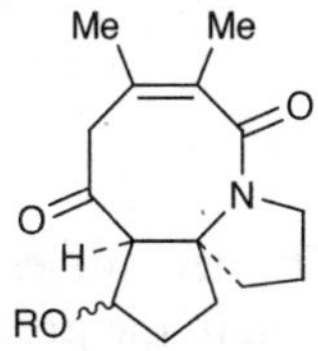
(39)

final products (39) are obtained in good yields (61-73%). The evaluation of several alkoxy protecting groups has shown that the best results are obtained from (38, R = H). This result suggests that hydrogen bonding is important in the alignment of the molecules. The intramolecular cycloaddition within the butenolides (40) affords the two products (41) and (42).[32] The products arise by two different cycloaddition modes and the outcome is dependent upon the substitution pattern. The control exercised is on the stability of the biradicals, and there is a preference for the biradical (43) to yield (41), while (44) affords (42). These suggestions have been substantiated by some simple theoretical calculations.

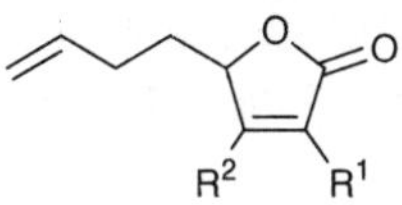
(40)

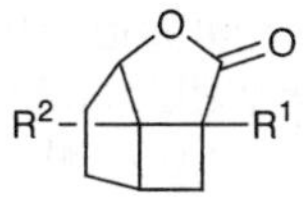
(41)

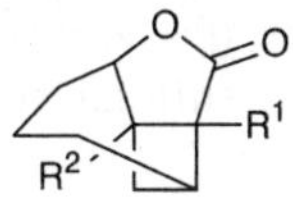
(42)

R^1	R^2	ratio of products
H	H	3 : 1
H	Me	7 : 1
Me	H	1.6 : 1

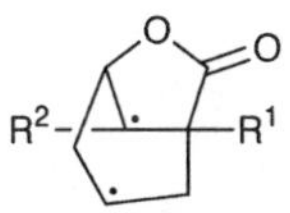
(43)

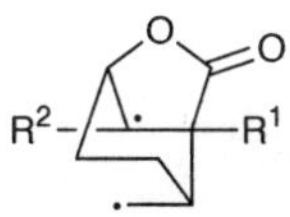
(44)

1.2.2 Additions to Cyclohexenones and Related Systems. Starting from (−)-α-pinene Mehta and Nandakumar[33] synthesized the enone (45). This compound is photochemically reactive and undergoes intramolecular (2 + 2)-cycloaddition to yield the (+)-adduct (46) and the (−)-adduct (47) in 45% and 15% yield, respectively. The former addition mode to yield (46) involves a 1,5-cyclization, while the latter to (47) is a 1,6-process. The related (+)-enone (48) is also photochemically reactive, and intramolecular cycloaddition yields (49, 70%). A detailed investigation of the intramolecular cycloaddition reactions within the enones (50) has been carried out,[34] but only a few of those reported are described here. The simplest of them, (50a), does not undergo intramolecular cycloaddition

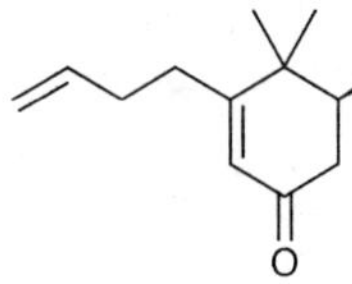
(45)

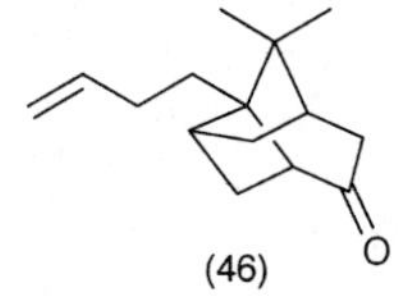
(46)

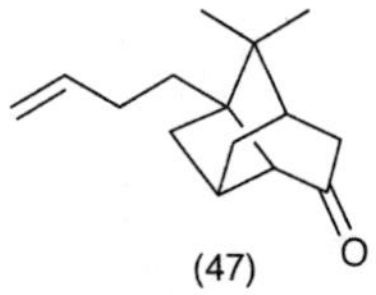
(47)

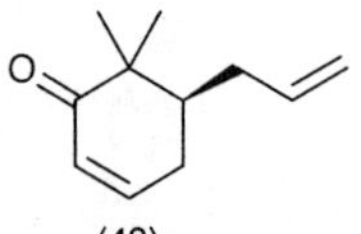
(48)

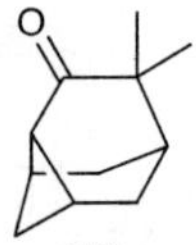
(49)

but does dimerize. The derivatives (50b-d) all undergo the intramolecular addition with the formation of the products shown in Scheme 4. The cycloadditions are diastereoselective, as can be seen from the *de* values reported. Mariano and his co-workers[35] have demonstrated that the photocyclizations of the perchlorate salts (51) provide a useful strategy for stereochemical control in the formation of the (2+2)-cycloadducts. The irradiation of (51) affords the cycloadduct (52) after work-up. The reactions do show some selectivity dependent on the nature of the R groups in (51). Thus with R = $MeOCH_2$, a 61% yield of the adduct can be obtained with an *ee* of 82%. When R = Me the reaction is poorer and the adduct is formed with an *ee* of 37%.

The enone (53) undergoes (2+2)-photocycloaddition to yield the adduct (54).[36] This compound is a key intermediate in an approach to the synthesis of spirolucidine, a lycopodium alkaloid. Intramolecular photoaddition has been studied within the enones (55), (56) and (59).[37] The additions of (55) and (56) follow the conventional reaction path to afford the adducts (57) and (58). In the more constrained system (59), addition also takes place and yields (60). The intramolecular photocyclization of 6-methyl-2-(4-methyl-3-cyclohexen-1-yl)-1,3-dioxin-4-one affords two isomeric eudesmane diols with a *cis*-decalin skeleton.[38] The enone (61) undergoes photochemical intramolecular (2+2)-cycloaddition to yield the highly strained cage compound (62).[39]

2 Rearrangement Reactions

2.1 α,β-Unsaturated Systems. – *2.1.1 Isomerization.* The sunlight-induced photochemical *trans,cis*-isomerism of the octylcinnamate (63) has been described.[40] A study of a new chiral dopant based on cinnamic acid and isosorbide has been reported.[41] The photochemical reactivity of β-phenylacryloyl compounds with chloromethyl substituents has been investigated.[42]

A report of the solvent dependence of electronic relaxation of the all *trans*-retinal (64) has been published.[43] The irradiation of the polyunsaturated allylic benzoates (65) brings about their selective conversion into the isomers (66).[44] In an effort to increase the yield of the *cis*-isomer (67) the *trans*-isomer (68) was irradiated in ether. This gave a 1:1 mixture of the isomers. This reaction was a key step in a synthetic approach to (−)-laulimalide.[45]

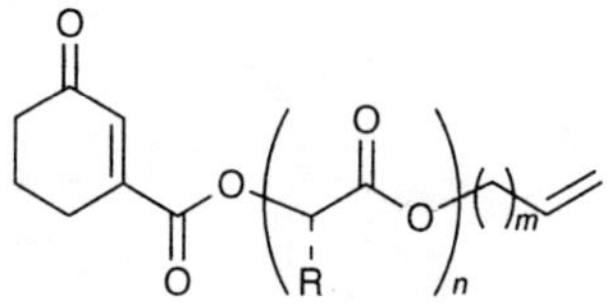

(50)

	n	m	R
a	1	1	Me
b	1	2	Me
c	2	1	Me
d	1	2	Ph

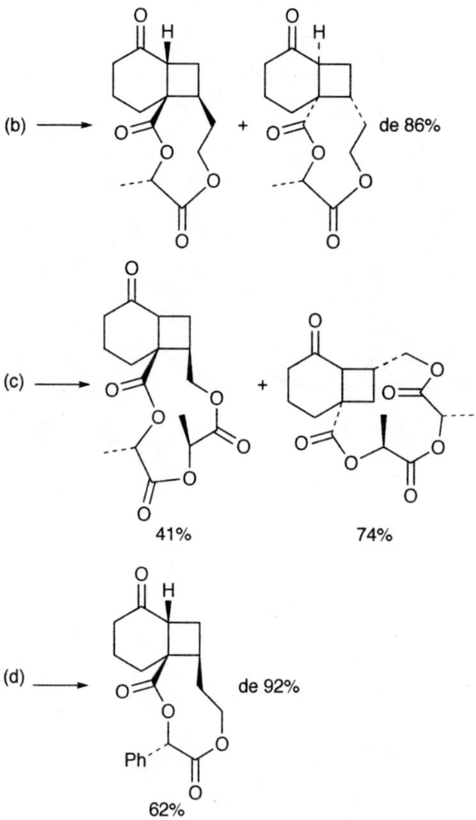

Scheme 4

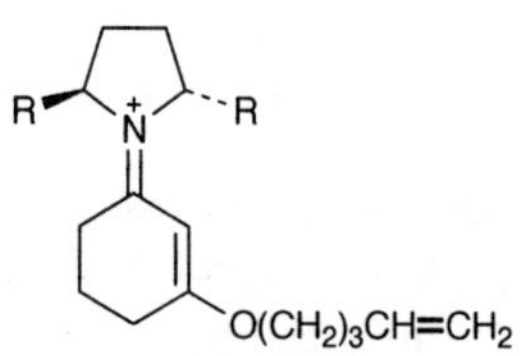

(51) R = $MeOCH_2$, Me

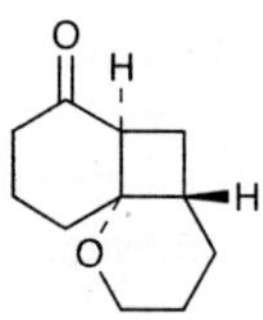

(52)

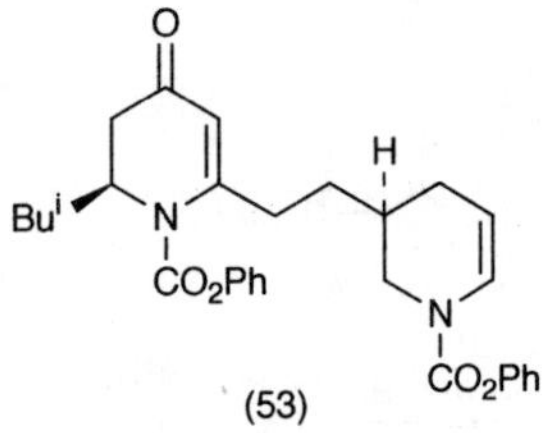

(53)

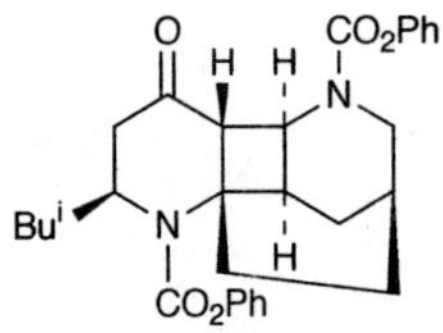

(54)

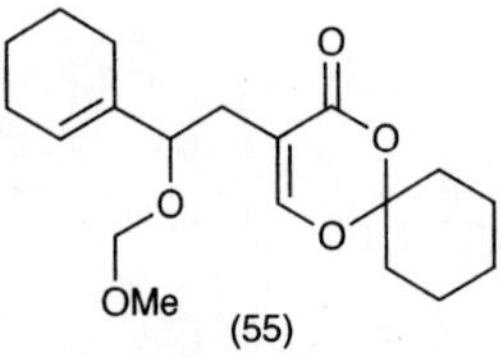

(55)

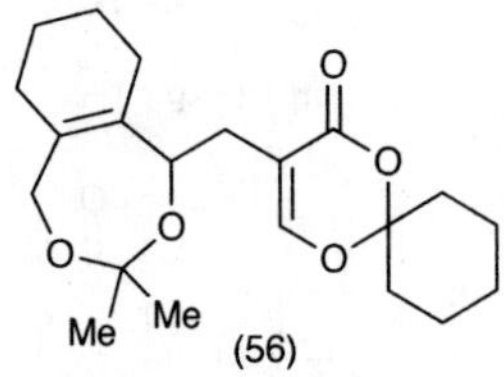

(56)

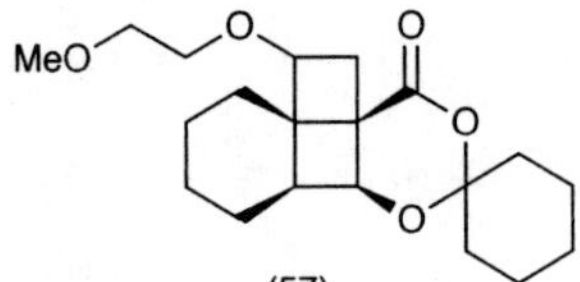

(57)

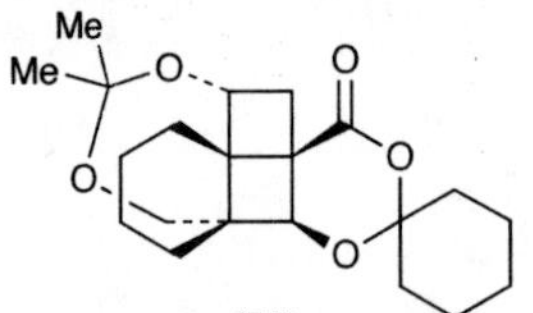

(58)

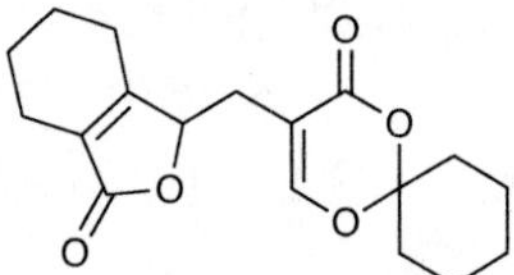

(59)

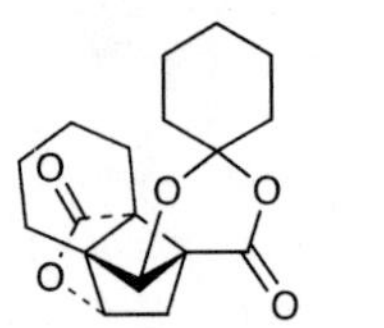

(60)

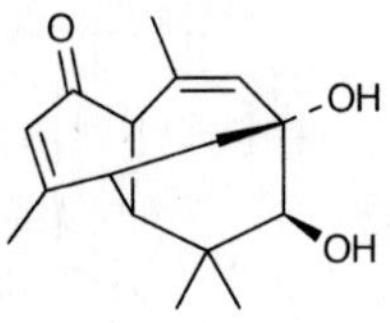

(61)

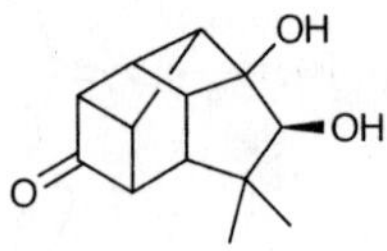

(62)

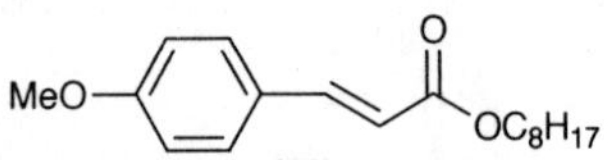

(63)

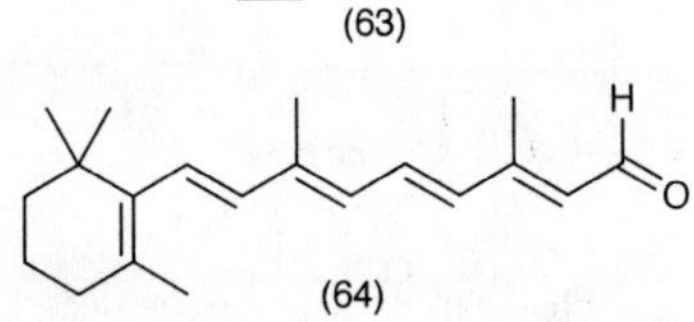

(64)

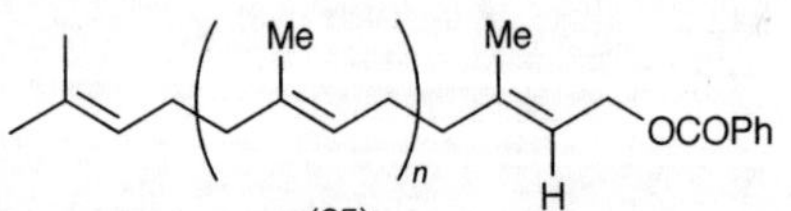

(65)

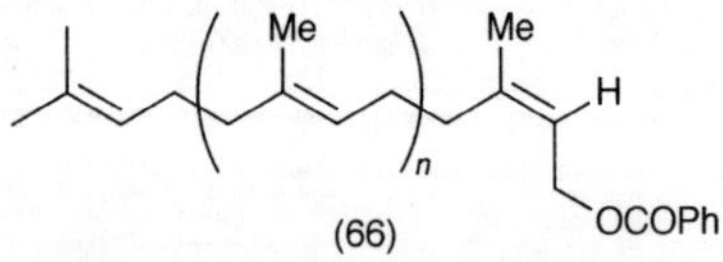

(66)

2.1.2 Hydrogen-abstraction Reactions. The enone (69) undergoes photoenolization to (70) when subjected to flash photolysis.[46] The rate to reversion to the enone was measured under a variety of conditions. Bach and Hofer[47] have investigated photochemical deconjugation within the enone derivatives shown in Scheme 5. All the examples exhibit reasonable chemical yields for the forma-

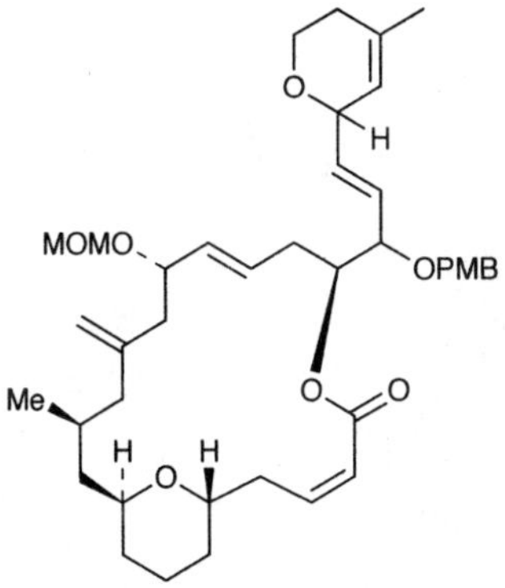

(67)

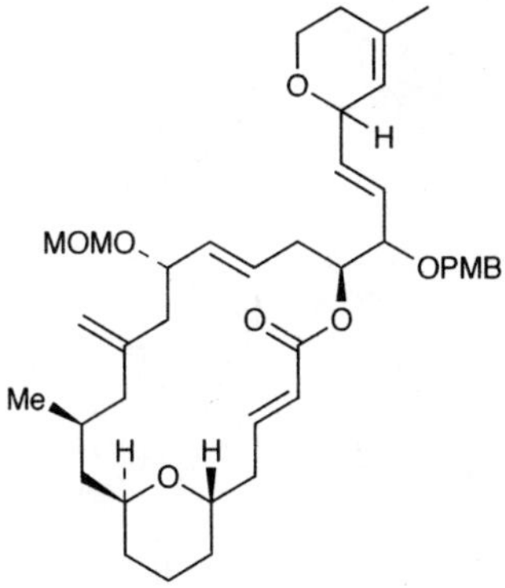

(68)

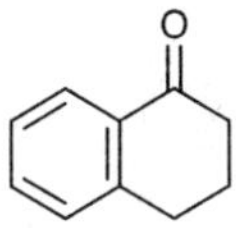

(69)

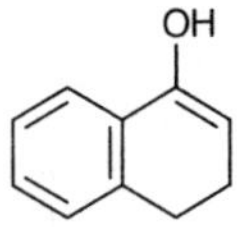

(70)

tion of the deconjugated product, but the best *de* of 95% is obtained with the diacetone L-fructopyranose derivative (71 DAG). Others[48] have studied the photochemical deconjugation reaction in α,β-unsaturated esters. These reactions are carried out in dichloromethane at −40°C. The results show that, for the esters using the chiral alcohols (72-75), the best overall yield was obtained using the optically active alcohol (74) as shown (Scheme 6). Having established the best chiral auxiliary to use, the ester (76) was converted into (77) using the (*R*)-lactone, while the (*S*)-lactone afforded (78). The *de* values from these two reactions were around 89%.

2.1.3 Rearrangement Reactions. A report has given details of the photoconversion (irradiation at 350 nm) of some bicyclic enones (79) into the corresponding phenols.[49] A further study of the photochemical cyclizations encountered with the deuteriated enone (80) has been reported.[50] The reaction paths were checked using this deuteriated ester, and the various rearrangement routes were identified. These afforded the cyclized products (81-83) as well as the *cis,trans* isomerized material (84). The involvement of a radical anion (85) was postulated. Other photocyclization reactions have been reported such as the conversion of (86) in cyclopentane solution into the quinolones (87) in the yields shown.[51] The cyclizations can be carried out in the presence of chiral inductors such as ephedrine, when enantiomeric excesses of 37% are achieved without a reduction in the overall yields. The photochemical synthesis of some quinol-2-ones, quinolines and coumarins has been described.[52] The key to the synthetic path is the *trans,cis* isomerization of *o*-aminocinnamoyl and of *o*-hydroxycinnamoyl derivatives.

Irradiation of (88) in anhydrous benzene affords the pyran derivatives (89).[53] These compounds are also photochemically reactive, and irradiation in aqueous methanol brings about their conversion into the bicyclic ketones (90). These are

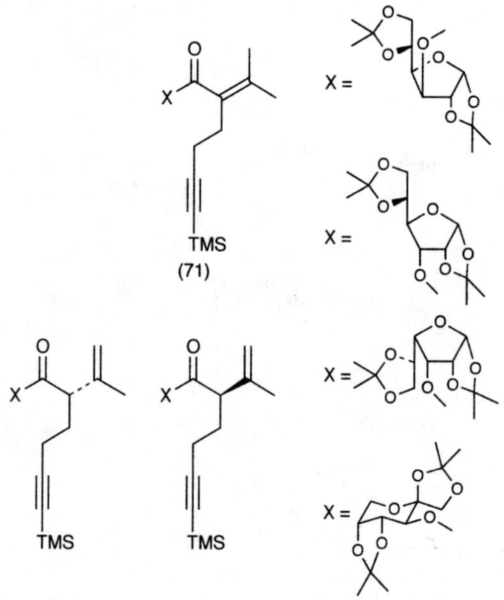

(71)

	%	de
DAG	87	>95
DAA	73	38
DAG*n*	80	66
DAF	70	>95

Scheme 5

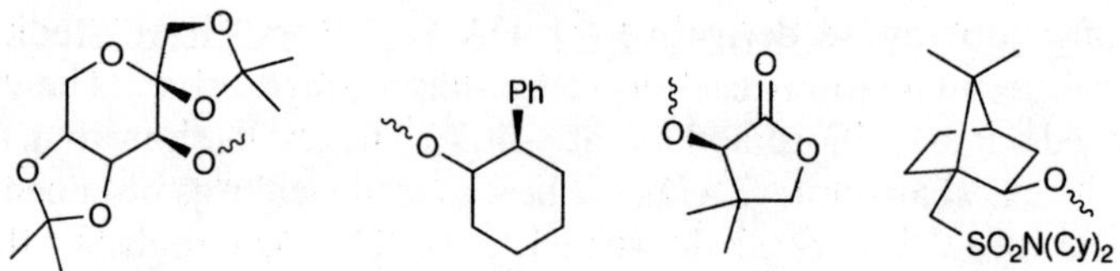

(72) (73) (74) (75)

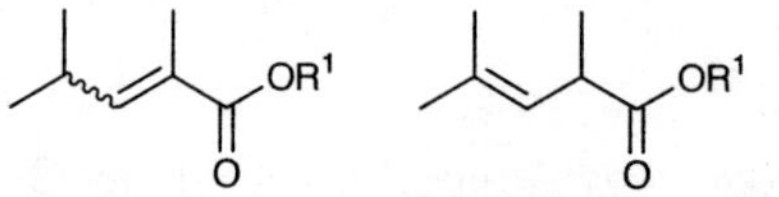

	yield (%)	de (%)
(72)	81	85
(73)	81	83
(74)	82	88
(75)	33	67

Scheme 6

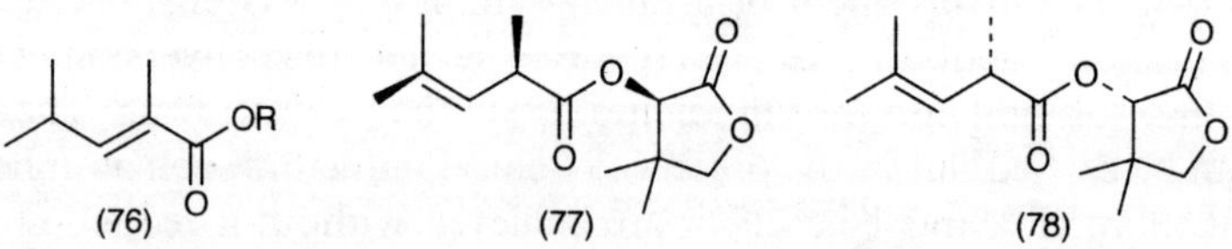

(76) (77) (78)

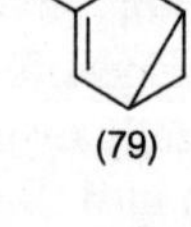

(79)

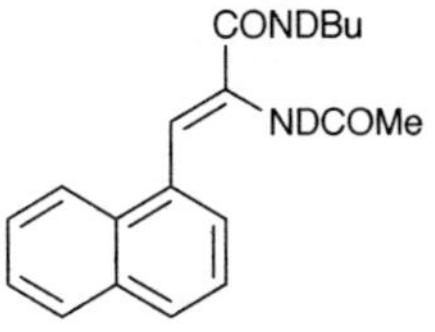
(80)

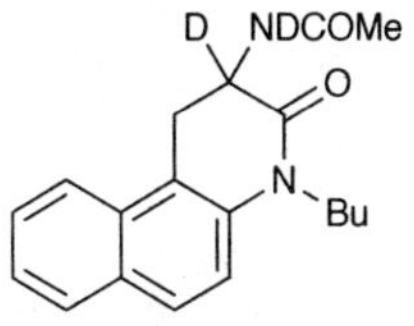
(81)

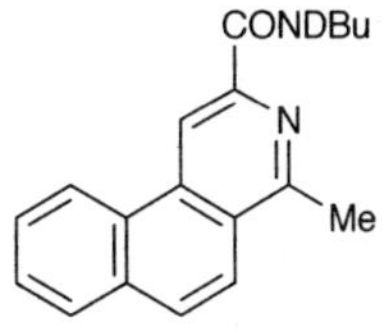
(82)

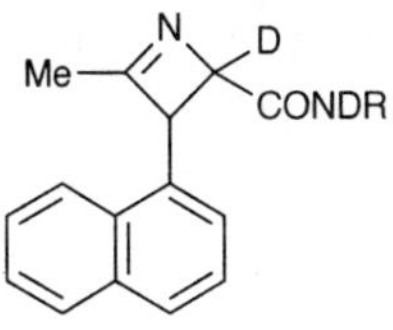
(83)

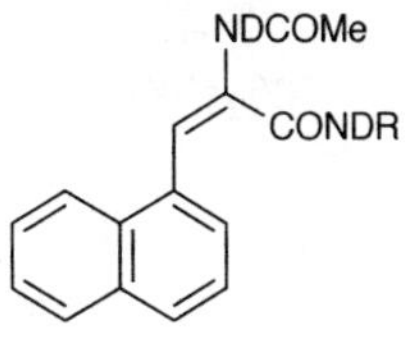
(84)

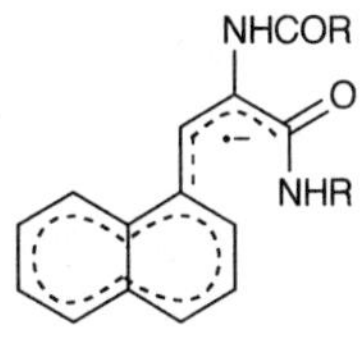
(85)

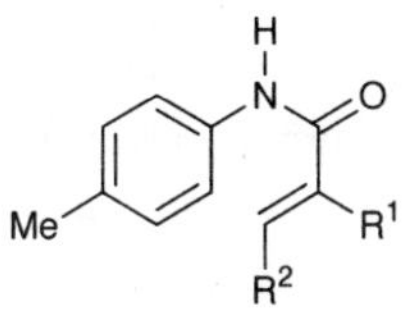
(86)

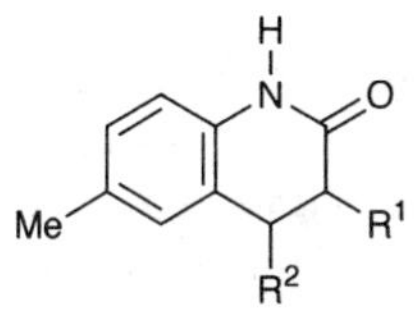
(87)

(86)	(87)
R^1 = Me, R^2 = H	96%
R^1–R^2 = $(CH_2)_4$	94% (1:1)

thermally labile and readily convert into the isomeric ketones (91). The yields of (91) are high except for the derivatives (90, Ar = *p*-$MeOC_6H_4$ or furan-2-yl) when the products (91) are accompanied by the tricyclic ketones (92).

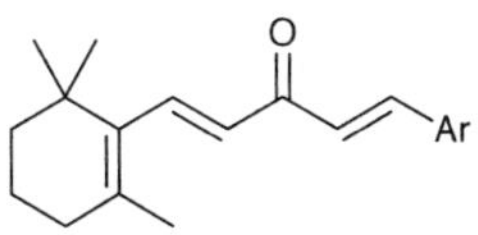
(88)

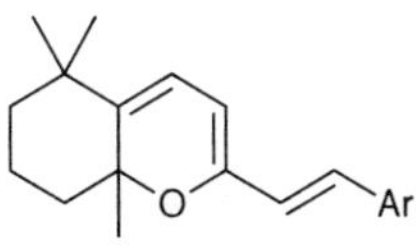
(89)

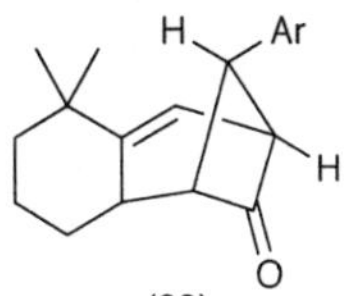
(90)

Ar = Ph, *p*-MeC_6H_4, *p*-$MeOC_6H_4$, *p*-BrC_6H_4, *p*-ClC_6H_4, furan-2-yl

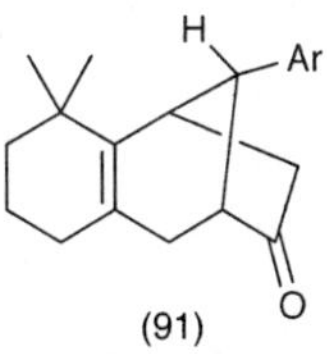
(91)

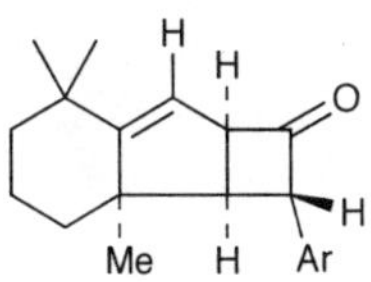
(92)

Ar = *p*-$MeOC_6H_4$, furan-2-yl

The influence of 4- and 4′-ether substituents on the photochemical isomerism of some benzylideneacetophenones has been studied.[54] The electron donating substituents influence the bond order of the single bonds and raise the barrier to rotational isomerism. The chalcones (93) are photochemically reactive and undergo isomerization around the double bond.[55] The quantum yields for this process are in the range 0.2 – 0.4 in neutral aprotic solvents. The intramolecular hydrogen transfer within (94) to yield (95) has been studied further.

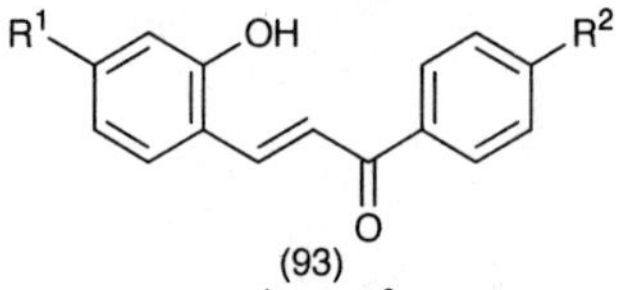

(93)

R^1	R^2
NEt_2	H
NEt_2	OMe
NEt_2	NEt_2
NMe_2	NMe_2

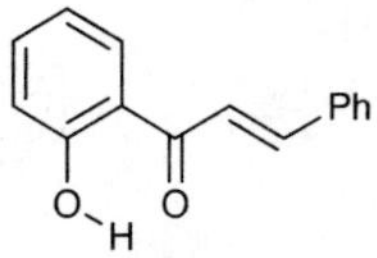

(94)

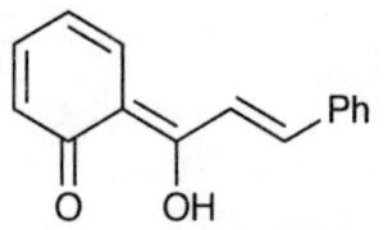

(95)

The *trans*-isomer shown does not undergo *trans,cis*-isomerization, but the *cis*-isomer of (94) undergoes a one-way *cis,trans*-isomerization.[56] Methoxynaphthyl derivatives of chalcones undergo *cis,trans* isomerization only.[57]

A theoretical study of the photochemical decomposition of cyclopropenones to CO and ethynes has shown that the decomposition occurs stepwise.[58] The irradiation of (96) leads to the formation of the free radicals (97) by what is essentially a Norrish Type I fission reaction.[59] An electron-transfer process is involved in the synthesis of the spongianone (98, 23%).[60] This has been prepared using the photocyclization of the lactone derivative (99) that is carried out by irradiation (λ = 300 nm) in acetonitrile/water at -25°C using 1,4-dicyano-2,3,5,6-tetramethylbenzene as the sensitizer.

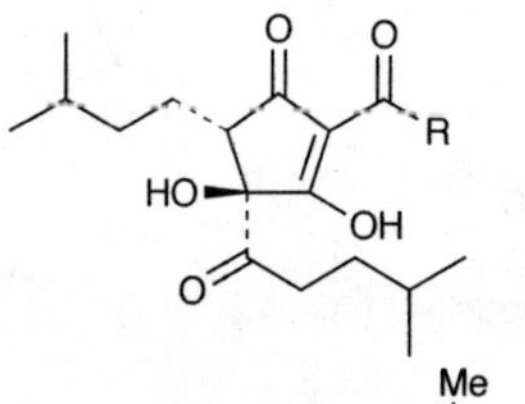

(96) R = $CH_2CH(Me)_2$, $CH(Me)_2$, $\overset{Me}{C}HCH_2Me$

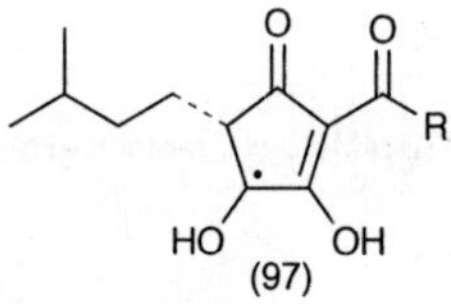

(97)

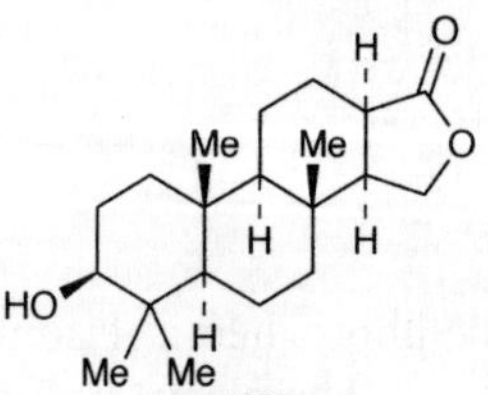

(98)

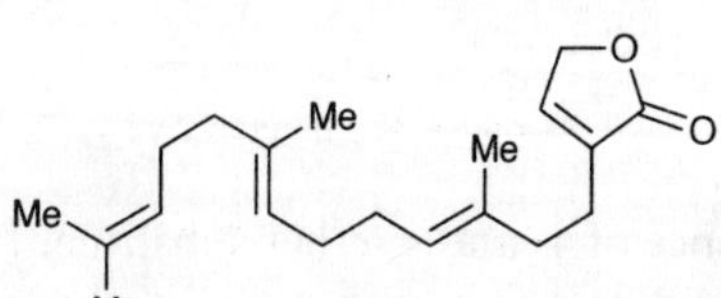

(99)

A detailed study of the photochemical reactivity of the enone (100) has been reported.[61-63] The prime photochemical reaction is the *cis,trans*-isomerization of (100) to yield (101). Subsequent cyclization yields the zwitterion (102) from which many of the products are formed.

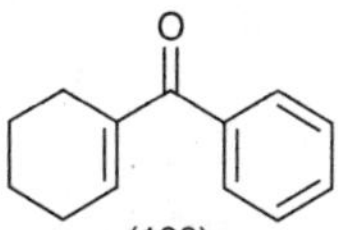
(100)

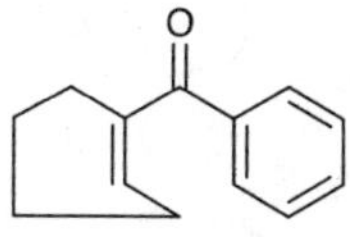
(101)

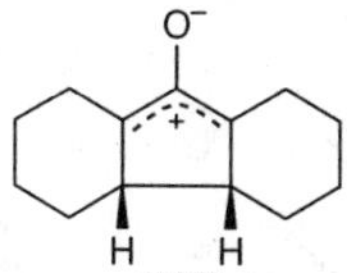
(102)

Further examples of the photochemical behaviour of some enones (103-105) in the solid state have been described by Zimmerman and Nesterov.[64] Interestingly the enone (103) undergoes formation of the cyclopropane (106) in solution. This is also formed in the solid state, but an additional reaction, the formation of (107), also occurs. The triphenyl-substituted enone (104) also behaves differently in the solid state from solution and yields the four products (108-111). The *t*-butyl derivative (105) follows an analogous reaction path and yields only the product (112). A detailed analysis of why such different photoreactions occur in the solid state was presented. The photochemical rearrangement of the enone (113) results in the formation of (114) and (115) as the two major products.[65] These are formed by 1,3-migrations, either of the 4,10-bond to afford (114, 15%) or of the 4,5-bond to yield (115, 10%). The former product undergoes an oxa-di-π-methane process to yield (116, 1%), while (115) undergoes decarbonylation to yield (117, 3%). Park and Oh[66] report that, upon irradiation, derivatives of 2-phenyl-4-alkylidene-5(4*H*)-oxazolone undergo 1,3-benzoyl migration. Irradiation at 436 nm of the N-acyl derivative (118) results in migration of the acyl group to the carbonyl oxygen to afford an enol acetate.[67]

The direct irradiation of certain allenyl alkenes brings about a series of photochemical reactions as illustrated in Scheme 7. (2 + 2)-Cycloaddition affords housane derivatives, while interaction with the ester group yields the dihydropyran derivatives. Di-π-methane reactions also compete, as does *trans-cis* isomerization of the α,β-unsaturated ester function. The dicyano derivatives also behave more or less in a like manner.[68]

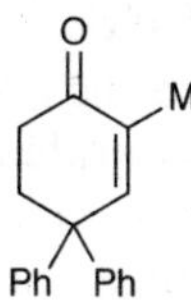
(103)

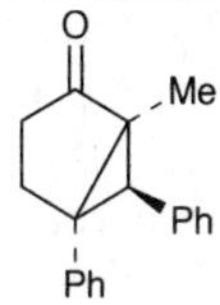
(106)

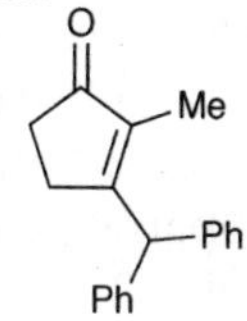
(107)

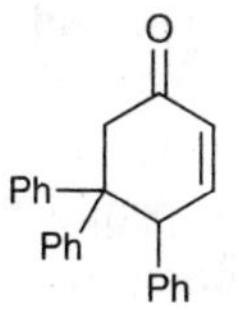
(104)

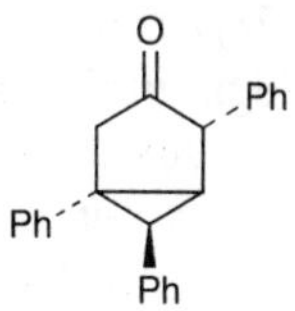
(108)

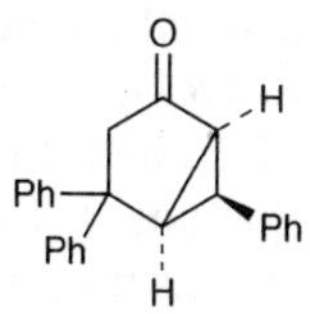
(109)

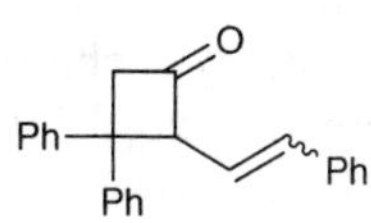
(110)

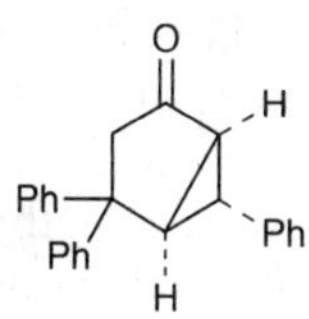
(111)

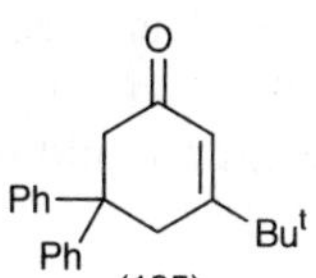
(105)

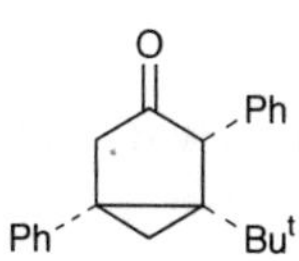
(112)

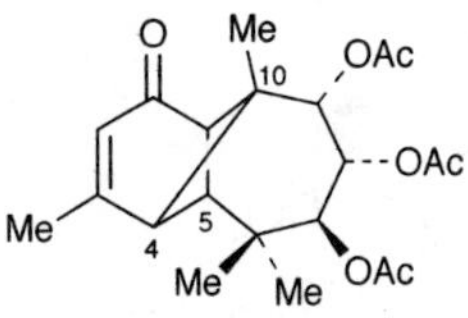

(113)

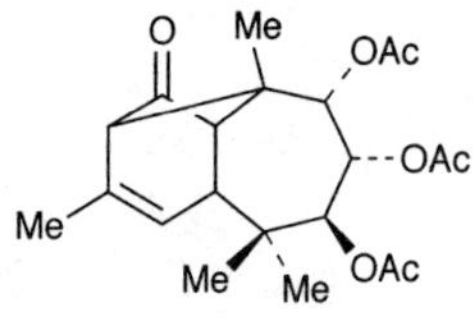

(114)

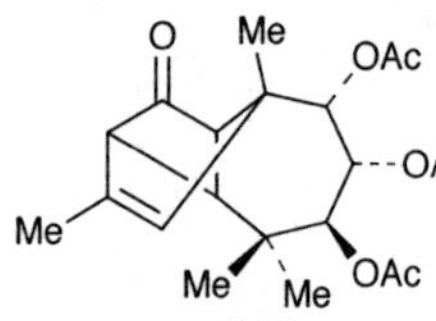

(115)

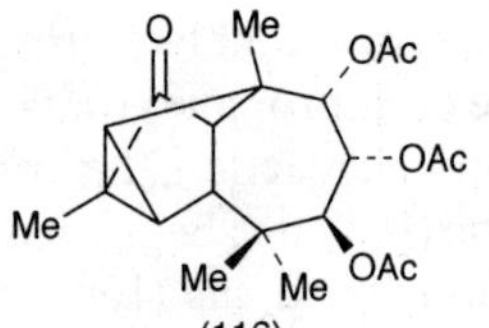

(116)

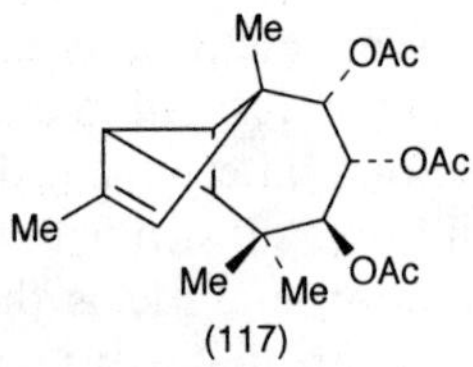

(117)

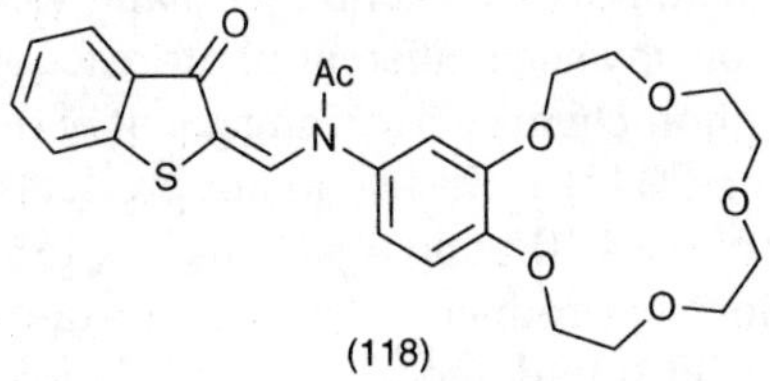

(118)

2.2 β,γ-Unsaturated Systems. – *2.2.1 The Oxa di-π-methane Reaction and Related Processes.* The dienone (119), as a complex in cyclodextrins, affords a single product when it is irradiated, identified as (120).[69] The structural rearrangement could be the result of an oxa-di-π-methane rearrangement followed by ring opening of the resultant cyclopropyl ketone. Further studies by Armesto and his co-workers[70] have examined the oxa-di-π-rearrangement of the unsaturated aldehyde (121). This compound undergoes the oxa-di-π-methane rearrangement on sensitized irradiation using *m*-methoxyacetophenone. The yields of the cyclopropyl products (122) are as shown under the appropriate structures. Decarbonylation and 1,3-acyl migration reactions were also detected and afford (123) and (124). The rearrangements of the aldehyde derivatives (125) into the cyclopropyl aldehydes (126) were also studied. Interestingly the β,γ-unsaturated aldehydes (127a) do not undergo rearrangement. Decarbonylation results on irradiation, with the formation of (127b). Several examples of this smooth photochemical conversion were reported.[71]

The photochemical rearrangement of the enone (128) has been used to provide a key intermediate in the synthesis of cyclopentanoid natural products.[72] Irradiation of 5-oxo-7-methylenebicyclo[2.2.2]oct-2-ene results in the formation of 3-oxo-6-methylenetricyclo[3.3.0.$0^{2,8}$]octane.[73] The formation of this product shows that only the oxa-di-π-methane rearrangement is operative in this system. The exploitation of the 1,2-acyl migration reactions within endo-tricyclo[5.2.2.$0^{2,6}$]undecanes as a synthetic approach to triquinanes has been reported.[74] The photochemical rearrangement of salannin has been described.[75] Apparently two reaction modes are observed, one involving a (4+2)-cycloaddition and another a 1,3-sigmatropic shift.

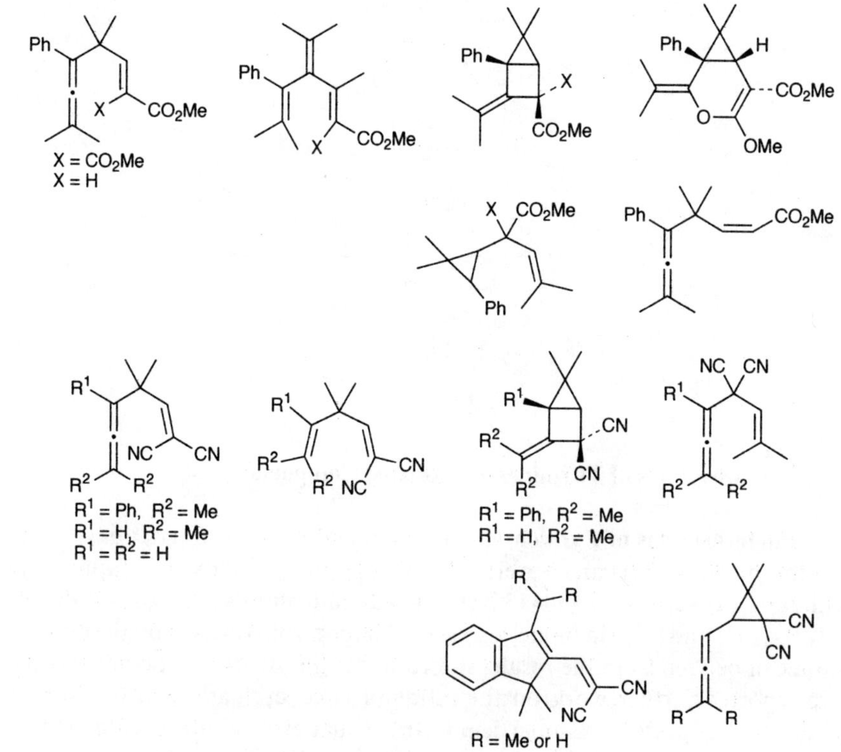

Scheme 7

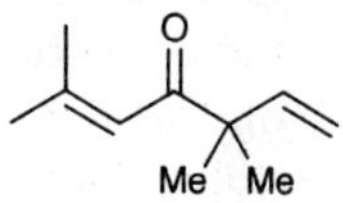

(119)

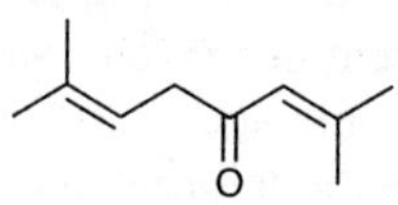

(120)

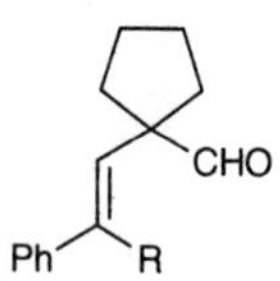

(121) R = H or Ph

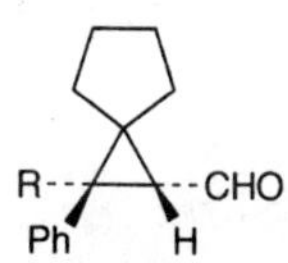

(122) R = H 36%
R = Ph 34%

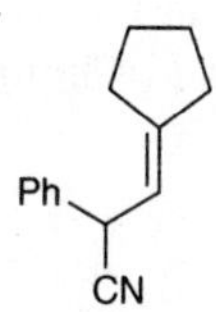

(123)
from R = H

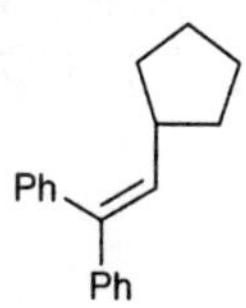

(124)
from R = Ph

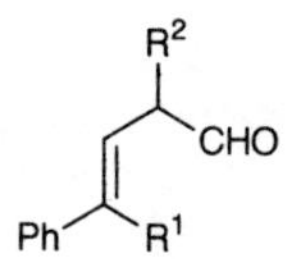

(125)

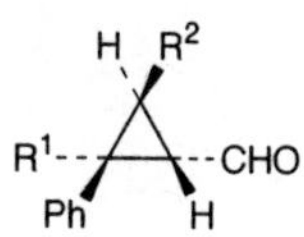

(126)

R^1	R^2	yield (%)
H	Pr^i	21
Ph	Pr^i	23
H	Et	22
Ph	Et	24

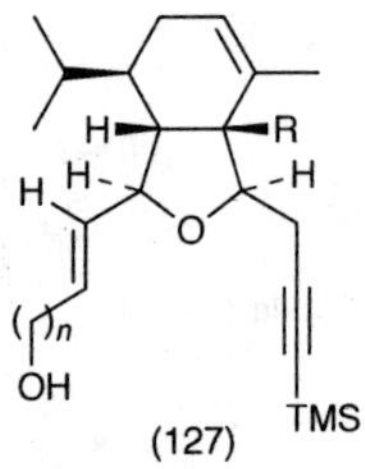

(127)

	R	n
a	CHO	1 or 2
b	H	1 or 2

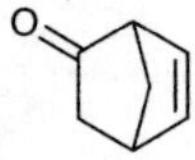

(128)

3 Photoreactions of Thymines and Related Compounds

3.1 Photoreactions of Pyridones. – Sieburth and his co-workers[76] have examined the influence that can be exerted by silyloxy groups on the (4+4)-photocycloaddition between (129) and (130). The irradiation affords a mixture of the two adducts (131) and (132) in moderate yields. The control exercised on the cycloaddition can be seen from the results shown under the structures. Seiburth *et al.*[77] have reported further work on the intramolecular cycloaddition reactions of pyridones. The present example demonstrates that cycloaddition within (133, R = Me) affords the product (134). When (133, R = H) is irradiated, a mixture of the two adducts (135) and (136) is obtained. The ratio of products has been demonstrated to be solvent dependent. Sieburth and co-workers[78] described the synthesis of (137) some time ago. Further work on this photoadduct has been reported, and reaction with chlorine affords the polyquinane compound (138). Irradiation of (139) in methanol at 340 nm brings about the formation of the hydroxypyridone (140) and the pyridone (141).[79] These products arise by either heterolysis of the CO bond or homolysis of the NO bond.

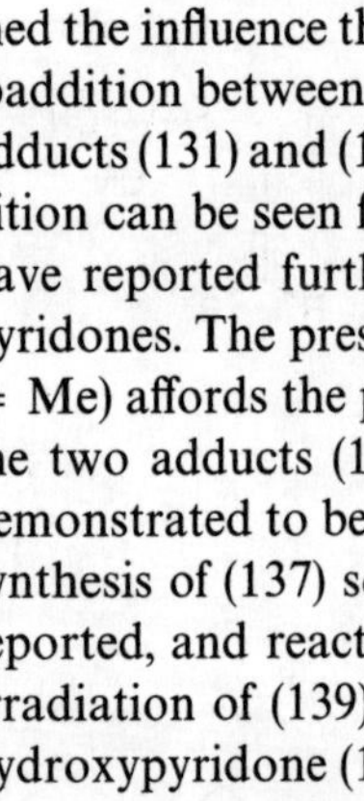

(129)

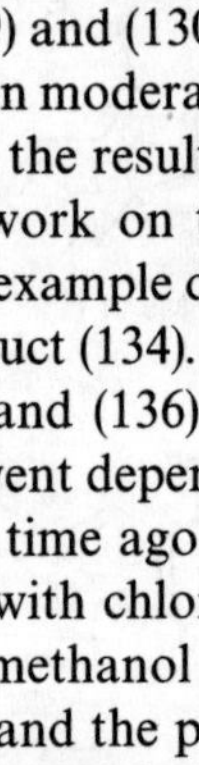

(130)

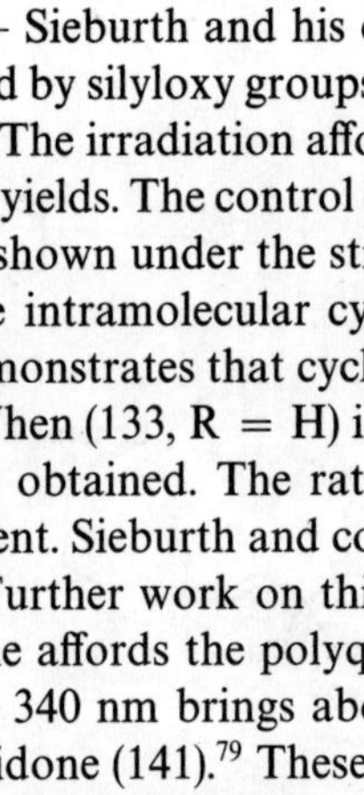

(131)

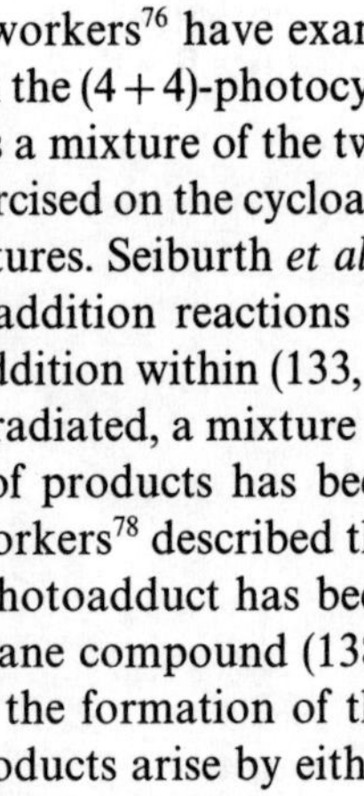

(132)

SiR_3		yield (%)
Bu^tMe_2Si	3 : 2	35
$(Pr^i)_3Si$	2 : 1	22
$(PhCH_2)_3Si$	10 : 1	49

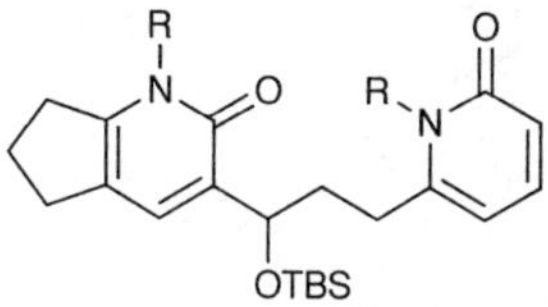

(133)

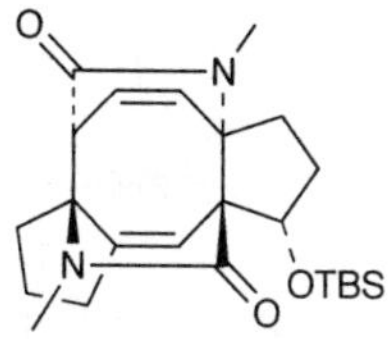

(134)

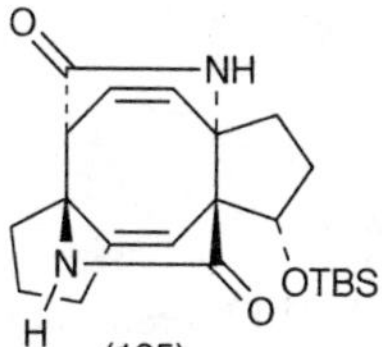

(135)

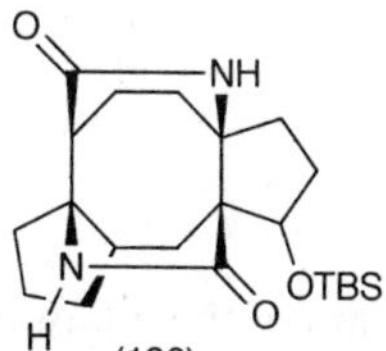

(136)

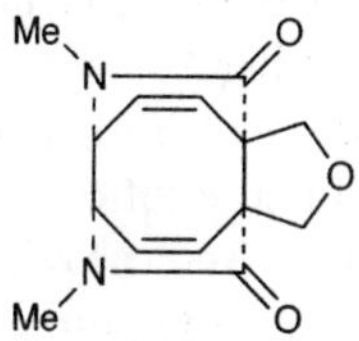

(137)

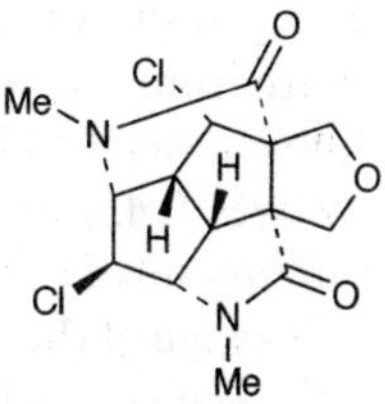

(138)

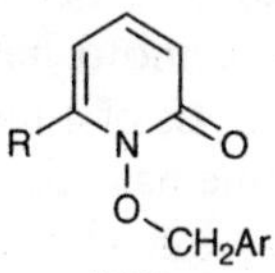

(139)

R = H, Ar = 9-anthryl
R = Me, Ar = 9-anthryl
R = H, Ar = 1-pyrenyl

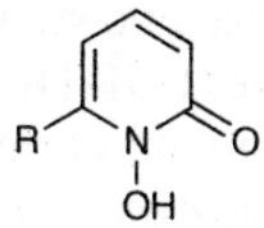

(140)

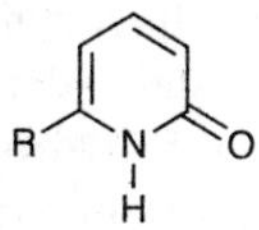

(141)

3.2 Photoreactions of Thymines *etc.* – The photochemical conversion of 1,3,6,8,10-pentamethylcyclooctapyrimidine-2,4-dione into 9, 11-diazapentacyclo[6.4.0.$0^{1,3}$.$0^{2,6}$.$0^{4,8}$]-dodecane and pentaleno[2,1-d]pyrimidine is carried out in the presence of trifluoroacetic acid.[80]

A study has demonstrated that clusters of thymine molecules undergo extensive dimerization when subjected to strong UV irradiation.[81] *p*-Aminobenzoic acid can act as a photosensitizer for the dimerization of thymine, but dimerization can be inhibited by carrying out the reactions in aqueous solutions of β-cyclodextrin.[82] The reason for the inhibition is thought to be the formation of an inclusion complex between the aminobenzoic acid and the cyclodextrin. Others have described the dimerization of thymine, cytosine and uracil,[83] and have also reported that a variety of products are formed when thymine, uracil or cytosine are irradiated in alkaline medium.[84] A polymer containing uracil units has been irradiated at 265nm.[85] This brings about changes that are considered to arise from dimerization of the uracil units.

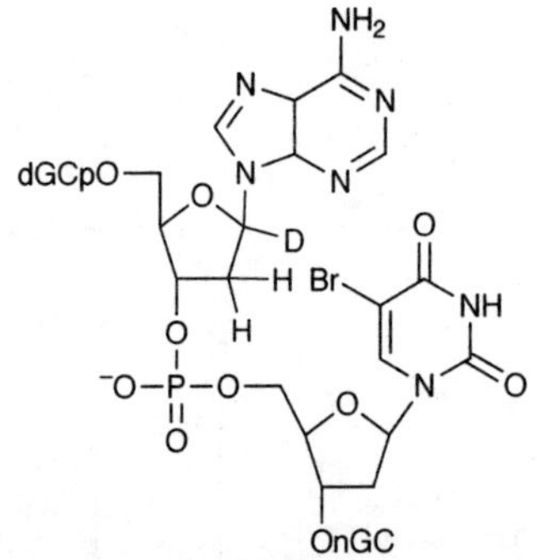

(142)

Irradiation at 302 nm of the oligonucleotide (142) results in C-Br bond fission, with the formation of the corresponding vinyl radical.[86] This radical exclusively abstracts the deuterium atom shown in the molecule. There is no evidence for a hydrogen-abstraction path from the neighbouring methylene group. The photolysis of N_1-retinoyl-5-fluorouracil has been studied.[87]

The coumarin derivatives (143) and (144) are photochemically reactive on irradiation in methanol/aqueous HEPES buffer solution.[88] The irradiation results in photochemical cleavage of the O-P bond and the release of the coumarin moiety. The photochemical behaviour of the benzophenone derivative (145) has been studied.[89] The synthesis of the benzophenone derivatives (146) has been carried out to ascertain their application in studies of protein prenyltransferases.[90] Kale *et al.*[91] have synthesized and studied the photochemical behaviour of the prenylated cysteine derivative (147). The synthesis of a photolabile group for the protection of 5′-hydroxy function of thymidine has been reported.[92] The stilbene moiety of (148) undergoes *trans,cis*-isomerization when it is irradiated in dichloromethane solution.[93] Prolonged irradiation brings about the

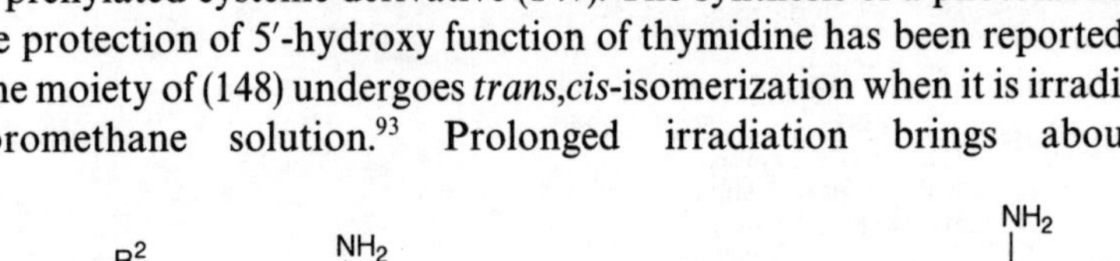

(143) (144)

$R^1 = R^2 = H$; $R^1 = H$, $R^2 = OMe$; $R^1 = OMe$, $R^2 = H$; $R^1 = R^2 = OMe$; $R^1 = N(Me)_2$ or $N(Et)_2$, $R^2 = H$

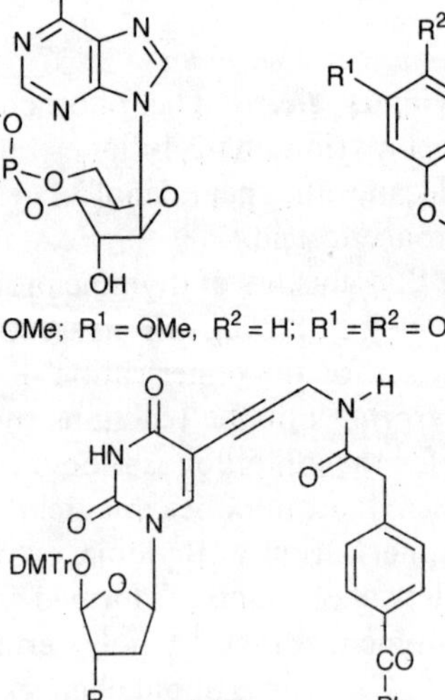

(145)

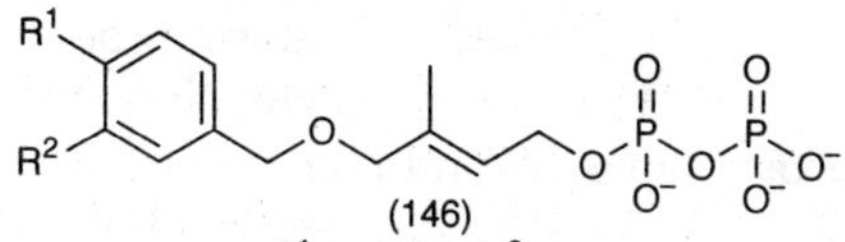

(146)

R^1 = PhCO, R^2 = H
R^1 = H, R^2 = PhCO

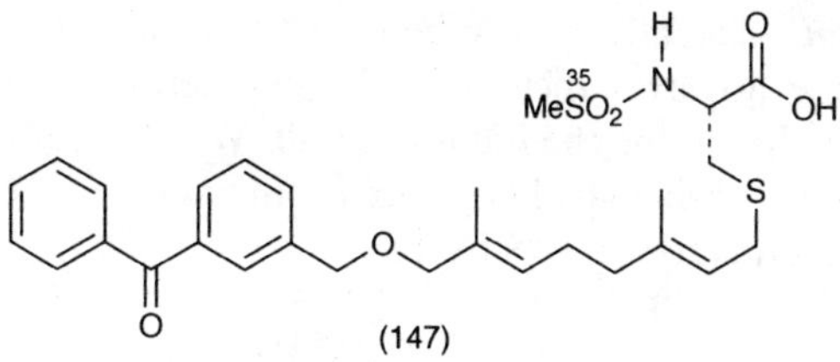

(147)

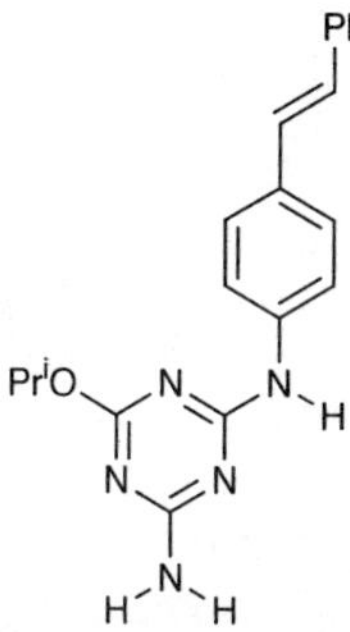

(148)

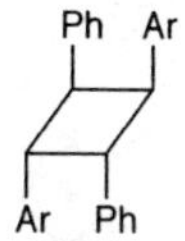

(149a)

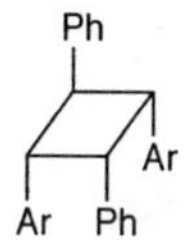

(149b)

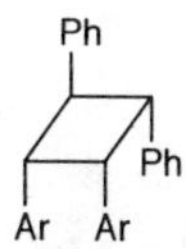

(149c)

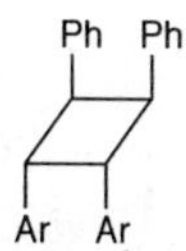

(149d)

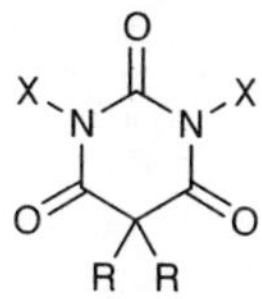

(150)

R	X
n-C_6H_{13}	H
Et	Me

formation of the four cyclobutane derivatives (149a-d), with (149a and 149d) as the major products. When (150, 0.5 equivalents) is added, the formation of (149b-d) is enhanced, with the suppression of (149a). The molecule (150) is considered to be the template on which the dimerization process takes place.

4 Photochemistry of Dienones

4.1 Cross-conjugated Dienones. – A further investigation by Uppili and Ramamurthy[94] of the control exercised by confinement has examined the outcome of the conversion of the dienone (151) into the cyclopentenones (152) and (153).

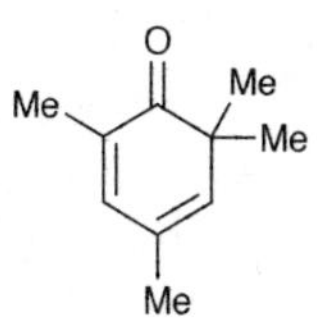

(151)

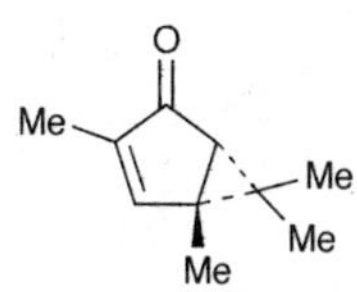

(152)

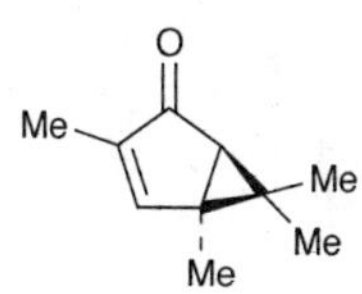

(153)

4.2 Linearly Conjugated Dienones. – Blay and his co-workers[95] have continued their study of the photochemical conversion of santonin (154) in ethanoic acid into lumi-santonin (155, 38%) and the use to which this product can be put. The principal aim is its use for the synthesis of natural products. The present work describes the elaboration of (155) into plagiochiline N (156). The photorearrangement of the cyclohexadienone (157) into the enone (158) is best carried out by irradiation in pentane/TFA (33%).[96] This treatment affords (158) in a yield of 82%. Other irradiation conditions such as acetic acid-TFA and 1,4-dioxane-TFA are less efficient. The photochemistry of the dienones (159) and (160) shows marked wavelength dependence.[97] When (159) is irradiated at 254 or 366 nm, where the dienone absorbs, then only the lumiketone (161) is formed. This behaviour is also shown for the dienone (160). When (159) and (160) are irradiated at 310 nm, Norrish Type I cleavage occurs at the C-17 position, with the formation of (162) and (163), respectively.

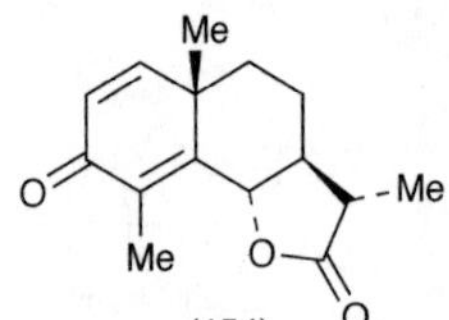

(154)

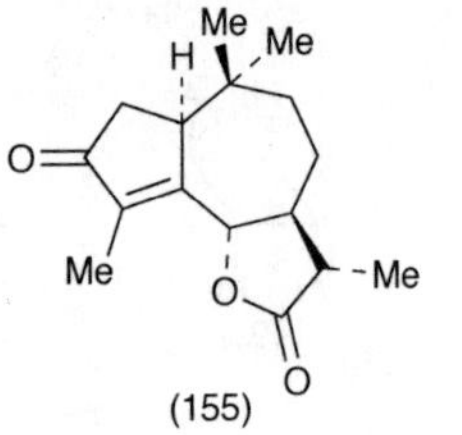

(155)

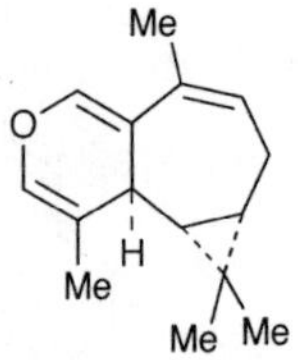

(156)

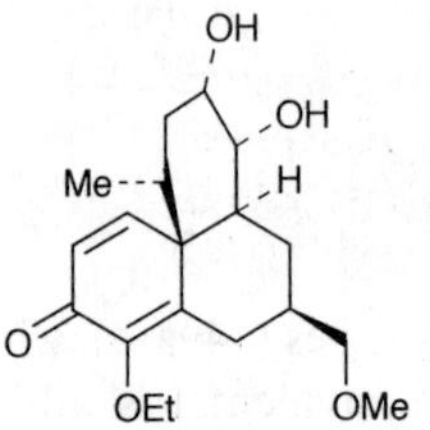

(157)

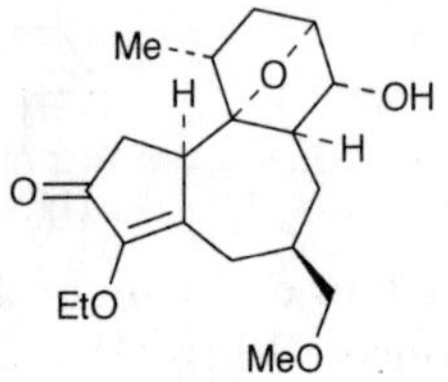

(158)

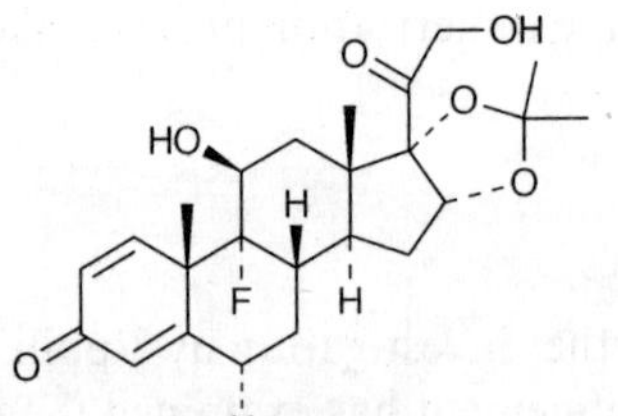

(159)

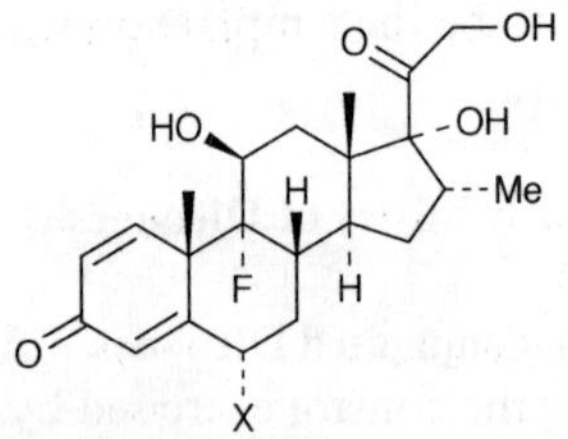

(160)

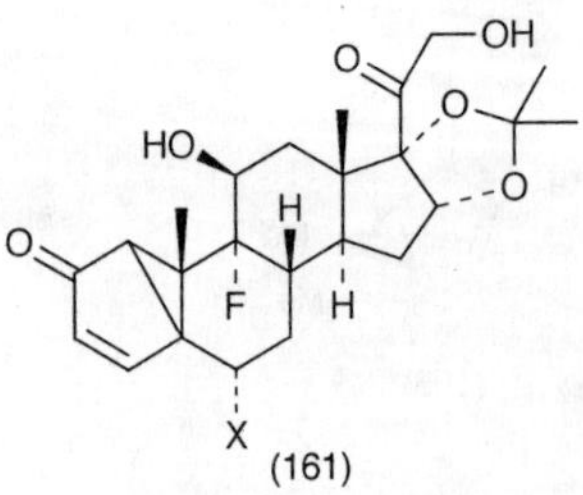

(161)

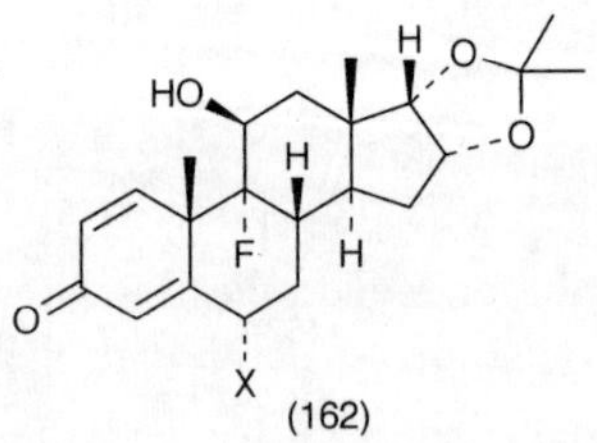

(162)

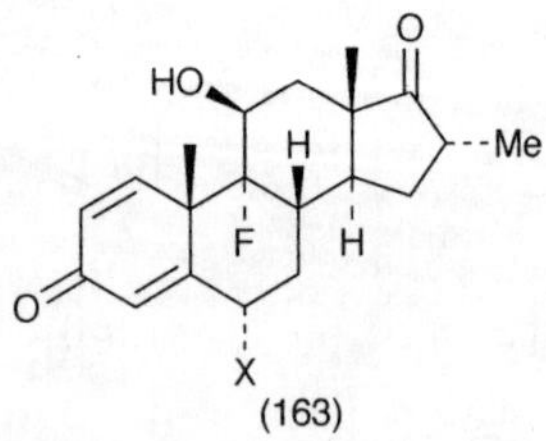

(163)

The triplet carbene (164) is formed by irradiation of the corresponding diazo compound.[98] This carbene adds to ethyne to afford the new triplet carbene (165), which in turn is photochemically reactive and cyclizes to (166) on irradiation at wavelengths > 515 nm.

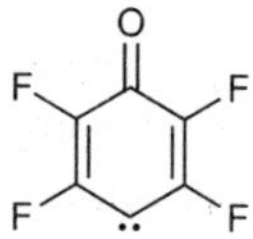
(164)

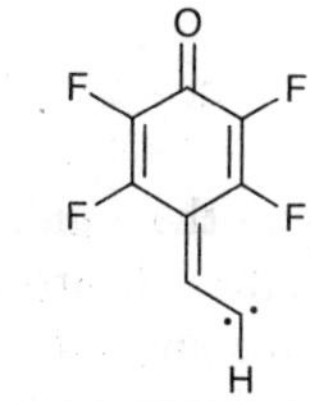
(165)

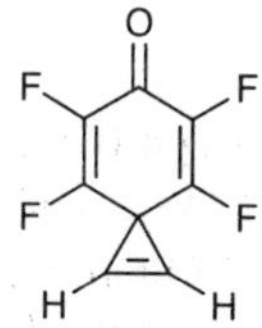
(166)

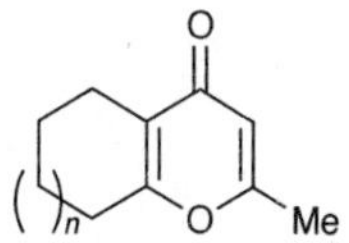
(167) $n = 1$, 2 or 3

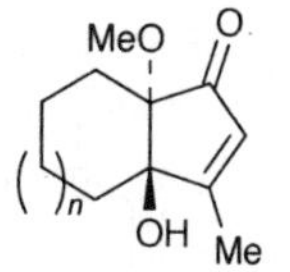
(168) $n = 1$, 2 or 3

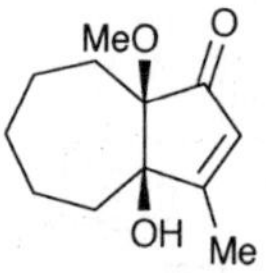
(169)

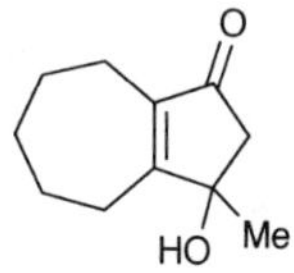
(170)

A further study of the photochemical rearrangement of pyrones such as (167) has been reported.[99] The reactions provide a useful synthetic path to the cyclopentenone products (168-170), when the reactions are carried out in methanol. In aqueous sulfuric acid, rearrangement and addition also occur. In this case, however, the addition to (167, n = 1) affords the isomeric products (171) and (172) (Scheme 8). The relative ratios of these are dependent upon the acid concentration, since the diols undergo a thermal retro-Aldol process with recyclization. Dimeric compounds are formed on the irradiation (300 nm) of derivatives of styryl-4-pyrones.[100] A photophysical study of some α- and γ-pyrones has examined the influence of ring substituents.[101]

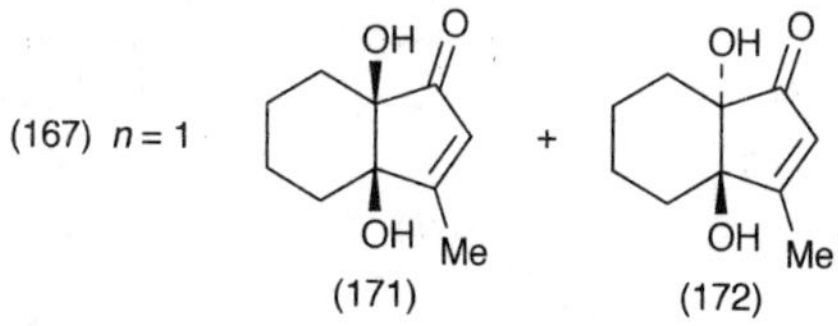

Scheme 8

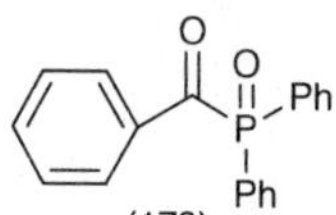
(173)

5 1,2-, 1,3- and 1,4-Diketones

5.1 Reactions of 1,2-Diketones and other 1,2-Dicarbonyl Compounds. – An ab initio study of the photodissociation of glyoxal has examined the extrusion of CO.[102] Other reports give details of the photochemical reactions open to both glyoxal and glycolaldehyde[103] and also of pyruvic acid.[104] The Norrish Type I fission of the keto derivative (173) and the subsequent addition of the resultant radicals to vinylacrylate have been studied.[105] Two-photon excitation of the acyl cyanides (174) brings about C-C fission, with the formation of a CN radical.[106]

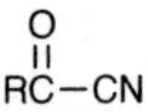

(174) R = Me or Bu^t

The Norrish Type II reactivity of the amide (175) in the crystalline state has been studied.[107] Crystals of salts of the acid group in (175) were prepared with the amines, (*R*)-(+)-1-phenylethylamine and L-(−)-prolinamine. Irradiation of the two crystalline compounds afforded single products that were converted to the esters (176). The products have an enantiomeric excess of > 99%. Griesbeck and

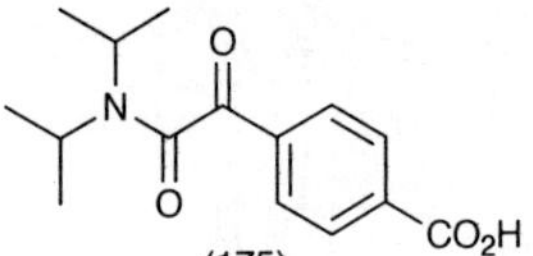

(175)

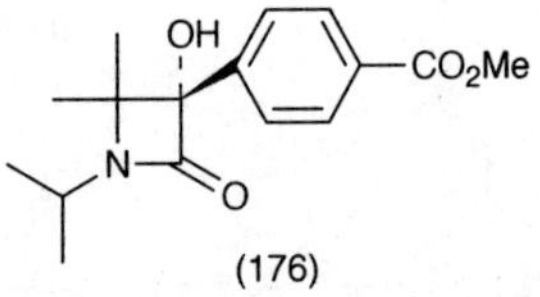

(176)

Heckroth[108] have also studied the photochemical cyclization of phenylglyoxylamides. They examined the cyclizations of (177) that were prepared from the enantiomerically pure α-amino acids identified below the structures. Irradiation of these derivatives at 300 nm follows the Norrish Type II hydrogen-abstraction path. Cyclization of the resultant biradicals afford excellent yields of racemic products (178) and (179). The diastereoselectivity for the product formation is shown beside the appropriate structures. A review has highlighted the solid-state enantioselective synthesis of β-lactams by the photocyclization of amides.[109] The reactions involved inclusion complexes between the amide and chiral host molecules.

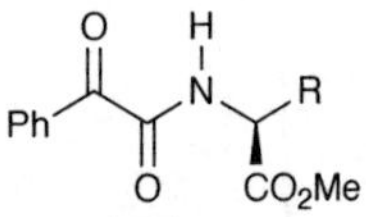

(177)

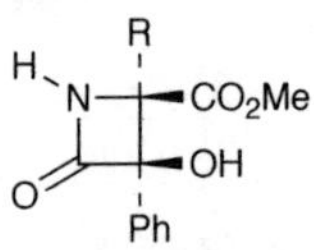

(178)

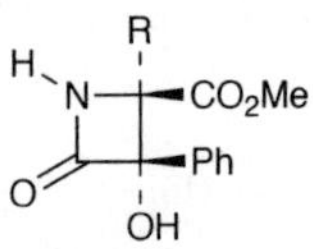

(179)

(177)	(178) : (179)
glycine	78 : 12
alanine	80 : 20
phenylalanine	80 : 20
valine	86 : 14
leucine	82 : 16
isoleucine	75 : 25
t-leucine	62 : 28
aspartic acid	41 : 59
glutamic acid	57 : 43

1-Acetylisatin (180) undergoes photochemical addition to a series of alkene systems such as furan, benzofuran and acetoxyethene.[110] The addition of the furans affords the pairs of oxetanes (181) and (182) in the yields shown. With the acetoxyethene, all modes of (2+2)-cycloaddition are followed and yield the mixture of oxetanes (183, 184).

Complexes of the (1*R*)-camphoroquinone (185) in α-, β- and γ-cyclodextrins have been studied spectroscopically.[111] The results indicate chiral discrimination within the complexes. Liao[112] has reviewed his work relating to the photochemical behaviour of bicyclo[2.2.2]octenones.

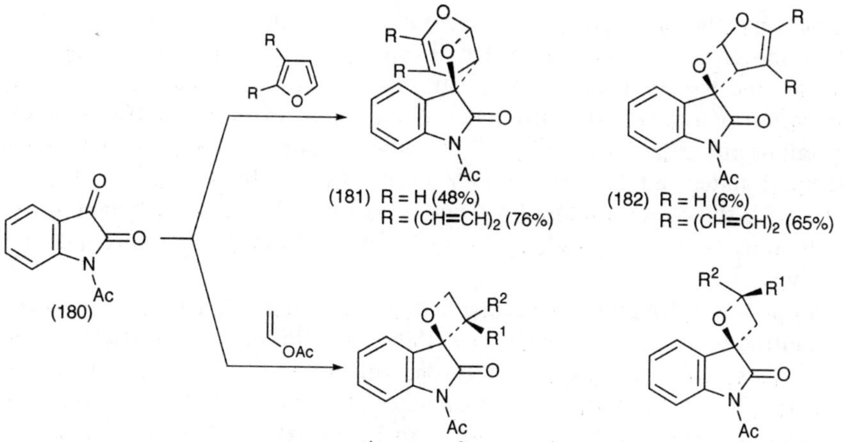

(180)

(181) R = H (48%)
R = (CH=CH)$_2$ (76%)

(182) R = H (6%)
R = (CH=CH)$_2$ (65%)

(183) R^1 = AcO, R^2 = H (55%)
R^1 = H, R^2 = AcO (5%)

(184) R^1 = H, R^2 = AcO (32%)
R^1 = AcO, R^2 = H (25%)

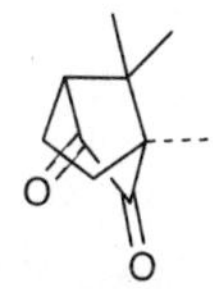

(185)

5.2 Reactions of 1,3-Diketones. – The irradiation (λ > 280 nm) of acetylacetone in a low temperature argon matrix has provided evidence for isomerization around the double bond of the enol form.[113] A laser flash study of the 1,3-diketone (186) has been carried out.[114] Irradiation at 355 nm brings about the

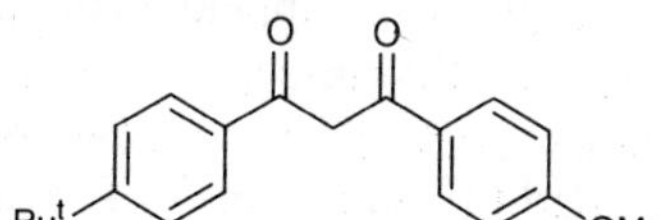

(186)

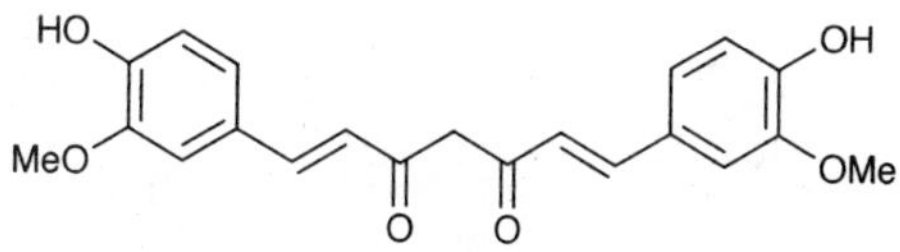

(187)

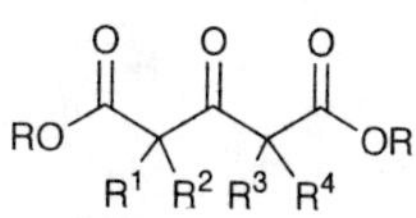

(188) R = adamantyl

	R^1	R^2	R^3	R^4	–CO in C_6H_3	–CO in crystal
a	Me	H	H	H	0	0
b	Me	H	Me	H	54	100
c	Me	Me	Me	H	19	100
d	Me	Me	Me	Me	20	100

formation of the enol form, and the decay kinetics of this species were measuered. Flash photolysis of curcumin (187) in dioxane-water leads to hydrogen abstraction by the triplet diketo form to afford the corresponding radical.[115] The generation of radical pairs from the triones (188) in both solution and the crystalline phase has been studied.[116] Decarbonylation is the principal process in the solid state, and follows Norrish Type I fission. In general the product obtained is the corresponding 1,4-dione. γ-Oxo-[1,1′-biphenyl]-4-butanoic acid reacts from its low lying ππ* triplet state and has an inefficient Norrish type I reactivity.[117]

Irradiation (through Pyrex) of the malononitrile derivative (189) in degassed acetonitrile with the dihydroacridine (190) results in an electron transfer process.[118] This is followed by a proton transfer and bonding to yield the adduct (191, 70% yield). The irradiation of 1,3-cyclohexadienone in methanol along with *p*-benzoquinone results in its conversion to 3-methoxycyclohex-2-en-1-one.[119]

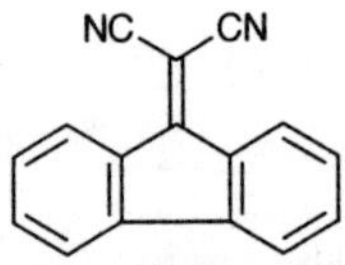

(189)

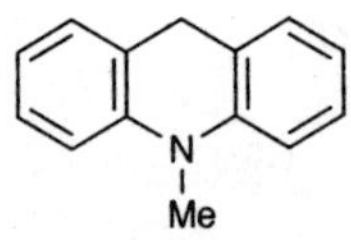

(190)

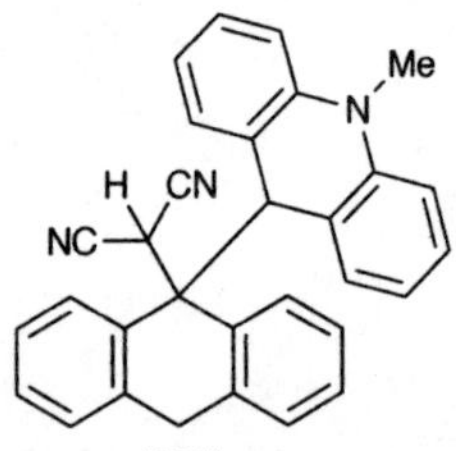

(191)

5.3 Reactions of 1,4-Diketones. – The *E,Z*-isomerization of (192) into (193) can be brought about photochemically.[120] Kokubo and his co-workers[121] have reported the photochemical addition of 1-phenylpropyne to the homoquinone (194). This yields the adduct (195) that undergoes thermal reactions with Lewis acid catalysts. The irradiation of (196) affords the cage compound (197) by way of a photochemical (2+2)-cycloaddition reaction.[122] The mechanism of this cyclization has been studied using ab initio calculations.

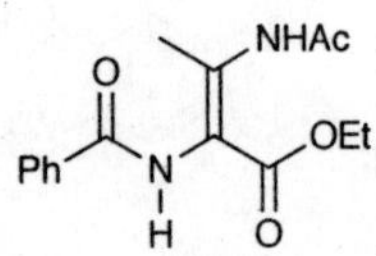

(192)

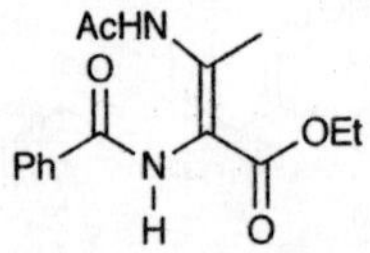

(193)

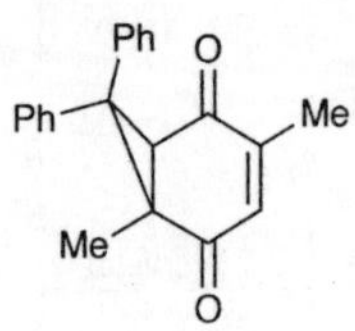

(194)

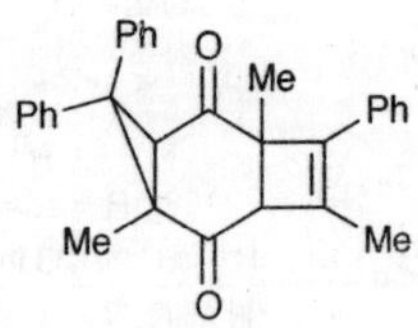

(195)

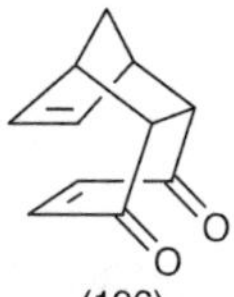

(196)

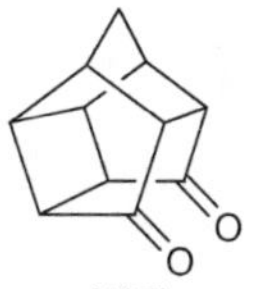

(197)

5.3.1 *Phthalimides and Related Compounds.* A detailed study of the synthesis of a series of tricyclic ring systems with a sulfur heteroatom has been reported.[123] This is an extension of earlier work by Griesbeck and his co-workers and makes use of the photochemical decarboxylation of potassium salts of acids such as (198) and (200). Irradiation brings about an intramolecular electron-transfer process followed by decarboxylation. Cyclization within the resultant species

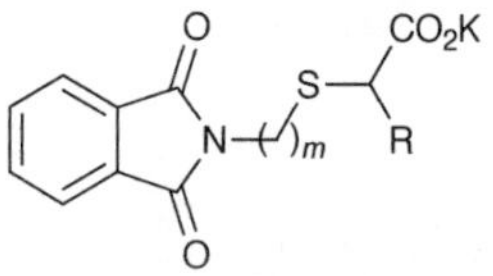

(198)

m	R	yield (%)
1	H	95
1	Me	98
2	H	61
2	Me	60
3	H	62
3	Me	93
4	H	66

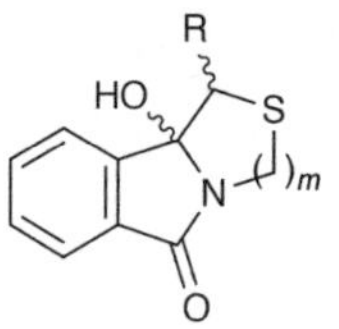

(199)

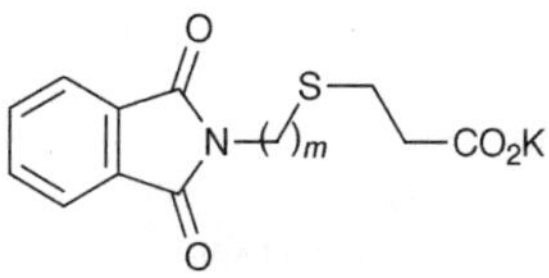

(200)

m	yield (%)
1	20
2	11
3	15
4	–

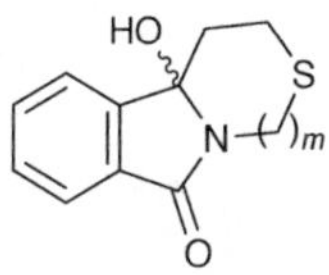

(201)

gives good yields of the products (199) and (201), respectively. Further examples of the photoinduced decarboxylation of anthranilic acid amides have been published.[124] The present work illustrates the conversion of the derivatives (202) into 1,4-dibenzodiazepines (203), in yields ranging from poor to excellent. With respect to this previous work, the decarboxylation of the phthalimides derivatives (204, 205) has been studied in detail.[125] The quantum yields for the decarboxylation process were measured. An example of the photodecarboxylation process occurring in ω-phthalimidoalkynoates has been used in the synthesis of a macroheterocyclic compound.[126] A study has been carried out on the photo-electron-transfer-induced cyclizations of ω-phthalimidoalkoxy acetic acids.[127] Photoelectron transfer has been studied in a new series of ferrocene-naphthalimide dyads.[128]

5.3.2 *Other 1,4-Dicarbonyl Compounds.* The photochemical decomposition of ethylenediaminetetraacetic acid using sunlight and using light in the 315-400 nm range has been studied.[129] Irradiation (at 254 nm) in an argon matrix at 9K of the phthalic anhydride (206) results in the formation of tetrafluorobenzyne.[130] The benzyne is photochemically reactive in the presence of CO, one of the by-products. Irradiation affords tetrafluorobenzocyclopropenone. UV irradiation of 1,3-diiodotetrafluorobenzene affords the corresponding tetrafluoro-1,3-benzyne. A study of the photochemical conversion of the anhydrides (207) and (208)

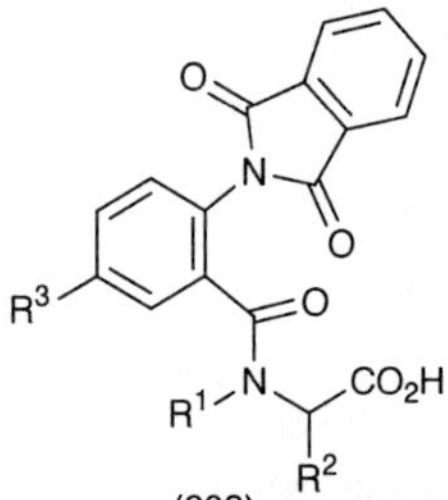

(202)

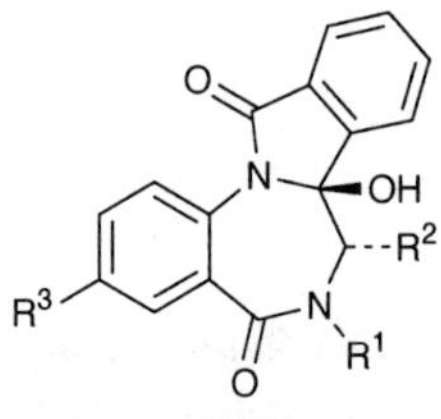

(203)

R^1	R^2	R^3	yield (%)
Me	H	H	56
H	Me	H	55
H	Pr^i	H	81
H	Bu^t	H	27
H	Bn	Cl	61

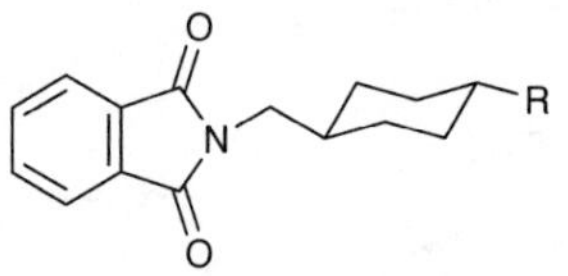

(204) R = CO_2H

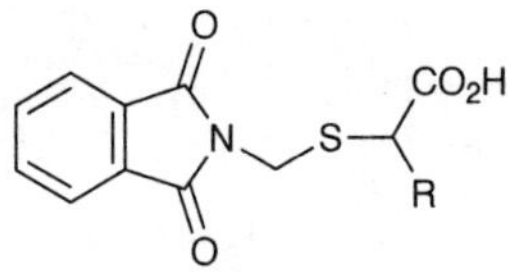

(205) R = H, Me, CH_2, CO_2H

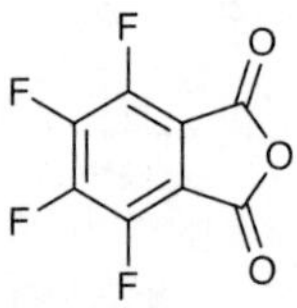

(206)

into the corresponding naphthynes has been reported.[131] Matrix photolysis of the dianhydrides, 1,2:5,6- and 2,3:6,7-naphthalenecarboxylic dianhydrides, has been examined as paths to 1,5- and 2,6-naphthodiynes.[132] The irradiation brought about decarboxylation and decarbonylation, with the formation of the desired products. These were identified by the usual spectroscopic techniques. The diynes are themselves photoreactive and undergo ring-opening processes. The irradiation using a XeCl laser of 3,4,9,10-perylenetetracarboxylic dianhyd-ride in the presence of Co powder results in enhanced elimination of the anhyd-ride groups.[133]

5.3.3 Fulgides and Fulgimides. A data-storage system based on pyrryl-substituted fulgides has been described.[134] The indolylfulgide derivative (209) has been synthesized and its photochromic behaviour has been studied.[135] The molecule (209) possesses a C_2 symmetric side chain. Irradiation at 313 nm brings about cyclization to the coloured form, with a *de* of 90%. The steric effects on the photocyclization of 2-indolylfulgides substituted with methyl and *iso*-propyl groups have been assessed.[136] Wolak and his co-workers[137] have described the photochromicity of a series of fluorinated indolylfulgides. A theoretical study of the CD spectra of indolylfulgides such as (210) has been carried out.[138]

The photochromic properties of (*E*)-dicyclopropylmethylene-(2,5-dimethyl-3-furylethyldiene)succinic anhydride have been studied.[139] Irradiation at 366 nm

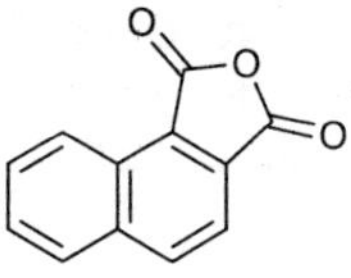

(207)

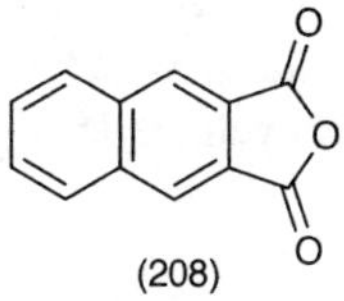

(208)

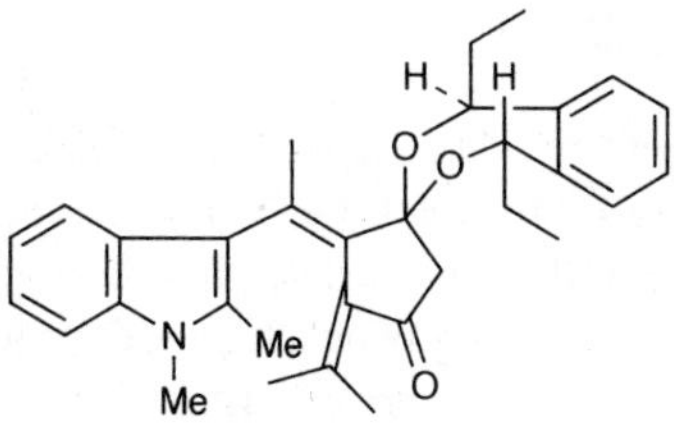

(209)

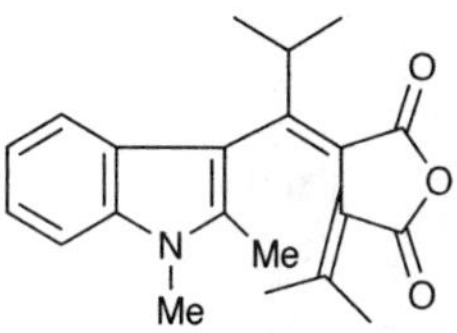

(210)

brings about conversion of the compound into the violet coloured cyclized form. The reverse process can be brought about by irradiation at 500 nm. Synthetic routes to photochromic compounds have been described in a short review.[140]

A study of the photochromic behaviour of the indolylfulgimide (211) in the presence of the pyridine derivative (212) has shown that association by hydrogen bonding is important in the outcome of the reaction.[141] The photochromism of the co-polymer formed between 2,3-bis(2,5-dimethyl-3-thienyl)-*N*-allylmaleimide and methyl methacrylate has been studied.[142] A report has dealt with the synthesis of new photochromic compounds that are based on 3,4-dimethylfuran-2,5-dione.[143]

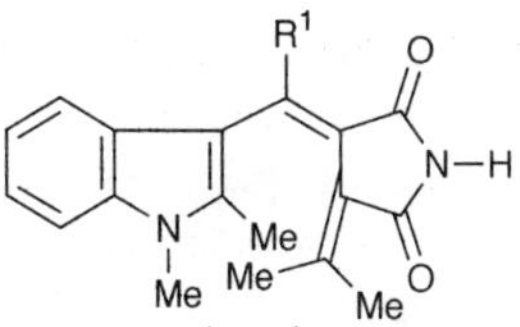

(211) $R^1 = Pr^i$ or Me

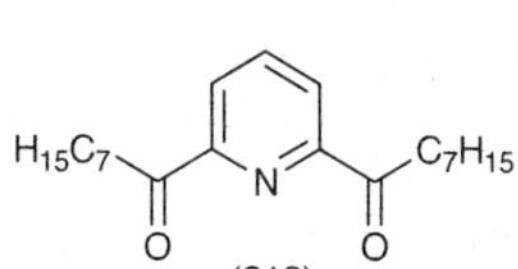

(212)

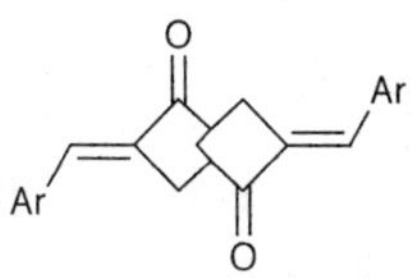

(213)

5.3.4 Other Diketones. Irradiation of 1,3,4,6-tetrathiapentalene-2,5-dione at 300 nm in a matrix at 10K shows the production of CO, CS_2 and SCO by competing pathways.[144] Irradiation of the dione-diene (213) in the presence of CuCl results in intramolecular cyclization to yield 7-*endo*-10-dihydroxy-2-phenyl-3,4-benzotetracyclo[4.3.3.1^{8,11}.0^{1,6}]-tridecane.[145]

6 Quinones

The quinone (214) undergoes photochemical addition to acenaphthylene to afford what is believed to be an oxetane.[146] The oxetane is either thermally or photochemically reactive and is converted into the two products (215, R = H) and (216, R = H) by a ring opening process. Evidence for such a mechanism was obtained when the reactions were carried out in methanol or ethanol, when the

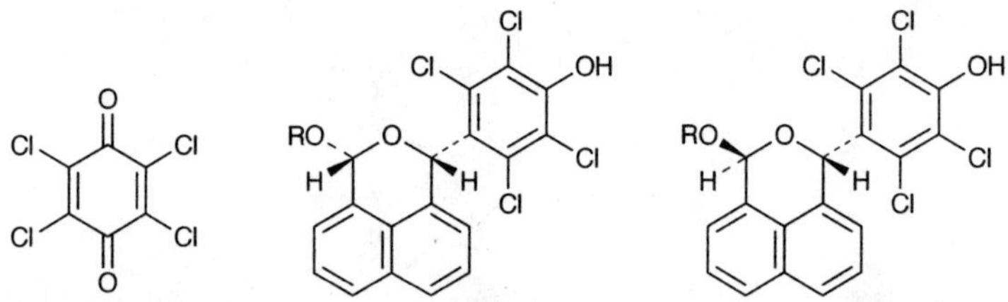

(214) (215) (216)

corresponding ethers (215, 216, R = OMe or OEt) were formed. A more detailed report of the scope of this reaction system has made recently.[147] The quinones (217) are all photoreactive in the presence of acenaphthylene. Irradiation of quinone (217a) sensitizes the (2 + 2)-cycloaddition to afford the acenaphthylene dimers (218) and (219). All the quinones bring about this dimerization, but other reactions are also observed. Thus, with (217b) the corresponding oxetane (220) is formed in 38% yield. With quinone (217c) the furan (221) is obtained in 19% yield. Quinone (217d) affords three products the oxetane (220, 9%), the furan (221, 26%) and the cyclobutane (222, 18%). The other quinones are also reactive. Both charge-transfer and direct irradiative techniques were employed.

Photochemical coupling between benzo-1,4-quinones and some Δ^5-steroids has been reported.[148] The radical cation of the di-(*p*-methoxyphenyl)methane (223) is formed on irradiation in the presence of quinones such as 1,4-benzoquinone, 2,5-dichloro-1,4-benzoquinone and chloranil.[149] A study of electron-transfer processes between carotenoids and benzoquinones such as 2,3-dichloro-5,6-dicyano-1,4-benzoquinone has been carried out.[150] The photochemical reaction of benzoquinone with benzodioxoles has been studied using the CIDNP technique.[151] Radical pairs were detected arising from hydrogen abstraction by the quinone from the benzodioxole. Irradiation of methoxy-*p*-benzoquinone in THF results in the formation of the corresponding hydroquinone.[152] The triplet state of fluoroanil can be generated by irradiation.[153] This species and the related radical anion and ketyl radical were studied using time resolved resonance Raman spectroscopy. Photoelectron transfer within the porphyrin-benzene-(bicyclo[2.2.2]octane)$_n$-quinone systems has been investigated.[154] The photo-

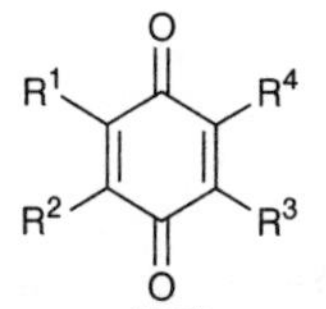

(217)

	R^1	R^2	R^3	R^4
a	Me	Me	Me	Me
b	Me	H	Me	H
c	H	H	H	H
d	Cl	H	H	H
e	Br	H	H	H
f	Cl	H	Cl	H
g	F	F	F	F
h	Br	Br	Br	Br

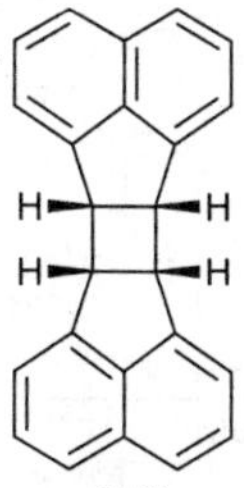

(218)

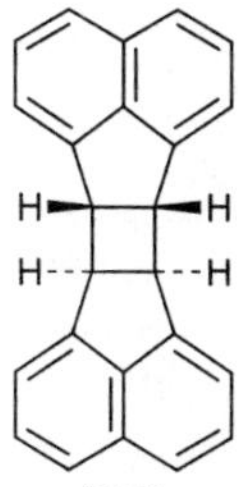

(219)

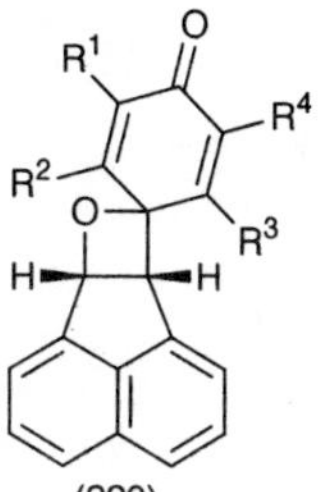

(220)

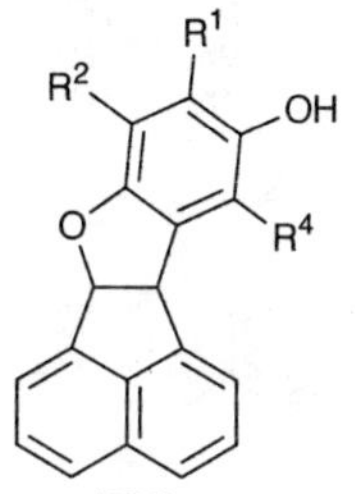

(221)

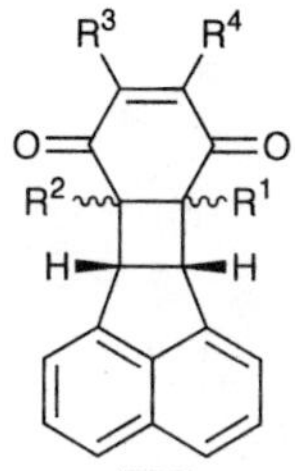

(222)

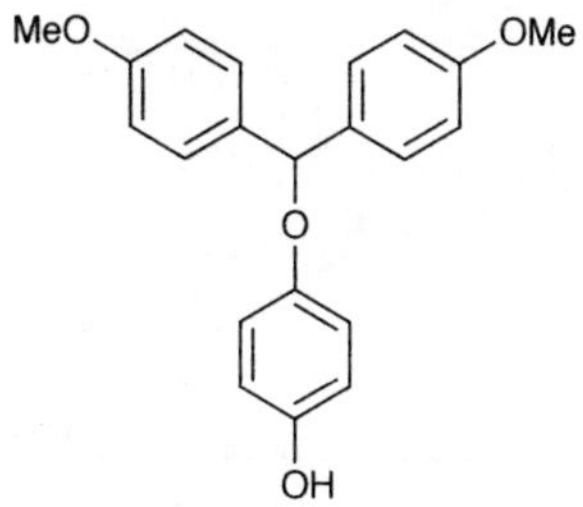

(223)

chromic behaviour of a series of quinonoid derivatives (224) has been studied.[155]

The formation of the radical cations of the naphthoquinones (225) occurs from the triplet state when the compounds are irradiated in β-cyclodextrins.[156] The photo-Friedel-Crafts acylation of 1,4-benzo- and naphthoquinones has been reported.[157] The reaction seems to be quite general and several examples of the process were cited. One will suffice here and this is shown in Scheme 9. In these cases the addition of aldehydes to the quinone (226) affords both the regioisomers (227) and (228). Regardless of the aldehyde used, however, there is always a preference for the formation of the 2,8-isomer (227). The solar initiated reactions between quinones and aldehydes have also been studied.[158]

1,4-Anthraquinone is photochemically reactive and irradiation affords the lowest triplet state.[159] In the presence of alkanes, products of hydrogen abstraction are formed, while with alkenes (2+2)-cycloaddition occurs. Furuta and his co-workers[160] have described the anthraquinone system (229) as a useful photochemically activated protecting group for alcohols. Blankespoor *et al.*[161] have

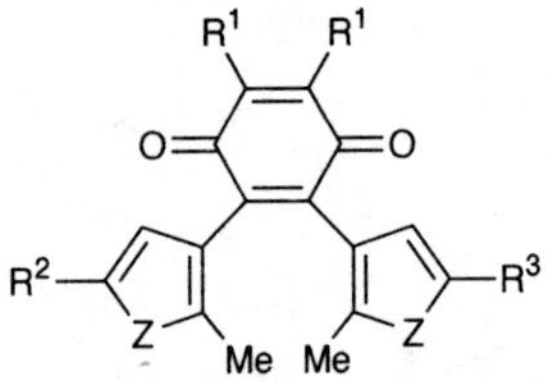

(224)

R^1–R^1	R^2	R^3	Z
benzo	H	H	S
	Me	Me	S
	Ph	Ph	S
	thienyl	thienyl	S
	Br	Br	S
	I	I	S
	I	H	S
	Ph	Ph	O
	Ph	Ph	S/NMe
Me Me	Ph	Ph	S

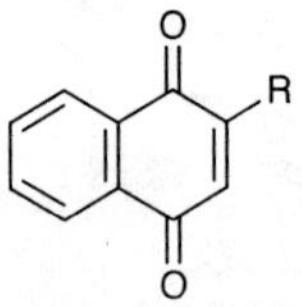

(225) R = H or Me

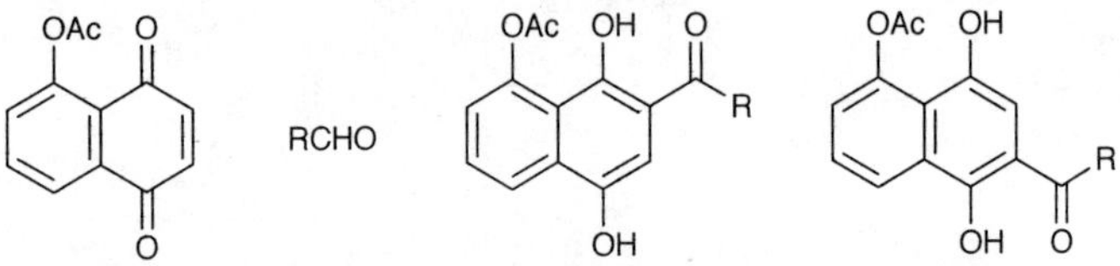

(226) (227) (228)

R	yield (%) (227)	(228)
Me	43	36
Et	41	18
Pr^n	42	34

Scheme 9

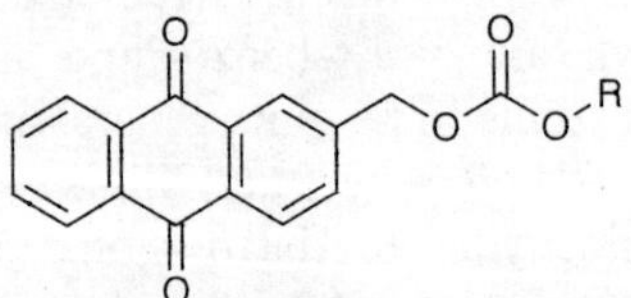

(229)

developed a method for the synthesis of benzaldehyde, using the photochemical reactivity of the quinone derivative (230). This is attached to a polymer that permits the reactions to be carried out in methanol. The yields of benzaldehyde are in the 50-55% range, and the reaction involves a Norrish Type II hydrogen-abstraction process. Wan and his co-workers[162] have examined the photochemical activity of the anthraquinone derivatives (231). Irradiation of (231a) gives no

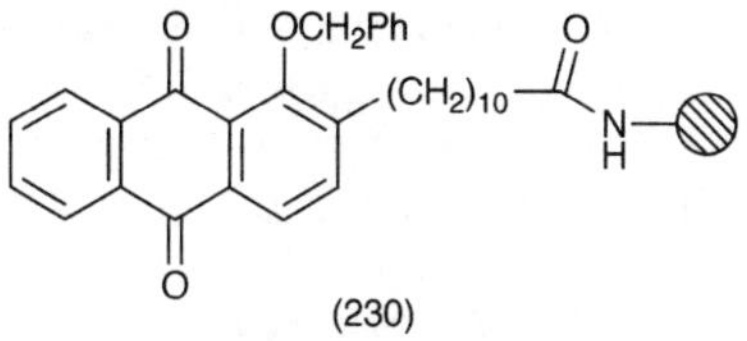

(230)

photochemical reaction on irradiation in water-acetonitrile. The derivatives (231b) and (231c), on the other hand, both produce a deep yellow-orange solution on irradiation. This colouration faded with time or exposure to oxygen. The colouration is due to the formation of the intermediate (232) formed by deprotonation. It is likely that this intermediate is the forerunner to the final product, 2-formylanthraquinone, formed in both cases. The authors[162] suggest that a highly polarized excited state is involved.

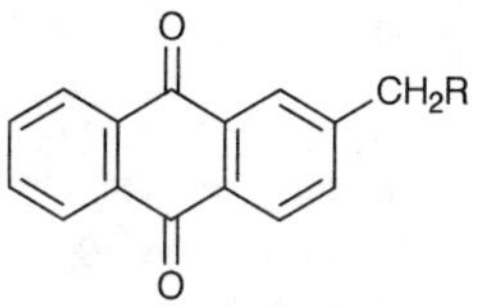

(231) a R = H
b R = OH
c R = OMe

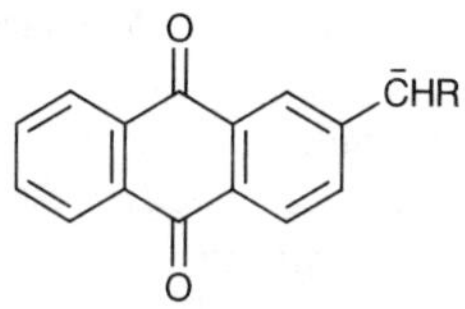

(232)

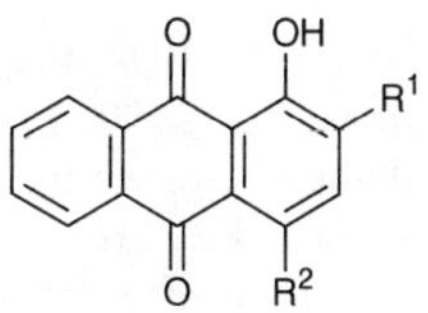

(233) R^1 = H, R^2 = $NHBu^n$
R^1 = $NHBu^n$, R^2 = H

The photoamination (using *n*-butylamine) of 1-hydroxyanthraquinone results in the formation of the two aminated products (233) in a ratio of 5:1 in acetonitrile solution without the exclusion of air.[163] Photolysis of aminoanthraquinones in cellulose and cellulose-like matrices results in reduction of the carbonyl groups.[164] This process is, however, dependent upon the nature of the substrate used and on the intensity of the irradiation.

Obermuller and Falk[165] have reported that intramolecular (2 + 2)-cycloaddition occurs on irradiation of ω,ω′-4-dimethylaminobenzal substituted hypericin derivatives.

References

1. R. S. Gopalan and G. U. Kulkarni, *Proc.- Indian Acad. Sci., Chem. Sci.*, 2001, **113**, 307.
2. M. M. S. Abdel-Mottaleb, S. de Feyter, A. Gesquiere, M. Sieffert, M. Klapper, K. Muellen and F. C. de Schryver, *Nano Lett.*, 2001, **1**, 353.
3. G. Kaupp, *Int. J. Photoenergy*, 2001, **3**, 55; *Chem. Abstr.*, 2001, **135**, 296042.

4. T. Hasegawa, K. Ikeda and Y. Yamazaki, *J. Chem. Soc., Perkin Trans. 1*, 2001, 3025.
5. S. Ohba, H. Hosomi and Y. Ito, *J. Am. Chem. Soc.*, 2001, **123**, 6349.
6. Y. Ito, S. Horie and Y. Shindo, *Org. Lett.*, 2001, **3**, 2411.
7. G. Andresen, A. B. Eriksen, B. Dalhus, L. L. Gundersen and F. Rise, *J. Chem. Soc., Perkin Trans. 1*, 2001, 1662.
8. I. Turowska-Tyrk, *Chem.-Eur. J.*, 2001, **7**, 3401.
9. E. Piers and A. Orellana, *Synthesis*, 2001, 2138.
10. R. Alibés, P. de March, M. Figueredo, J. Font and M. Racamonde, *Tetrahedron Lett.*, 2001, **42**, 6695.
11. C. Bertrand, J. Bouquant, J. P. Pete and S. Humbel, *J. Mol. Struct., THEOCHEM*, 2001, **538**, 165; *Chem. Abstr.*, 2001, **134**, 340171.
12. R. Mosca, M. Fagnoni, M. Mella and A. Albini, *Tetrahedron*, 2001, **57**, 10319.
13. U. Brule and N. Hoffmann, *Tetrahedron Lett.*, 2002, **43**, 69.
14. S. Marinkovic and N. Hoffmann, *J. Chem. Soc., Chem. Commun.*, 2001, 1576.
15. G. Mehta and K. Sreenivas, *Tetrahedron Lett.*, 2002, **43**, 703.
16. B. Hatano, S.-Y. Hirano, T. Yanagihara, S. Toyota and F. Toda, *Synthesis*, 2001, 1181.
17. D. Schwebel, M. Soltau and P. Margaretha, *Synthesis*, 2001, 1111.
18. A. Hilgeroth and U. Baumeister, *Chem.-Eur. J.*, 2001, **7**, 4599.
19. M. Ghosh, S. Chakrabarti and T. N. Misra, *Indian J. Phys., B*, 2001, **75B**, 341.
20. J. Fang, C. Whitaker, B. Weslowski, M.-S. Chen, J. Naciri and R. Shashidhar, *J. Mater. Chem.*, 2001, **11**, 2992.
21. M. Nowakowska, M. Smoluch and D. Sendor, *J. Inclusion Phenom. Macrocyclic Chem.*, 2001, **40**, 213.
22. A. P. Shkumat, *Visn. Khark. Nats. Univ.*, 1999, **454**, 150.
23. E. S. Greenwood and P. J. Parsons, *Synlett*, 2002, 167.
24. H. R. Memarian, M. Nasr-Esfahani, H. Görner and D. Döpp, *Indian J. Chem., Sect. B: Org. Chem. Incl. Med. Chem.*, 2001, **40B**, 1187.
25. D. Döpp, S. K. Mohamed and A. El-Khwaga, *Helv. Chim. Acta*, 2001, **84**, 3673.
26. D. Cesarin-Sobrinho and J. C. Netto-Ferreira, *Quimica Nova*, 2002, **25**, 62.
27. M.A. Kinder, L. Meyer and P. Margaretha, *Helv. Chim. Acta*, 2001, **84**, 2373.
28. Y. S. Kwak and J. D. Winkler, *J. Am. Chem. Soc.*, 2001, **123**, 7429.
29. B. K. Joseph, B. Verghese, C. Sudarsanakumar, S. Deepa, D. Viswam, P. Chandran and C. V. Asokan, *J. Chem. Soc., Chem. Commun.*, 2002, 736.
30. P. C. K. Lo and M. L. Snapper, *Org. Lett.*, 2001, **3**, 2819.
31. K. I. Booker-Milburn, L. F. Dudin, C. E. Anson and S. D. Guile, *Org. Lett.*, 2001, **3**, 3005.
32. F. Busque, P. De March, M. Figueredo, J. Font, P. Margaretha and J. Raya, *Synthesis*, 2001, 1143.
33. G. Mehta and J. Nandakumar, *Tetrahedron Lett.*, 2001, **42**, 7667.
34. S. Faure, S. Piva-Le-Blanc, C. Bertrand, J. P. Pete, R. Faure and O. Piva, *J. Org. Chem.*, 2002, **67**, 1061.
35. C. F. Chen, V. Chang, X. L. Cai, E. Duesler and P. S. Mariano, *J. Am. Chem. Soc.*, 2001, **123**, 6433.
36. D. L. Comins and A. L. Williams, *Org. Lett.*, 2001, **3**, 3217.
37. R. H. Blaauw, J. C. J. Benningshof, A. E. van Ginkel, J. H. Maarseveen and H. Hiemstra, *J. Chem. Soc., Perkin Trans. 1*, 2001, 2250.
38. T. Hatsui, K. Li, A. Mori and H. Takeshita, *Heterocycles*, 2001, **54**, 765; *Chem. Abstr.*, 2001, **134**, 340568.
39. L. U. Roman, N. R. Morales, J. D. Hernandez, C. M. Cerda-Garcia-Rojas, L. G.

Zepeda, C. A. Flores-Sandoval and P. Joseph-Nathan, *Tetrahedron*, 2001, **57**, 7269.
40. S. Pattanaargson and P. Limphong, *Int. J. Cosmet. Sci.*, 2001, **23**, 153; *Chem. Abstr.*, 2002, **136**, 221490.
41. A. Yu. Bobrovsky, N. I. Boiko and V. P. Shibaev, *Mol. Cryst. Liq. Cryst. Sci. Technol., Sect. A*, 2001, **363**, 35.
42. E. Rusu, V. Barboiu, A. Airinei, E. Comanita and G. Rusu, *Rev. Roum. Chim.*, **2000**, **45**, 451; *Chem. Abstr.*, 2001, **135**, 46829.
43. E. J. Larson, S. J. Pyszcysynski and C. K. Johnson, *J. Phys. Chem. A*, 2001, **105**, 8136.
44. T. Hu and E. J. Corey, *Org. Lett.*, 2001, **3**, 3547.
45. A. K. Ghosh, Y. Wang and J. T. Kim, *J. Org. Chem.*, 2001, **66**, 8973.
46. Y. Chiang, A. J. Kresge, Q. Meng, R. A. M. O'Ferrall and Y. Zhu, *J. Am. Chem. Soc.*, 2001, **123**, 11562.
47. T. Bach and F. Hofer, *J. Org. Chem.*, 2001, **66**, 3427.
48. F. Bargiggia and O. Piva, *Tetrahedron: Asymmetry*, 2001, **12**, 1389.
49. I. Marchueta, S. Olivella, L. Sola, A. Moyano, M. A. Pericas and A. Riera, *Org. Lett.*, 2001, **3**, 3197.
50. K. Maekawa, T. Igarashi, K. Kubo and T. Sakurai, *Tetrahedron*, 2001, **57**, 5515.
51. P. Formentin, M. J. Sabater, M. N. Chreetien, H. Garcia and J. C. Scaiano, *J. Chem. Soc., Perkin Trans. 2*, 2002, 164.
52. T. Horaguchi, N. Hosokawa, K. Tanemura and T. Susuki, *J. Heterocycl. Chem.*, 2002, **39**, 61.
53. M. P. S. Ishar, R. Singh, K. Kumar, G. Singh, D. Velmurugan, A. S. Randi, S. S. S. Raj, and H. K. Fun, *J. Org. Chem.*, 2002, **67**, 2234.
54. D.-M. Shin, D.-M. Song, K.-H. Jung and J.-H. Moon, *J. Photosci.*, 2001, **8**, 9.
55. R. Matsushima, S. Fujimoto and K. Tokumura, *Bull. Chem. Soc. Jpn.*, 2001, **74**, 827.
56. Y. Norikane, H. Itoh and T. Arai, *J. Phys. Chem. A*, 2002, **106**, 2766.
57. S. Kar and S. Lahiri, *Indian J. Chem., Sect. B: Org. Chem. Incl. Med. Chem.*, 2001, **40B**, 1121; *Chem. Abstr.*, 2002, **136**, 316797.
58. L. T. Nguyen, F. De Proft, M. T. Nguyen and P. Geerlings, *J. Chem. Soc., Perkin Trans. 2*, 2001, 898.
59. C. S. Burns, A. Heyerick, D. De Keukeleire and M. D. E. Forbes, *Chem.-Eur. J.*, 2001, **7**, 4554.
60. F. Goeller, C. Heinemann and M. Demuth, *Synthesis*, 2001, 1114.
61. J. Leitich, I. Heise, J. Rust and K. Schaffner, *Eur. J. Org. Chem.*, 2001, 2719.
62. J. Leitich, I. Heise and K. Schaffner, *J. Photochem. Photobiol. A- Chem.*, 2001, **141**, 207.
63. A. E. A. Gaber, I. Heise, J. Leitich and K. Schaffner, *J. Photochem. Photobiol. A-Chem.*, 2001, **141**, 133.
64. H. E. Zimmerman and E. E. Nesterov, *J. Am. Chem. Soc.*, 2002, **124**, 2818.
65. M. Melendez-Rodriguez, C. M. Cerda-Garcia-Rojas and P. Joseph-Nathan, *Tetrahedron*, 2002, **58**, 2331.
66. B. S. Park and C. M. Oh, *J. Photosci.*, 2000, **7**, 109; *Chem. Abstr.*, 2002, **136**, 19797.
67. V. P. Rybalkin, A. D. Dubonosov, E. N. Shepelenko, L. L. Popova, V. A. Bren' and V. I. Minkin, *Russ. J. Org. Chem.*, 2001, **37**, 1034; *Chem. Abstr.*, 2002, **136**, 294813.
68. T. Tsuno, H. Hoshino, R. Okuda and K. Sugiyama, *Tetrahedron*, 2001, **57**, 4831.
69. O. A. Luzina, S. M. Adekenov, E. M. Suleimenov, G. A. Atazhanova, N. F. Salakhutdinov and V. A. Barkhash, *Russ. J. Org. Chem.*, 2001, **37**, 740; *Chem. Abstr.*, 2001, **135**, 336803.

70. D. Armesto, M. J. Ortiz, A. R. Agarrabeitia and S. Aparicio-Lara, *Synthesis*, 2001, 1149.
71. D. W. C. MacMillan, L. E. Overman and L. D. Pennington, *J. Am. Chem. Soc.*, 2001, **123**, 9033.
72. H.-M. Tai, C.-C. Yang, M.-L. Yeh and N.-C. Chang, *J. Chin. Chem. Soc. (Taipei, Taiwan)*, 2001, **48**, 901.
73. N. Tanaka, *Fukuoka Kyoiku Daigaku Kiyo*, 2002, **51**, 31; *Chem. Abstr.*, 2002, **136**, 269193.
74. V. Singh and S. Prathap, *J. Ind. Inst. Sci.*, 2001, **81**, 75.
75. G. Gopalakrishnan, N. D. P. Singh and V. Kasinath, *Molecules [online computer file]*, 2001, **6**, 551; *Chem. Abstr.*, 2002, **136**, 118597.
76. S. M. Sieburth, C. B. Madsen-Duggan and F. N. Zhang, *Tetrahedron Lett.*, 2001, **42**, 5155.
77. K. F. McGee, Jnr. T. H. Al-Tel and S. McN. Sieburth, *Synthesis*, 2001, 1185.
78. T. A. Ader, C. A. Champey, L. V. Kuznetsova, T. Li, Y. H. Lim, D. Rucando and S. M. Sieburth, *Org. Lett.*, 2001, **3**, 2165.
79. N. Yoshioka, C. Andoh, K. Kubo, T. Igarashi and T. Sakurai, *J. Chem. Soc., Perkin Trans. 2*, 2001, 1927.
80. K. Ohkura, K.-I. Nishijima, S. Uchiyama, A. Sakushima and K.-I. Seki, *Heterocycles*, 2001, **55**, 1015; *Chem. Abstr.*, 2001, **135**, 218555.
81. N. J. Kim, H. Kang, G. Jeong, Y. S. Kim, K. T. Lee and S. K. Kim, *J. Chem. Phys.*, 2001, **115**, 7002.
82. M. Nowakowska, M. Grebosz, M. Smoluch and W. Tatara, *Photochem. Photobiol.*, 2002, **75**, 92.
83. S. S. Chawada and S. Jain, *Asian J. Chem.*, 2001, **13**, 1006.
84. S. S. Chawada and S. Jain,*Asian J. Chem.*, 2001, **13**, 1231; *Chem. Abstr.*, 2001, **135**, 226955.
85. Y. Ohtani, Y. Inaki and M. Miyata, *J. Photopolym. Sci. Technol.*, 2001, **14**, 295; *Chem. Abstr.*, 2001, **135**, 331791.
86. K. Fujimoto, Y. Ikeda, S. Ishihara and I. Saito, *Tetrahedron Lett.*, 2002, **43**, 2243.
87. Z. R. Zhang, W. Z. Luo and T. Nagai, *S.T.P. Pharma Sci.*, 2001, **11**, 243.
88. T. Eckardt, V. Hagen, B. Schade, R. Schmidt, C. Schweitzer and J. Bendig, *J. Org. Chem.*, 2002, **67**, 703.
89. K. Nakatani, T. Yoshida and I. Saito, *J. Am. Chem. Soc.*, 2002, **124**, 2118.
90. T. C. Turek, I. Gaon, M. D. Distefano and C. L. Strickland, *J. Org. Chem.*, 2001, **66**, 3253
91. T. A. Kale, C. Raab, N. Yu, D. E. Dean and M. D. Distefano, *J. Am. Chem. Soc.*, 2001, **123**, 4373.
92. C. Muller, P. Even, M. L. Viriot and M. C. Carre, *Helv. Chim. Acta*, 2001, **84**, 3735.
93. D. M. Bassani, X. Sallenave, V. Darcos and J. P. Desvergne, *J. Chem. Soc., Chem. Commun.*, 2001, 1446.
94. S. Uppili and V. Ramamurthy, *Org. Lett.*, 2002, **4**, 87.
95. G. Blay, L. Cardona, B. Garcia, L. Lahoz and J. R. Pedro, *J. Org. Chem.*, 2001, **66**, 7700.
96. S. R. Jackson, M. G. Johnson, M. Mikami, S. Shiokawa and E. M. Carreira, *Angew. Chem. Int. Edn.*, 2001, **40**, 2694.
97. A. Ricci, E. Fasani, M. Mella and A. Albini, *J. Org. Chem.*, 2001, **66**, 8086.
98. H. H. Wenk, R. Hubert and W. Sander, *J. Org. Chem.*, 2001, **66**, 7994.
99. M. Fleming, P. V. Fisher, G. U. Gunawardena, Y. Jin, C. Zhang, W. Zhang, A. M. Arif and F. G. West, *Synthesis*, 2001, 1268.

100. Z. Stiplosek, M. Sindler-Kulyk, K. Jakopcic, A. Visnjevac and B. Kojic-Prodic, *J. Heterocycl. Chem.*, 2002, **39**, 37.
101. J. Seixas de Melo, G. Quinteiro, J. Pina, S. Breda and R. Fausto, *J. Mol. Struct.*, 2001, **565-566**, 59; *Chem. Abstr.*, 2001, **135**, 195269.
102. X. Li, J. M. Millam and H. B. Schlegel, *J. Chem. Phys.*, 2001, **115**, 6907.
103. I. Magneron, A. Horowitz, J. Tadic, K. Wirtz, M. Pons and G. K. Moortgat, *Transport and Chemical Transformation in the Troposphere, Proceedings of EUROTRAC Symposium, 6th, Garmisch-Partenkirchen, Germany, Mar. 27-31*, 2000, 414; *Chem. Abstr.*, 2002, **136**, 102052.
104. R. Winterhalter, N. R. Jensen, I. Magneron, K. Wirtz, W. Mellouki, M. Yujing, J. Tadic, A. Horowitz, G. K. Moortgat and J. Hjorth, *Transport and Chemical Transformation in the Troposphere, Proceedings of EUROTRAC Symposium, 6th, Garmisch-Partenkirchen, Germany, Mar. 27-31*, 2000, 464; *Chem. Abstr.*, 2002, **136**, 102054.
105. M. Weber, I. V. Khudyakov and N. J. Turro, *J. Phys. Chem. A*, 2002, **106**, 1938.
106. I-R. Lee, Y.-C. Chung, W.-K. Chen, X.-P. Hong and P.-Y. Cheng, *J. Chem. Phys.*, **115**, 10656.
107. J. R. Scheffer and K. Wang, *Synthesis*, 2001, 1253.
108. A. G. Griesbeck and H. Heckroth, *Synlett*, 2002, 131.
109. H. Miyamoto, Z. Urbanczyk-Lipkowska and F. Toda, *Trends Org. Chem.*, 2000, **8**, 93; *Chem. Abstr.*, 2002, **136**, 199782.
110. Y. Zhang, J. Xue, Y. Gao, H.-K. Fun and J.-H. Xu, *J. Chem. Soc., Perkin Trans. 1*, 2002, 345.
111. P. Bortolus, G. Marconi, S. Monti and B. Mayer, *J. Phys. Chem. A*, 2002, **106**, 1686.
112. C.-C. Liao, *Huaxue*, 2001, **59**, 163.
113. N. Nagashima, S. Kudoh, M. Takayanagi and M. Nakata, *J. Phys. Chem. A*, 2001, **105**, 10832.
114. A. Cantrell and D. J. McGarvey, *J. Photochem. Photobiol. B-Biol.*, 2001, **64**, 117.
115. F. Ortica and M. A. J. Rodgers, *Photochem. Photobiol.*, 2001, **74**, 745.
116. Z. Yang, D. Ng and M. A. Garcia-Garibay, *J. Org. Chem.*, 2001, **66**, 4468.
117. S. Sortino, L. J. Martinez and G. Marconi, *New J. Chem.*, 2001, **25**, 975.
118. H. Jiang, Y. C. Liu, J. Li, G. W. Wang, Y. D. Wu, Q. M. Wang and T. C. W. Mak, *J. Chem. Soc., Chem. Commun.*, 2002, 882.
119. S. S. Kim, J. A. Chang, A. R. Kim, Y. J. Mah, H. J. Kim and C. Kang, *J. Photosci.*, 2000, **7**, 111.
120. A. J. Robinson, P. Stanislawski, D. Mulholland, L. N. He and H. Y. Li, *J. Org. Chem.*, 2001, **66**, 4148.
121. K. Kokubo, T. Koizumi, H. Yamaguchi and T. Oshima, *Tetrahedron Lett.*, 2001, **42**, 5025.
122. A. P. Marchand, T. D. Power and H. G. Kruger, *Croat. Chem. Acta*, 2001, **74**, 265; *Chem. Abstr.*, 2001, **135**, 195192.
123. A. G. Griesbeck, M. Oelgemoller, J. Lex, A. Haeuseler and M. Schmittel, *Eur. J. Org. Chem.*, 2001, 1831.
124. A. G. Griesbeck, W. Kramer and J. Lex, *Synthesis*, 2001, 1159.
125. H. Gorner, M. Oelgemoller and A. G. Griesbeck, *J. Phys. Chem. A*, 2002, **106**, 1458.
126. D. J. Yoo, E. Y. Kim, M. Oelgemoller and S. C. Shim, *Heterocycles*, 2001, **54**, 1049; *Chem. Abstr.*, 2001, **134**, 353245.
127. U. C. Yoon, C. W. Lee, S. W. Oh, H. J. Kim and S. J. Lee, *J. Photochem.*, 2000, **7**, 143; *Chem. Abstr.*, 2002, **136**, 5626.
128. J. Gan, H. Tian, Z. Wang, K. Chen. J. Hill, P. A. Lane, M. D. Rahn, A. M. Fox and

D. D. C. Bradley, *J. Organomet. Chem.*, 2002, **645**, 168.
129. S. Metsarinne, T. Tuhkanen and R. Aksela, *Chemosphere*, 2001, **45**, 949; *Chem. Abstr.*, 2002, **136**, 123070.
130. H. H. Wenk and W. Sander, *Chem.-Eur. J.*, 2001, **7**, 1837.
131. T. Sato, H. Niino and A. Yabe, *J. Phys. Chem. A*, 2001, **105**, 7790.
132. T. Sato, H. Niino and A. Yabe, *J. Photochem. Photobiol. A-Chem.*, 2001, **145**, 3
133. S. Nishio, K. Tamura, Y. Tsujine, T. Fukao, M. Nakano, A. Matsuzaki, H. Sato and T. Yamabe, *J. Photochem. Photobiol., A: Chem.*, 2001, **145**, 165.
134. N. Liao, M. Gong, D. Xu, G. Qi and K. Zhang, *Chin. Sci. Bull.*, 2001, **46**, 1856.
135. Y. Yokoyama, T. Okuyama, Y. Yokoyama and M. Asami, *Chem. Lett.*, 2001, 1112.
136. Y. C. Kiang, A. S. Dvornikov and P. M. Rentzepis, *J. Photochem. Photobiol. A-Chem*, 2001, **146**, 83.
137. M. A. Wolak, N. B. Gillespie, C. J. Thomas, R. R. Birge and W. J. Lees, *J. Photochem. Photobiol., A: Chem* 2001, **144**, 83.
138. E. Ankai, K. Sakakibara, S. Uchida, Y. Uchida, Y. Yokoyama and Y. Yokoyama, *Bull. Chem. Soc. Jpn.*, 2001, **74**, 1101.
139. A. M. Asiri, *J. Photochem. Photobiol. A-Chem.*, 2001, **146**, 133.
140. M. M. Krayushkin, *Chem. Heterocycl. Compd. (N. Y., NY, U. S.)*2001, **37**, 15.
141. T. Okuyama, Y. Yokoyama and Y. Yokoyama, *Bull. Chem. Soc. Jpn.*, 2001, **74**, 2181.
142. H. Tian and H.-Y. Tu,*Adv. Mater. (Weinheim, Ger.)*, 2000, **12**, 1597; *Chem. Abstr.*, 2001, **135**, 19982.
143. V. Z. Shirinyan, M. M. Krayushkin, L. I. Belen'kii, L. G. Vorontsova, Z. A. Starikova, A. Yu. Martynkin, V. L. Ivanov and B. M. Uzhinov, *Chem. Heterocycl. Compd. (N. Y., NY, U. S.)* 2001, **37**, 77.
144. F. Risi, Florence and J.-P. Aycard, *Spectrosc. Lett.*, 2001, **34**, 167.
145. N. V. Averina, G. S. Borisova, A. A. Borisenko and N. S. Zefirov, *Russ. J. Org. Chem.*, 2001, **37**, 957.
146. N. Haga, H. Takayanagi and K. Tokumaru, *Chem. Lett.*, 2001, 448.
147. N. Haga, H. Takayanagi and K. Tokumaru, *J. Chem. Soc., Perkin Trans. 2*, 2002, 734.
148. C.-G. Yang, Z.-Q. Jiang, D.-R. Zhu, S.-P. Wu, L. Yang and L.-M. Wu, *Chin. J. Chem.*, 2001, **19**, 1211; *Chem. Abstr.*, 2002, **136**, 247738.
149. T. Del Giacco, E. Baciocchi, O. Lanzalunga and E. Elisei, *Chem.-Eur. J.*, 2001, **7**, 3005.
150. N. E. Polyakov, V. V. Konovalova, T. V. Leshina, O. A. Luzina, N. F. Salakhutdinov, T. A. Konovalova and L. D. Kispert, *J. Photochem. Photobiol. A-Chem.*, 2001, **141**, 117.
151. T. Wang and B.-z. Yan, *Beijing Huagong Daxue Xuebao*, 2001, **28**, 72.
152. S. Bearnais-Barbry, R. Bonneau and A. Castellan, *Photochem. Photobiol.*, 2001, **74**, 542.
153. G. Balakrishnan, P. Mohandas and S. Umapathy, *J. Phys. Chem. A*, 2001, **105**, 7778.
154. S. A. do Monte, *Chem. Phys. Lett.*,2001, **336**, 462; *Chem. Abstr.*, 2001, **135**, 26748.
155. X. H. Deng and L. S. Liebeskind, *J. Am. Chem. Soc.*, 2001, **123**, 7703.
156. D. Takamori, T. Aoki, H. Yashiro and H. Murai, *J. Phys. Chem. A*, 2001, **105**, 6001.
157. C. Schiel, M. Oelgemoller and J. Mattay, *Synthesis*, 2001, 1275.
158. C. Schiel, M. Oelgemoeller, J. Ortner and J. Mattay, *Green Chem.*, 2001, **3**, 224.
159. T. Yoshihara, M. Yamaji, T. Itoh, J. Nishimura, H. Shizuka and S. Tobita, *J. Photochem. Photobiol. A-Chem.*, 2001, **140**, 7.

160. T. Furuta, Y. Hirayama and M. Iwamura, *Org. Lett.*, 2001, **3**, 1809.
161. R. L. Blankespoor, T. DeVries, E. Hansen, J. M. Kallemeyn, A. M. Klooster, J. A. Mulder, R. P. Smart and D. A. Vander Griend, *J. Org. Chem.*, 2002, **67**, 2677.
162. M. Lukeman, M. S. Xu and P. Wan, *J. Chem. Soc., Chem. Commun.*, 2002, 136.
163. M. Tajima, K. Kato, K. Matsunaga and H. Inoue, *J. Photochem. Photobiol. A-Chem.*, 2001, **140**, 127.
163. R. P. Ponomareva and O. P. Studzinskii, *Russ. J. Gen. Chem.*, 2001, **71**, 759.
165. R. A. Obermuller and H. Falk, *Monatsh. Chem.*, 2001, **132**, 1519.

3
Photochemistry of Alkenes, Alkynes and Related Compounds

BY WILLIAM M. HORSPOOL

1 Reactions of Alkenes

1.1 *cis,trans*-Isomerization. – The photoisomerization of *p*-dimethylamino-β-chlorostyrene has been studied in a low temperature matrix.[1] The research has shown that the *trans* form is present immediately after deposition in the argon matrix. Irradiation populates the S_1 state and it is this that brings about the isomerization to the *cis* form of the alkene. The influence that *trans,cis*-isomerism of styryl groups can exercise on the spectral luminescent properties of molecules has been studied.[2] The photoisomerization of 1,4-bis[(*N*-methylquinolinium-4-yl)vinyl]benzene affords the *E,E* and the *Z,E* isomers.[3] The influence of group I and group II metal ions on the photochemical behaviour of the crown ethers *trans,trans*-1,4-bis[2-(3′,4′-benzo[15]crown-5)ethenyl]-2,3,5,6-tetrafluorobenzene and *trans,trans*-1,4-bis[2-(3′,4′-benzo[18]crown-6)ethenyl]-2,3,5,6-tetrafluorobenzene has been investigated.[4] Steady state irradiation shows that both *trans,cis*-isomerization and (2+2)-cycloaddition occur. Ions such as Li^+, Na^+ and Ca^{2+} influence the isomerism, while the larger ions K^+, Rb^+ and Ba^{2+} bring about the cycloaddition. The isomerization of (1) into (2) can be brought about by irradiation at 300 nm using I_2 in tetrachloromethane as solvent.[5] The molecule (3) has been tested as a potential molecular switch.[6] Its irradiation at 0°C brings about conversion to the *Z* isomer (4). The photostationary state achieved was 1:1. Photochemical isomerism has been used in a new synthesis of (−)-bauhinin.[7]

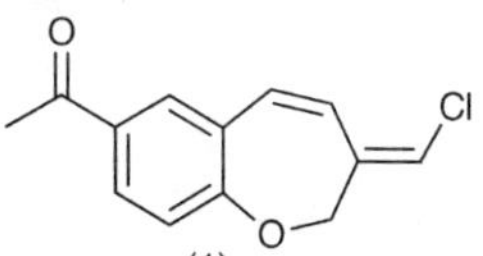

(1)

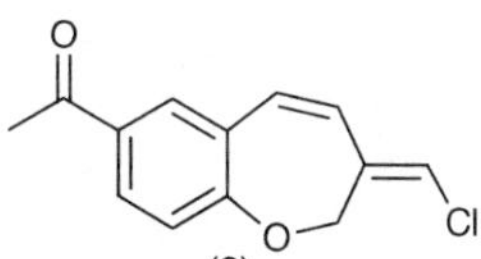

(2)

Some optically active symmetric phosphoramides have been used as sensitizers for the isomerism of cyclooctene.[8] The reactions were carried out at low temperature and ees of 20-28% were achieved. A further account by Matsuyama *et al.*[9] has reported studies into the self-sensitized photoisomerizations of the cyclic alkenes (5). They have demonstrated that this self-sensitized process is dia-stereodifferentiating, and irradiation of the *Z* isomer was shown to be

Photochemistry, Volume 34

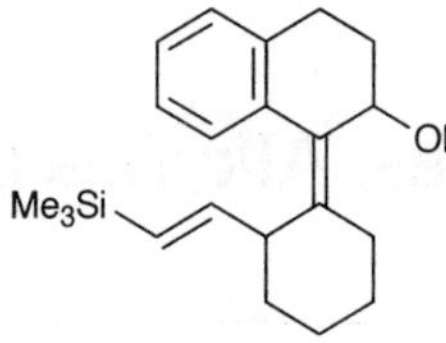

(3)

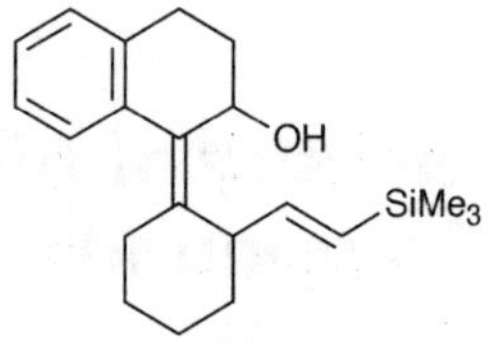

(4)

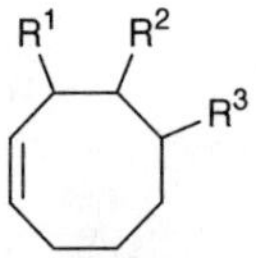

(5) $R^1 = PhCO_2$, $R^2 = R^3 = H$
$R^1 = R^3 = H$, $R^2 = PhCO_2$
$R^1 = R^2 = H$, $R^3 = PhCO_2$

concentration dependent. Thus, the major diastereoisomer obtained could be switched from the (1*R*,3*R*)-*E* isomer with a *de* of 26% to (1*R*,3*S*)-*E* with 11% *de*.

1.1.1 Stilbenes and Related Compounds. The photoisomerization of *trans*-stilbene can be brought about in the ionic liquid 1-butyl-3-methylimidazolium hexafluorophosphate.[10] The rate for the isomerization is greater than that expected. The results of a study of the solvent and temperature dependent Raman spectral changes of singlet-excited *trans*-stilbene has been published.[11] The study focused on the use of the dynamic polarization model, and both temperature and solvent effects were studied. A theoretical study of the iodine-catalysed photoisomerism of stilbene has been carried out.[12] An essential step in the process is the formation of a complex between iodine atoms and stilbene. The rate determining step for the isomerization is the conversion from the complex to the 1,2-diphenyl-2-iodoethyl radical. A series of stilbene dendrimers based on the example (6) has been synthesized for the first time.[13] Irradiation of (6) at 330 nm brings about *trans,cis*-isomerization of the double bond. Interestingly isomerization also occurs in the heavily substituted derivatives that were also studied.

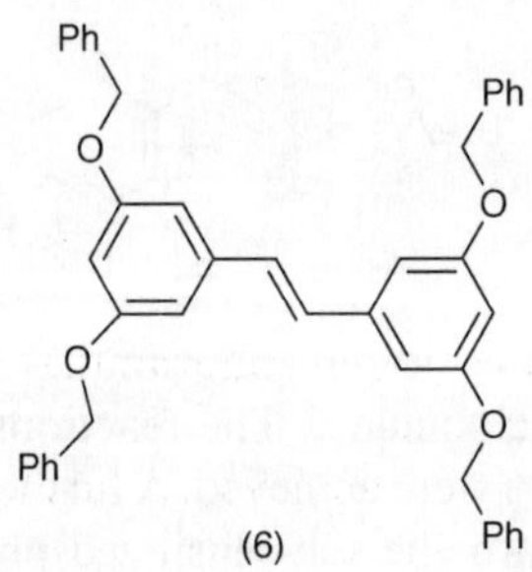

(6)

Pulsed and steady state irradiation of 4-nitrostilbene, 4,4′-dinitrostilbene and 4-methoxy-4′-nitrostilbene in the presence of amines such as DABCO, diethylamine and triethylamine has been carried out.[14] Triplet states of the stilbenes are involved, and all the amines used have a quenching effect on these

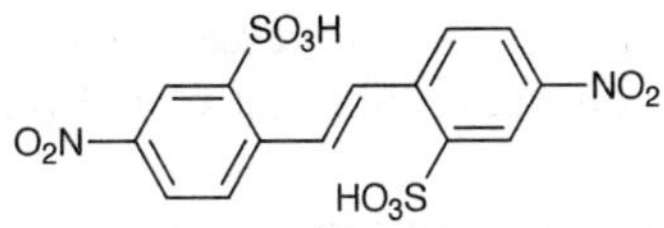

(7)

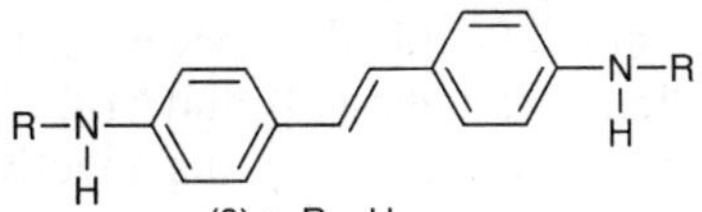

(8) a R = H

b R = C(=S)–N(H)–Ph

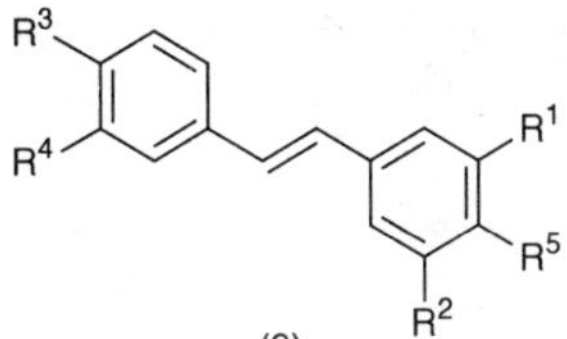

(9)

R^1	R^2	R^3	R^4	R^5
NH_2	H	H	H	H
NMe_3	H	H	H	H
NH_2	NH_2	H	H	H
H	NMe_2	NMe_2	H	H
H	NH_2	H	NH_2	H
H	H	NMe_2	H	NMe_2
H	H	NH_2	H	NH_2

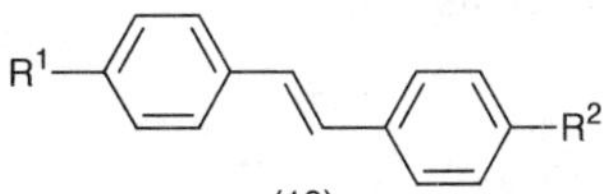

(10)

R^1	R^2
NMe_2	CN
NMe_2	NO_2
$N(CH_2CH_2OH)_2$	NO_2

excited states. The radical anions of the stilbenes are formed in all cases. The water-soluble stilbene (7) has been studied using laser-induced optoacoustics.[15] The work demonstrated the influence of monovalent cations on the *trans,cis*-isomerism of the stilbenes. A short review has focused attention on the new areas of applications to which substituted stilbenes can be put.[16] Anomalous temperature effects affecting the excited states of the stilbenes (8) have been examined.[17] The authors point out that in the case of donor-substituted stilbenes it is necessary to measure the quantum yields for both the isomerization and the fluorescence. This guarantees a true picture of the excited states involved. A further examination of the *meta*-amino effect in the stilbenes (9) has been published.[18] The influence of torsional barriers in non-polar solvents was assessed. A high-resolution spectroscopic study of the stilbenes (10) has pointed out the dominance of low frequency vibrational modes in the excitation spectra.[19] Results from irradiation and calculations upon the photochemical activity of 4-dimethylamino-4′-cyanostilbene indicate that twisted intramolecular charge transfer does not contribute to the fluorescence of the system.[20] The results of SCF calculations on the photochemical activity of the same stilbene have been published.[21]

cis-Stilbene cyclization to phenanthrenes is a well-known photochemical reaction and follows the path shown in scheme 1. Here irradiation of (11) brings about a 6e cyclization to afford the dihydro species (12).[22] This then undergoes oxidation to afford the final products (13) and (14). The current study has examined push-pull systems and especially the reactivity of the derivatives

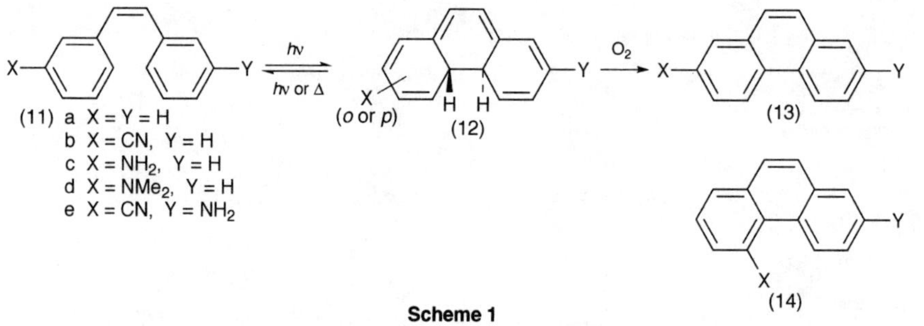

Scheme 1

(11c)-(11e). The irradiation of all these derivatives demonstrates cyclization, with the formation of (13) or (14). Apparently the dihydrophenanthrenes (12c-e) all show relatively large ground-state and excited-state dipoles. The authors[22] propose that, in the amino cyano derivative, the likelihood is the formation of a quinonoid like compound such as (15).

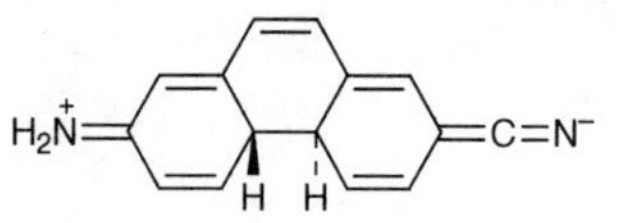

(15)

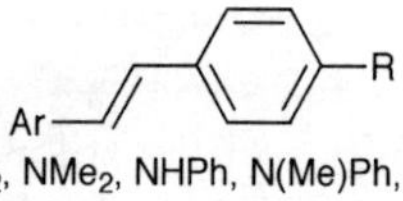

(16) Ar = Ph, R = Br, NH_2, NMe_2, NHPh, N(Me)Ph, NPh_2, NH(2,5-diMeC_6H_3)
Ar = 3-pyridyl or 2-naphthyl

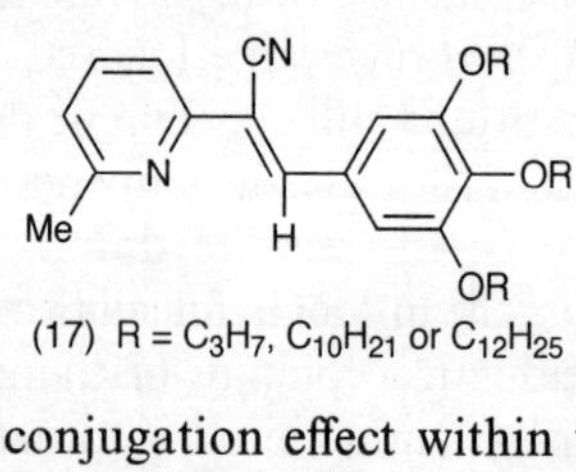

(17) R = C_3H_7, $C_{10}H_{21}$ or $C_{12}H_{25}$

A study of the amino conjugation effect within the stilbenes (16) has been studied in some detail.[23] Interestingly, the *N*-substituted aryl groups have low photoisomerization yields but high quantum yields for fluorescence. The stilbene derivatives (17) are photochemically reactive and undergo *trans,cis*-isomerization on irradiation.[24] The ferrocenyl derivative (18) undergoes *E,Z*-isomerization when irradiated.[25] Karatsu[26] has reviewed the work relating to the study of

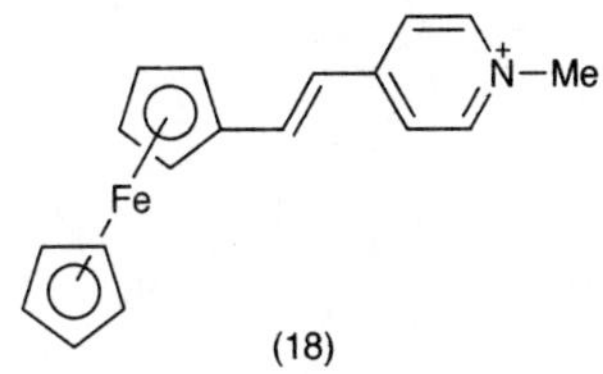

(18)

the potential energy surfaces and rotational isomerism of 1-(2-anthryl)-2-phenylethene and 1-(2-anthryl)-3,3-dimethyl-1-butene.

1.1.2 The Dithienylethene System and Related Compounds. The interest in these photochromic compounds continues to attract considerable attention. Many different studies have been reported throughout the year, dealing with reactions in the crystal and in solution of variously substituted derivatives. The stacking of compound (19a) in the crystal is such that 17% of the molecules undergo photochromic ring closure.[27] The derivative (19b), however, does not stack correctly and shows no photochromism in the crystalline state. In a co-crystal, however, there is evidence that (19b) is slightly photochemically reactive. Photochemical decomposition is brought about by the irradiation of (20a) and (20b) on porous Al_2O_3.[28] The other derivative (20c) does not decompose and exhibits the usual photochromism observed for such compounds. A Raman spectroscopic study of the photochromic reactivity of 1,2-bis(3-methyl-2-thienyl)perfluorocyclopentene has been reported.[29]

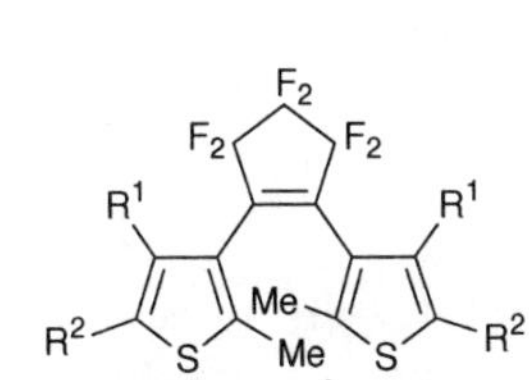

(19) a R = H
b R = Me

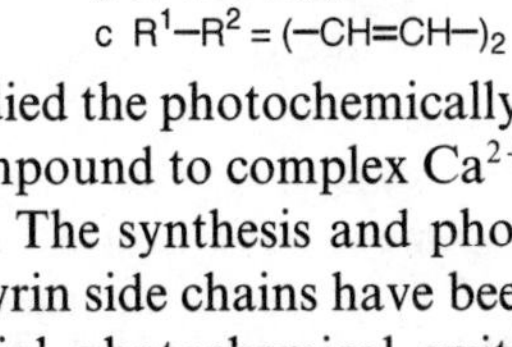

(20) a R^1 = H, R^2 = Me
b R^1 = R^2 = Me
c R^1–R^2 = (–CH=CH–)$_2$

Malval *et al.*[30] have studied the photochemically active crown ether derivative (21). The ability of the compound to complex Ca^{2+} increased by a factor of >10^3 on irradiation at 330 nm. The synthesis and photoisomerization of a series of alkenes (22) with diporphyrin side chains have been reported.[31] These molecules are of interest as potential photochemical switches. The dimer, trimer and tetramers (23) are all photochemically reactive and exhibit photochromism.[32] The quantum yields for the photocyclizations increase from 0.21 to 0.40 with an increase in the number of thienyl units.

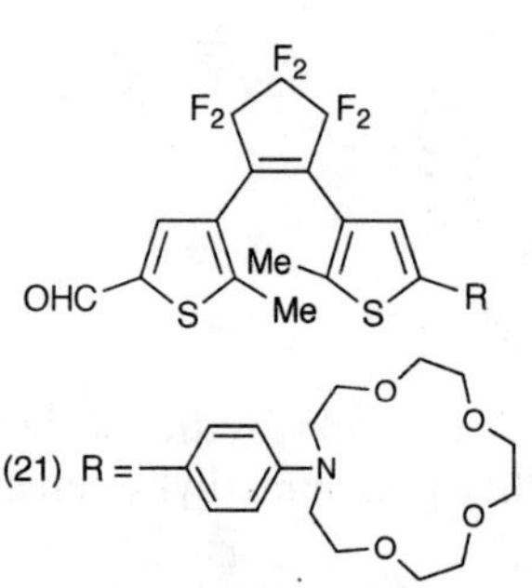

(21)

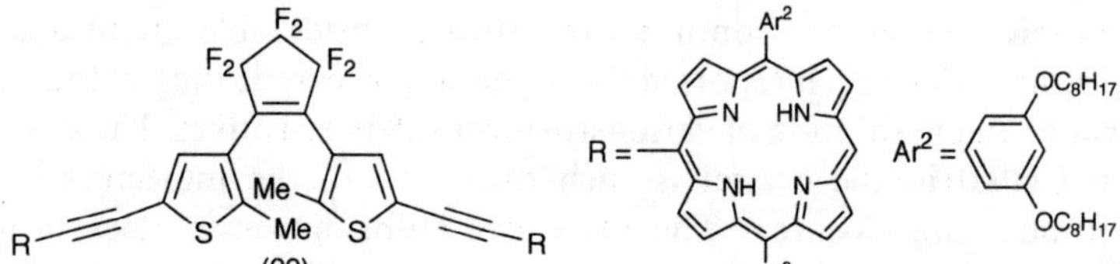

(22)

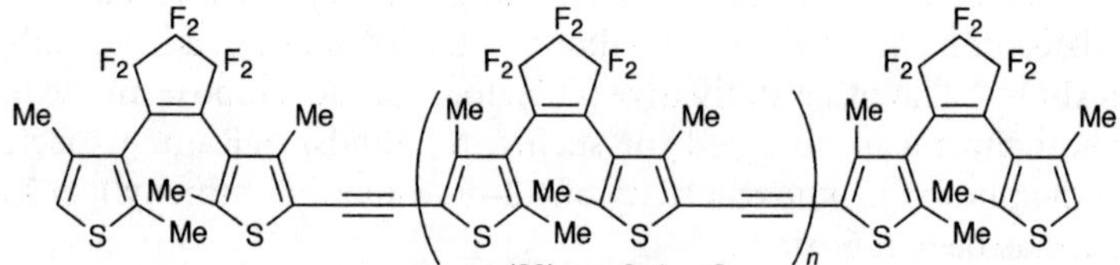

(23) $n = 0$, 1 or 2

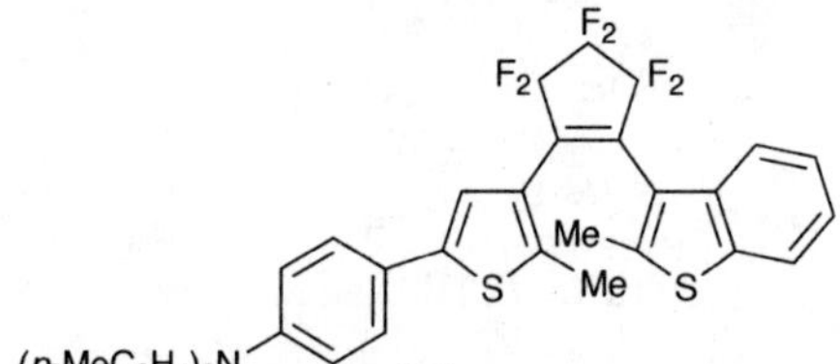

(24)

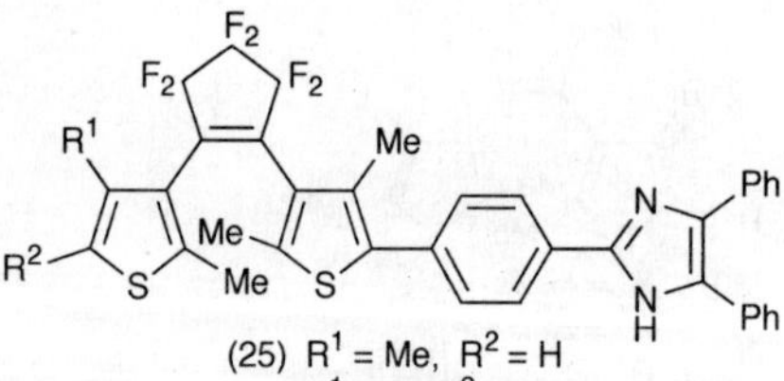

(25) $R^1 = Me$, $R^2 = H$
$R^1 = H$, $R^2 = Me$

The photochromic properties of some amorphous materials using dithienylethene (24) as a component have been studied.[33] The imidazole substituted diarylethene system (25) has been synthesized.[34] The photochromic behaviour of this is similar to that of the parent system, and irradiation at 366 nm brings about ring closure. Ring opening occurs on irradiation using wavelengths > 480 nm.

Yamada *et al.*[35] have carried out an X-ray study of the alkene (26). They have found that there are two different molecular conformations in the asymmetric unit. The difference in photocyclization of the two conformations has been examined, and the molecules with the shorter distance between the reactive centres undergo the usual cyclization mode. The influence of substituents other than methyl groups on the ring closure and cycloreversion of the diarylethenes (27) has been studied.[36] While the compounds in the report all cyclize with

(26)

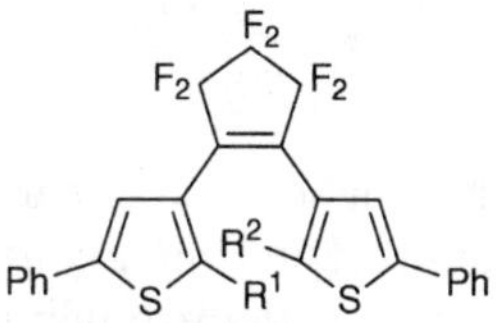

(27)

R^1	R^2	$\phi_{o\to c}$	$\phi_{c\to o}$
OMe	OMe	0.44 (309nm)	2.0×10^{-5} (525nm)
OMe	Me	0.44 (298nm)	1.6×10^{-3} (600nm)
Me	Me	0.59 (280nm)	1.3×10^{-2} (492nm)

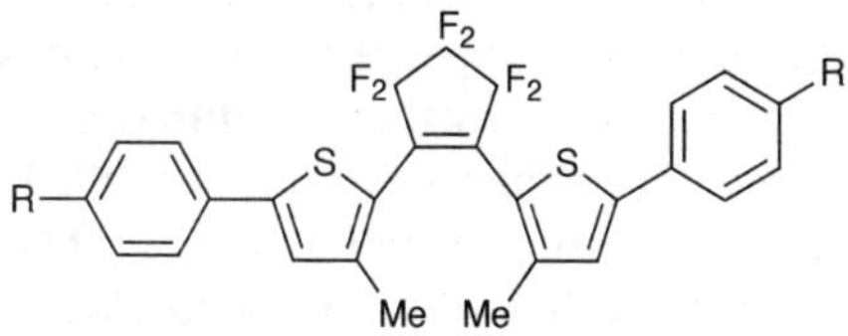

(28)

R	ϕ_{cyc}	ϕ_{open}
H	0.17 (366nm)	0.48 (440nm)
OMe	0.45 (366nm)	0.40 (440nm)
NMe_2		

comparable quantum yields, the reverse is not so: the dimethoxy and the monomethoxy derivatives are powers of 10 slower in the photo-reversion. The influence of aryl substituents on the photoreactivity of the bis-thienyl-perfluorocyclopentenes (28) has been measured.[37] The molecule (28, R = H) and the methoxy substituted compound (28, R = MeO) both undergo cyclization to afford the corresponding cyclized compounds. Interestingly, with R = NMe_2, cyclization does not occur. The quantum yields for the forward and backward reactions were also measured, and again a substituent effect was observed. The quantum yield for the photochemical cyclization of (29) in the crystal has been measured as unity.[38] This high quantum yield is attributed to the conformational constraint within the crystal. 1,2-Bis[(2-methyl-5-(4-methoxyl-phenyl)thiophene-3-yl)]perfluorocyclopentene exhibits photochromism and has been used for a re-writeable storage system.[39] The photodynamics of the photochromic molecule (30) have been measured.[40] Again the quantum yield for ring closure is high.

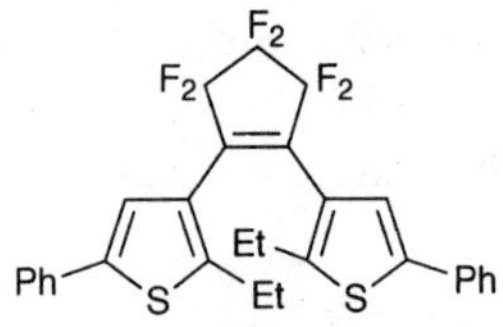

(29)

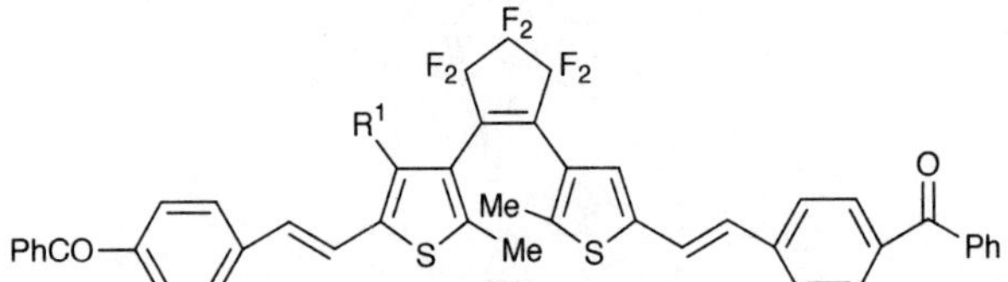

(30)

The photochromic behaviour of the series of nitronylnitroxide derivatives (31) has been studied.[41] Irie and his co-workers[42] have continued their examination of the photochromic alkenes (32), again with nitronylnitroxide substituents. In the present examples they have compared the magnetic interactions between the nitroxide radicals. The biradical molecules (33), where the number of thienyl spacers is varied, are photochromic in the usual cyclization mode.[43] A study of the interaction between the spins through the spacers was carried out. The alkene 1,2-bis[6-(1-oxyl-3-oxide-4,4,5,5-tetramethylimidazolin-2-yl)-2-methylbenzo[*b*]thiophen-3-yl]hexafluorocyclopentene shows photoswitching of intramolecular magnetic interaction between the two nitroxide radicals. The compound is also photochromic.[44] A further paper has described the photochromic reactivity of nitronyl nitroxide substituted diarylethenes.[45] The exchange interaction between the two radicals both in the open and closed forms was investigated. A short review has also highlighted this area of study.[46]

The photochromic asymmetric alkene (34) has been studied to examine the asymmetry of the cyclization process.[47] The compound undergoes diastereoselective cyclization in the amorphous state. The predominant diastereoisomer formed was identified as *R*,*R* and, when R = menthyl, the *de* was 37.4%. The amorphous fluorescent diarylethenes (35) have been synthesized and studied.[48]

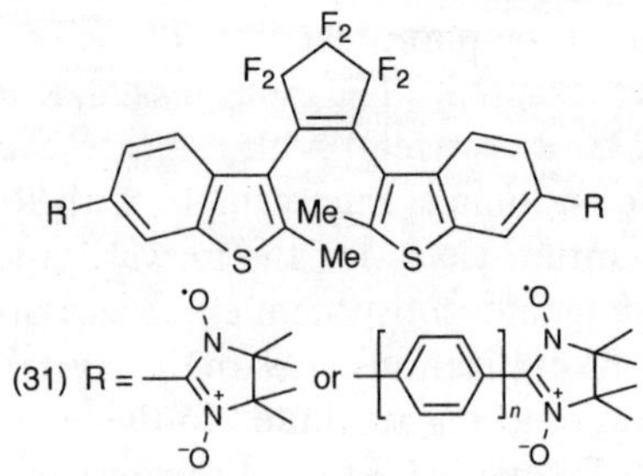

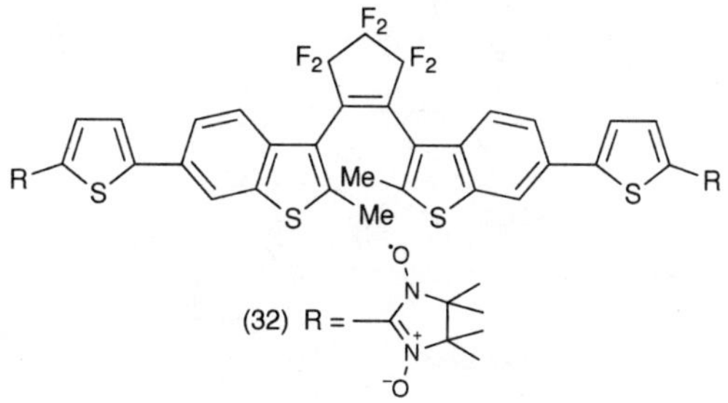

(32)

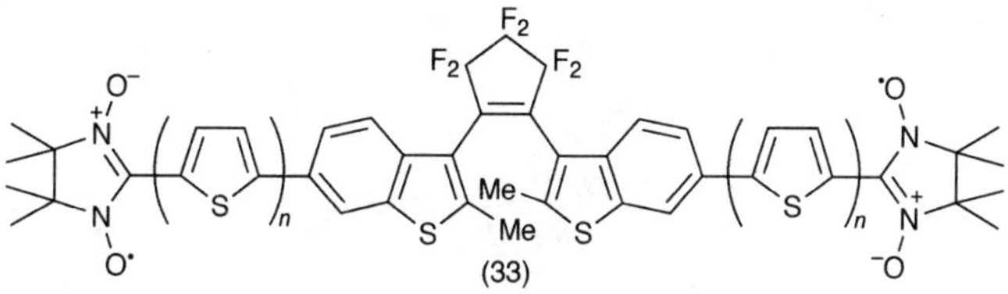

(33)

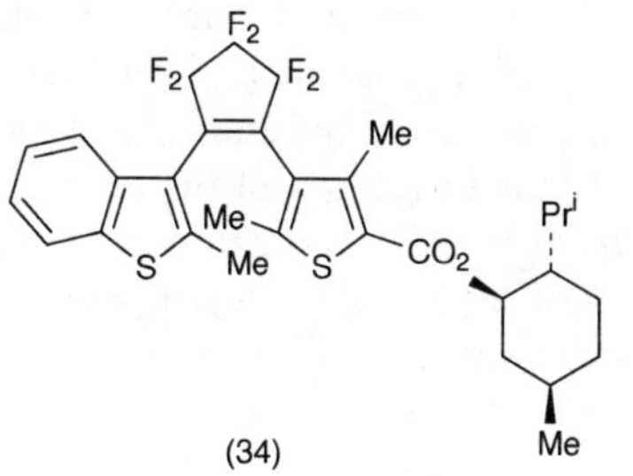

(34)

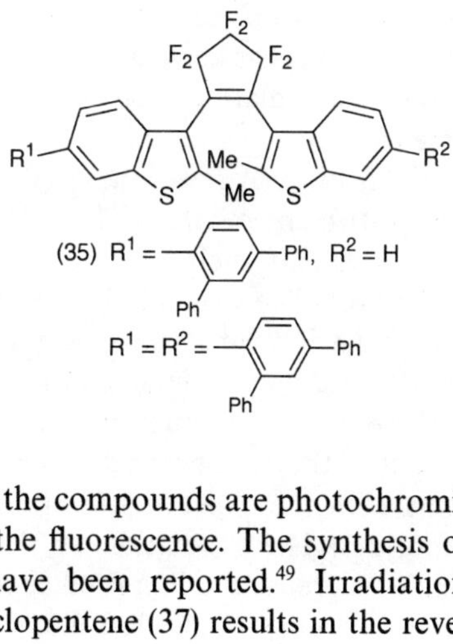

(35)

Like others of their kind, the compounds are photochromic and show dramatic wavelength changes for the fluorescence. The synthesis of (36) and its photochromism in solution have been reported.[49] Irradiation at 410 nm in dichloromethane of the cyclopentene (37) results in the reversible formation of a [7]-thiahelicene.[50]

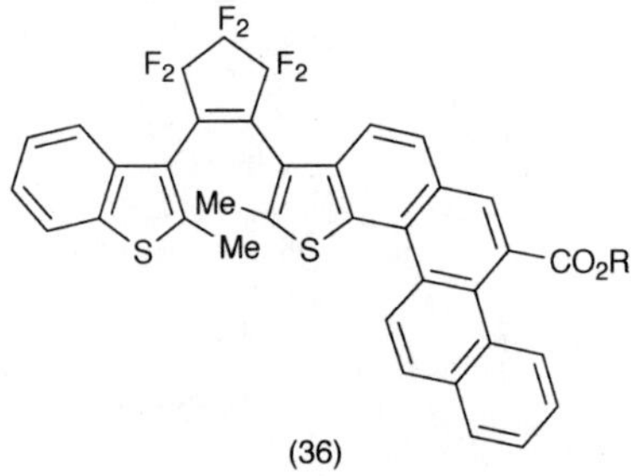

(36)

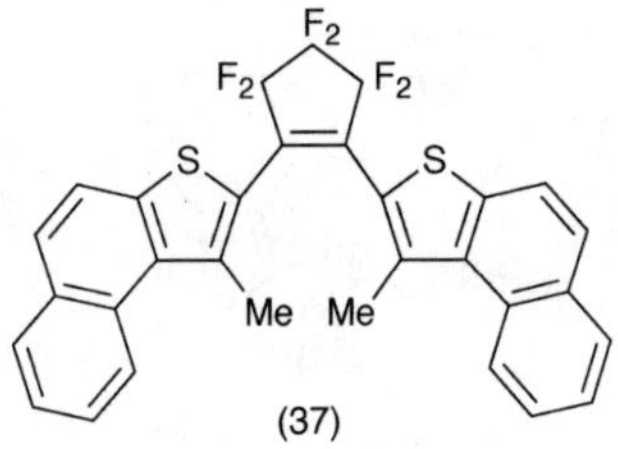

(37)

A short review has dealt with the photochromism of thiophene derivatives and focussed attention on the effect that such photochemistry has on crystal structure.[51] Refractive index changes in an amorphous bisbenzothienylethene have been observed following the formation of the coloured species on irradiation.[52] NMDO calculations have been performed on the open and closed forms of the photochromic 1,2-di(3-thienyl)perfluorocyclopentenes.[53] A theoretical study dealing with the photochromism of diarylethenes has been published.[54] Irie and his co-workers[55] have published the results of ab initio calculations related to the understanding of the temperature dependent effect on the quantum yields of the ring opening of diarylethene derivatives. Bouas-Laurent and Dürr[56] have published a general introduction to photochromism. The influence of temperature, pressure, solvent and concentration of the substrate on some enantio- and diastereo-differentiating reactions has been investigated.[57] The effect of the incorporation of chiral photochromic diarylethene compounds as dopants in liquid crystals has been assessed.[58] A review dealing with the photochromicity of solid diarylethenes has been published.[59] A review has highlighted the special properties of the photochromic dithienylethenes.[60] Details of the X-ray diffraction analysis of photochromic systems such as 1,2-di(3-thienyl)perfluorocyclopentenes have been summarized.[61] The reaction and relaxation dynamics of some photochromic dithienylethene derivatives have been studied.[62]

Considerable interest is being shown in the photochromic alkenes such as (38). These compounds have use in the construction of optical devices and since the closed form is optically active they can be used for chiroptical devices. Murguly *et al.*[63] have shown that both the (*R,R*) and (*S,S*) forms of (38) undergo self-assembly as Cu(I) complexes. The irradiation at 313 nm of these affords the cyclized forms with high diastereoselectivity. A patent has been lodged detailing the synthetic approach to molecules of the class illustrated as (39), in which X = H, F, Cl, Br, I, OH *etc.*, R = a heterocyclic substituent and R^1 = an electroactive substituent.[64] Compounds of this type become highly coloured on irradiation. Some photochromic bisthienylazoles have been synthesized and studied.[65] 1,2-Bis{5-(2-benzothiazolyl)-2-methyl-6-trifluoromethylthieno[3,2-*b*]thiophen-3-yl}hexafluorocyclopentene has been shown to have a high fatigue resistance.[66]

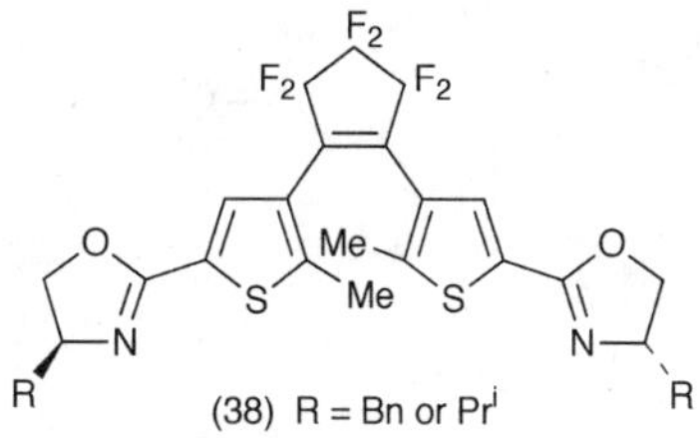

(38) R = Bn or Pr^i

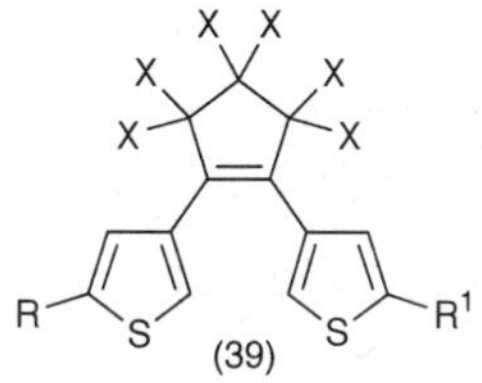

(39)

The efficiency of the cyclization of the photochromic alkenes *cis*-1,2-dicyano-1, 2-dithienylethene, 2,3-bis(2,4,5-trimethyl-3-thienyl)maleic anhydride and 1,2-bis (2,4,5-trimethyl-3-thienyl) perfluorocyclopentene has been studied in polymers and organic solvents[67] and semi-empirical calculations have been used to study the conformational aspects of the compounds.[68]

Ivanov and co-workers[69] have reported the synthesis of 5,6-bis(2,5-dimethyl-3-thienyl)-1,2,4-triazines that are photochromic analogues of the 1,2-diarylcyclopentenes. The amorphous spirobifluorene derivative (40) has been synthesized.[70] This molecule is photochemically reactive and undergoes cyclization of the dithienylethene moiety on irradiation at 365 nm. This brings about ring closure in a yield of 25.4%.

1.2 Miscellaneous Reactions. – *1.2.1 Addition Reactions.* The irradiation of the alkenes (41) in α-, β- and γ-cyclodextrin cavities results in photohydration in a dramatically modified manner.[71] The irradiation in solution results mainly in the formation of the dihydrofuran cyclization product. Irradiation of the alkenes in the cyclodextrins, however, affords the products shown in Scheme 2. One of the key processes in the formation of the products is thought to be an intramolecular charge-transfer exciplex. This would permit the activation of the usually inactive allyl double bond. Clearly, proton transfer is not the key step under these reaction conditions. The efficient photohydration of a series of *m*-hydroxy-1,1-diarylalkenes has been studied.[72] The reactions take place readily in water-acetonitrile (1:1). Two distinct mechanisms were observed, one involving a water-mediated process, as previously reported for the parent molecule, and the other a direct protonation at the β-carbon of the styryl group. The choice of which mechanism is operative is substituent dependent. The addition of 2-aminoethanethiol to oligosaccharide allyl glycosides such as (42) can be

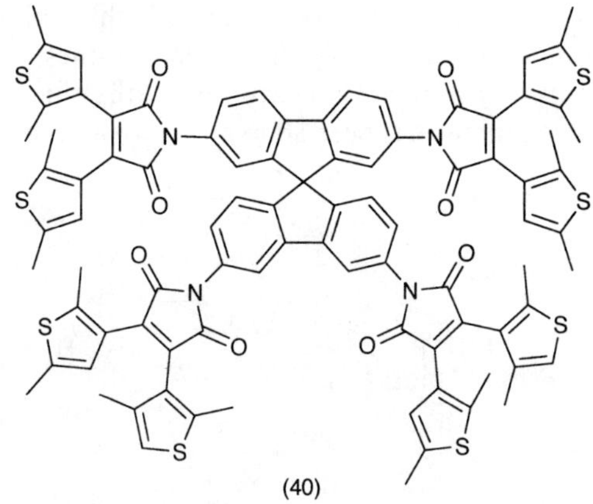

(40)

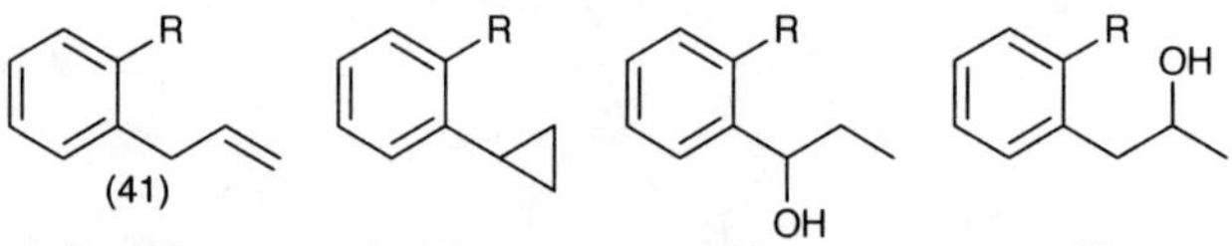

R = HO	11	20	69
R = MeO	27	36	37
R = NH_2	–	–	55

Scheme 2

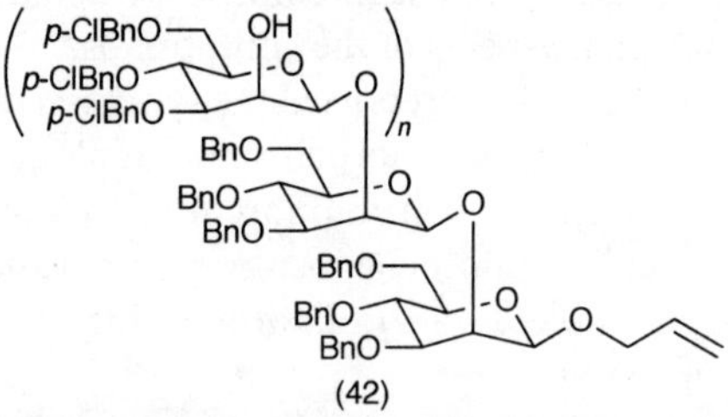

(42)

brought about photochemically.[73] The reaction affords yields >72% by irradiation at 365 nm in methanol-dichloromethane solution. The thiol adds to the double bond in the conventional manner for photoaddition to alkenes. Fulton and Stoddart[74] have also reported the photoaddition of thiols such as (43) to the allyl ether moieties in the cyclodextrin-based carbohydrate cluster (44).

Further work by Miranda and his co-workers[75] has been devoted to the photochemical transformations of the bichromophoric systems (45). They have observed the presence of an exciplex for such systems, a good indication of the excited state interaction between the chromophores. Acetone-sensitized irradiation brings about a quantitative *trans,cis*-isomerization, but there is a solvent dependency for the reactions encountered. In hexane, for example, the predominant reaction is the formation of the products (46) and (47) in the yields shown. In acetonitrile only the product (47) is formed from (45a-c). Sensitized addition of methanol to limonene has been examined in some detail.[76] In general the additions occur using both triplet (xylene) and singlet sensitizers (methyl benzoate

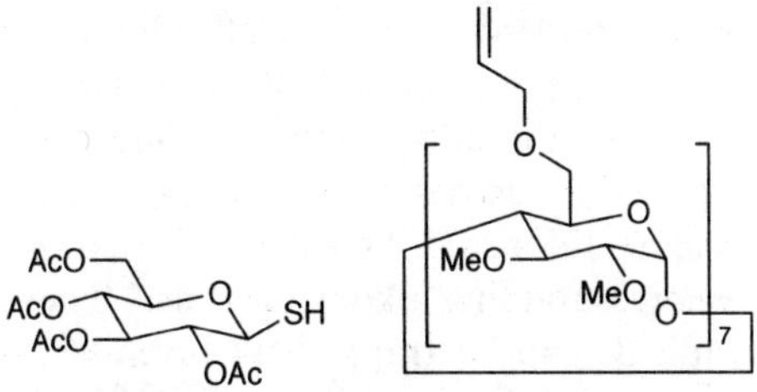

(43) (44)

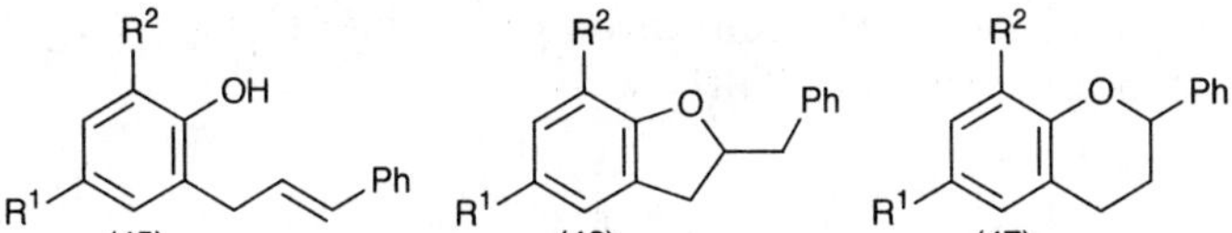

(45) (46) (47)

	R^1	R^2	yield (%) hexane	acetonitrile	hexane	acetonitrile
a	MeO	H	10	–	86	99
b	Me	H	18	–	45	80
c	Me	Me	5	–	65	99
d	Bu^t	Bu^t	9	8	91	70

and dimethyl phthalate). There is a dependence of the diastereoselectivity on the polarity of the medium and on the temperature at which the reactions are carried out. The highest values of *de* obtained are found with low concentrations of methanol in ether. Some of the results from the neat methanol reactions are shown below the structures. Here the temperature and sensitizer effects can be seen (Scheme 3).

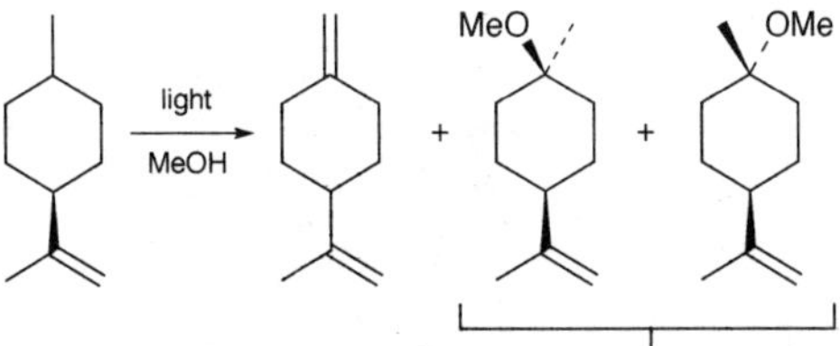

		temp (°C)	%		de
triplet	Sensitiser	25	32.8	40.4	28.1
	xylene	−75	27.4	40.9	46.7
singlet	Me Benzoate	25	53.1	30.0	31.6
		−75	27.5	22.4	48.7
	di Me phthalate	25	18.4	22.4	33.4
		−75	11.5	15.8	51.1

Scheme 3

1.2.2 Other Processes. The irradiation of a series of bromo or iodopyran derivatives has been studied.[77] The primary photochemical reaction is fission of the CBr or the CI bond and the formation of a radical. Cyclization of the radical to the pendant alkyne or alkene moiety affords the bicyclic products shown. Several examples of this reaction are shown in Scheme 4 and the yields of products are good to excellent. More complex reactivity is seen with the bicyclic ether (48), where cyclization of the type described above affords (49) and (50). Ring opening, however, also occurs in competition with the cyclization and yields (51). The intramolecular addition of the alkyl radical, formed by irradiative CI bond fission in (52), to the double bond has provided a path to a macrocyclic lactone.[78] Irradiation of the thiazole thiones (53) results in the fission of the NO bond and the formation of the corresponding alkoxy radicals.[79] These undergo cyclization and the resultant radicals can be trapped by bromine to afford (54). Several examples of this process were examined. Wipf *et al.*[80] have reported a stereoselective spiroketalization as part of a sequence aimed at the synthesis of a bioactive polyether bistramide C. The conversion of (55) into (56) is brought about by irradiation in the presence of $PhI(OAc)_2$-I_2.

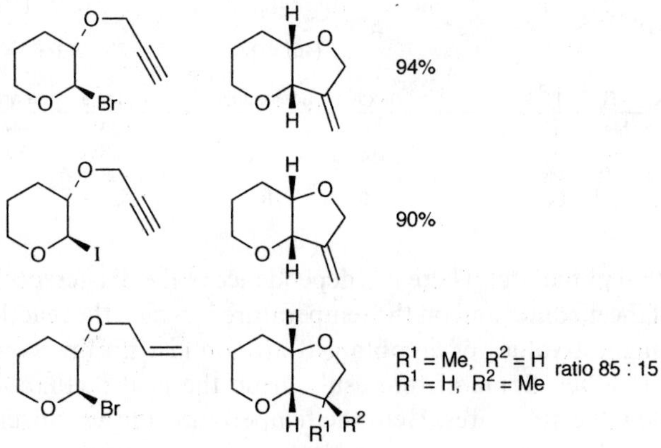

Scheme 4

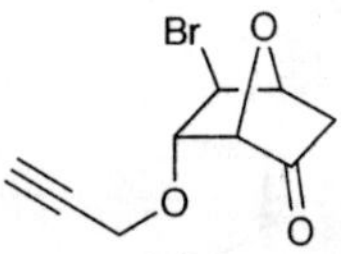
(48)

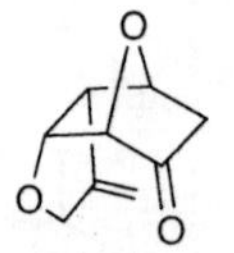
(49)

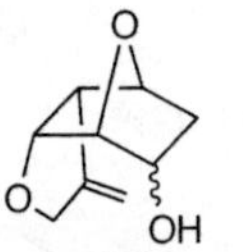
(50)

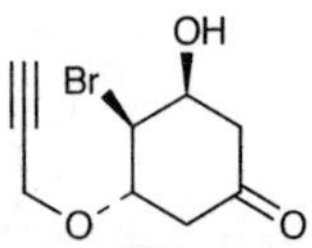
(51)

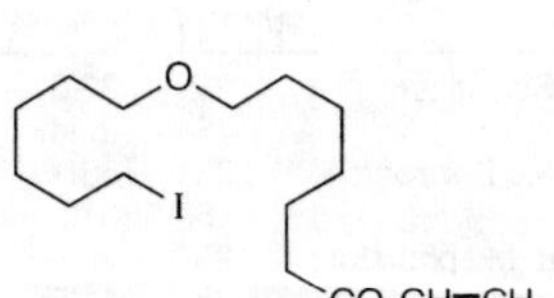
(52)

The electron-transfer-induced cyclization of a variety of unsaturated amino acid derivatives has been studied.[81] The electron accepting sensitizer used is the well-known DCA-biphenyl system. Irradiation of (57) yields the cyclized derivative (58). This is the result of loss of the silyl group to yield the radical (59) as the key intermediate, which then undergoes cyclization. Conversions of (60) into (61) and (62) into (63) are shown.

Other radical-addition reactions involving intermolecular processes have also been described. Photochemically excited ketones can readily abstract hydrogen

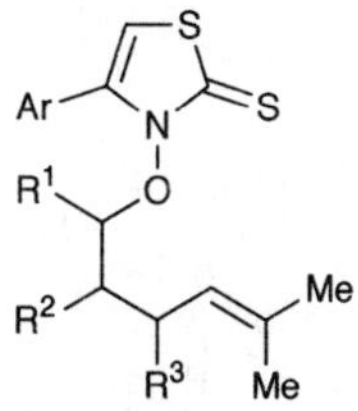

(53)

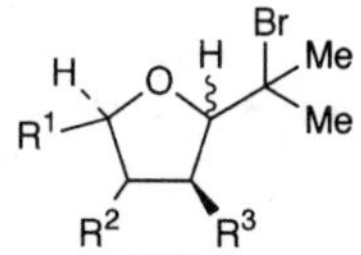

(54)

Ar = *p*-ClC_6H_4, R^1 = Ph, $R^2 = R^3$ = H
Ar = *p*-ClC_6H_4, $R^1 = R^3$ = H, R^2 = 1-propenyl
Ar = *p*-ClC_6H_4, $R^1 = R^2$ = H, R^3 = Ph

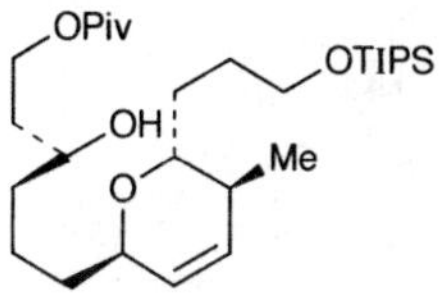

(55)

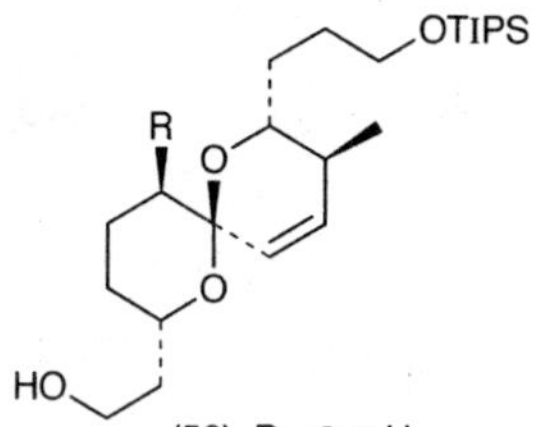

(56) R = I or H

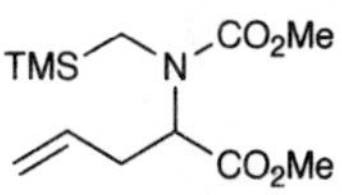

(57)

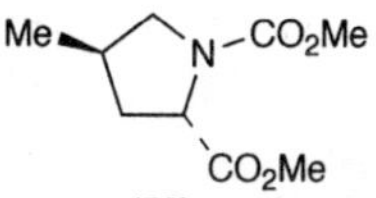

(58)

via

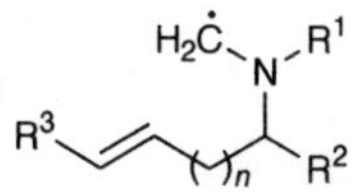

(59)

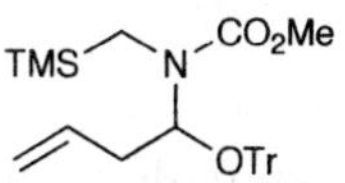

(60)

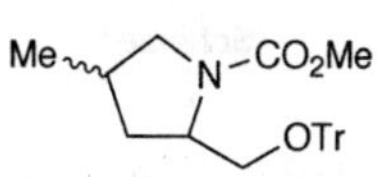

(61)

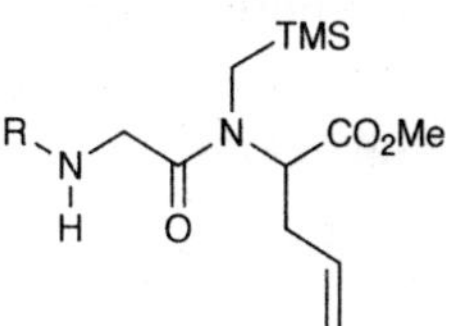

(62)

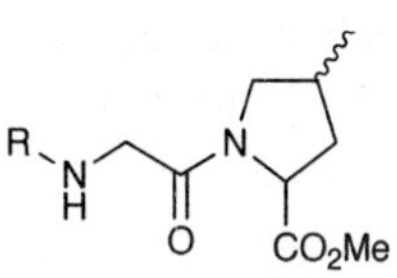

(63) R = PhCO 24%

from cycloalkanes. In one example, the resultant cycloalkanyl radicals undergo addition to the vinylnitriles (64) to afford low to moderate yields of the addition products (Scheme 5).[82] Srivastava *et al.*[83] have reported that various alkenes undergo photoaddition of nitroarenes when they are irradiated in the presence of the iron catalyst $[(C_5Me_5)Fe(CO)_2]_2$. The yields of products are variable, as can be seen from the examples cited in Scheme 6. The successful photocarbonylation of alkenes can be carried out using cobalt catalysts.[84] The efficiency of the process depends to a large extent on the nature of the catalyst, but yields of methyl esters can be obtained in the range of 69-85%. Irradiation of phenylacetylene in the presence of PCl_3 results in the formation of 1-phenyl-2-chlorovinyldichlorophosphine as a mixture of isomers (E:Z = 97:3).[85] The reaction is highly regioselective.

The irradiation of either of the epimeric 2-[(2-methyl-cyclohexyl)-methylene]-malononitriles brings about the formation of a 1:1 mixture of epimeric 4-methyl-

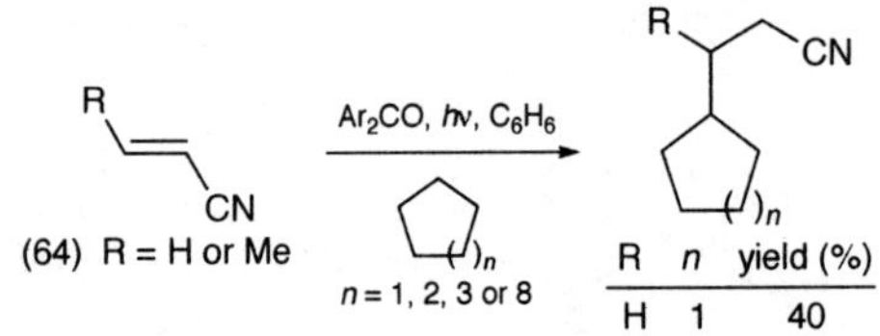

R	n	yield (%)
H	1	40
H	2	37
H	3	50
H	8	50
Me	1	22
Me	2	39

Scheme 5

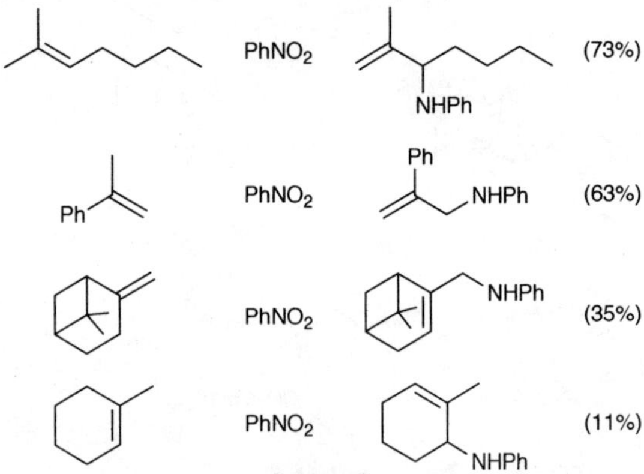

Scheme 6

spiro[2.5]octane-1,1-dicarbonitriles.[86] The general aspects of the photochemical reactivity of a series of 1,1-dicyano-1-alkenes have been reported in some detail.[87] Leitich and Heise[88] report that irradiation of 2-(2,2-dimethyl-furan-3-ylidene)-malononitrile brings about its conversion into a 2-alkoxy-1,1-dicyanocyclopropane.

2 Reactions involving Cyclopropane Rings

2.1 The Di-π-methane Rearrangement and Related Processes. – A detailed re-examination of the photochemical behaviour of the 1,3-diphenylpropenes (65) has been carried out.[89] The singlet excited state is reactive in the di-π-methane process by vinyl-aryl bonding. Double bond isomerization also occurs from the singlet state and yields the *cis* isomer (66). The cyclopropanes formed by the di-π-methane rearrangement were identified as (67). Substitution at the 3-position had little effect on the quantum yield. There is evidence, however, that the activation energies decrease with increasing substitution. Miranda and his co-workers[90] have studied the photochemical reactivity of the styryl derivatives (68) and (69). These differ from those studied in earlier work by the same authors in that the molecules have an ethyl group on the methylene spacer between the alkene and the aryl unit. The compounds undergo the usual cyclizations to

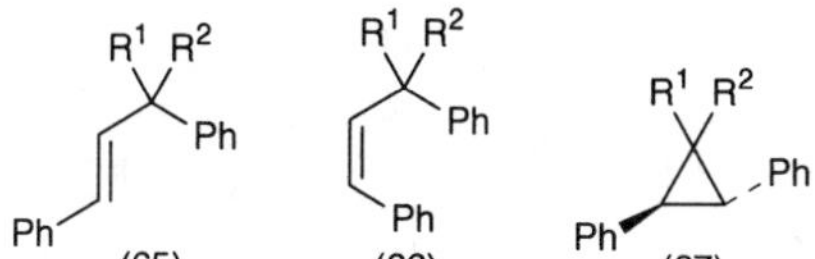

(65) (66) (67)

R^1	R^2	$\phi_{di\text{-}\pi}$
H	H	0.005
Me	H	0.21
Me	Me	0.42
		0.30

afford dihydrofurans and pyrans, but the presence of the ethyl group changes the path of the reaction, and a di-π-methane process occurs. Irradiation in acetonitrile at 254 nm of the *E* isomer (68) brings about the formation of the two cyclopropyl derivatives (70) and (71) in 26% and 37% yield, respectively. The *Z* isomer (69) is also reactive in the same mode and yields both products in 22% and 36% yield, respectively.

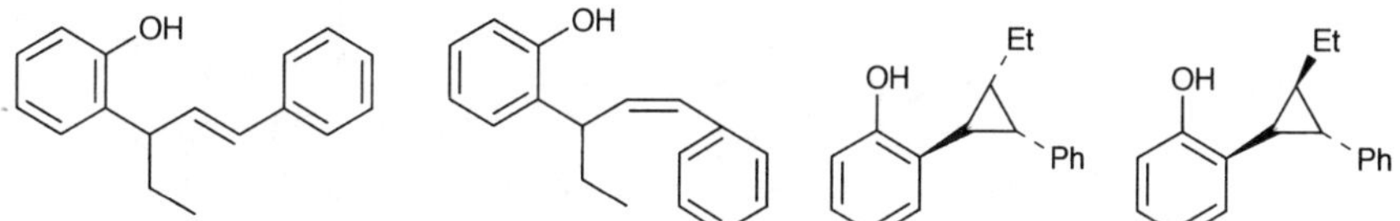

(68) (69) (70) (71)

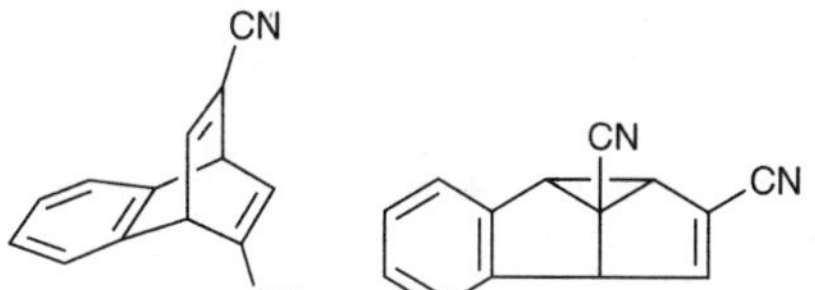

(72) (73)

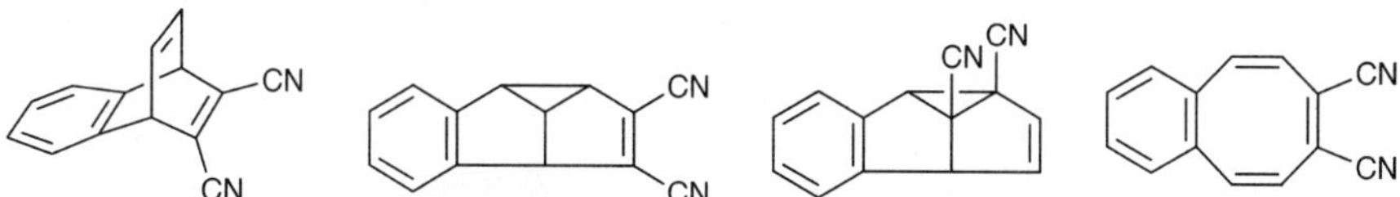

(74) (75) 82% (76) 12% (77) 6%

The triplet-sensitized reaction of the barrelene derivative (72) affords the semibullvalene (73) in 90% yield.[91] The reaction occurs exclusively by way of benzovinyl bridging. Interestingly the isomeric barrelene (74) also undergoes benzovinyl bridging, yielding (75) as the predominant reaction mode. In this example there is some vinyl-vinyl bridging that yields (76). The cyclooctatetraene (77) is formed also. In another related study[92] the influence of substituents on the photochemical reactions of the barrelene (78) has been evaluated. Sensitized irradiation of (78, R = Me) affords the two products (79) and (80) in the yields shown. The major product arises by bonding between the vinyl group and the methylvinyl site, while the minor product involves the interaction between the vinyl group and the carbomethoxy-substituted position. Similar irradiation of (78, R = CN) fails to yield di-π-methane products and instead the cyclooctatetraene (81) is formed. Di-π-methane reactivity is also absent in the benzonorbornadiene derivative (82). This is photochemically reactive, however, and yields the

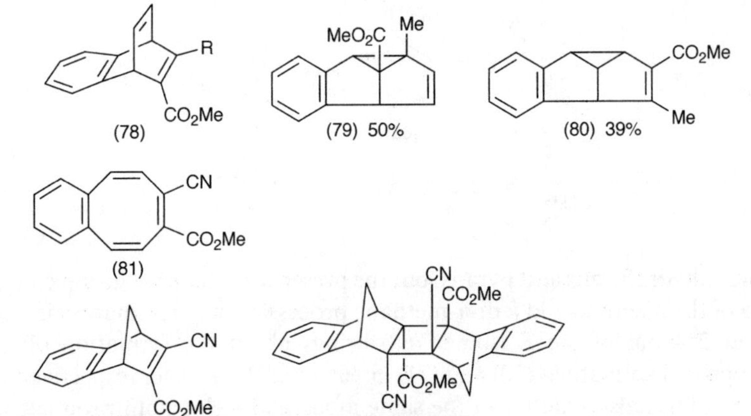

(78) (79) 50% (80) 39% (81) (82) (83) 68% yield

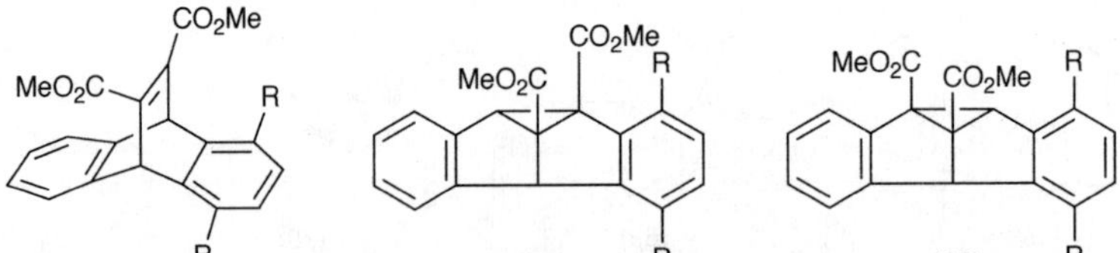

(84) (85) (86)

	R	solvent	ratio	COT
a	Me	C_6D_6	79 : 20	: 1
		CD_3CN	59 : 14	: 27
		THF_{d8}	71 : 21	: 8
		Crystal	74 : 26	: 0
b	Bu^tCH_2O	C_6D_6	78 : 13	: 9
		CD_3CN	86 : 8	: 6
		Crystal	10 : 90	: 0

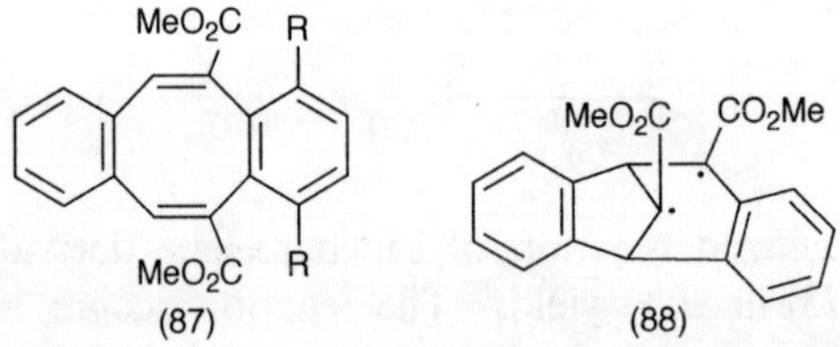

(87) (88)

(2+2)-dimer (83). Others[93] have postulated that the stability of the intermediate free radicals produced on irradiation ($\lambda > 280$ nm) of the dibenzobarrellenes (84) controls the regioselectivity of the di-π-methane rearrangement to yield the semibullvalenes (85) and (86). The yields obtained are shown below the appropriate structures. The influence of substituents was also investigated, as illustrated. Prolonged irradiation brings about a secondary process with the formation of a cyclooctatetraene (87). This is thought to arise via the biradical (88) formed on fission of a cyclopropyl ring bond. In the solid state the irradiation of (84, R = *t*-BuCH$_2$O) affords the semibullvalene that is also formed to a minor extent in the solution phase irradiations.

The dihydrofuran derivative (89) is formed on irradiation of the di-π-methane system (90).[94] This diversion from normal reactivity arises when one of the alkene moieties is substituted with either one or two acyl groups.

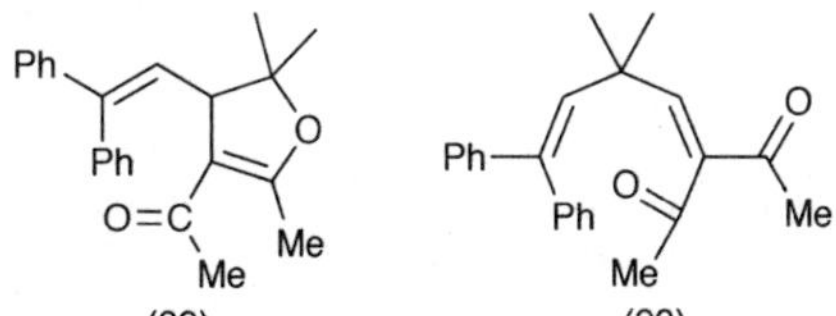

(89) (90)

2.2 Other Reactions Involving Cyclopropane Rings. – A two-photon process is involved in the photochemical conversion of (91) into the radical (92).[95] This involves the intermediacy of the benzylic radical (93). Laser-flash photolysis of (91) produces the cation (94) by a photoheterolysis route. Irradiation of the cyclopropane derivatives (95) in the presence of phenanthrene at wavelengths >280 nm in benzene solution results in the formation of the four products (96), (97), (98) and (99).[96] The formation of the principal products is thought to involve a singlet exciplex between the aryl groups of the cyclopropanes and the phenanthrene. The irradiation affords the biradical (100) that is responsible for the formation of the products by paths involving cyclization or hydrogen transfer.

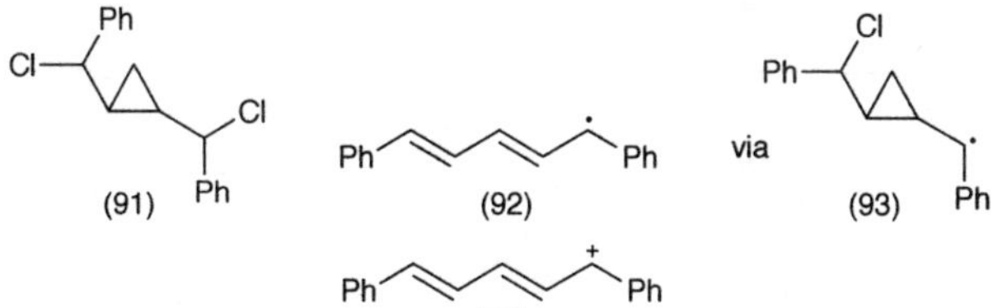

(91) (92) (93) (94)

Scheffer and co-workers[97] have studied the reactions of the cyclopropyl-ketones (101, Scheme 7) confined in zeolites and the influence that such constraint has on the *ee* of products formed. The photochemical reaction affords the isomerized products (102) and (103). The highest *ee* values were observed with the salts (101d,e) and the results are as shown. The amides (104) also undergo *cis,trans*-isomerization on irradiation within alkali-exchanged Y-zeolites.[98] The report states that the introduction of the carboxamide group overcomes many of the difficulties experienced previously. The results obtained indicate that the cations within the zeolite control the extent of the diastereoselectivity and also influence the isomer that is being enhanced. This can be exemplified by the results obtained from (104a,b,d), where the *de* values obtained are in the 80-83% range. The irradiation of either of the isomers of (105) brings about conversion to the phenanthrene derivative (106) as well as isomerization of the cyclopropane ring.[99] The mechanism of the transformation into (106) involves cyclopropane ring bond fission followed by a hydrogen transfer.

Irradiation of (107) brings about extrusion of ethene, with the formation of the carbene (108).[100] The irradiation of 1,6-methano[10]annulene in a supersonic jet results in the extrusion of singlet methylene.[101] The elimination of the carbene has been shown to involve the S_2 excited state.

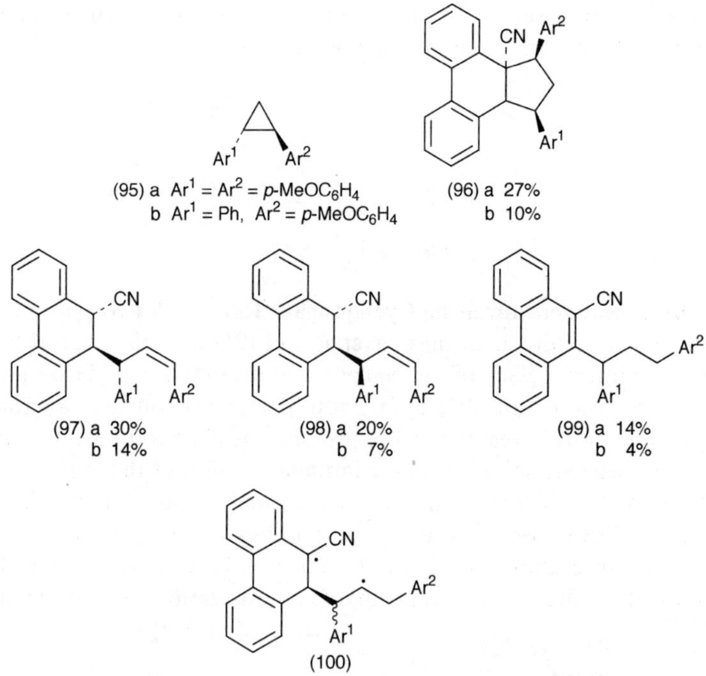

(95) a $Ar^1 = Ar^2 = p\text{-}MeOC_6H_4$
b $Ar^1 = Ph$, $Ar^2 = p\text{-}MeOC_6H_4$

(96) a 27%
b 10%

(97) a 30%
b 14%

(98) a 20%
b 7%

(99) a 14%
b 4%

(100)

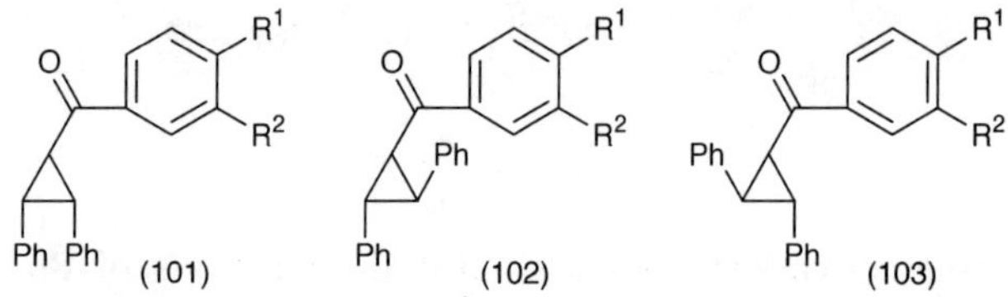

(101) (102) (103)

	R^1	R^2			
a	H	H			
b	CO_2Me	H			
c	H	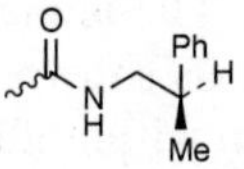 or vice versa			
d	CO_2^-, 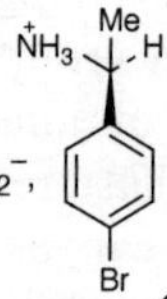	H	25% conversion	99%	99% ee +
			35% conversion	95%	96% ee +
e	CO_2^-, 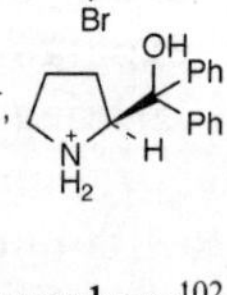	H	25% conversion	99%	94% ee −
			38% conversion	85%	96% ee −

Scheme 7

Campos and co-workers[102] have reported an interesting example of aza-vinylcyclopropane-cyclopentene rearrangement, converting (109) into (110). A review has dealt with the thermal and photochemical behaviour of adducts formed by the addition of cyclopropenes to a variety of substrates.[103] The azirine (111) is converted into the nitrile oxide (112) on irradiation in the presence of O_2

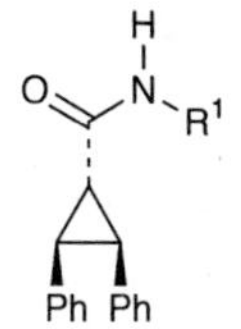

(104) R^1 =

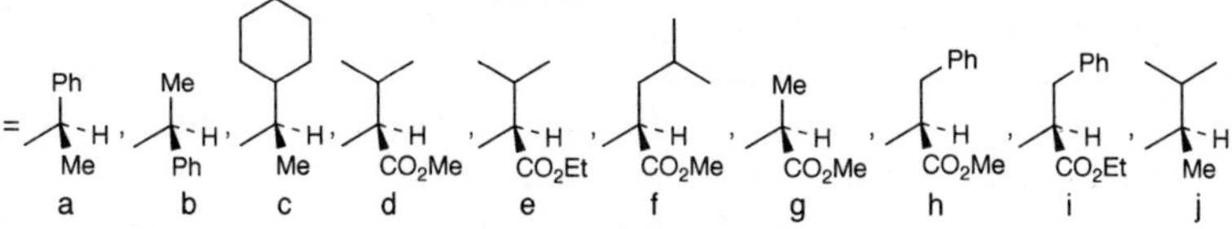

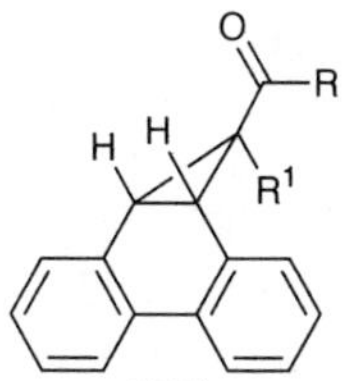

(105)

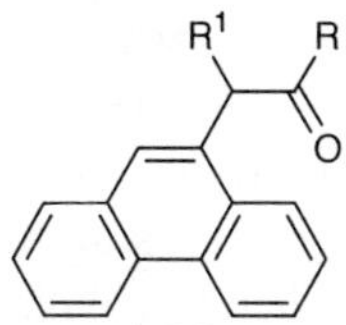

(106)

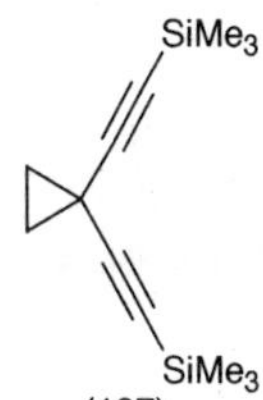

(107)

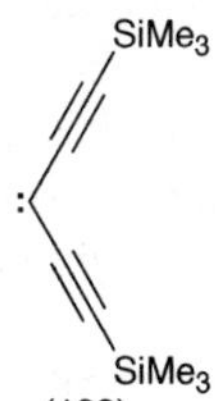

(108)

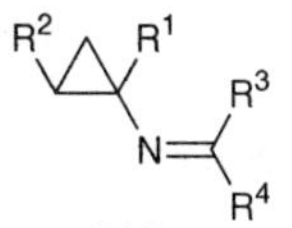

(109)

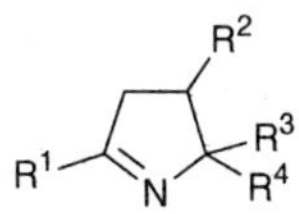

(110)

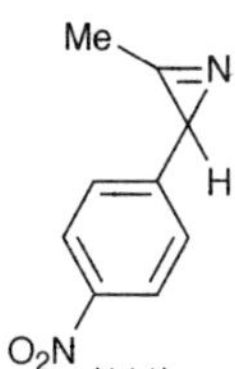

(111)

$Me-C\equiv\overset{+}{N}-O^{-}$

(112)

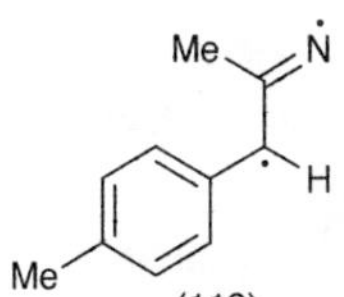

(113)

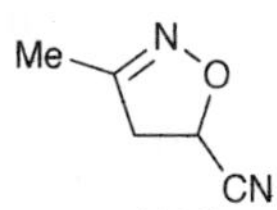

(114)

in fluid solutions.[104] The reaction proceeds by fission of a CN bond to yield the biradical (113). When the reaction is carried out in the presence of acrylonitrile, the products are *p*-nitrobenzaldehyde and the adduct (114). This is unprecedented behaviour, since the usual reactivity is fission of the CC bond within the azirine rather than fission of the CN single bond.

3 Reactions of Dienes and Trienes

The photofragmentation dynamics of allene and propyne have been studied at a variety of wavelengths (203.3, 209.0 and 213.3 nm).[105] Two reactions have been identified for the photodecomposition of buta-1,2-diene at 193 nm: the paths produce methyl and allenyl radicals and butadienyl radicals and hydrogen atoms.[106] Non-adiabatic molecular dynamic computations have been carried out on the photochemical reactivity of *cis*-buta-1,3-diene.[107] An ab initio study of the potential energy surfaces of *s-trans*-buta-1,3-diene has been reported.[108] Robb and his co-workers[109] have reviewed the available quantum chemical packages for such calculations.

Previously Leigh and his co-workers[110] have studied the conrotatory ring opening of the cycloalkenes (115) and (116) to yield the products shown. In the present study they have examined the behaviour of (117) and (118) using Rydberg excitation of electrons [π,R(3s) excitation]. Their results indicate that clean conrotatory ring opening occurs along with reverse (2+2)-addition, affording the products shown in Scheme 8. Thus, they conclude that the conrotatory ring opening is a true photochemical process. A decision has yet to be reached concerning competition between the ππ* excited state process and the Rydberg excitation. Fokialakis *et al.*[111] report the isolation of the alkaloid (119). They suggest that the formation of this compound in the plant may arise from oxidation of (120) to the diene (121). This diene will readily undergo UV transformation, a standard diene-cyclobutene transformation, into the final product (119).

An examination of the formation of an intramolecular charge transfer complex in the dienes (122) has been reported.[112] The dienes undergo *E*,*Z*-isomerization from the singlet excited state with a quantum efficiency of 0.1. Both stationary and pulsed photochemical techniques have been used to study a series of 1-aryl-4-phenylbutadienes (Ar = pyridyl, thienyl, naphthyl, phenanthryl, anthryl and pyrenyl).[113] The cyclobutane derivative (123) undergoes electron

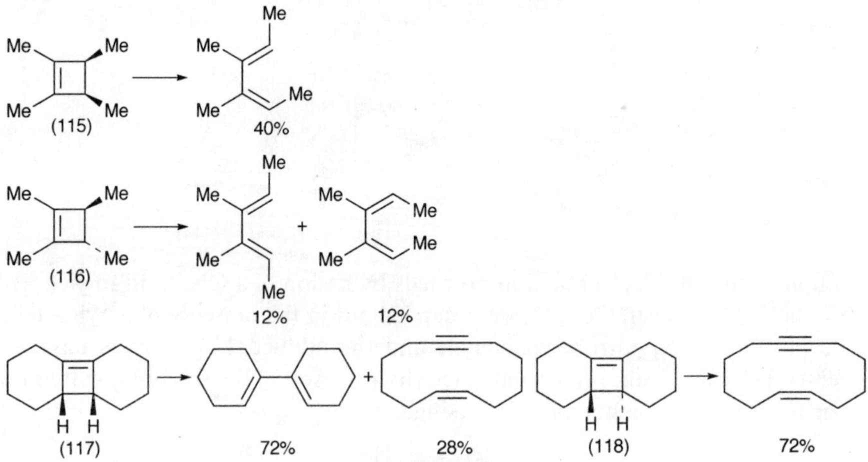

Scheme 8

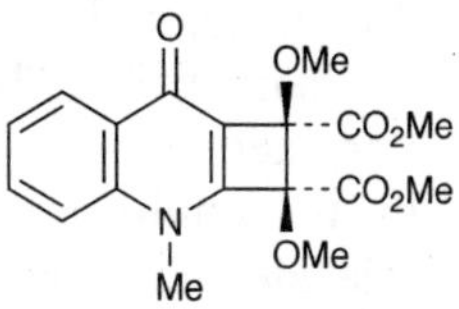

(119)

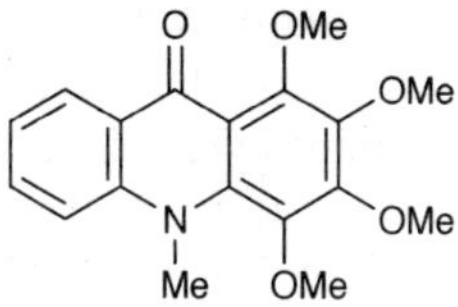

(120)

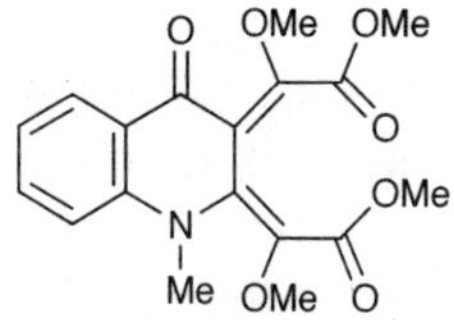

(121)

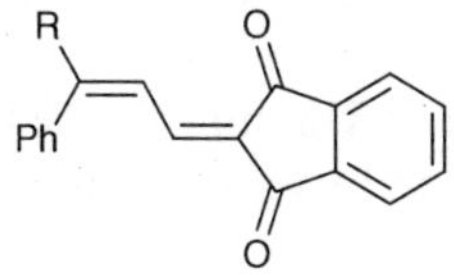

(122) R = 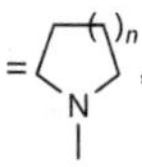, $n = 1$ or 2

R = NEt_2

transfer when irradiated in dichloromethane or benzene with $\lambda > 360$ nm, in the presence of DCA as the electron-accepting sensitizer.[114] This produces the corresponding radial cation, which undergoes dimerization to yield the *syn*-(124) and *anti*-(125) dimers. The radical cation also adds to the DCA to yield the adduct (126). The cyclopentadiene derivative (127, n = 1) undergoes irradiative conversion into the adduct (128).[115] The other derivatives (127, n = 2,3) are unreactive under such conditions. A photodimer is obtained when 1-(*o*-hydroxyphenyl)-2-pentamethyldisilanylethyne is irradiated in benzene solution.[116,117] Different reactions occur in methanol, and irradiation converts 1-(*o*-hydroxyphenyl)-2-pentamethyldisilanylethyne into 1-(*o*-hydroxyphenyl)-2-trimethylsilylethyne. The transformation takes place *via* a silacyclopropene intermediate.

Studies with circularly polarized light have demonstrated that it is possible to obtain enantiomeric enrichment. Thus the irradiation of the chiral norbornadiene ester (129) affords the corresponding chiral quadricyclane derivative (130).[118-120] Dubonosov *et al.*[121] report the conversion of the diene (131) and

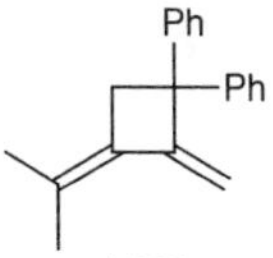

(123)

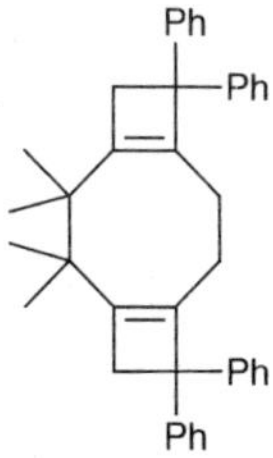

(124)

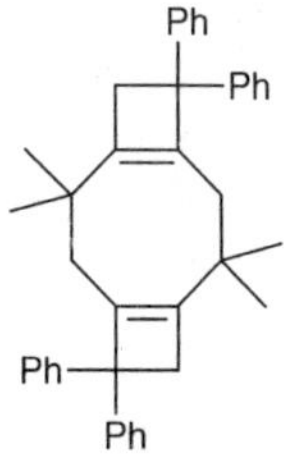

(125)

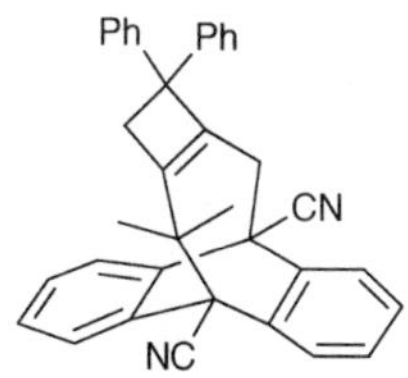

(126)

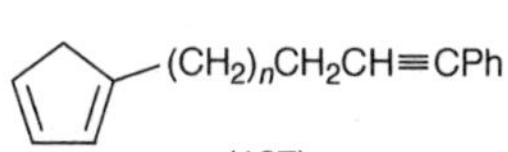

(127)

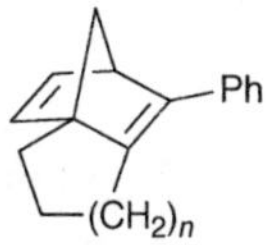

(128)

derivatives thereof into the quadricyclane (132) by irradiation. The acetophenone-sensitized irradiation of the norbornadiene derivative (133) in THF at -30°C affords the quadricyclane (134).[122] Warrener and his co-workers[123] have reported the formation of aziridines by the photochemical extrusion of nitrogen from adducts formed between benzylazide and di-methyl 3,6-di(2-pyridyl)-4,5-diazatetracyclo[6.4.1.$0^{2,3}$.$0^{9,12}$]trideca-3,5,7,10-tetraene-10,12-dicarboxylate. The aziridines are accompanied by hitherto unknown pyridazinonorbornadienes. Norbornadiene can be readily brominated by irradiation in the presence of bromine.[124] This affords the hexabromo derivative (135). The same product can be obtained by photobromination of (136).

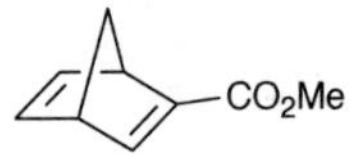

(129)

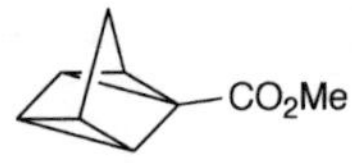

(130)

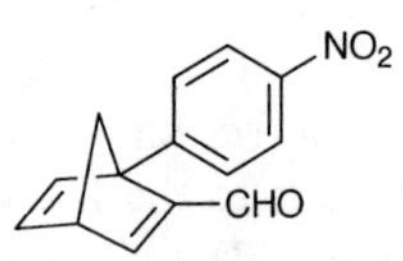

(131)

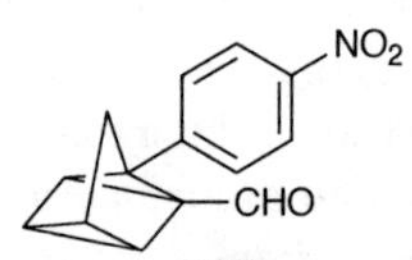

(132)

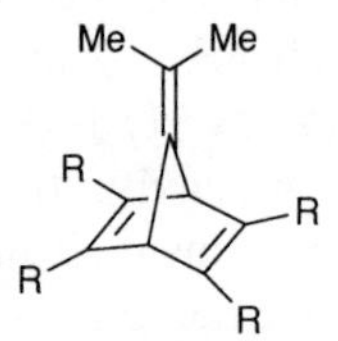

(133) R = $CH_2OCH_2CH_2OCH_2$

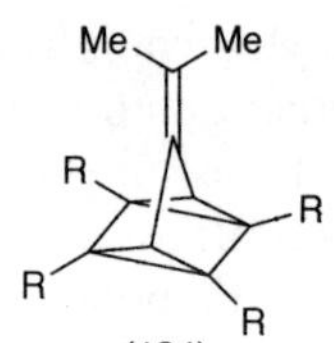

(134)

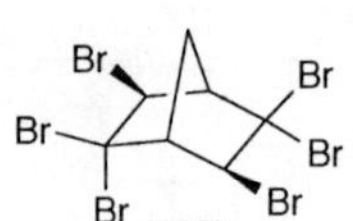

(135)

(136)

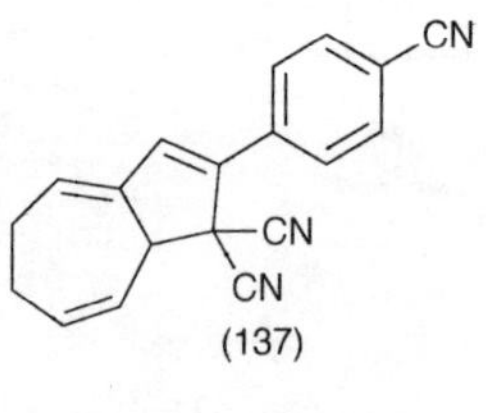

(137)

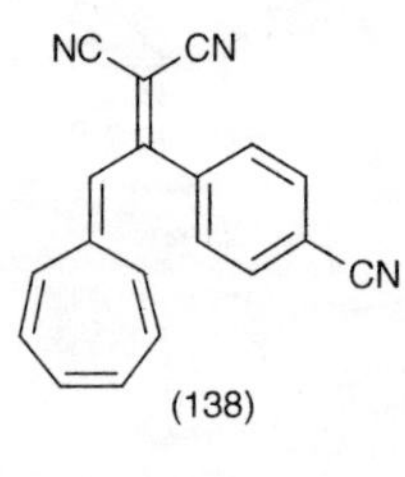

(138)

Irradiation of (137) at 340 nm using a 30fs pulse in ethanol brings about ring opening and the formation of (138).[125] Calculations have dealt with the photocyclization of the vinylheptafulvene (139) into the cyclized form (140).[126] The

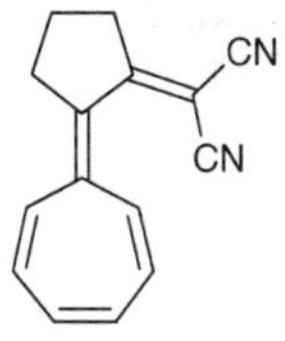
(139)

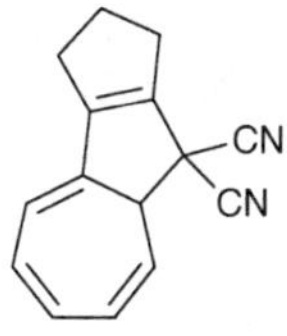
(140)

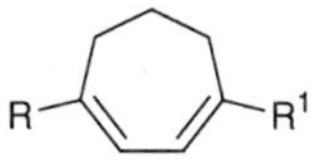
(141)

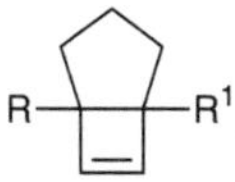
(142)

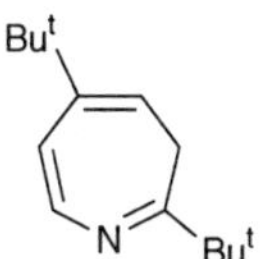
(143)

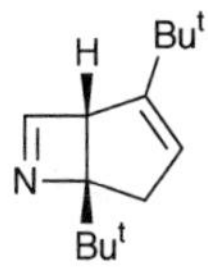
(144)

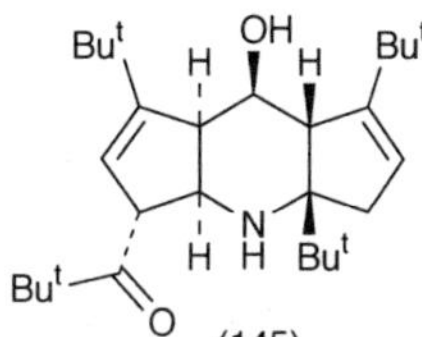
(145)

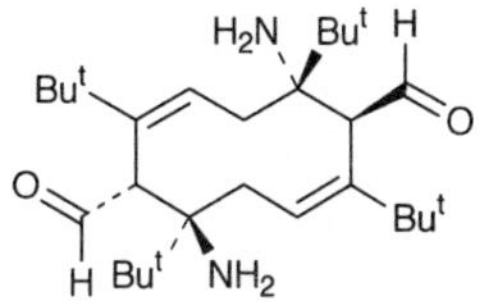
(146)

results obtained explain why the quantum yield for the forward reaction approaches unity.

The cycloheptadiene derivatives (141, R and R′ = aryl groups) undergo direct irradiative conversion (XeCl laser at 308 nm) into the bicycloheptenes (142).[127] The photophysical properties of 5*H*-dibenzo[*a,d*]cycloheptene have been measured and the production of a triplet state verified.[128] Irradiation of the azepine (143) results in the formation of the primary product identified as the bicyclic compound (144).[129] This compound could not be isolated, and work-up in water afforded the two products (145, 61%) and (146, 5%). The second product (146) is thought to arise from a dimer of the starting material presumed to have the structure (147). The route to the major product remains speculative, but a possible path involves rearrangement of (144) into (148) and cycloaddition of these two bicyclic compounds. Subsequent ring opening of this adduct would afford (145).[129] Tanaka and his co-workers[130] have demonstrated that the irradiation of the inclusion complex composed of (−)-(149), the tropone (150) and chloroform (ratio 2:1:1) brings about the formation of the two products (151) and (152) in 96% and 90% *de*, respectively.

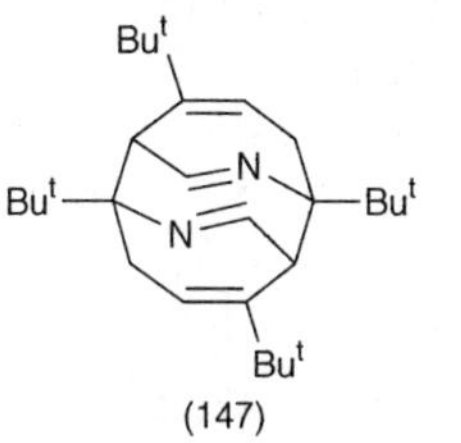
(147)

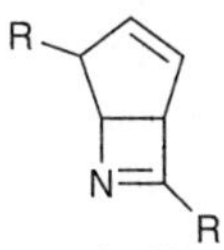
(148)

Irradiation (254 nm in hexane solution at 0°C) of the bicyclo[4.3.1]decatrienes (153) results in the formation of the triynes (154).[131] These products arise via formation of the carbenes (155), which then undergo rearrangement. Another product was identified as (156). This arises by a 1,5-rearrangement of the starting material (153) to afford the intermediate (157). Ring opening of this yields (156).

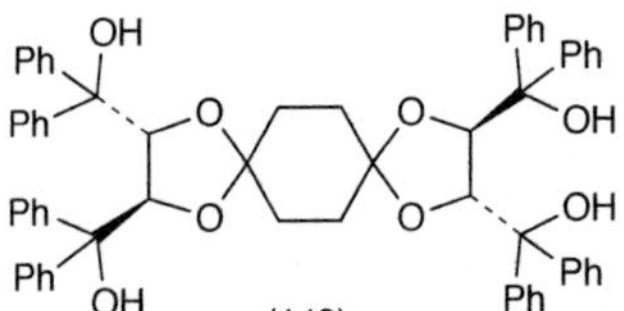

(149)

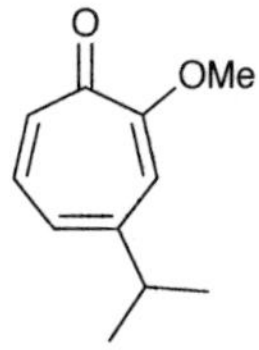

(150)

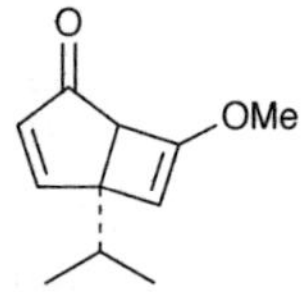

(151) (–)

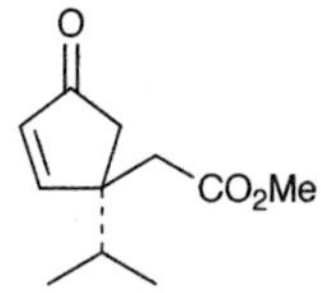

(152) (–)

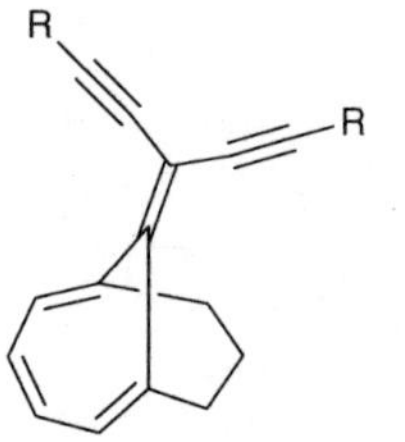

(153) R = TMS or ≡–TIPS

(154) R = TMS 46%
R = ≡–TIPS 59%

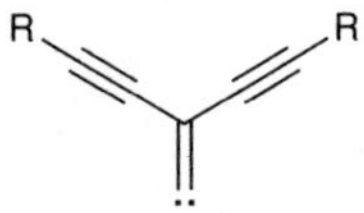

(155)

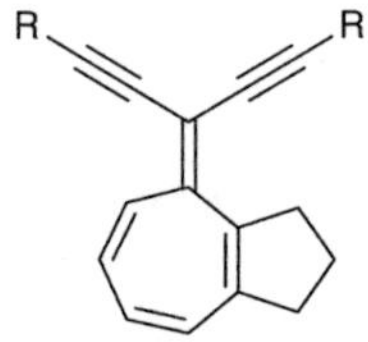

(156) R = TMS 30%
R = ≡–TIPS 24%

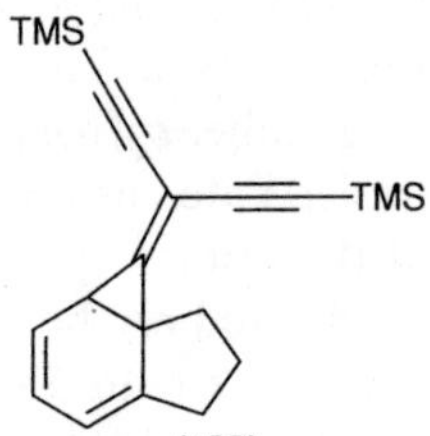

(157)

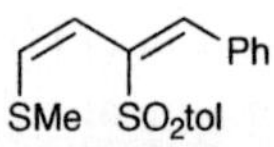

(158)

The photocyclization of the butadiene (158) in the presence of iodine affords 2-*p*-tolylsulfonylnaphthalene.[132] The cyclization is retarded when electron-withdrawing groups are introduced into the phenyl substituent at C-1. A discussion of the use of a variety of stilbene derivatives as an approach to novel heterocyclic compounds has been published.[133] The irradiation of 2-[2-(2-vinylphenyl)ethenyl]furan brings about its conversion into an intermediate identified as 3a,9-dihydro-4,9-methano-4*H*-benzo[4,5]cyclohepta[1,2-*b*] furan.[134] The chromenes (159-161, X = O) are photochemically reactive, and irradiation brings about a marked colour change.[135] This is not the case when the substituent is X = C(CN)$_2$. These derivatives have been shown to be photochemically inert under the same conditions. Some derivatives of [2*H*]-chromene have been investigated by semi-empirical quantum methods.[136] The pyran derivative 2,2-di-*t*-butyl-6-(4,4-di-*t*-butylbuta-1,3-dienyl)-2*H*-pyran undergoes photochemical ring opening.[137]

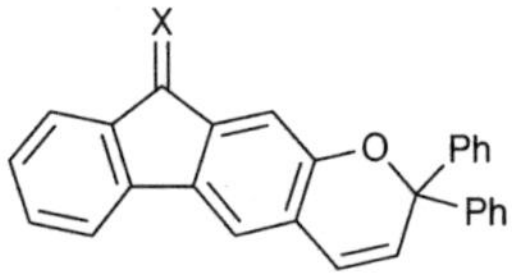

(159) X = O or $C(CN)_2$

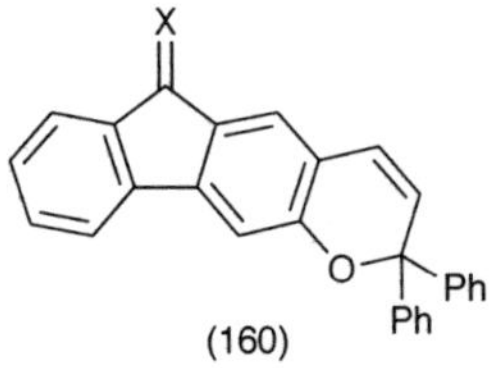

(160)

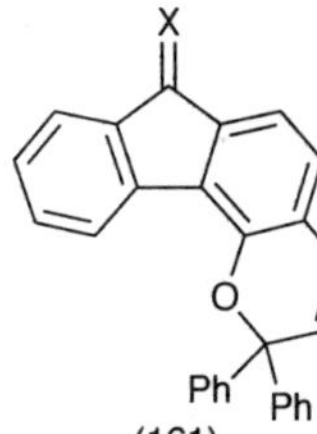

(161)

A further example of the photo-Bergman reaction has shown that irradiation of (162) affords the cyclized products (163) in 5% yield.[138] Ab initio calculations have been used to examine the ring opening of cyclohexa-1,3-diene.[139] Others[140] have calculated that a single conical intersection accounts for all the products obtained from the irradiation of cyclohexa-1,4-diene in solution. Provitamin D_3 has been subjected to photoisomerization on a silica gel-hexane matrix.[141] Using these conditions a high yield of previtamin D_3 is obtained with the formation of undesirable by-products at a minimum. An example of the so-called Hula-twist isomerization has been reported for the conversion of (164) into (165).[142]

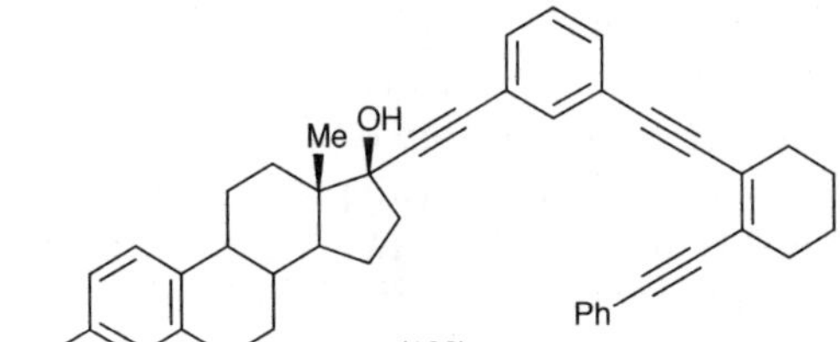

(162)

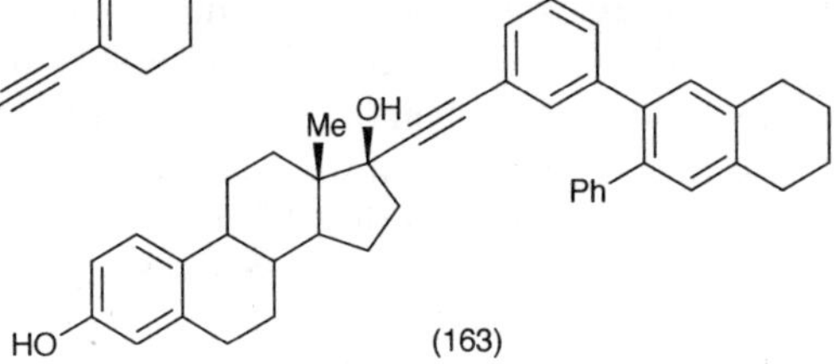

(163)

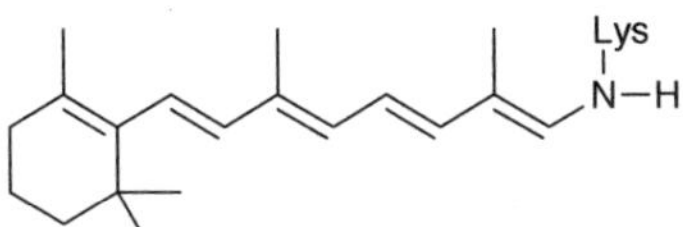

(164)

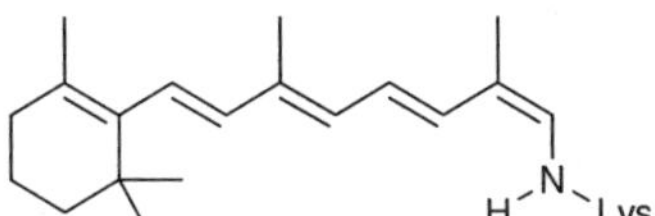

(165)

4 (2+2)-Intramolecular Additions

The Cu(I)-catalysed cycloaddition of the *trans* bis-alkene (166, n = 1) proceeds with excellent facial selectivity and affords the cycloalkane (167, n = 1) in 80% yield with a *dr* of 98:2.[143] The *cis* isomer (168) is also reactive and gives the cycloalkane (169, 77%, *dr* 75:25). The corresponding cyclohexane derivatives (166 and 168, n = 2) are also reactive and yield the cycloalkanes in similar yield and diastereoselectivity. The Cu(I)-catalysed intramolecular cycloadditions of compounds such as (170) results in the formation of the adducts (171).[144] The authors suggest that the formation of this *cis,syn,cis* adduct is unusual. The intramolecular photocycloaddition of (172) can be brought about in 70% yield

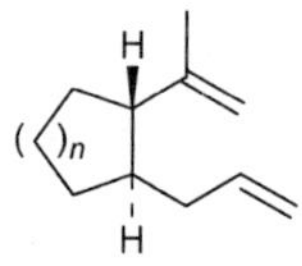
(166)

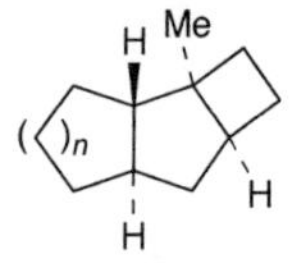
(167)

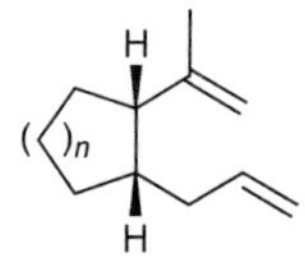
(168)

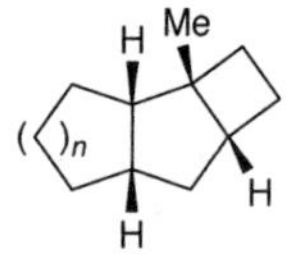
(169)

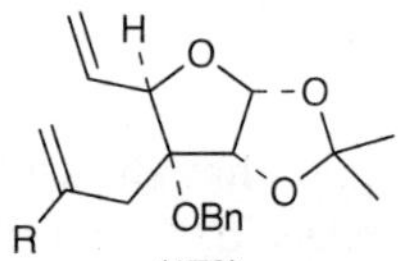
(170)

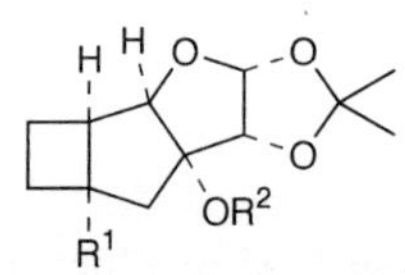
(171) R^1 = H, R^2 = Bn yield 73%
R^1 = Me, R^2 = Bn yield 72%

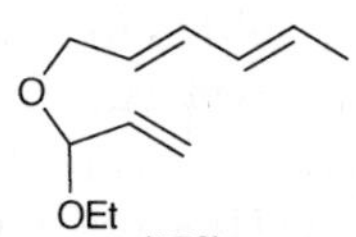
(172)

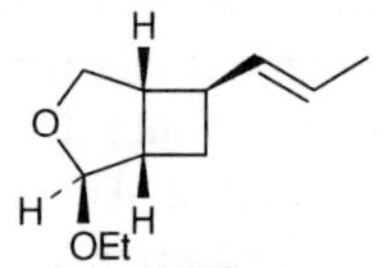
(173)

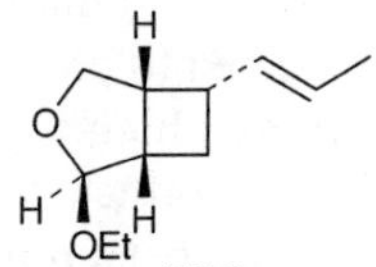
(174)

by irradiation in ether solution with added CuOTf.[145] This treatment affords a mixture of the two cycloadducts (173) and (174), and was used as a key step in an approach to the synthesis of cyclobut-A (175). Irradiation of ipsdienol (176) in hexanes using benzophenone as the sensitizer affords the two products (177) and (178) in the yields shown.[146] The products are formed by two modes of (2+2)-cycloaddition. The phenanthrene derivative (179) undergoes photochemical cycloaddition on irradiation through Pyrex in benzene solution.[147] This treatment brings about cycloaddition of the vinyl groups to afford a mixture of two cyclobutyl-substituted phenanthrenophanes in a total yield of 40%. Neckers and his co-workers[148] report that polymers incorporating the distyrylbenzene moiety undergo (2+2)-photocycloaddition. The cage compound (180) can be formed by irradiation of (181).[149]

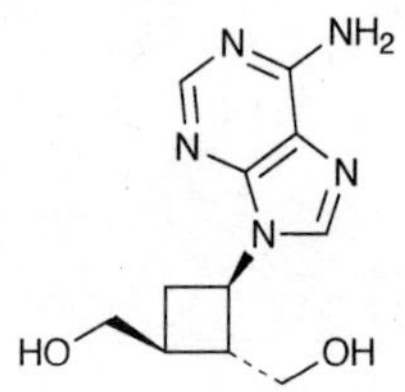
(175)

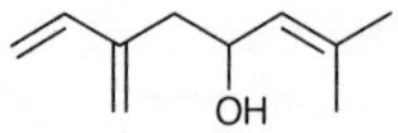
(176)

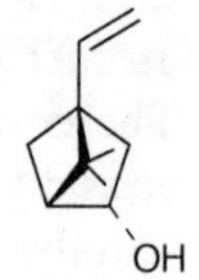
(177) 85%

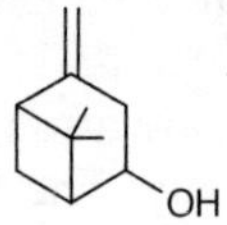
(178) 5%

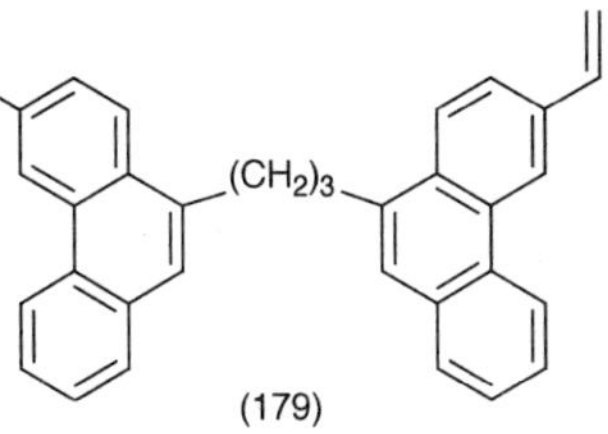

(179)

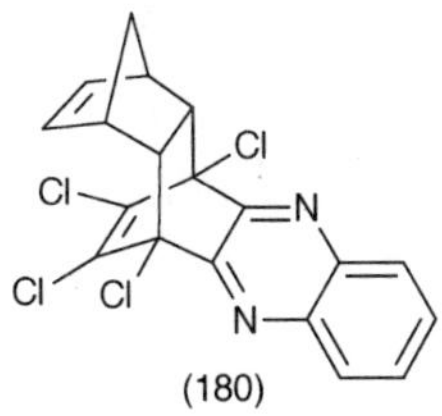

(180)

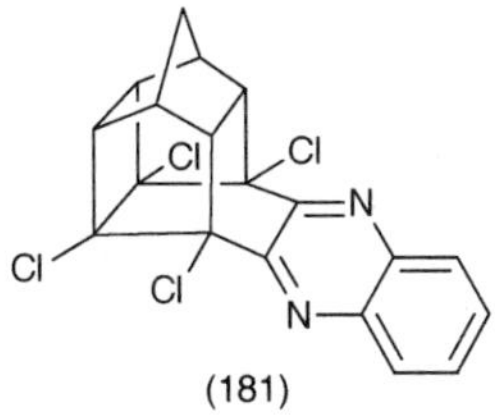

(181)

5 Dimerization and Intermolecular Additions

Dimerization is the result of irradiation of 4-methoxystyrene in a NaY zeolite.[150] The co-crystals obtained from 1,8-naphthalenedicarboxylic acid and the stilbene derivative (182) are arranged in such a manner that the two ethylene units lie in close proximity.[151] The arrangement is shown in (183), where the diacid behaves as a linear template with hydrogen bonding to the pyrimidine nitrogens. Irradiation at 300 nm results in 100% stereospecific conversion into the corresponding cyclobutane derivative. Single crystals of the stilbene (184) co-crystallized with bis-*p*-phenylen[34]crown-10 have been irradiated.[152] This affords the dimer (185) in 80% yield. The photochemical dimerization of the *trans*-1-[2-(5-*R*-benzoxazolyl)]-2-(4-R′-phenyl)ethenes (R = R′ = H; R = Me, R′ = H; R = Me, R′ = OMe) affords cyclobutane derivatives by head-to-tail dimerization.[153] Photochemical isomerization is brought about when the trienes (186) are irradiated in solution.[154] In the solid phase, however, irradiation using wavelengths > 370nm induces (2 + 2)-cycloaddition, to yield the dimers (187, R = CHO, 16%) and (187, R = CN, 21%).

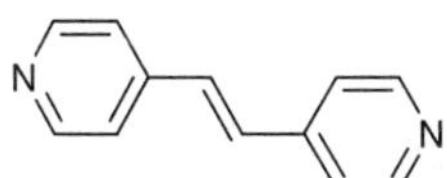

(182)

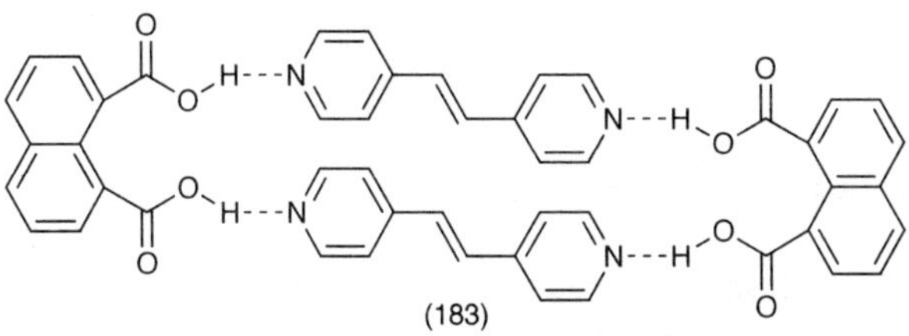

(183)

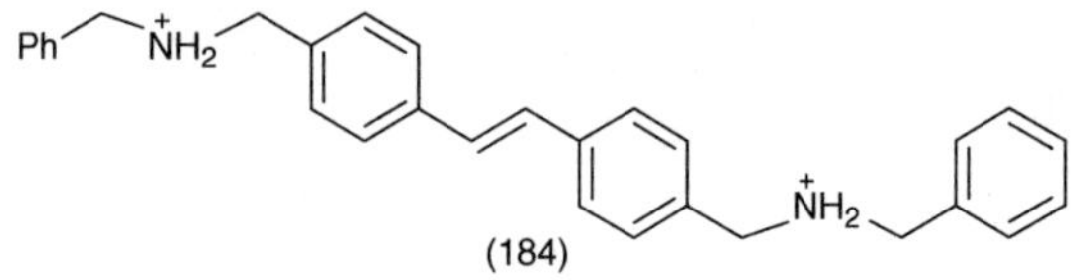

(184)

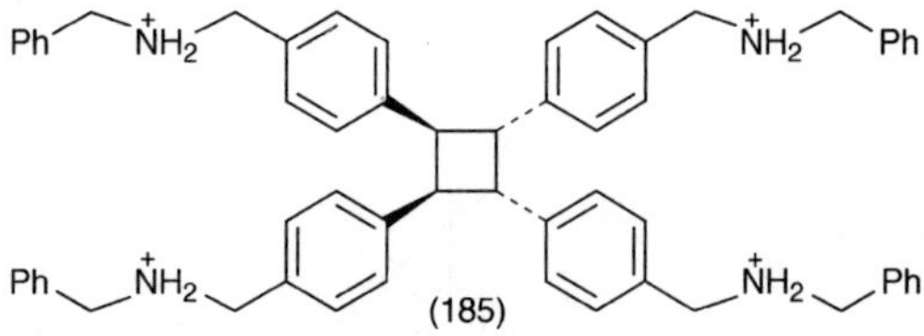

(185)

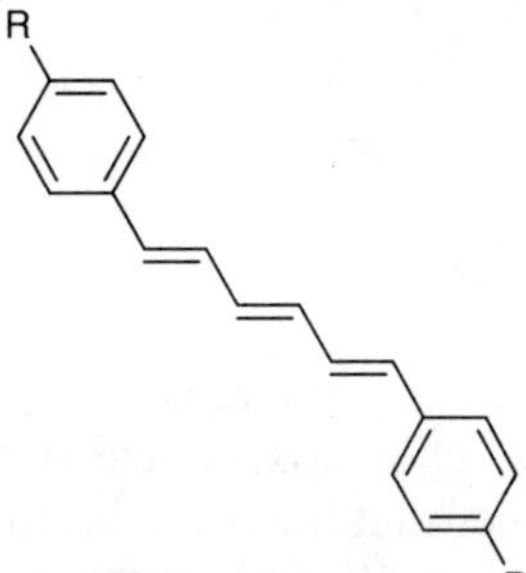

(186) R = CHO or CN

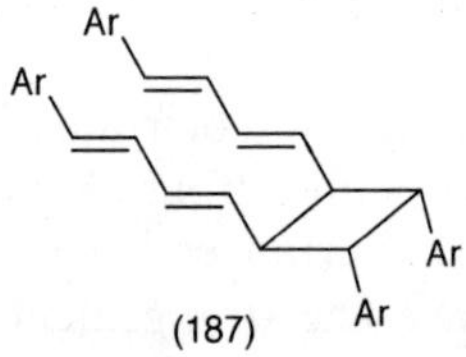

(187)

Yoshizawa and co-workers[155] have described the photochemical dimerization process observed within co-ordination cages. The nano-cage is of the M_6L_4 type and is based on Pd complexed with tripyridyltriazine. When acenaphthylene (188, R = H) is complexed within the cage and the system irradiated in water, a quantitative conversion into the dimer (189, R = H) is observed. The methyl derivative (188, R = Me) is also reactive and affords (189, R = Me) exclusively. The dimerization of other compounds such as naphthoquinone (190) was also studied and this affords the dimer (191).

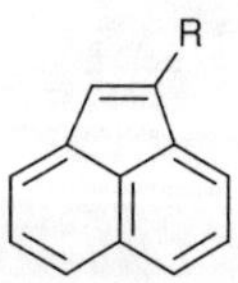

(188)

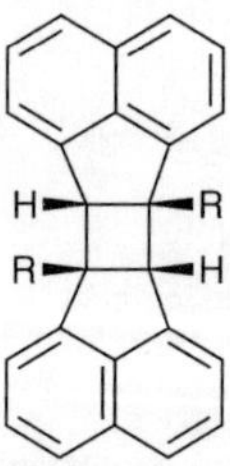

(189)

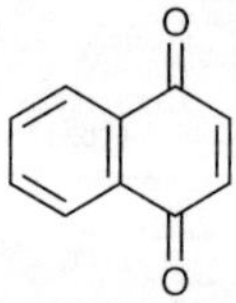

(190)

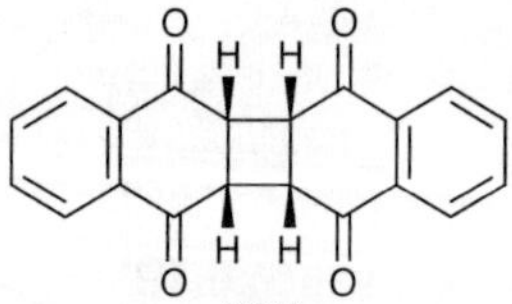

(191)

6 Miscellaneous Reactions

The photodissociation of jet-cooled methyl chloride using 235 nm irradiation has been studied and the fragments produced studied by resonance enhanced multiphoton ionization.[156] The transient resonance Raman spectra obtained following the irradiation of iodomethane in cyclohexane suggests that the species formed is CH_3I-I.[157]

The formation of *iso*-CH_2BrI has been demonstrated following irradiation of bromoiodomethane.[158] The iodomethylene radical was found to be the most energetic following the photodissociation of diiodomethane in the wavelength range 277 nm to 305 nm.[159] A detailed study of the photochemical reaction between diiodomethane and alkenes has indicated that the likely precursor of cyclopropanation of alkenes is isodiiodomethane (192).[160] Further studies on the isodihalomethanes (CH_2X_2, X = Cl, Br or I) have been reported.[161] A detailed study of the mechanism of the reaction of isoiodomethane with cyclohexene has shown that an iodine-cyclohexene complex is formed in the 5-10ns timescale.[162] The authors have presented a mechanism that accounts for the cyclopropanation of alkenes by this reagent. A Raman study has sought to obtain the vibrational spectrum of the species produced on the irradiation of diiodomethane.[163]

CH_2—I—I
(192)

The fission of CCl bonds is brought about by irradiation of chlorodifluoromethane under jet-cooled conditions.[164] The photodissociation at 266 nm of bromoform has been studied using transient frequency modulation spectroscopy.[165] The reactions encountered involve a two photonprocess with the formation of a CHBr fragment. The dynamics for the photodissociation of CF_3Br at 234 nm have been recorded, and the prime event is the fission of the CBr bond.[166] The photoionization of jet-cooled trifluorobromomethane has been used to determine the ionization energy of the trifluoromethyl radical.[167] A study of the photodecomposition of trifluoromethyl iodide at 304 nm has been reported.[168] Two primary photodissociation channels have been observed for the dissociation of CF_2ClBr at 234 nm.[169] The two processes involve either CBr or CCl bond fission, with a preference shown for the CBr process. A study of the photochemical decomposition of trichlorofluoromethane in ice on a Ru surface has been carried out.[170] The wavelength effects on the photodissociation dynamics of dibromodifluoromethane have been studied.[171]

Irradiation of dibromoacetonitrile in cyclohexane has been shown to yield significant amounts of the isomeric Br-NCCHBr.[172] Irradiation of dibromocinnamide derivatives affords the corresponding cinnamides in good yield.[173] Irradiation of 1,1-dichloro-1-fluoroethane at 235 nm brings about its photodissociation.[174] The photodissociation of perfluoroethyliodide can be brought about by irradiation at 266 nm.[175]

A study of the conformation-specific paths for the photodissociation of 1-iodopropane has been reported.[176] The photochemical reactivity of 1,n-diiodoalkanes has been reviewed.[177]

The decomposition of bromoethene at 193 nm affords ethyne and HBr as the primary reaction products.[178] The photodissociation of 1,2-dibromoethene using 248 nm light has shown that decomposition to bromine and ethyne is a major reaction path.[179] A study of the decomposition of trichloroethene in the gas phase has been reported.[180] The photosensitized irradiation (acetone as sensitizer) of trichloroethene in solution with a surfactant brings about photodegradation.[181] The sensitized degradation is more efficient than direct irradiation.

The primary dissociation processes following irradiation of allyl chloride at 193 nm have been studied.[182] This work reports a previously undiscovered low energy path for the CCl bond fission. Elimination of HCl is also observed. The formation dynamics of bromine and chlorine atoms from irradiation of allyl bromide and allyl chloride at 234 nm have been measured.[183] The multiphoton dissociation of C_3F_6 has been studied.[184] Allyl cyanide undergoes dissociation on irradiation at 193 nm to yield cyano and allyl radicals.[185] The photodissociation of acrylonitrile brought about by irradiation at 157.6 nm has been used to assess the energy disposal in the resultant CN radical.[186]

The photodissociation of the chloromethyl radical in the range 235 to 243 nm results in CCl bond fission.[187] The long-lived radical perfluoro-2,4-dimethyl-3-ethylpent-3-yl is photochemically reactive in the solid phase at 77K and liberates trifluoromethyl radicals.[188] Irradiation of the radical 1-(dimethylethyl)-2,2-dimethylpropyl at 254 nm brings about its conversion into the 2,2,4,4-tetramethylpentyl radical.[189] Detection of a vinyl radical has been reported following the irradiation of ethyne at 230 nm.[190]

The singlet excited state is involved in the photochemical free radical decomposition of 3-*t*-butyl-3-methyl-1-butyne.[191] A quantum chemical study of the photochemistry of heptane has shown that the two central CC and CH bonds are the most labile in the S_1 state.[192] Selective photochlorination of alkanes can be brought about using irradiation in the presence of chlorine in carbon disulfide, and a study has examined the bonding in the chlorine atom-CS_2 complex that is thought to be involved in the process.[193]

The photodissociation of the fluorenyl cation (193) has been examined.[194] Density functional calculations have been carried out to explain the photodissociation of this cation (193).[195] It has been shown to undergo sequential loss of one to five hydrogens. Itoh *et al.*[196] have demonstrated that benzylic and allylic alcohols such as (194) and (195) can be oxidized by irradiation in the presence of iodine (Scheme 9).

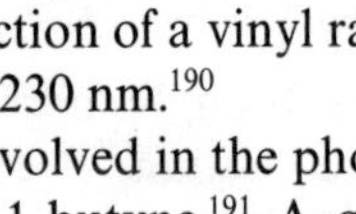

(193)

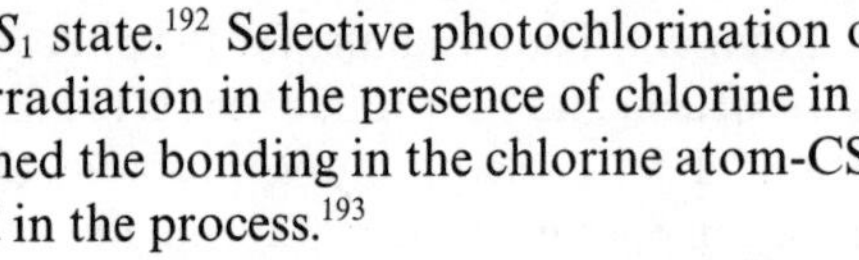

Scheme 9

Miller and Salvador[197] have investigated the photochemical substitution reactions of 1-haloadamantanes, 1-halonorbornanes and menthyl chloride with a series of amines and alcohols. This treatment affords the corresponding substituted amines and ethers. The authors[197] suggest that a novel photoinduced electron-transfer mechanism is involved. The stereochemical implications were probed using conversion of the optically active halide (196). Irradiation in methanol gives (197) with complete retention of configuration. Pandey and co-workers[198] have described an elegant approach to the synthesis of phenanthridone alkaloids. This involves the electron-transfer induced photocyclization of (198), affording (199). The reaction was brought about by irradiation ($\lambda > 280$ nm) in acetonitrile-water with dicyanonaphthalene as the sensitizer. The yield of cyclized product (199) is 68%. The photochemical formation of the radical cations of enol ethers such as (200) has been studied.[199]

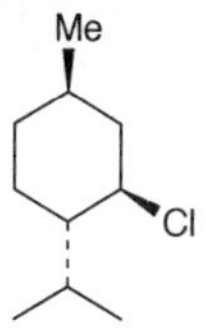

(196)

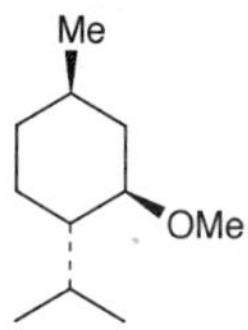

(197)

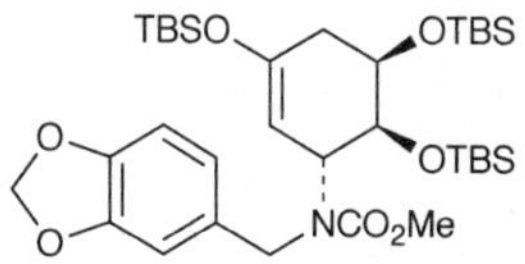

(198)

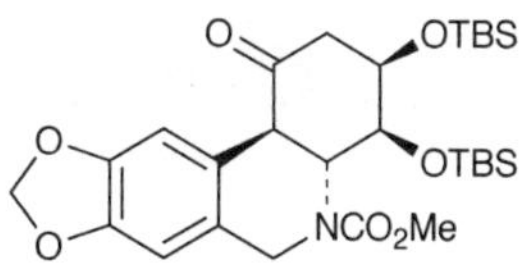

(199)

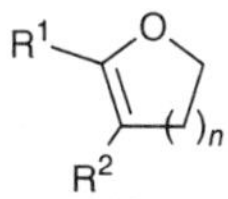

(200)

R^1	R^2	n
H	H	1
Me	H	1
H	Me	1
Me	Me	1
H	H	2

A study of the photodissociation of 1-chloroadamantane has shown that loss of chloride occurs at photon energies < 13.0 eV.[200] Rossi and his co-workers[201] have described the photochemical reactivity of dibromoadamantane (201), among other compounds, with the trimethyltin anion in ammonia. Under these conditions the adamantane derivative (202) is formed. Irradiation of (203) in DMSO in the presence of diphenylphosphide ions results in an $S_{RN}1$ reaction, with the formation of (204).[202] Reduction to the hydrocarbon also occurs in competition with the substitution reaction.

Ionization of the phenolic group is the principal step in the formation of the corresponding *o*-quinonemethide from the alcohol (205).[203] A laser-flash study of the generation of *o*-quinonemethide from the starting materials (206) has been reported.[204] The reactivity of the quinonemethide was also studied. Wan and his

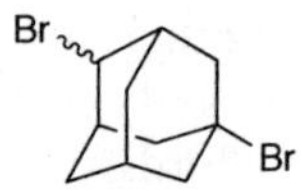

(201)

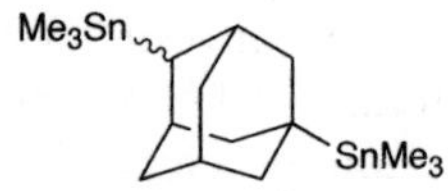

(202)

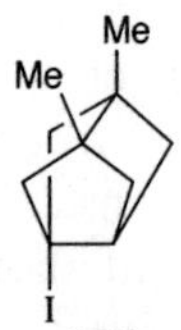

(203)

(204)

co-workers[205] have reviewed the synthesis of quinonemethides by photochemical means. A brief review has supplied details of the photogeneration of *o*-, *m*- and *p*-quinonemethides from phenol derivatives.[206] The formation of the corresponding xylylenes by irradiation of the cyclophanes (207) and (208) has been reported.[207]

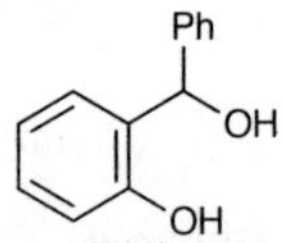

(205)

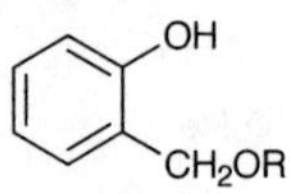

(206) R = H or *p*-CNC_6H_4, OR = $Me_3N^+ I^-$

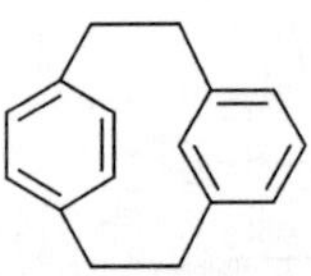

(207)

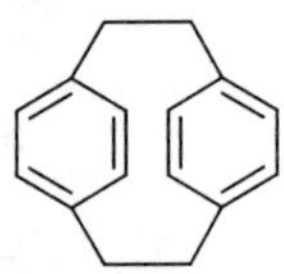

(208)

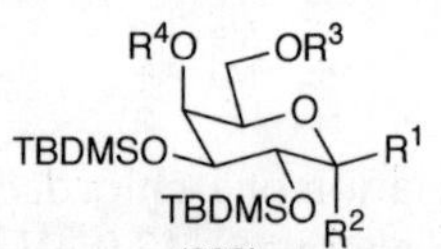

(209)

	R^1	R^2	R^3	R^4	
a	SPhMe	H	TBDMS	TBDMS	
b	SPhMe	H	H	TBDMS	75%
c	SPhMe	H	TBDMS	H	
d	SPhMe	H	H	H	87%
e	H	OMe	TBDMS	TBDMS	
f	H	OMe	H	TBDMS	94%

A method for the selective deprotection of primary *t*-butyldimethylsilyl ethers has been reported.[208] Some of the examples cited in the paper are shown above. The ethers (209) are irradiated in Pyrex vessels in CBr_4. There is some confusion over the irradiating wavelength since the authors report the use of a TLC lamp emitting at 245nm. Regardless of this, the conversions give high yields of products as can be seen from the yields quoted.

References

1. N. Chattopadhyay, I. Reva, L. Lapinski, R. Fausto, L. G. Arnaut and S. J. Formosinho, *J. Phys. Chem. A*, 2002, **106**, 3722.
2. V. Ya. Artyukhov and E. I. Sinchenko, *Russ. Phys. J.*, 2001, **44**, 718.
3. B. Juskowiak, T. Ichihara, S. Tanaka and M. Takagi, *Pol. J. Chem.*, 2001, **75**, 1377; *Chem. Abstr.*, 2001, **135**, 357590.
4. B. Strehmel, K. B. Henbest, A. M. Sarker, J. H. Malpert, D. Y. Chen, M. A. J. Rodgers and D. C. Neckers, *J. Nanosci. Nanotechnol.*, 2001, **1**, 107.
5. P. Kahnberg and O. Sterner, *Tetrahedron*, 2001, **57**, 7181.
6. J. F. Tietze, K. Kahle and T. Raschke, *Chem.-Eur. J.*, 2002, **8**, 401.
7. D. Josien-Lefebvre, G. Desmares and C. Le Drian, *Helv. Chim. Acta*, 2001, **84**, 890.
8. M. Shi and Y. Inoue, *Aust. J. Chem.*, 2001, **54**, 113.
9. K. Matsuyama, T. Inoue and Y. Inoue, *Synthesis*, 2001, 1167.
10. R. Ozawa and H. Hamaguchi, *Chem. Lett.*, 2001, 736.
11. K. Iwata, R. Ozawa and H. O. Hamaguchi, *J. Phys. Chem. A*, 2002, **106**, 3614.
12. T. S. Zyubina and V. F. Razumov, *High Energy Chem.*, 2001, **35**, 100.
13. T. Mizutani, M. Ikegami, R. Nagahata and T. Arai, *Chem. Lett.*, 2001, 1014.
14. H. Goerner, *Phys. Chem. Chem. Phys.*, 2002, **4**, 482.
15. I. Michler, A. Feis, M. A. Rodriguez and S. E. Braslavsky, *J. Phys. Chem. A*, 2001, **105**, 4814.
16. V. Papper and G. I. Likhtenshtein, *J. Photochem. Photobiol. A- Chem.*, 2001, **140**, 39.
17. M. Seydack and J. Bendig, *J. Phys. Chem. A*, 2001, **105**, 5731.
18. F. D. Lewis, W. Weigel and X. B. Zuo, *J. Phys. Chem. A*, 2001, **105**, 4691.
19. R. A. Rijkenberg, D. Bebelaar, W. J. Buma and J. W. Hofstraat, *J. Phys. Chem. A*, 2002, **106**, 2446.
20. S. A. Kovalenko, R. Schanz, T. A. Senyushkina and N. P. Ernsting, *Phys. Chem. Chem. Phys.*, 2002, **4**, 703.
21. Y. Amatatsu, *Chem. Phys.*, 2001, **274**, 87.
22. F. D. Lewis, T. L. Kurth and R. S. Kalgutkar, *J. Chem. Soc., Chem. Commun.*, 2001, 1372.
23. J.-S. Yang, S.-Y. Chiou and K.-L. Liau, *J. Am. Chem. Soc.*, 2002, **124**, 2518.
24. H. Meier, C. Schnorpfeil, M. Fetten and S. Hinnenschiedt, *Eur. J. Org. Chem.*, 2002, 547.
25. H. Kunkely and A. Vogler, *J. Organomet. Chem.*, 2001, **637-639**, 777.
26. T. Karatsu, *Kokagaku*, 2001, **32**, 10.
27. T. Yamada, S. Kobatake and M. Irie, *Bull. Chem. Soc. Jpn.*, 2002, **75**, 167.
28. K. Uchida, M. Fujita, Y. Aoi, M. Saito and M. Irie, *Chem. Lett.*, 2001, 366.
29. C. Okabe, N. Tanaka, T. Fukaminato, T. Kawai, M. Irie, S. Nibu and H. Sekiya, *Chem. Phys. Lett.*, 2002, **357**, 113.
30. J.-P. Malval, I. Gosse, J.-P. Morand and R. Laypouyade, *J. Am. Chem. Soc.*, 2002, **124**, 904.
31. A. Osuka, D. Fujikane, H. Shinmori, S. Kobatake and M. Irie, *J. Org. Chem.*, 2001, **66**, 3913.

32. T. Kaieda, S. Kobatake, H. Miyasaka, M. Murakami, N. Iwai, Y. Nagata, A. Itaya and M. Irie, *J. Am. Chem. Soc.*, 2002, **124**, 2015.
33. H. Utsumi, D. Nagahama, H. Nakano and Y. Shirota, *J. Mater. Chem.*, 2000, **10**, 2436; *Chem. Abstr.*, 2000, 756295.
34. K. Yagi, C. F. Soong and M. Irie, *J. Org. Chem.*, 2001, **66**, 5419.
35. T. Yamada, K. Muto, S. Kobatake and M. Irie, *J. Org. Chem.*, 2001, **66**, 6164.
36. K. Shibata, S. Kobatake and M. Irie, *Chem. Lett.*, 2001, 618.
37. K. Uchida, T. Matsuoka, S. Kobatake, T. Yamaguchi and M. Irie, *Tetrahedron*, 2001, **57**, 4559.
38. K. Shibata, K. Muto, S. Kobatake and M. Irie, *J. Phys. Chem. A*, 2002, **106**, 209.
39. H. Guo, F. Zhang, Y. Tang, F. Zhao and G. Wu, *Proc. SPIE-Int. Soc. Opt. Eng.*, 2001, **4085**(Optical Storage [ISOS 2000]), 158.
40. J. Ern, A. T. Bens, H. D. Martin, K. Kuldova, H. P. Trommsdorff and C. Kryschi, *J. Phys. Chem. A*, 2002, **106**, 1654.
41. K. Matsuda and M. Irie, *Chem.-Eur. J.*, 2001, **7**, 3466.
42. K. Matsuda, M. Matsuo and M. Irie, *Chem. Lett.*, 2001, 436.
43. K. Matsuda, M. Matsuo and M. Irie, *J. Org. Chem.*, 2001, **66**, 8799.
44. K. Matsuda and M. Irie, *Polyhedron*, 2001, **20**, 1391.
45. K. Matsuda and M. Irie, *Kyushu Daigaku Chuo Bunseki Senta Hokoku*, 2001, **19**, 9.
46. K. Matsuda, *Kagaku to Kogyo (Tokyo)*, 2002, **55**, 141; *Chem. Abstr.*, 2002, **136**, 224066.
47. T. Yamaguchi, Y. Tanaka, H. Nakazumi and M. Irie, *Shikizai Kyokaishi*, 2001, **74**, 8; *Chem. Abstr.*, 2001, **135**, 19570.
48. M. S. Kim, T. Kawai and M. Irie, *Chem. Lett.*, 2001, 702.
49. T. Yamaguchi, H. Kashiyama, H. Nakazumi, T. Yamada and M. Irie, *Chem. Lett.*, 2002, 58.
50. T. B. Norsten, A. Peters, R. McDonald, M. T. Wang and N. R. Branda, *J. Am. Chem. Soc.*, 2001, **123**, 7447.
51. S. Kobata and M. Irie, *Kagaku (Kyoto, Jpn.)*, 2001 **56**, 19; *Chem. Abstr.*, 2002, **136**, 239250.
52. J. Chauvin, T. Kawai and M. Irie, *Jpn. J. Appl. Phys., Part 1*, 2001, **40**, 2518; *Chem. Abstr.*, 2001, **135**, 99710.
53. M. M. Krayushkin, A. Yu. Martynkin and N. D. Chuvylkin, *Russ. Chem. Bull.*, 2001, **50**, 381.
54. K. Uchida, D. Guillaumont, E. Tsuchida, G. Mochizuki, M. Irie, A. Murakami and S. Nakamura, *J. Mol. Struct., THEOCHEM*, 2002, **579**, 115; *Chem. Abstr.*, 2002, 100657.
55. S. Nakamura, K. Kanda, D. Guillaumont, K. Uchida and M. Irie, *MCLC S&T, Sect. B; Nonlinear Opt.*, 2000, **26**, 201.
56. H. Bouas-Laurent and H. Dürr, *Pure Appl. Chem.*, 2001, **73**, 639.
57. Y. Inoue, N. Sugahara and T. Wada, *Pure Appl. Chem.*, 2001, **73**, 475; *Chem. Abstr.*, 2001, **135**, 202851.
58. T. Yamaguchi, T. Inagawa, H. Nakazumi, S. Irie and M. Irie, *Mol. Cryst. Liq. Cryst. Sci. Technol. Section A: Mol. Cryst. Liq. Cryst.*, 2001, **365**, 861; *Chem. Abstr.*, 2002, **136**, 142500.
59. M. Irie and T. Kawai, *Mirai Zairyo*, 2002, **2**, 30; *Chem. Abstr.*, 2002, **136**, 109969.
60. M. M. Krayushkin, L. G. Vorontsova and B. M. Uzhinov, *Int. J. Photoenergy*, 2001, **3**, 25; *Chem. Abstr.*, 2001, **135**, 280304.
61. M. M. Krayushkin, L. G. Vorontsova and B. M. Uzhinov, *Int. J. Photoenergy*, 2001, **3**, 25.

62. A. T. Bens, J. Ern, K. Kuldova, H. P. Trommsdorff and C. Kryschi, *J. Lumin.*, 2001, **94-95**, 51.
63. E. Murguly, T. B. Norsten and N. R. Branda, *Angew. Chem. Int. Ed.*, 2001, **40**, 1752.
64. P. Guetlich, R. Lapouyade, Y. Garcia and V. Ksenofontov, *Ger. Offen. DE*, **10,039,903**; *Chem. Abstr.*, 2002, **136**, 209642.
65. M. M. Krayushkin, S. N. Ivanov, A. Yu. Martynkin, B. V. Lichitsky, A. A. Dudinov and B. M. Uzhinov, *Russ. Chem. Bull.*, 2001, **50**, 116.
66. M. M. Krayushkin, F. M. Stoyanovich, O. Yu. Zolotarskaya, I. V. Murav'ev, A. Yu. Martynkin, L. G. Vorontsova, Z. A. Starikova, V. L. Ivanov and B. M. Uzhinov, *Russ. Chem. Bull.*, 2001, **50**, 110.
67. K. Kasatani and T. Fujiwaki, *ITE Lett. Batteries, New Technol. Med.*, 2001, **2**, 215.
68. K. Kasatani, *ITE Lett. Batteries, New Technol. Med.*, 2001, **2**, 220.
69. S. N. Ivanov, B. V. Lichitskii, A. A. Dudinov, A. Yu. Martynkin and M. M. Krayushkin, *Chem. Heterocycl. Compd. (N. Y., NY, U. S.)*, 2001, **37**, 85; *Chem. Abstr.*, 2001, **135**, 242209.
70. H. Tian, B. Z. Chen and P. H. Liu, *Chem. Lett.*, 2001, 990.
71. O. Benali, M. C. Jiminez, M. A. Miranda and R. Tormos, *J. Chem. Soc., Chem. Commun.*, 2001, 2328.
72. J. G. Cole and P. Wan, *Can. J. Chem.*, 2002, **80**, 46.
73. M. Nitz and D. R. Bundle, *J. Org. Chem.*, 2001, **66**, 8411.
74. D. A. Fulton and J. F. Stoddart, *J. Org. Chem.*, 2001, **66**, 8309.
75. M. C. Jiminez, M. A. Miranda and R. Tormos, *Tetrahedron*, 2002, **58**, 115.
76. D. S. Kim, S. C. Shim, T. Wada and Y. Inoue, *Tetrahedron Lett.*, 2001, **42**, 4341.
77. J. Cossy, V. Bellosta, J. L. Ranaivosata and B. Gille, *Tetrahedron*, 2001, **57**, 5173.
78. T. Kurata and R. Kinoshita, *J. Oleo Sci.*, 2001, **50**, 759; *Chem. Abstr.*, 2001, **135**, 344466.
79. J. Hartung, R. Kneuer, T. M. Kopf and P. Schmidt, *Compt. Rend.: Chim.*, 2001, **4**, 649; *Chem. Abstr.*, 2002, **136**, 69706.
80. P. Wipf, Y. Uto and S. Yoshimura, *Chem.-Eur. J.*, 2002, **8**, 1670.
81. M. Jonas, S. Blechert and E. Steckhan, *J. Org. Chem.*, 2001, **66**, 6896.
82. A. M. Cardarelli, M. Fagnoni, M. Mella and A. Albini, *J. Org. Chem.*, 2001, **66**, 7320.
83. R. S. Srivastava, M. Kolel-Veetil and K. M. Nicholas, *Tetrahedron Lett.*, 2002, **43**, 931.
84. V. Dragojlovic, D. B. Gao and Y. L. Chow, *J. Mol. Catal. A: Chem.*, 2001, **171,** 43; *Chem. Abstr.*, 2001, 419456.
85. K. I. Borovinski, M. N. Krivchun and B. I. Ionin, *Russ. J. Gen. Chem.*, 2001, **71**, 813; *Chem. Abstr.*, 2002, **136**, 61382.
86. J. Leitich, U. Ritter-Thomas and I Heise, *J. Photochem. Photobiol. A-Chem.*, 2002, **147**, 177.
87. J. Leitich, U. Ritter-Thomas, I Heise, Y. H. Tsay and J. Rust, *J. Photochem. Photobiol. A-Chem.*, 2002, **147**, 157.
88. J. Leitich and I. Heise, *J. Photochem. Photobiol. A -Chem.*, 2002, **147**, 137.
89. F. D. Lewis, X. B. Zuo, R. S. Kalgutkar, J. M. Wagner-Brennan, M. A. Miranda, E. Font-Sanchis and J. Perez-Prieto, *J. Am. Chem. Soc.*, 2001, **123**, 11883.
90. M. C. Jiminez, M. A. Miranda, T. Tormos and S. Gil, *Eur. J. Org. Chem.*, 2002, 297.
91. N. S. Unaldi and M. Balci, *Tetrahedron Lett.*, 2001, **42**, 8365.
92. R. Altundas, A. Dastan, N. S. Unaldi, K. Guven, O. Uzun and M. Balci, *Eur. J. Org. Chem.*, 2002, 526.
93. H. Ihmels, C. J. Mohrschladt, J. W. Grimme and H. Quast, *Synthesis*, 2001, 1175.
94. H. E. Zimmerman and W. S. Chen, Org. *Lett.*, 2002, **4**, 1155.

95. M. A. Miranda, E. Font-Sanchis, J. Perez-Prieto and J. C. Scaiano, *J. Org. Chem.*, 2002, **67**, 1162.
96. H. Maeda, Y. Miyata and K. Mizuno, *Tetrahedron Lett.*, 2002, **43**, 1481.
97. K. C. W. Chong, J. Sivaguru, T. Schichi, Y. Yoshimi, V. Ramamurthy and J. R. Scheffer, *J. Am. Chem. Soc.*, 2002, **124**, 2858.
98. J. Sivaguru, J. R. Scheffer, J. Chandarasekhar and V. Ramamurthy, *J. Chem. Soc., Chem. Commun.*, 2002, 830.
99. A. Bogdanova and V. V. Popik, *Org. Lett.*, 2001, 3, 1885.
100. A. de Meijere, S. I. Kozhushkov, R. Boese, T. Haumann, D. S. Yufit, J. A. K. Howard, L. S. Khaikin and M. Traetteberg, *Eur. J. Org. Chem.*, 2002, 485.
101. W.-H. Park, S. H. Cho, H. Cho, S. K. Kim and Y. S. Choi, *Bull. Korean Chem. Soc.*, 2001, **22**, 1030; *Chem. Abstr.*, 2002, **136**, 69547.
102. P. J. Campos, A. Soldevilla, D. Sampedro and M. A. Rodriguez, *Org. Lett.*, 2001, **3**, 4087.
103. K. Ito and K. Saito, *Rec. Res. Dev. Pure. Appl. Chem.*, 2000, **4**, 81; *Chem. Abstr.*, 2002, **136**, 69478.
104. H. Inui and S. Murata, *J. Chem. Soc., Chem. Commun.*, 2001, 1036.
105. R. F. Qadiri, E. J. Feltham, E. E. H. Cotrill, N. Taniguchi and M. N. R. Ashfold, *J. Chem. Phys.*, 2002, **116**, 906.
106. J. C. Robinson, W. Sun, S. A. Harris, F. Qi and D. M. Neumark, *J. Chem. Phys.*, 201, **115**, 8359.
107. M. Garavelli, F. Bernardi, M. Olivucci, M. J. Bearpark and M. A. Robb, *J. Phys. Chem. A*, 2001, **105**, 11496.
108. B. Ostojic and W. Domcke, *Chem. Phys.*, 2001, **269**, 1; *Chem. Abstr.*, 2001, **135**, 226600.
109. M. A. Robb, M. Garavelli, M. Olivucci and F. Bernardi, *Rev. Comput. Chem.*, 2000, **15**, 87; *Chem. Abstr.*, 2002, **136**, 309486.
110. B. H. O. Cook, W. J. Leigh and R. Walsh, *J. Am. Chem. Soc.*, 2001, **123**, 5188.
111. N. Fokialakis, P. Magiatis, A. Terzis, F. Tillequin and A. L. Skaltsounis, *Tetrahedron Lett.*, 2001, **42**, 5323.
112. R. Davis, S. Das, M. George, S. Druzhinin and K. A. Zachariasse, *J. Phys. Chem. A*, 2001, **105**, 4790.
113. G. Bartocci, G. Galiazzo, G. Gennari, E. Marri, U. Mazzucato and A. Spalletti, *Chem. Phys.*, 2001, **272**, 213 (*Chem. Abstr.*, 2002, **136**, 102059).
114. H. Ikeda, N. Aburakawa, F. Tanaka, T. Fukushima and T. Miyashi, *Eur. J. Org. Chem.*, 2001, 3445.
115. H. J. Yoon, J. M. Lee, S. S. Im and W. K. Chae, *Bull. Korean Chem. Soc.*, 2001, **22**, 543; *Chem. Abstr.*, 2001, **135**, 241934.
116. S. K. Park, *J. Photochem. Photobiol. A-Chem.*, 2001, **144**, 167.
117. S. K. Park, *Bull. Korean Chem. Soc.*, 2001, **22**, 1202.
118. H. Nishino, A. Nakamura and Y. Inoue, *J. Chem. Soc., Perkin Trans. 2*, 2001, 1693.
119. H. Nishino, A. Nakamura and Y. Inoue, *J. Chem. Soc., Perkin Trans. 2*, 2001, 1701.
120. H. Nishino, A. Nakamura, H. Shitomi, H. Onuki and Y. Inoue, *J. Chem. Soc., Perkin Trans. 2*, 2001, 1706.
121. A. D. Dubonosov, S. V. Galichev, V. A. Chernoivanov, V. A.; Bren' and V. I. Minkin, *Russ. J. Org. Chem.*, 2001, **37**, 67; *Chem. Abstr.*, 2001, **135**, 180458.
122. R. Herges and T. Winkler, *Eur. J. Org. Chem.*, 2001, 4419.
123. D. Margetic, D. N. Butler and R. N. Warrener, *Aust J. Chem.*, 2000, **53**, 959.
124. W. Adam, O. Cakmak, C. R. Saha-Moller and A. Tutar, *Synlett*, 2002, 49.
125. V. De Waele, U. Schmidhammer, T. Mrozek, J. Daub and, R. T. Eberhard, *J. Am.*

Chem. Soc., 2002, **124**, 2438.

126. M. Boggio-Pasqua, M. J. Bearpark, P. A. Hunt and M. A. Robb, *J. Am. Chem. Soc.*, 2002, **124**, 1456.
127. K. Nakabayashi, H. Enami, H. Tsukamoto and A. Ouchi, *Nippon Kagaku Kaishi*, 2001, 433; *Chem. Abstr.*, 2001, **135**, 241841.
128. F. Bayrakceken, B. B. Jomehri and A. Yaman, *J. Indian Chem. Soc.*, 2001, **78**, 234; *Chem. Abstr.*, 2001, **135**, 195276.
129. K. Satake, S. Takami, T. A. Yuko and M. Kimura, *J. Chem. Soc., Chem. Commun.*, 2001, 1382.
130. K. Tanaka, R. Nagahiro and Z. Urbancyzk-Lipkowska, *Org. Lett.*, 2001, **3**, 1567.
131. Y. Tobe, N. Iwasa, R. Umeda and M. Sonoda, *Tetrahedron Lett.*, 2001, **42**, 5485.
132. S. Matsumoto, S. Takahashi and K. Ogura, *Heteroat. Chem.*, 2001, **12**, 385.
133. M. Sindler-Kulyk and N. Basaric, *Kemija u Industriji*, 2002, **51**, 169.
134. I. Skoric, N. Basaric, Z. Marinic and M. Sindler-Kulyk, *Heterocycles*, 2001, **55**, 1889; *Chem. Abstr.*, 2002, **136**, 102246.
135. P. J. Coelho, L. M. Carvalho, S. Rodrigues, A. M. F. Oliveira-Campos, R. Dubest, J. Aubard, A. Samat and R. Guglielmetti, *Tetrahedron*, 2002, **58**, 925.
136. I. I. Chuev, S. M. Aldoshin, A. Samat, F. Maurel and J. Aubar, *J. Mol. Struct., THEOCHEM*, 2001, **548**, 123; *Chem. Abstr.*, 2001, **135**, 357605.
137. P. Kilickiran, S. Sankararaman and H. Hopf, *Indian J. Chem., Sect. B: Org. Chem. Incl. Med. Chem.*, 2001, **40B**, 781.
138. A. Purohit, J. Wyatt, G. Hynd, J. Wright, A. El-Shafey, N. Swamy, R. Ray and G. B. Jones, *Tetrahedron Lett.*, 2001, **42**, 8579.
139. M. Garavelli, C. S. Page, P. Celani, M. Olivucci, W. E. Schmid, S. A. Trushin and W. Fuss, *J. Phys. Chem. A*, 2001, **105**, 4458.
140. S. Zilberg and Y. Haas, *Phys. Chem. Chem. Phys.*, 2002, **4**, 34.
141. Y. Y. Gao, Y. Y. Liu, B. J. Wang, W. Bao and Y. Li, *Chin. Chem. Lett.*, 2001, **12**, 309; *Chem. Abstr.*, 2001, 354018.
142. R. S. H. Liu and G. S. Hammond, *Chem.-Eur. J.*, 2001, **7**, 4536.
143. T. Bach and A. Spiegel, *Eur. J. Org. Chem.*, 2002, 645.
144. S. Ghosh, S. Banerjee, K. Chowdhury, M. Mukherjee and J. A. K. Howard, *Tetrahedron Lett.*, 2001, **42**, 5997.
145. J. Panda, S. Ghosh and S. Ghosh, *J. Chem. Soc., Perkin Trans. 1*, 2001, 3013.
146. A. Barbero, C. Garcia and F. J. Pulido, *Synlett*, 2000, 824.
147. Y. Nakamura, T. Fujii and J. Nishimura, *Chem. Lett.*, 2001, 970.
148. V. Strehmel, B. Stiller, B. Strehmel, A. Sarker and D. C. Neckers, *Polym. Prepr. (Am. Chem. Soc., Div. Polym. Chem.) [computer optical disk]*, 2001, **42**, 749; *Chem. Abstr.*, 2001, **135**, 358503.
149. M. R. Johnston, M. J. Latter, R. N. Warrener, M. Golic, D. McKavanagh and D. Margetic, *Proc. ECSOC-3, Proc. ECSOC-4, 1999, 2000 [computer optical disk]* 1999-2000 (Pub. **2000**), 858; *Chem. Abstr.*, 2001, **135**, 226960.
150. C. Matsubara and M. Kojima, *Res. Chem. Intermed.*, 2001, **27**, 975; *Chem. Abstr.*, 2002, **136**,295726.
151. G. S. Papaefstathiou, A. J. Kipp and L. R. MacGillivray, *J. Chem. Soc., Chem. Commun.*, 2001, 2462.
152. D. G. Amirsakis, M. A. Garcia-Garibay, S. J. Rowan, J. F. Stoddart, A. J. P. White and D. J. Williams, *Angew. Chem. Int. Edn.*, 2001, **40**, 4256.
153. W.-Q. Zhang, J.-P. Zhuang, C.-B. Li, H. Sun and X.-N. Yuan, *Chin. J. Chem.*, 2001, **19**, 695.
154. Y. Sonoda, A. Miyazawa, S. Hayashi and M. Sakuragi, *Chem. Lett.*, 2001, 410.

155. M. Yoshizawa, Y. Takeyama, T. Usukawa and M. Fujita, *Angew. Chem. Int. Edn.*, 2002, **41**, 1347.
156. C. Tao and P. J. Dagdigian, *Chem. Phys. Lett.*, 2001, **350**, 63.
157. Y.-L. Li and D. L. Phillips, *Chem. Phys. Lett.*, 2001, **349**, 291.
158. W. M. Kwok, C. Ma, D. Phillips, A. W. Parker, M. Towrie, P. Matousek, P. and D. L. Phillips, *Chem. Phys. Lett.*, 2001, **341**, 292.
159. H.-F. Xu, S.-L. Liu, X.-X. Ma, D.-X. Dai, J.-C. Xie and G.-H. Sha, *Wuli Xuebao*, 2002, **51**, 240.
160. D. L. Phillips, W. H. Fang and X. M. Zheng, *J. Am. Chem. Soc.*, 2001, **123**, 4197.
161. D. L. Phillips and W. H. Fang, *J. Org. Chem.*, 2001, **66**, 5890.
162. Y. L. Li, K. H. Leung and D. L. Phillips, *J. Phys. Chem. A*, 2001, **105**, 10621.
163. Y. L. Li, D. Q. Wang, K. H. Leung and D. L. Phillips. *J. Phys. Chem. A*, 2002, **106**, 3463.
164. L. Li, G. Dorfman, A. Melchior, S. Rosenwaks and I. Bar, *J. Chem. Phys.*, 2002, **116**, 1869.
165. W.-L. Liu and B.-C. Chang, *J. Chin. Chem. Soc. (Taipei, Taiwan)*, 2001, **48**, 613.
166. T. K. Kim, M. S. Park, K. W. Lee and K.-H, Jung, *J. Chem. Phys.*, 2001, **115**, 10745.
167. G. A. Garcia, P. M. Guyon and I. Powis, *J. Phys. Chem. A*, 2001, **105**, 8296.
168. G.-H. Xiao, Z. Yu and J.-c. Xie, *Harbin Ligong Daxue Xuebao*, 2001, **6**, 81.
169. S.-H. Lee and K.-H. Jung, *Chem. Phys. Lett.*, 2001, **350**, 306.
170. H. Ogasawara and K. Hirohito, *Riken Review*, 2002, **44** 502; *Chem. Abstr.*, 2002, **136**, 285792.
171. M. S. Park, T. K. Kim, S. H. Lee, K. H. Jung, H. R. Volpp and J. Wolfrum, *J. Phys. Chem. A*, 2001, **105**, 5606.
172. Y.-L. Li, C. W. Lee, D. Wang, K. H. Leung and D. L. Phillips, *Chem. Phys. Lett.*, 2001, **350**, 78.
173. S. Aruna, R. Kalyanakumar and V. T. Ramakrishnan, *Synth. Commun.*, 2001, **31**, 3125; *Chem. Abstr.*, 2001, 735245.
174. T. Einfeld, C. Maul, K.-H. Gericke, R. Marom, S. Rosenwaks and I. Bar, *J. Chem. Phys.*, 2001, **115**, 6418.
175. A. V. Baklanov, G. A. Bogdanchikov, M. Aldener, U. Sassenberg and A. Persson, *J. Chem. Phys.*, 2001, **115**, 11157.
176. S. T. Par, S. K. Kim and M. S. Kim, *Nature*, 2002, **415**, 306.
177. M. A. Miranda, J. Perez-Prieto, E. Font-Sanchis and J. C. Scaiano, *Prog. React. Kinet. Mech.*, 2001, **26**, 139.
178. D.-K. Liu, L. T. Letendre and H.-L. Dai, *J. Chem. Phys.*, 2001, **115,** 1734.
179. Y. R. Lee, C. C. Chou, Y. J. Lee, L. D. Wang and S. M. Lin, *J. Chem. Phys.*, 2001, **115**, 3195.
180. M. Murabayashi, K. Itoh and K. Kawamura, *Yokohama Kokuritsu Daigaku Kankyo Kagaku Kenkyu Senta Kiyo*, 2001, **27**, 1.
181. W. K. Choy and W. Chu, *Chemosphere*, 2001, **44**, 943.
182. M. L. Morton, L. J. Butler, T. A. Stephenson and F. Qi, *J. Chem. Phys.*, 2002, **116**, 2763.
183. M. S. Park, W. Keon and K.-H. Jung, *J. Chem. Phys.*, 2001, **114**, 10368; *Chem. Abstr.*, 2001, **135**, 195252.
184. L. Rubio, M. Santos and J. A. Torresano, *J. Photochem. Photobiol. A-Chem.*, 2001, **146**, 1.
185. C. Y. Oh, S. K. Shin, H. L. Kim and C. R. Park, *Chem. Phys. Lett.*, 2001, **342**, 27.
186. J. Guo, T. Carrington and S. V. Filseth, *J. Chem. Phys.*, 2001, **115**, 8411.
187. A. B. Potter, V. Dribinski, A. V. Demyanenko and H. Reisler, *Chem. Phys. Lett.*,

2001, **349**, 257.
188. S. R. Allayarov and A. I. Mikhailov, *Russ. Chem. Bull.*, 2001, **50**, 1198.
189. H. Koizumi, T Takada, T.Ichikawa and A. Lund, *Chem. Phys. Lett.*, 2001, **340**, 256; *Chem. Abstr.*, 2001, **135**, 218545.
190. D. L. Osborn and J. H. Frank, *Chem. Phys. Lett.*, 2001, **349**, 43.
191. T. G. Shutova, E. D. Skakovskii, V. E. Agabekov, V. L. Murashko and E. A. Dikusar, *Russ. J. Gen. Chem.*, 2001, **71**, 627.
192. S. V. Nikolaev, *Opt. Atmos. Okeana*, 2001, **14**, 1038.
193. D. Q. Wang, Y. L. Li, W. S. Ho, K. H. Leung and D. L. Phillips, *J. Org. Chem.*, 2002, **67**, 747.
194. M. J. Dibben, D. Kage, J. Szczepanski, J. R. Eyler and M. Vala, *J. Phys. Chem. A*, 2001, **105**, 6024.
195. J. Szczepanski, M. J. Dibben, W. Pearson, J. R. Eyler and M. Vala, *J. Phys. Chem. A*, 2001, **105**, 9388.
196. A. Itoh, T. Kodama and Y. Masaki, *Chem. Lett.*, 2001, 686.
197. J. B. Miller and J. R. Salvador, *J. Org. Chem.*, 2002, **67**, 435.
198. G. Pandey, A. Murugan and M. Balakrishnan, *J. Chem. Soc., Chem. Commun.*, 2002, 624.
199. K. Bernhard, J. Geimer, M. Canle-Lopez, J. Reynisson, D. Beckert, R. Gleiter and S. Steenken, *Chem.-Eur. J.*, 2001, **7**, 4640.
200. Y. Li and T. Baer, *J. Phys. Chem. A*, 2002, **106**, 272.
201. A. N. Santiago, C. A. Toledo and R. A. Rossi, *J. Org. Chem.*, 2002, **67**, 2494.
202. P. Camps, A. E. Lukach and R. A. Rossi, *J. Org. Chem.*, 2001, **66**, 5366.
203. Y. Chiang, A. J. Kresge and Y. Zhu, *J. Am. Chem. Soc.*, 2002, **124**, 717.
204. Y. Chiang, A. J. Kresge and Y. Zhu, *J. Am. Chem. Soc.*, 2001, **123**, 8089.
205. P. Wan, D. W. Brousmiche, C. Z. Chen, J. Cole, M. Lukeman, and M. Xu, *Pure Appl. Chem.*, 2001, **73**, 529.
206. M. Yasuda, *Kokagaku*, 2000, **31**, 30.
207. Y. Hosoi, T. Yatsuhashi, K. Ohtakeyama, S. Shimizu, Y. Sakata and N. Nakashima, *J. Phys. Chem. A*, 2002, **106**, 2014.
208. M. Y. Chen, K. C. Lu, A. S. Y. Lee and C. C. Lin, *Tetrahedron Lett.*, 2002, **43**, 2777.

4

Photochemistry of Aromatic Compounds

BY ANDREW GILBERT

1 Introduction

Several reviews have been published within the year, which provide an overview of various areas relevant to the photochemistry of aromatic compounds. The subjects of these reviews include hexafluorobenzene photochemistry,[1] spin-orbit coupling induced electron-spin polarization in photoinduced electron-transfer reactions,[2] advances in organic photochromics,[3] photochemical isomerization of penta-atomic heterocycles,[4] and the development of experimental and theoretical crystal-lattice organic photochemistry.[5]

2 Isomerization Reactions

The photoinduced *cis-trans* interconversions of 1,2-diarylethenes are reviewed in Part II, Chapter 3 of this Volume, but it is relevant to note here studies into the geometrical isomerizations of 1,2-diarylcyclopropanes. In particular, the confined space and cations in alkali-exchanged Y-zeolites are reported to enhance markedly the asymmetric induction of the photoisomerization of several 1,2-diphenylcyclopropane derivatives (1) which have a distant chiral auxiliary.[6]

The photoinduced positional isomerization of substituted heteroarenes has been the subject of a number of reports over the years.[7] The photoisomerization channels of dimethylpyridines on the S_1 and T_1 potential energy surfaces have been theoretically investigated using density function theory, CASSCF, CASPT2 and MRCI methods.[8] From these studies it is suggested that the Möbius (2) and the azabenzvalene (3) intermediates are likely respective precursors to the Dewar isomers such as (4) from 2,3-dimethylpyridine and to phototransposition products from the singlet state. For 3,5-dimethylpyridine, the precursors which give

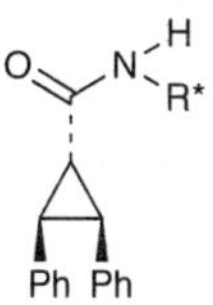

(1) for example R* = 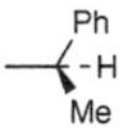

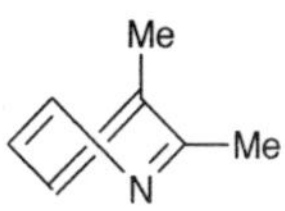

(2)

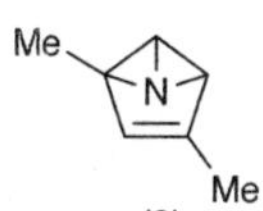

(3)

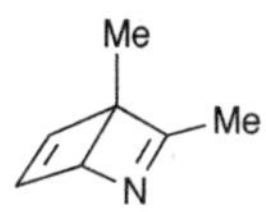

(4)

Photochemistry, Volume 34

the 2,4-positional isomer are proposed to have an out-of-plane distorted geometry arising from vibrational relaxation of the T_1 state and a further out-of-plane distorted structure originating from a triplet azaprefulvene.

In 1996, Mariano and co-workers described the sequential photoisomerization-aziridine ring opening of pyridinium salts,[9] and now report this as a successful strategy towards both stereochemically defined enantiomerically enriched aminocyclopentendiols such as (5),[10] and to provide the aminocyclopentane core intermediate (6) in a concise synthesis of (−)-allosamizoline (7).[11] The

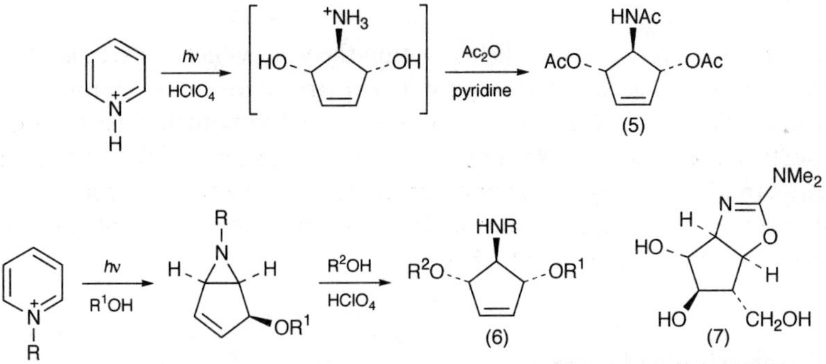

basic photochemistry of this synthetically useful process has been further investigated theoretically and experimentally by other workers.[12] From computational studies, 6-azabicyclo[3.1.0]hexenyl ions and azabenzvalenes are proposed as intermediates, and the results from irradiation of the deuteriated compound (8) allow the authors to advance the mechanism outlined in Scheme 1, in which the nitrogen atom migrates around the ring by way of the azabenzvalene ion (9). The observed preference for *exo* products is in accord with the computed relative energies of the ions and transition states.

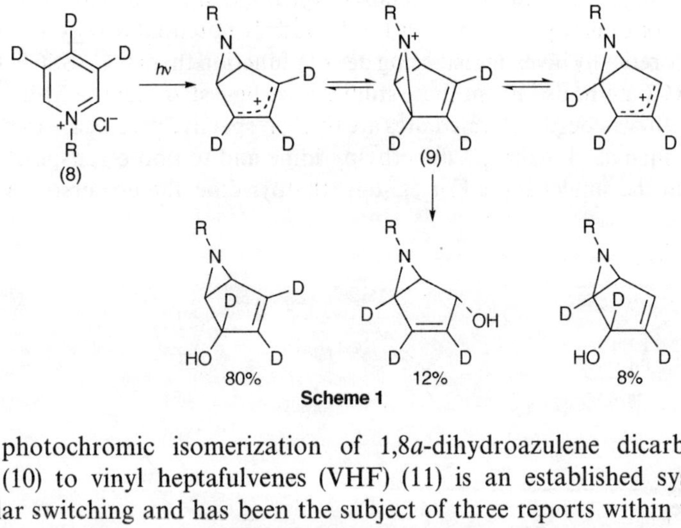

Scheme 1

The photochromic isomerization of 1,8*a*-dihydroazulene dicarbonitriles (DHA) (10) to vinyl heptafulvenes (VHF) (11) is an established system for molecular switching and has been the subject of three reports within the year. The photochromism of 1,2,3,8*a*,9-pentahydrocyclopent[a]azulene (12) has been

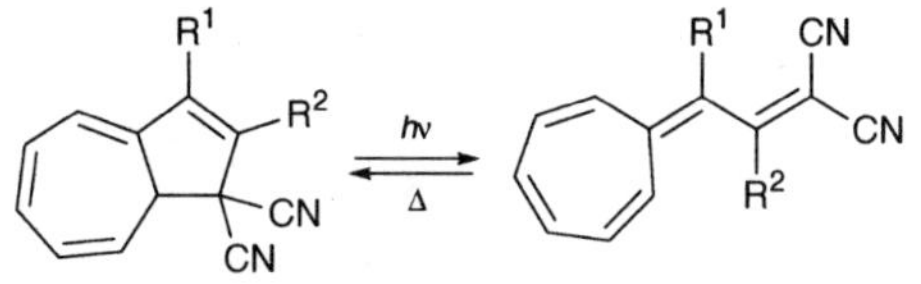

(10) (11)

investigated through CASSCF calculations of the S_0 and S_1 states, and the S_1 reaction co-ordinate is characterized by a transition structure for adiabatic ring opening, connecting a DHA-like intermediate to a VHF-like structure of greater stability.[13] The latter species is a conical intersection between the S_1 and S_0 potential energy surfaces, and the presence of such a crossing is consistent with the previously measured lifetime of *ca.* 600fs for the DHA-like intermediate.[14] In previous studies, the reversion process in these systems (*i.e.*, (11) to (10)) has been thermally induced and has a much longer time scale than the photoisomerization (ps). The possibility of photoreversal of the process has been investigated using the cyanophenyl dihydroazulene derivative (13).[15] Irradiation of (13) initially

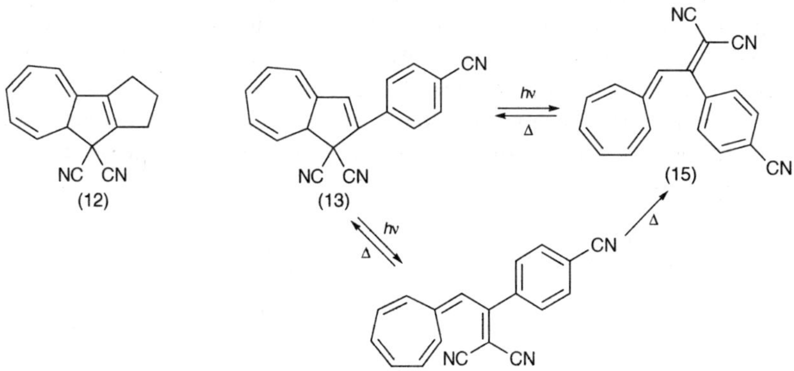

forms the s-*cis* vinyl heptafulvene (14), which converts into the 'photostable' s-*trans* isomer (15) on a nanosecond time scale. By following the initial UV pulse with a visible femtosecond pulse on a 25ps delay, however, ring closure of the primary photoisomer (14) back to (13) has been achieved. The authors thus consider that the DHA → VHF → DHA sequence can be a very fast photoreversible molecular switch as well as a photochemical-thermal switch which has a thermal lock. The complex hybrid systems (16) which have both photoswitchable dihydroazulene and diethynylhex-3-en-1,5-diyne moieties have been synthesized and their switching properties investigated.[16] Although the dihydroazulene unit underwent photoinduced ring opening and thermal closure (half life of 3.9-5.8 h at 25°C in CH_2Cl_2), the reaction of the second switch involved solely a comparatively slow photoinduced $E \rightarrow Z$ isomerization, and, for the derivative of (16) having R = *N*,*N*-dimethylanilino, the only photochemistry was the geometrical isomerization.

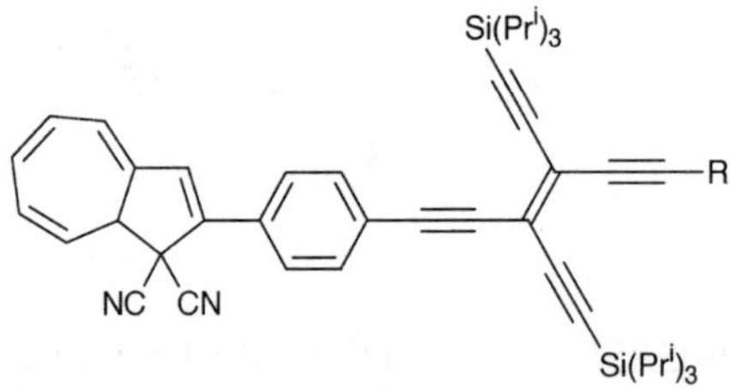

(16) R = 2-thienyl, 4-substituted phenyl and Si(Pr^i)$_3$

3 Addition Reactions

An excellent and comprehensive review of the *ortho* photocycloadditions of alkenes and alkynes to the benzene ring has been published,[17] and this complements the similarly extensive review of the corresponding *meta* process from the same research group.[18] The applicability of the latter process to complex molecule synthesis has been well illustrated over the past decade and a half, and more fundamental systems continue to be reported. The ethenyl-phenyl bichromophore (17), as may be predicted, undergoes intramolecular 2,6-addition to give the 1,6-bridged dihydrosemibullvalene (18) (52% yield),[19] and in the same report it is noted that the absence of cyclopropane ring opening during the formation of (19) from the photocycloaddition of cyclopentene to phenylcyclopropane, suggests that an intermediate such as (20) is either not involved or has a very short lifetime.

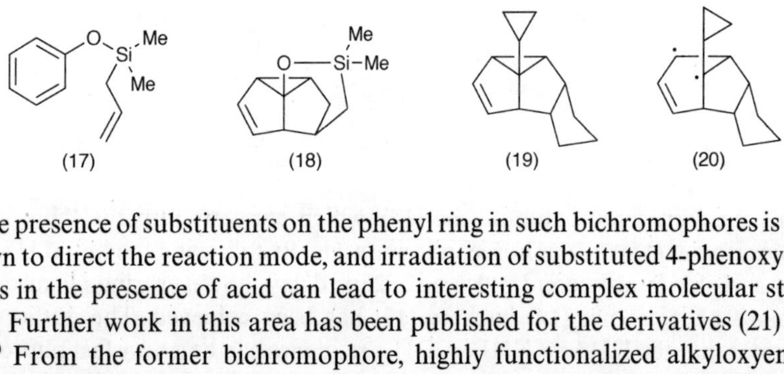

(17) (18) (19) (20)

The presence of substituents on the phenyl ring in such bichromophores is well known to direct the reaction mode, and irradiation of substituted 4-phenoxybut-1-enes in the presence of acid can lead to interesting complex molecular structures. Further work in this area has been published for the derivatives (21) and (22).[20] From the former bichromophore, highly functionalized alkyloxyenone compounds (23) result, while (22) displays a higher reactivity and yields the product (24) even in the absence of acid. Interestingly, a side reaction of (22) yields (25) by a formal (2 + 3) photocycloaddition involving, it is suggested, the diradical (26).

The intermolecular ($2\pi + 2\pi$) photocycloaddition of captodative ethenes to acenaphthones has been well documented and exploited by Döpp and co-workers, and further reports of these processes have appeared within the year. The reaction arises from the T_1 state of the arene, which is quenched by the ethene with rate constants which are approximately two orders of magnitude below that of a diffusion controlled process.[21] For 1-acenaphthones (27) having substituents at the 2- or 4- positions, the photochemistry can be more complex,

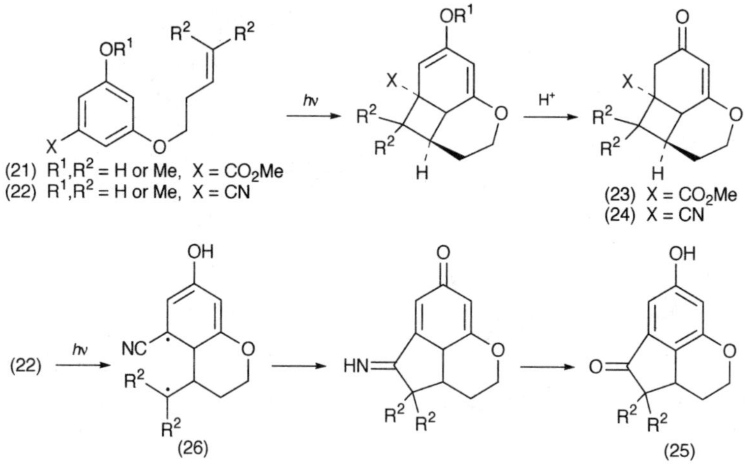

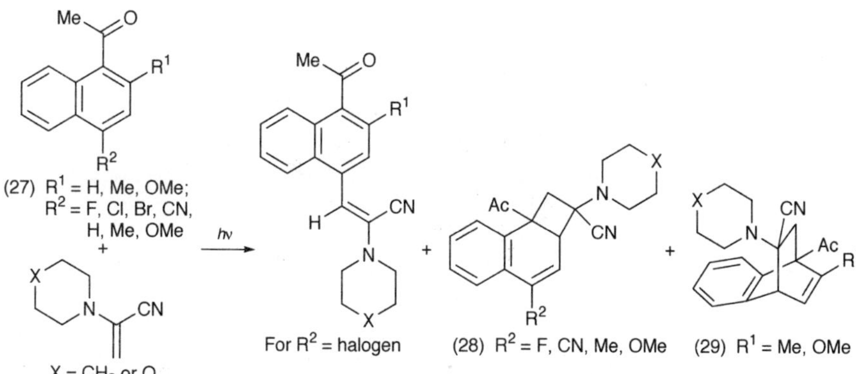

Scheme 2

and for R^2 = Cl or Br substitution occurs, while 1,2- or 1,4-cycloadditions affording (28) or (29), respectively, are observed in other cases, as outlined in Scheme 2.[22] The addition of 2-morpholinoprop-2-enenitrile to perinaphthenone (30), which has a high intersystem crossing efficiency to the T_1 state, gives the one stereoisomer (31) of the head-to-tail photoadduct, but low conversions are beneficial and the absence of oxygen is essential, since (30) is a sensitizer for singlet oxygen.[23] The products observed from irradiation of methoxylated naphthalenes in the presence of acrylonitrile are reported to be solvent- and substituent-dependent.[24] Thus $(2\pi + 2\pi)$ cycloaddition of (32) occurs in benzene solution for R = H to give (33), which then undergoes ring opening and intramolecular $(4\pi + 2\pi)$ addition affording (34), but irradiation of (32, R = OMe) in methanol or acetonitrile as solvent, yields (35) or (36), respectively, by way of an electron-transfer process.

Intramolecular photocycloaddition of naphthalene-ethene bichromophores occurs for a variety of derivatives. Two reports describe the reactions of substituted alk-2-enyl 1-cyano-2-naphthylmethyl ethers (37), which give the products, (38) and (39), of both 1,2- and 3,4- attack, respectively, in yields of up to

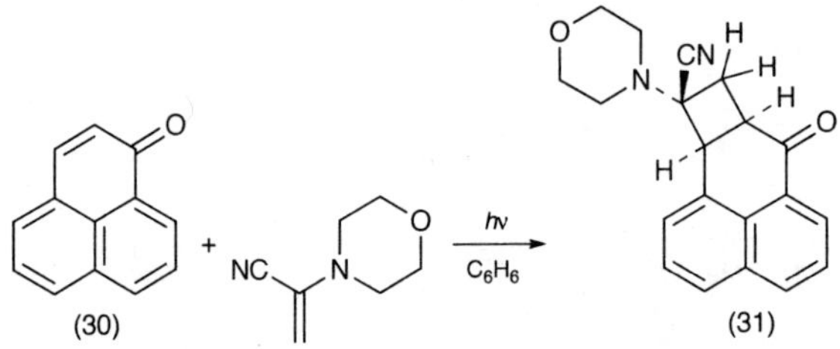

(30) (31)

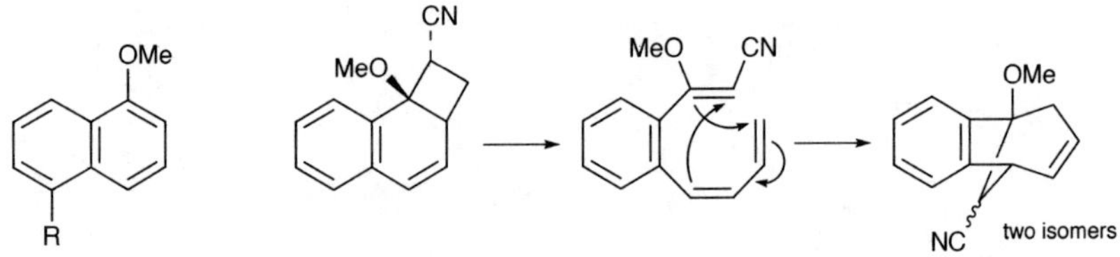

(32) R = H or MeO (33) (34)

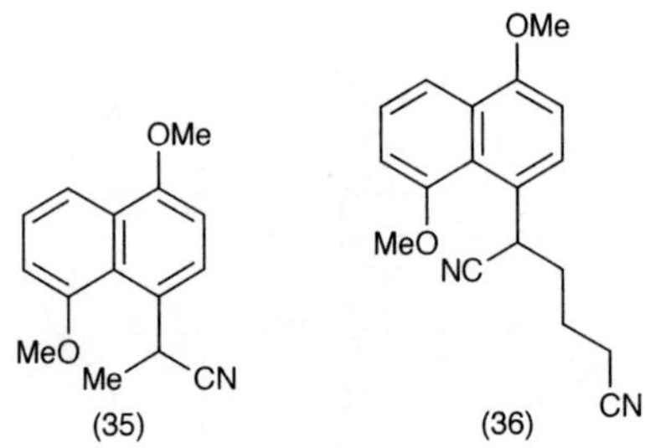

(35) (36)

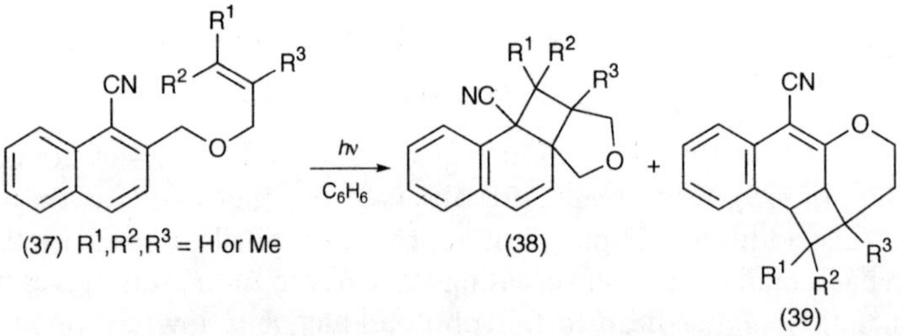

(37) R^1,R^2,R^3 = H or Me (38) (39)

90%.[25] The ratio of the two adducts is solvent-polarity dependent, with (38) predominating in benzene solution and (39) becoming more favoured in acetonitrile. The addition of a europium compound to the irradiation inhibits the formation of (39) and the cycloreversion of (38), and the presence of magnesium perchlorate accelerates the formation of (39). The latter feature is rationalized in terms of (39) arising from polar intermediates. *N*-(1-Naphthylethyl)prop-2-enamides (40) are reported to undergo intramolecular ($2\pi + 2\pi$) and ($4\pi + 2\pi$) cycloadditions, as well as Type II cyclization, yielding (41), (42) and (43), respectively.[26] Both substituents and experimental conditions influence the type of reaction which is favoured, with (41) and (43) arising from the acryloyl derivatives (40 a-c), and ($4\pi + 2\pi$) intramolecular adducts (42) being formed in 75-97% yield from the cinnamoyl series (40 d-i). The latter bichromophores also yield only (42) on sensitization with benzophenone or with 254 nm radiation in benzene, while the formation of (41) is subject to similar benzophenone sensitization, and (43) is formed almost exclusively from (40 a-c) in benzene with 254 nm radiation. The naphthalene-furan bichromophores (44) exhibit exciplex emission and reduced arene fluorescence.[27] Irradiation ($\lambda > 280$ nm) of these systems in benzene solution under argon induces intramolecular photocycloaddition to give the two ($4\pi + 4\pi$) adducts (45) and (46) for R = H, and (45) for R = CN (100% yield) or CO_2Me (76% yield). As may be expected, irradiation of (46) gives the caged compound (47).

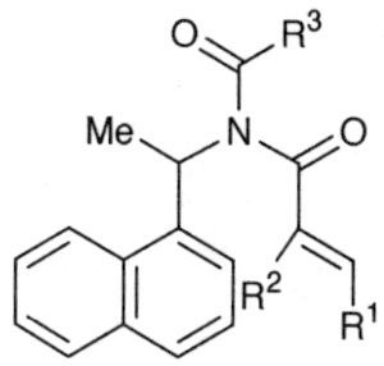

(40) a $R^1 = R^2 = H$, $R^3 = Me$
b $R^1 = H$, $R^2 = R^3 = Me$
c $R^1 = H$, $R^2 = Me$, $R^3 = Ph$
d $R^1 = C_6H_5$, R^2 and R^3 = H or Me
e R^1 = 4-MeC_6H_4, R^2 and R^3 = H or Me
f R^1 = 4-$NO_2C_6H_4$, R^2 and R^3 = H or Me
g R^1 = 4-ClC_6H_4, R^2 and R^3 = H or Me
h R^1 = 4-$MeOC_6H_4$, R^2 and R^3 = H or Me
i R^1 = 3,4,5-$(MeO)_3C_6H_2$, R^2 and R^3 = H or Me

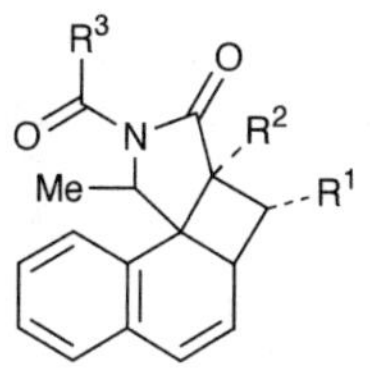

(41)

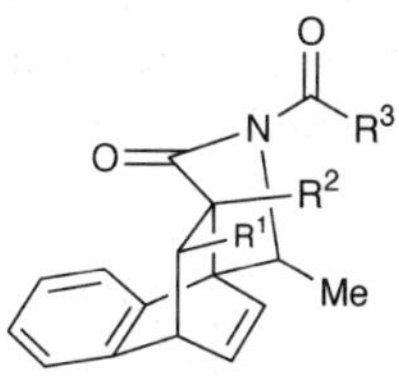

(42)

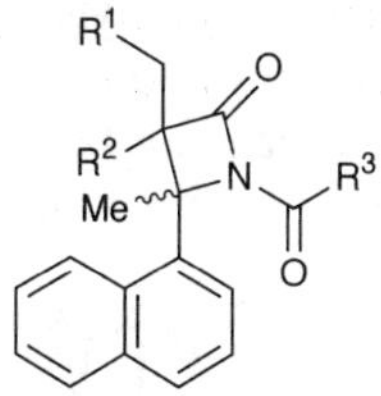

(43)

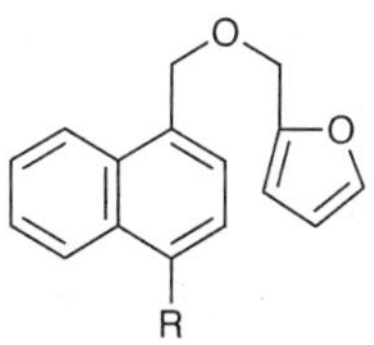

(44) R = H, CN or CO_2Me

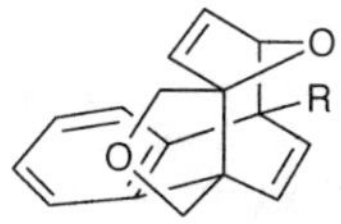

(45) R = H, CN or CO_2Me

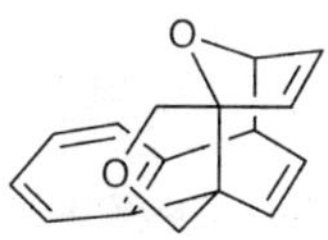

(46)

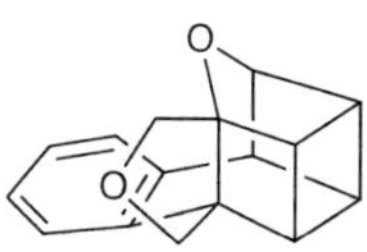

(47)

Irradiation of benzotriazole in the presence of various maleimide derivatives is reported to afford the ($2\pi + 2\pi$) adduct (48) regio- and stereo-selectively.[28] The reaction is considered to proceed by an unprecedented tautomer-selective addition to the 2-*H* tautomer (49) to yield (50, R = H), which rearranges to (48), and this is supported by the isolation of (50, R= alkyl) from reactions with 2-alkylbenzotriazoles.

The photochemical reaction of the homochiral anthracene (51) with maleic anhydride occurs with excellent diastereoselectivity (>95:5) forming (52), and

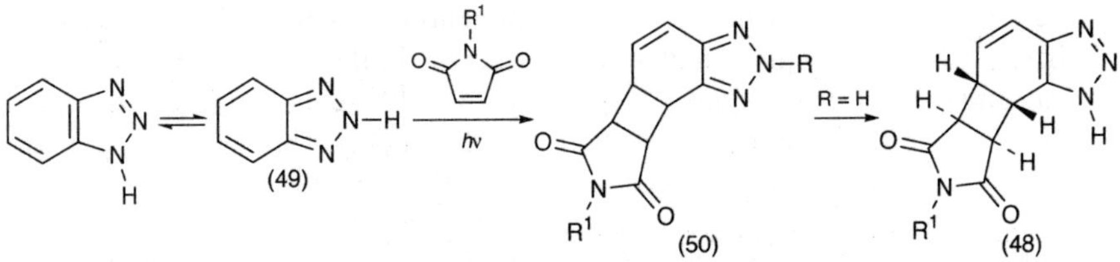

(49) (50) (48)

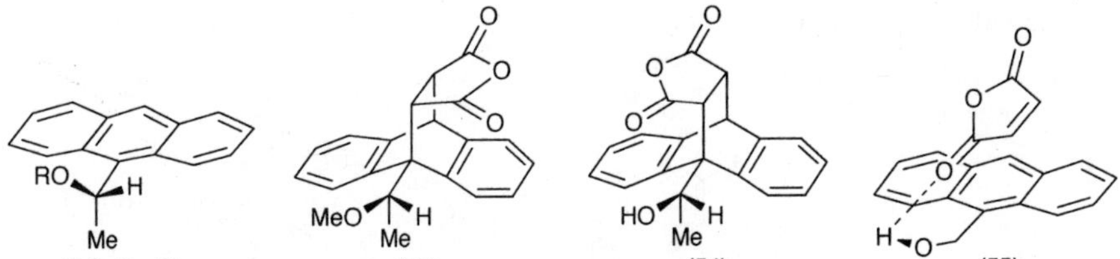

(51) R = Me
(53) R = H
(52) (54) (55)

more rapidly than the corresponding thermal process.[29] In contrast, with 1-anthracen-9-yl-ethanol (53), there is a complete reversal of the sense of selectivity, and (54) is the major product; this is rationalized in terms of hydrogen-bonding influences, which orient the molecules as shown in (55).[30] $(2\pi+2\pi)$ Photocycloaddition of ethenes to the 9,10-positions of phenanthrenes has been long established. This process has now been used with phenanthrene-9-carboxylate, in the presence of excess of the methyl orthoformate of *cis* cyclohexa-3,5-diene-1,2-diol, to obtain the precursor adducts (56) and (57), following acid treatment of the reaction mixture, which can be dehydroxylated to give the phenanthrene-benzene dimers (58) and (59), respectively.[31] The formation of (56) and (57) is

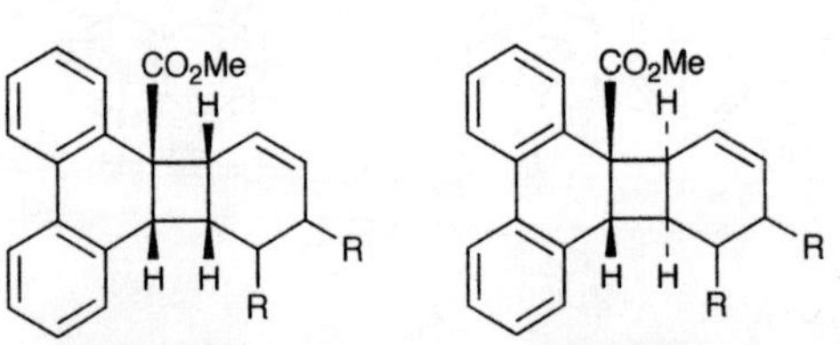

(56) R = OH
(58) R–R = bond
(57) R = OH
(59) R–R = bond

sensitized by Michler's ketone, but only (56) results from direct irradiation, and, in contrast to previously reported studies on dibenzenes and heterodimers, the *endo* isomer (58) is observed to be kinetically less stable than (59). The novel (3 + 2) photoadduct (60) has been reported from the irradiation of benzene solutions of 9-cyanophenanthrene in the presence of *trans* 1,2-dianisylcyclopropane.[32] The mechanism for the formation of (60) is suggested to involve collapse of an exciplex of the addends to the 1,5-diradical (61), which is also the precursor of the other products (62)-(64). A zwitterionic species rather than the diradical (61) is considered to be an unlikely intermediate, as yields of all products are reduced in reactions in methanol and, furthermore, none are formed in which the solvent is incorporated.

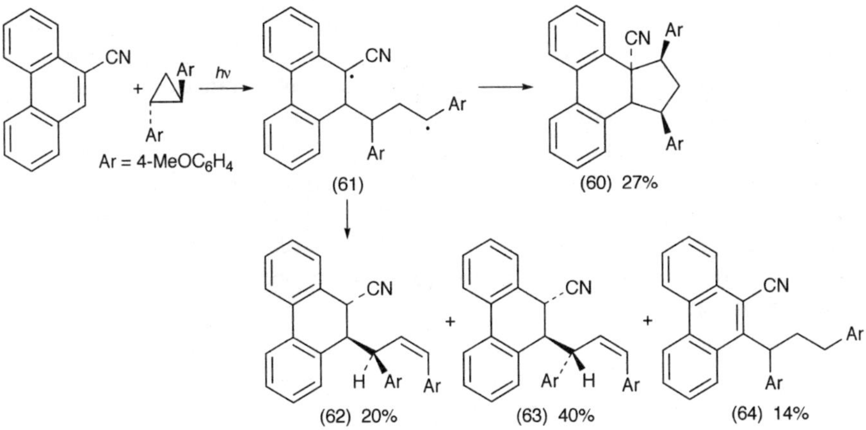

4 Substitution Reactions

There continues to be a growing interest in the photochemical reactions of halogenoarenes. The focus of many studies lies in the oxidative degradation of such compounds, and the publications coming from these are considered in Part II Chapter 5 of this review. In this section, the photochemistry of halogenoarenes, and of other substrates, which leads to substitution is presented. Direct irradiation of 1,4-dichlorobenzene in air-saturated aqueous solution is reported to result in the formation of 4-chlorophenol, quinol, hydroxybenzo-1,4-quinone and 2,5-dichlorophenol.[33] The substitution products are suggested to arise by direct attack of water onto the chlorine bearing carbon atom rather than from cleavage of the C–Cl bond, since no appropriate transient for phenyl radicals was observed. The presence of oxygen was essential for the formation 2,5-dichlorophenol, and a mechanism involving electron transfer from the arene to oxygen and reaction of the subsequent radical cation with water is proposed for this hydroxylated product without dechlorination. The formation of this phenol is, however, enhanced by the presence of nitrate ions or ferric salts. In the absence of oxygen, the unexpected formation of 4,4′-dichlorobiphenyl, 2,4′,5-trichlorobiphenyl and a terphenyl derivative is also observed. In aqueous solution, 3,6-dichloro-2-methoxybenzoic acid undergoes photochemical substitution of the chlorine atoms by the hydroxyl group as the major route, but, from irradiation of the acid in the solid state or supported on a synthetic clay, several products are formed.[34] 4-Chloroanilines can be converted into 4-(2′-chloroalkyl)anilines by their irradiation in the presence of alkenes.[35] The reaction is suggested to proceed by heterolytic cleavage of the C–Cl bond and, in the presence of $NaBH_4$, 4-alkylanilines are obtained directly. In contrast to these reports of reactions in aqueous solution, the irradiation of 1,4-dichlorobenzene, bromobenzene and 1,4-dibromobenzene in ice leads to processes of dehalogenation, coupling and rearrangement, and no solvolysis products were observed.[36]

Polychlorinated dibenzo-1,4-dioxins have high toxicity and bioaccumulatory properties and hence are of considerable concern in the environment. Their photochemistry is thus of great importance. Wan and co-workers have studied the phototransformation of 2,3,7,8-tetrachlorodibenzo-1,4-dioxin (65) in order to identify all the products and hence complete a mass balance of the system and improve the understanding of its photochemistry.[37] In aqueous solution and with 302 nm radiation, (65) undergoes a novel process of C-O bond homolysis and formation of 2,2′-dihydroxy-4,4′,5,5′-tetrachlorobiphenyl (66). Conversion of (65) to its dechlorinated derivatives or chlorinated phenoxy-phenols is a minor process, and from the results it is suggested that the photoreaction of chlorinated dioxins to the chlorinated dihydroxy-biphenyls may be an important reaction in the environment.

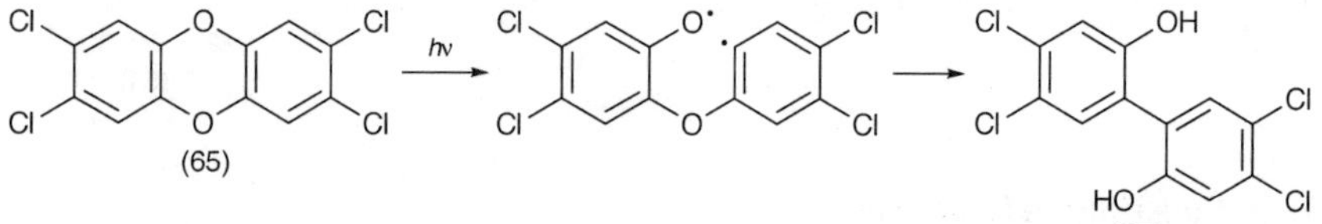

(65) (66)

The formation of phenyl-substituted derivatives by photoinduced substitution of 4-iodo-nitrobenzene, 2-bromo- and 2-iodo-5-cyanothiophenes and of 2-iodo-5-nitrothiophene in benzene solution has been studied by steady state and pulsed techniques, as well as analysed using semi-empirical quantum mechanical calculations.[38] The use of sensitizers with a range of triplet energies has indicated that these substitutions involve an upper triplet state in the C-halogen bond cleavage. Photochlorination of α,α-difluoroanisole has been reported;[39] and the formation of nitrophenols from irradiation of phenol in the presence of nitrate ion in aqueous solution or in aqueous suspensions of TiO_2 has been evaluated over the pH range 1-12.[40] 1-Hydroxyanthraquinone undergoes photoamination in the presence of *n*-butylamine.[41] Both the 2- and 4-butylamino derivatives are formed, in a respective ratio of 0.2, under air in acetonitrile or aqueous acetonitrile, but, from nitrogen-degassed solutions, the ratio increases to 0.32 as the water content is increased from zero to 50%. In contrast, only 4-photoamination occurs with 1-aminoanthraquinone, while 1-hydroxy-2-bromo- and 1-hydroxy-2,4-dibromoanthraquinones gave only 1-hydroxy-2-bromo-4-(butylamino)anthraquinone. Cyanation of pyrene has been achieved in 83% yield by irradiation of the arene in an oil-in-water emulsion in the presence of 1,4-dicyanobenzene and sodium cyanide.[42] A photoinduced electron-transfer pathway is suggested in which the pyrene radical cation is attacked by CN^- and the resulting radical is oxidized to 1-cyanopyrene in air.

2-Substituted phenols are quantitatively formed from either exposure to sunlight or 350 nm irradiation of the readily available 3-substituted bicyclo[3.1.0]hex-3-en-2-ones (67).[43] A photocatalysed procedure has been developed for the production of 2-substituted pyridines.[44] The pyridines result from a (2 + 2 + 2) cycloaddition of two equivalents of acetylene with one equivalent of a nitrile using 350-500nm radiation in the presence of a precatalyst such as η^5-cyclopentadienyl-η^4-cyclo-oct-1,5-diene-cobalt and at room temperature and atmospheric pressure. The solvents can be toluene or cyclohexane, as well as water, and

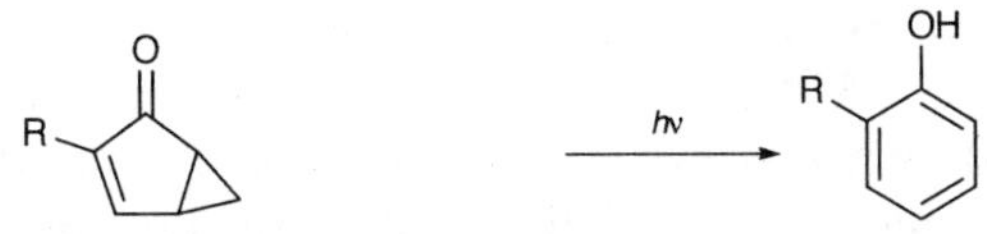

(67) R = CMe_3, 4-FC_6H_4, n-hexyl, $SiPh_3$, $C(OH)Me_2$, C(OH)HPh

in view of these mild conditions, which are in sharp contrast to the drastic requirements of the thermally-initiated variant of this process, the authors' claim that the reaction is 'a valuable extension to common synthetic methods' seems wholly justified.

5 Cyclization Reactions

Photoinduced 6π-electrocyclizations of a wide variety of systems incorporating aromatic moieties remains an area of considerable research activity, which has led to numerous commercial applications particularly in the areas of photochromism and molecular switches. Lewis and co-workers have examined the photocyclization of *cis* aminostilbenes and report differing activities for the isomers.[45] While the 6π-electrocyclization to give the phenanthrene occurs for the *meta* isomer with a quantum efficiency greater than any reported for a mono-substituted *cis* stilbene, and the *cis-trans* isomerization has an abnormally low efficiency, the *ortho* and *para* isomers photoreact similarly to many substituted stilbenes and undergo predominantly the *cis-trans* conversion. In the case of the 'push-pull' stilbene (68), the 4*a*,4*b*-dihydrophenanthrene (69)

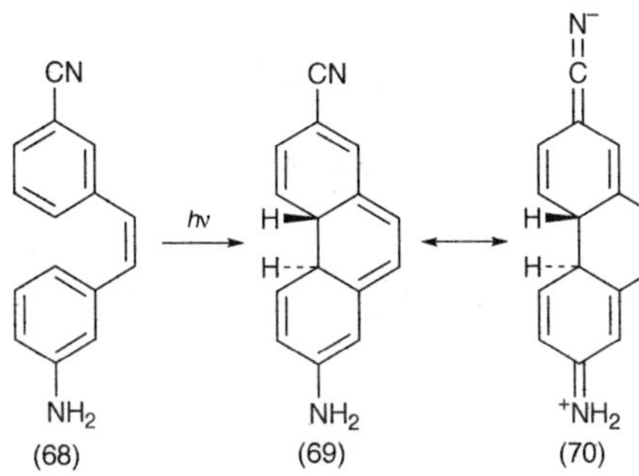

(68) (69) (70)

formed in the absence of oxygen, is deduced from the solvent dependence of the absorption spectrum (λ_{max} 578 nm and 530 nm in acetonitrile and cyclohexane, respectively) to have appreciable quinonoid character as in (70).[46] Similar features are also noted for the 3-amino- and *N*,*N*-dimethylaminostilbenes. Methyl α-phenylcinnamate derivatives which have halogen substituents at both the *ortho* positions of the β-phenyl group yield 9-phenanthroates in an oxygen-insensitive photocyclization and dehydrohalogenation procedure;[47] and the 1,2-dicyano[2.n]metacyclophan-1-enes (71) undergo photocyclization with 366 nm radiation to give (72), which can be readily reversed with wavelengths longer than 500 nm or in the dark at room temperature.[48] 2,2′-Dihydroxystilbenes undergo a photoinduced enantioselective aerobic oxidative cyclization in the presence of (nitrosyl)Ru(salen) to give the *cis*-4*b*,9*b*-dihydrobenzofuro[3,2-b] benzofurans (73).[49] The reaction is considered to proceed by way of the oxygen cation radical and gives enantiomeric excesses from 34% in chlorobenzene up to 89% in *t*-butanol, although the chemical yield is described as only 'modest'.

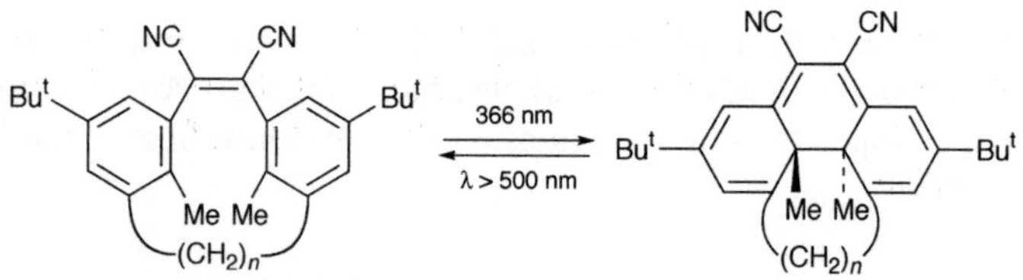

(71) *n* = 2 or 3, pale yellow (72) violet

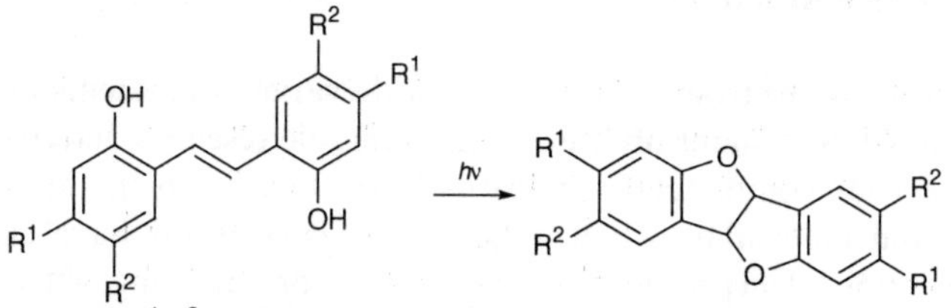

R^1,R^2 = H or Me (73)

cis-1,2-Di-(3′-methoxynaphthyl-2′-yl)ethene (74) undergoes photocyclization,[50] and Mallory and co-workers have used this photochemical approach to gain access to the phenacene family of arenes.[51] These polynuclear arenes have extended phenanthrene-like structures, and an example, [11] phenacene (75), from the present report has been formed in 43% yield by the irradiation of (76) in

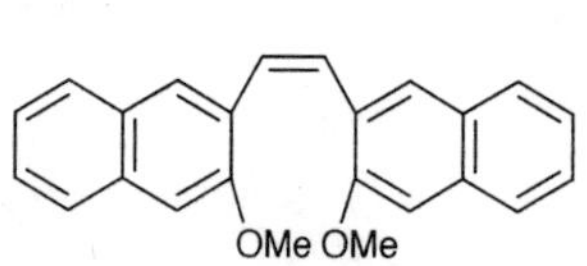

(74)

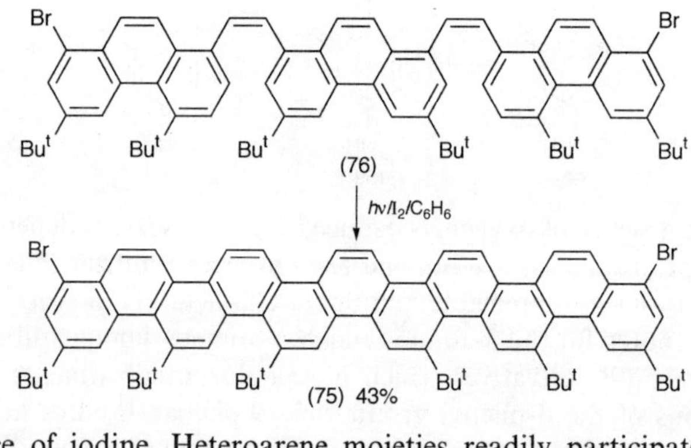

(75) 43%

the presence of iodine. Heteroarene moieties readily participate in the 6π-cyclization process and, in the case of the 3-styrylpyridines (77), the 2-azaphenanthrenes (78) are formed highly regioselectively under anaerobic conditions.[52] The formation of (78) under these conditions is rationalized in terms of the 4*a*,4*b*-dihydro compound (79) undergoing a [1,7] shift to give the 1,4-dihydropyridine

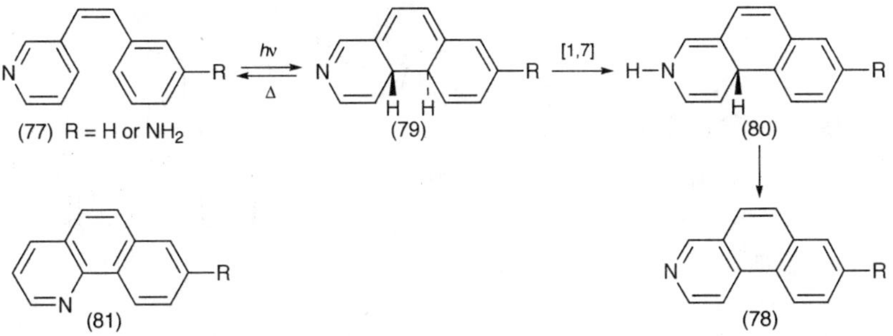

(77) R = H or NH_2 (79) (80) (81) (78)

(80), which readily aromatizes. In the presence of oxygen, mixtures of both (78) and the 4-azaphenanthrene (81) are formed, and this is considered to arise as a consequence of the relative rate constants of ring opening and sigmatropic shifts in the intermediates as well as their rates of reaction with oxidants. The photocyclization of the 3-styrylpyridine (82) has been investigated as a key step in a first total synthesis of the alkaloid toddaquinoline (83).[53] Unfortunately, although the desired 4-azaphenanthrene (84) was formed, the major product was

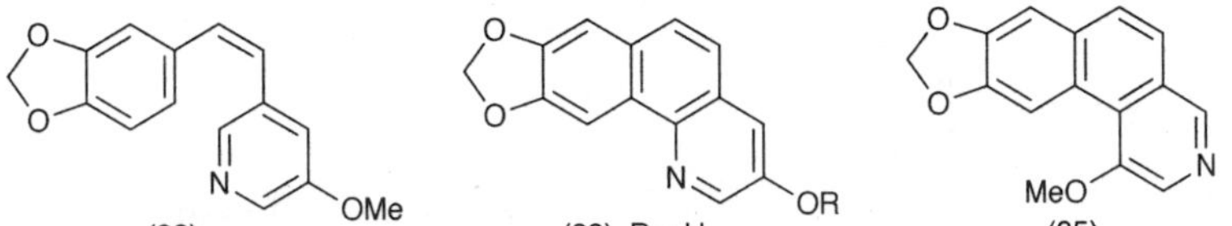

(82) (83) R = H (84) R = Me (85)

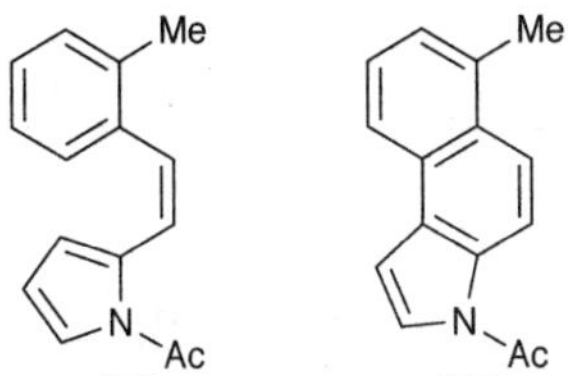

(86) (87)

the 2-isomer (85) in 53% yield. This type of process with 2-styrylpyrroles provides a useful access to benz[*e*]indoles, and using this approach Sindler-Kulyk and co-workers have converted the *N*-acetyl-2-methyl compound (86) into 2-acetyl-7-methylbenz[*e*]indole (87).[54] The same research group has also investigated the bifunctional system (88), which may be expected to undergo a double cyclization, but instead intramolecular cyclization within the *o*-divinylbenzene moiety occurs to give a 1:1 mixture of (89) and (90).[55] The latter is considered to arise from a [1,7] shift in the intermediate (91). The primary intermediates (92) and (93), from irradiation of the *o*-vinylstyrylfurans (94) and (95), respectively, have been characterized spectroscopically.[56] Both intermediates are trapped by oxygen to give the hydroperoxides (96) and (97) and the ketones (98) and (99), but only (93) reacts with methanol and this yields (100).

1,2-Diheteroarylethenes have been the subject of numerous photochemical publications in recent years, with the interest being principally focused on their photochromic properties and characteristics. This year has been no exception,

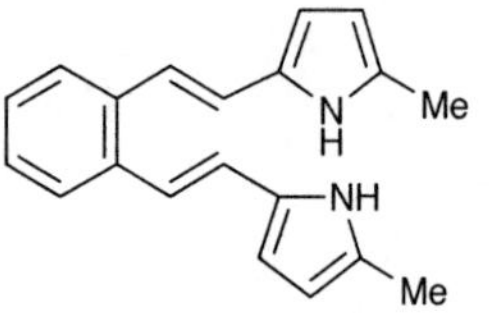

(88)

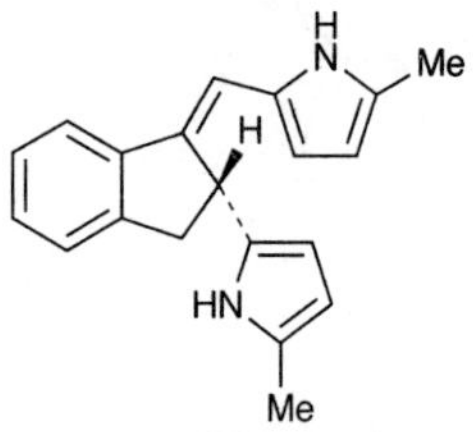

(89)

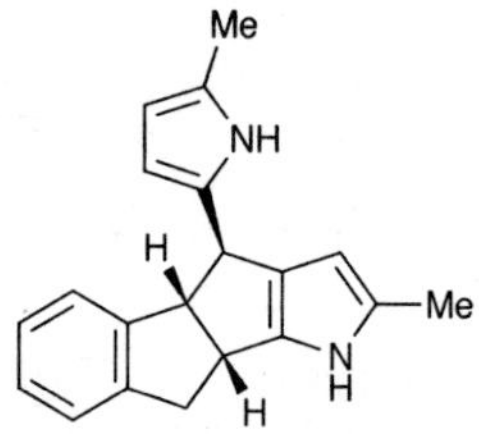

(90)

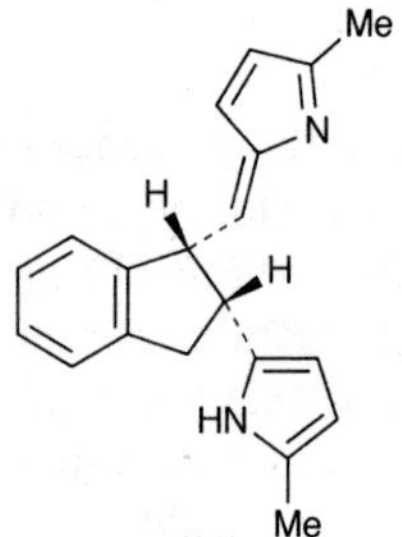

(91)

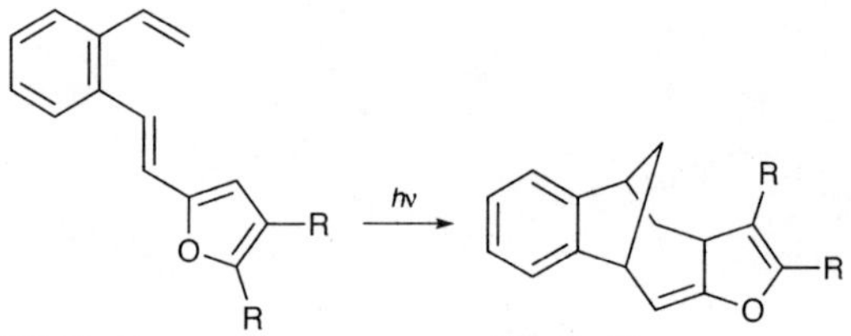

(94) R = H
(95) R–R = –(CH=CH)$_2$–

(92) R = H
(93) R–R = –(CH=CH)$_2$–

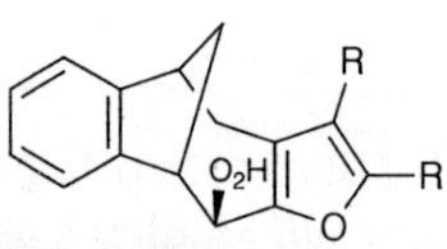

(96) R = H
(97) R–R = –(CH=CH)$_2$–

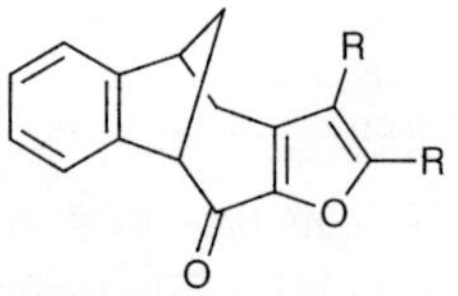

(98) R = H
(99) R–R = –(CH=CH)$_2$–

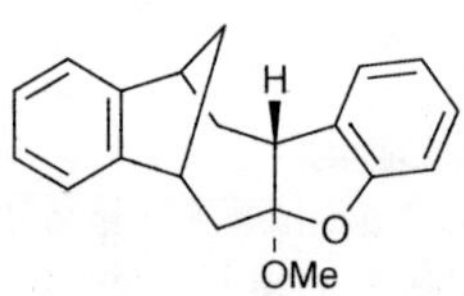

(100)

and again a variety of aspects of these systems, particularly those incorporating the 1,2-dithienylperfluorocyclopentene unit, have been described, and the use of such compounds in non-destructive erasable memory devices has been reviewed.[57] Compounds which show photochromic activity in the crystalline state are rare and x-ray data are limited. Recently, *in situ* crystallography has been carried out for several photochromic systems including diarylethenes,[58] and such data for diheteroarylethenes with perfluorocyclopentene, maleic anhydride and cyclobutanedione moieties between the aryl units have been discussed.[59] The photochromism of 1,2-bis(2-methyl-5-phenyl-3-thienyl)perfluorocyclopentene (101) has been monitored by *in situ* x-ray crystallography, from which it is noted that the molecule with the shorter distance between the reactive centres reacts more efficiently, and a quantum yield of unity is reported.[60] In crystals of the 2,4-dimethyl-3-thienyl compound (102), 17% of the molecules have the photoactive anti-parallel conformation and are photochromic, whereas the 2,4,5-trimethyl derivative (103) does not undergo photocyclization, since it has the photochemically inactive conformation.[61] In contrast, all three 1,2-dithienylperfluorocyclopentenes (104)-(106) undergo cyclization in single crystals with quantum yields twice those in solution, a feature which is attributed to only the

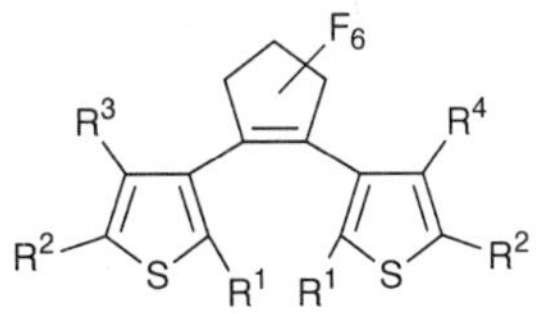

(101) R^1 = Me, R^2 = Ph, R^3 = R^4 = H
(102) R^1 = R^3 = R^4 = Me, R^2 = H
(103) R^1 = R^2 = R^3 = R^4 = Me
(104) R^1 = Et, R^2 = Ph, R^3 = R^4 = H
(105) R^1 = R^2 = Me, R^3 = R^4 = H
(106) R^1 = R^3 = Me, R^2 = Ph, R^4 = H
(107) R^1 = OMe, R^2 = Ph, R^3 = R^4 = H
(108) R^1 = Me, Et, Pr^i, C_6H_{12}, R^2 = Ph, R^3 = R^4 = H
(109) R^1 = Me, R^2 = —CH=CH—$C_6H_4$4-COPh, R^3 = R^4 = H

photoactive anti-parallel conformers being packed in the crystal, whereas both conformers are present in solution.[62] The very high quantum yields of cyclization and cycloreversion in the crystals are explained in terms of constrained conformations in the crystal lattice. The introduction of methoxy groups to replace methyl groups in these systems, as in (107), does not markedly affect the cyclization quantum yields in solution but does decrease the photoreversion efficiency by a factor of 10^3.[63] Indeed, for (108) with bulky alkoxy groups, the photodecolouration efficiency is very low ($\phi<10^{-3}$) but the thermal stability of the ring-closed isomer at high temperature is decreased; for example, the fade-out of the colour of the dicyclohexyloxy derivative was less than 1min at 160°C.[64] Complex derivatives of the 1,2-dithienylperfluorocyclopentene system have been synthesized in order to modify the properties of these photochromes and to assess various influences on their photochemical behaviour. The effect of bulky strongly-coupled substituents on the photocyclization and photoreversion of these systems has been investigated with (109), from which it is deduced that following relaxation of the S_1 state, the cyclization occurs with high quantum efficiency through a conical intersection, which also acts as a relaxation funnel for the S_1 state of the ring-closed isomer.[65] Furthermore, branching in the conical intersection favours the ring-closed isomer, with the result that the efficiency of cycloreversion is low (10^{-3}-10^{-4}). The usefulness of these diheteroaryl perfluorocyclopentene systems as thermally stable photoswitchable acceptors in photochromic fluorescence resonance energy transfer has been studied with (110), as only the corresponding cyclized forms have absorption bands which overlap with the emission band of the donor (111).[66] Only the (M*, R*, R*) diastereoisomer of the [7]thiahelicene (112) is reported to be formed from irradiation of (113) at 410 nm.[67] The process is reversed with wavelengths longer

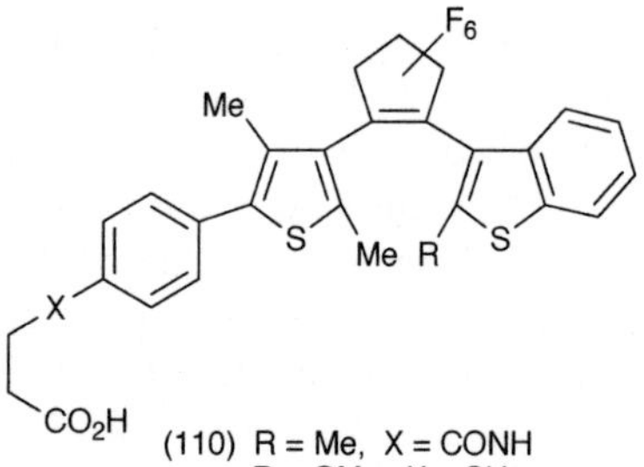

(110) R = Me, X = CONH
or R = OMe, X = CH_2

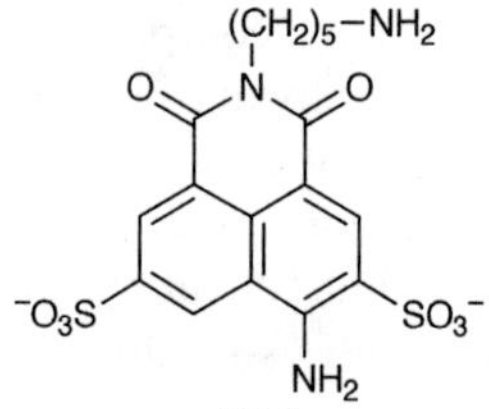

(111)

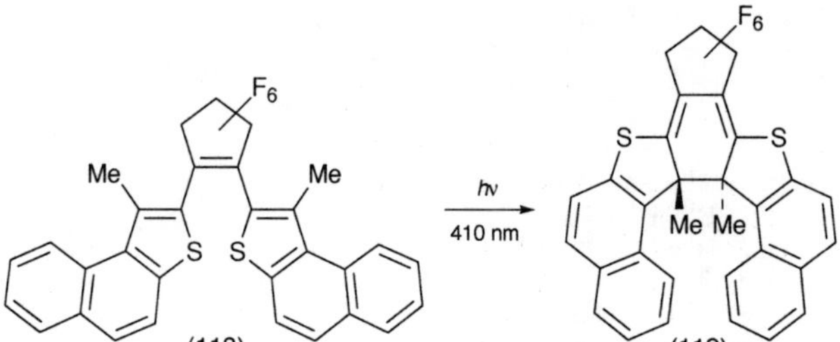

(113) (112)

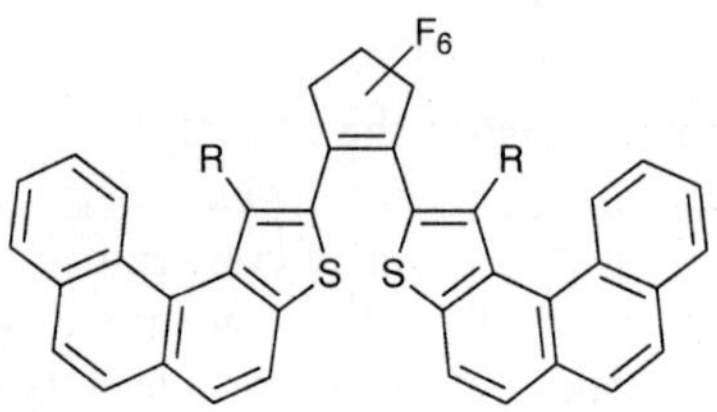

(114) R = H or Me

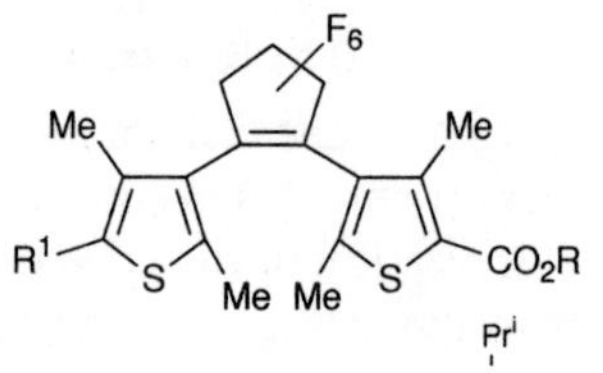

(115) R^1 = Me R =

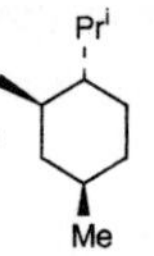

(116) R^1 = CO_2R

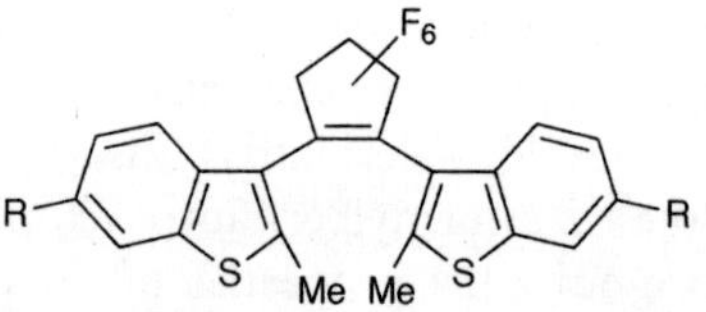

(117) R =

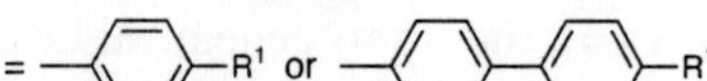

R^1 = H or

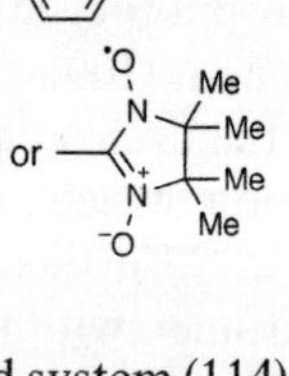

than 458 nm, but the extended system (114) is photoreactive only for R = H, and it then undergoes irreversible oxidative cyclization in the presence of iodine to give an aromatic dithia[9]helicene. The dithienylcyclopentenes (115) and (116), which have one or two chiral 1-menthyl substituents, are reported to undergo diastereoselective photocyclization in solution and in the amorphous state.[68] In solution, the two diastereomers are formed in an approximately 1:1 ratio, but the (*R*,*R*) isomer predominates in the amorphous state and for (116) has a *de* of

37.4%. The quantum yields have been determined for the cyclization and reversion of (117) which has *p*-phenylene-substituted benzothiophene units with and without nitronyl nitroxide radical moieties at each end of the molecule.[69,70] The efficiencies of both processes are found to increase for the compounds with the radicals which have an increase in the π-conjugation, and the photoswitching of the magnetic interaction between the two radicals was determined by ESR spectroscopy to be more than 30-fold for the change in exchange between the two photoisomers for the *p*-phenylene compound. High fatigue resistance is reported for 1,2-bis{5-(2-benzothiazolyl)-2-methyl-6-trifluoromethylthieno[3,2-*b*] thiophen-3-yl}perfluorocyclopentene,[71] and the introduction of porphyrin units at the 5- and 5′-positions of the basic 1,2-bis(2-methylthiophen-3-yl)perfluorocyclopentene structure results in the loss of photoreactivity, probably as a result of efficient charge-transfer quenching of the dithienylethene unit by the porphyrin.[72]

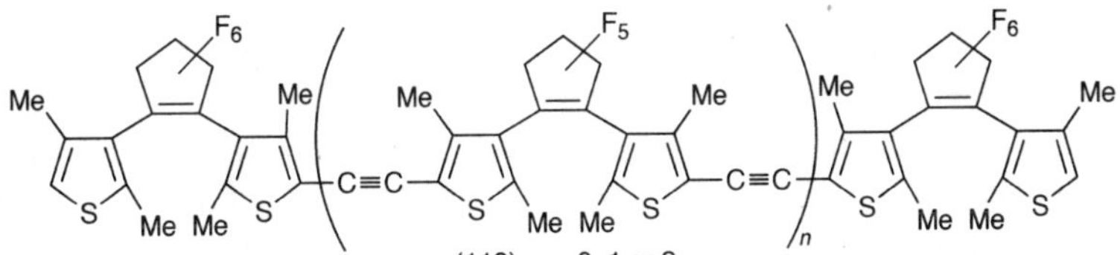

(118) $n = 0$, 1 or 2

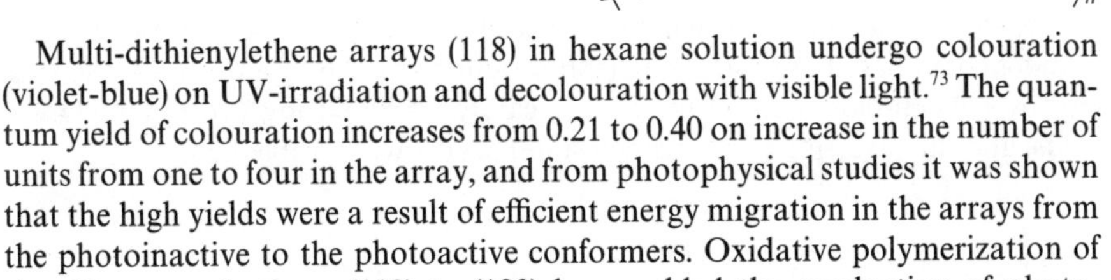

(119) (120)

Multi-dithienylethene arrays (118) in hexane solution undergo colouration (violet-blue) on UV-irradiation and decolouration with visible light.[73] The quantum yield of colouration increases from 0.21 to 0.40 on increase in the number of units from one to four in the array, and from photophysical studies it was shown that the high yields were a result of efficient energy migration in the arrays from the photoinactive to the photoactive conformers. Oxidative polymerization of the diheteroaryl ethene (119) to (120) has enabled the production of photochromic films which undergo colouration and decolouration cycles on UV and visible light irradiation, respectively.[74] A refractive index change of 0.064 at 817 nm has been observed between the open and cyclized isomers of a 1,2-bis benzothiophenyl ethene in an amorphous thin film,[75] and calculated UV-visible absorption spectra of dithienylethene derivatives linked by pyridinium betaines indicate that reading information from these systems above 700 nm will not destroy the recorded data.[76] Other novel dithienyl systems, involving a linked tetra-azaporphyrin, have been investigated for a molecular device which may be useful for non-destructive readout,[77]and the results have been reported of semi-empirical molecular orbital calculations on the optimized ground and S_1 states of dithienyl derivatives with perfluorocyclopentene, maleic anhydride and 1,2-dicyanoethene linking units.[78] Other theoretical studies suggest that it is more appropriate to consider the cyclized isomer than the open form for correlation analysis and to predict the photochromism of dithienyl perfluorocyclopentenes,[79] and a joint experimental and theoretical study of the mechanism of photochromic diarylethenes has been published by prominent workers in this field.[80]

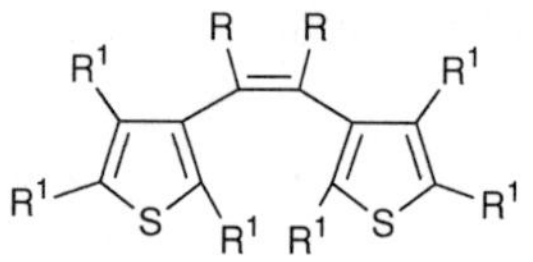

(121) R = CN or R–R = –CO–O–CO–, R^1 = H or Me

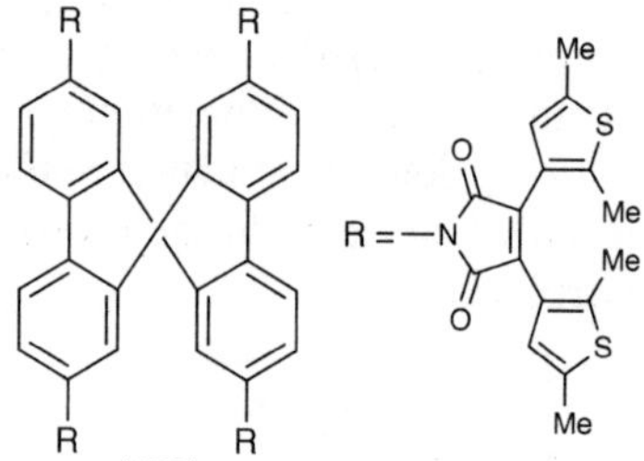

(122)

A number of reports have been published which describe the photocyclization of dithienyl systems having linking units other than the perfluorocyclopentene moiety. The photochemistry of the 1,2-bisdithienyl dicyanoethene and maleic anhydride compounds (121) has been studied in a variety of polymer films, and, although the efficiency of the dicyano compound is little affected by the polymer type, the anhydride compound is more photoactive in polymer matrices which have low glass-transition temperatures and low polarity.[81] The novel tetra bis-thienylethene (122), with a spiro linker, is reported to have high cyclization quantum yields and good switching stability (>200 cycles), and, as the authors suggest, provides a new approach towards amorphous photochromic materials.[82] Photochromism is also observed in dithienyl compounds which are linked through azole rings,[83] and a most interesting variant on this type of process for use in molecular switches has been reported by Deng and Liebeskind, with a system based on bis(heteroaryl)1,4-quinones (123).[84] In this case, the ring closure (switch-closed state) to (124) is catalysed by strong protic and Lewis acids, and the reversion (switch-open state) occurs with visible light.

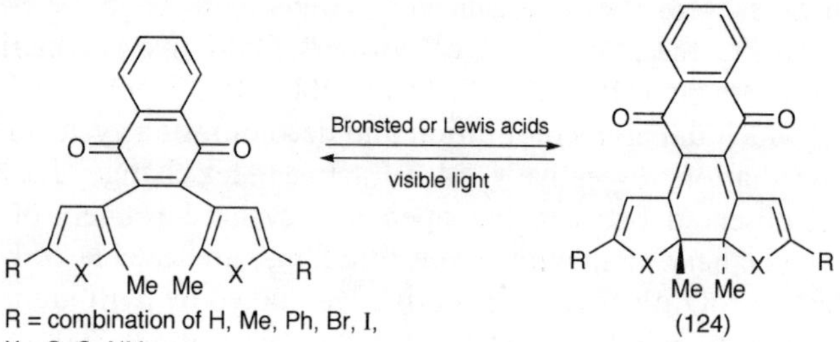

(123) R = combination of H, Me, Ph, Br, I,
X = S, O, NMe

(124)

Ar–CH=CH–CH=CH–Ph

(125) Ar = 1-naphthyl, 9-anthryl, 9-phenanthryl, 1-pyrenyl, 2′,3′and 4′-pyridyls, 2′ and 3′-thienyl

It is well known that 1-arylbuta-1,3-dienes undergo photoinduced 6π-electrocyclization to give naphthalene derivatives. The excited-state properties of the 1-aryl-4-phenylbuta-1,3-dienes (125) have been investigated by stationary and pulsed techniques in order to determine the influence of the size and characteristics of the aryl group on the photoprocess of the 6π-system,[85] and this type of cyclization has been used with 4-methylthio-1-phenyl-2-(*p*-tolylsulfonyl)-1,3-diene (126) to synthesise the 2-substituted naphthalene (127).[86] The differences in the structures and conformations between the imide moiety in the open (128) and

(126) (127)

closed (129) isomers of the thermally-irreversible indolylfulgimide have been determined by their respective association constants of 885 ± 63 and 156 ± 11 $mol^{-1}dm^3$ with 2,6-bis(octanoylamino)pyridine at 21°C in toluene;[87] and the photochromic reactivity of the fulgides, Aberchrome 670 and 540, in poly(methyl methacrylate) and in Langmuir-Blodgett films has been reported.[88] The first report of the 6π-photoelectrocyclization of aryl-1-methoxy-2-aza-1,3-dienes (130a and b) has been published and reasonable yields (38-69%) of the isoquinolines (131a and b, respectively) are formed in neutral methanol solution.[89] Arylenynes such as (132) also undergo this type of cyclization, in this case by way of the allene intermediate (133), which in non-polar solvents yields the phenanthrene (134) by a [1,5] sigmatropic shift or, as shown by the reaction with MeOD, undergoes protonation at the central allenic carbon atom.[90]

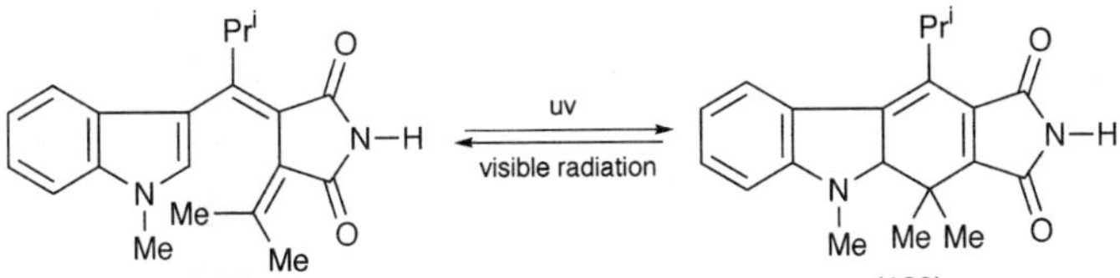

(128) (129)

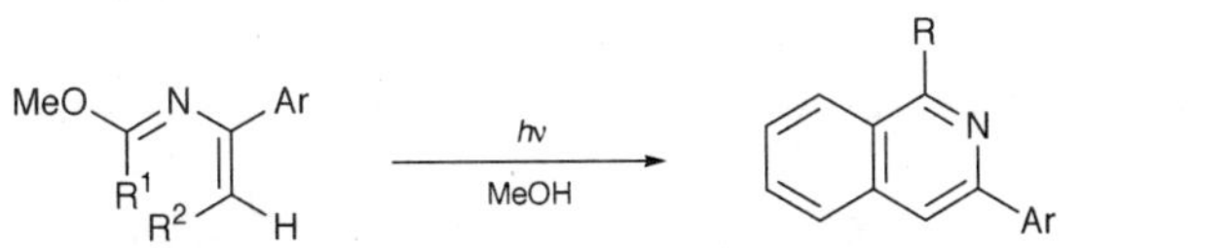

(130) a R^1 = Ph, R^2 = H,
Ar = Ph, $C_6H_4$4-Me, $C_6H_4$4-OMe
b R^1 = Ph or Me, R^2 = Ar = Ph

(131) a R = OMe,
Ar = Ph, $C_6H_4$4-Me, $C_6H_4$4-OMe
b R = Ph or Me, Ar = Ph

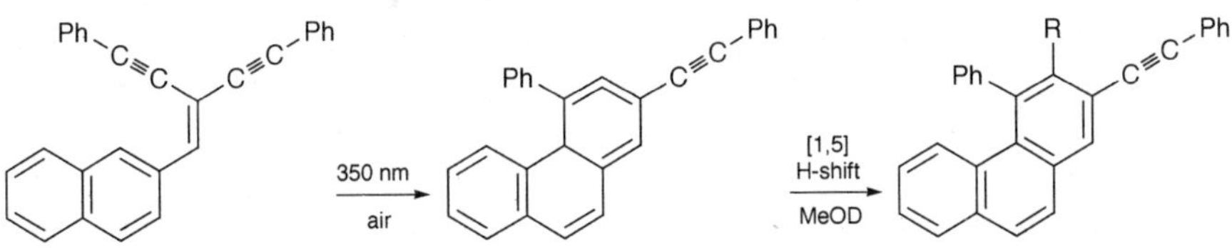

(132) (133) (134) R = H or D

N,N-Diarylamines yield carbazoles on irradiation, but for the *N*-alkyl-(*p*-methoxyphenyl)arylamines (135), in the presence of acid, a novel photochemical transformation is reported to occur.[91] As outlined in Scheme 3 for Ar = Ph, a sequence of protonation of the dihydrocarbazole intermediate (136), followed by deprotonation and formal [1,5] and [1,3] hydrogen shifts, leads to the enol ether (137), which is hydrolysed to (138) with overall yields of 60-96% dependent on

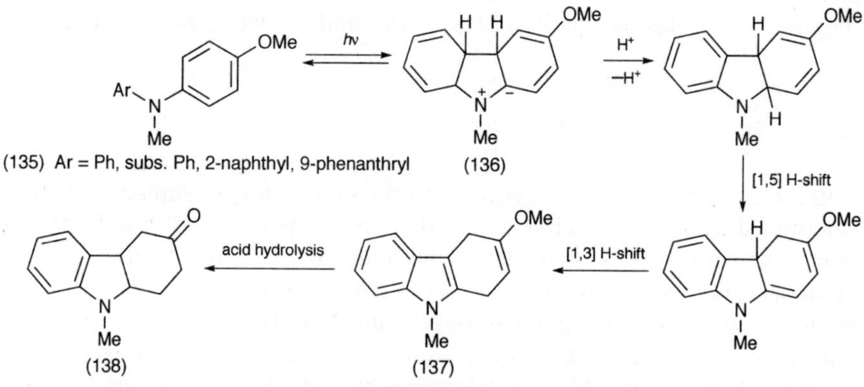

Scheme 3

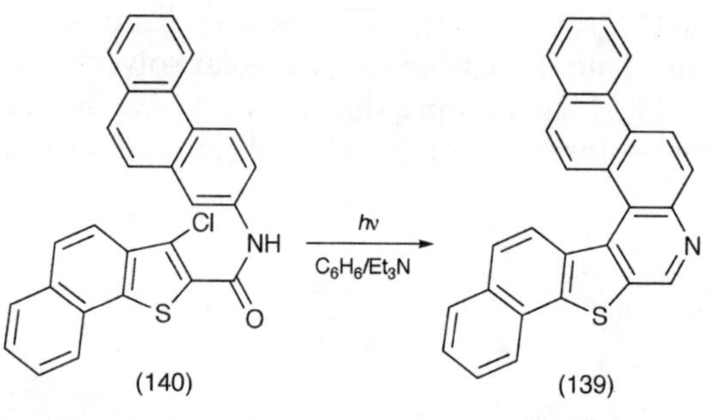

the nature of the aryl group. Castle and co-workers have again used the photocyclization-dehydrohalogenation of diaryl carboxamides to good synthetic effect. Thus the previously unknown polycyclic heterocyclic ring system (139) has been obtained in 87% yield by irradiation of 3-chloro-*N*-(2-phenanthryl)naphtho[1,2-*b*]thiophene-2-carboxamide (140) in benzene solution in the presence of triethylamine followed by further elaboration.[92] 2-Quinolones can be readily obtained by the photocyclization of derivatives of anilides of acrylic acid, and this process with (141) gives 21-26% yields of (142), which is a key synthetic intermediate for a novel benzophenanthridine structure (143), needed to assess if the inclusion of nitrogen in certain estrogen receptor modulators is effective.[93] By a similar process, irradiation of nitrogen-degassed pentane solutions of (144) yields the corresponding 2-quinolones (145) in over 90% yield at close to complete conversion.[94] The reaction also gives enantiomeric excesses up to 37% in the presence of substoichiometric amounts of chiral compounds (*e.g.* ephedrine), and the weakly absorbing transient at 400 nm, which appears on laser

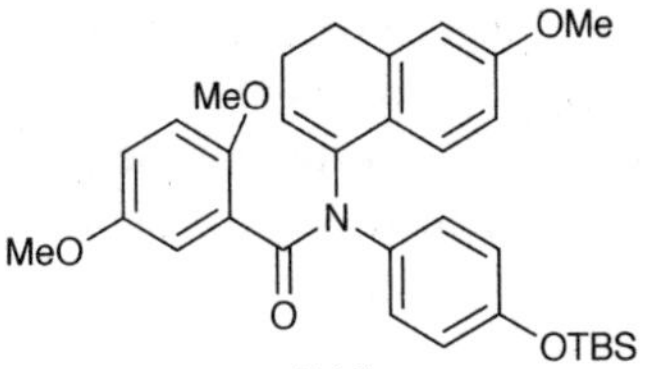

(141)

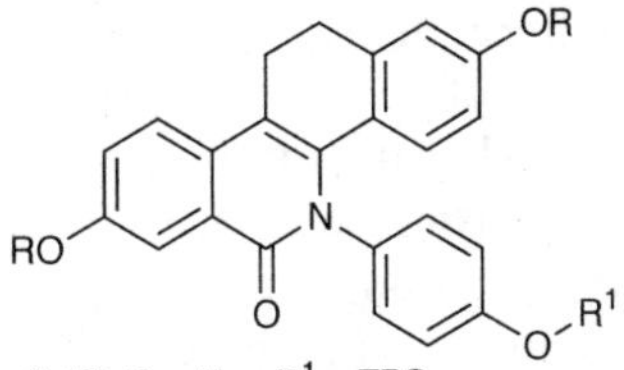

(142) R = Me, R^1 = TBS
(143) R = H, R^1 = —(CH$_2$)$_2$—N

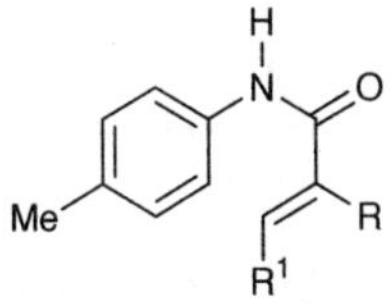

(144) a R = Me, R^1 = H
b R—R^1 = —(CH$_2$)$_4$—

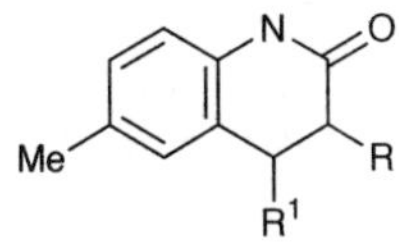

(145)

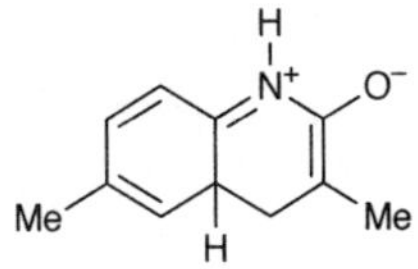

(146)

excitation, is assigned to the primary zwitterionic enolate intermediate (146) from concerted electrocyclization of (144a). Interestingly, this intermediate is not affected by oxygen but is quenched by ephedrine ($k_q = 1.19 \times 10^7$ $M^{-1}s^{-1}$). The 2-quinolones (147) or quinolines (148) and coumarin (149) or 2*H*-benzopyran hemiacetals (150) are formed by irradiation of the 2-aminocinnamoyl (151) and 2-hydroxycinnamoyl (152) derivatives, respectively.[95] These reactions can be carried out in a variety of solvents and yields up to 100% are claimed, depending on the particular derivative and reaction media. In methanol solution containing triethylamine, the *N*-acetyl-α-dehydro (1-naphthyl)alanines (153) undergo photocyclization to give 1,2-dihydrobenzo[*f*]quinolinones (154) in preference to benzo[*f*]isoquinolines (155); the 1-azetines (156) are also formed in this reaction.[96] The selectivity for (154) formation is observed to decrease with increasing

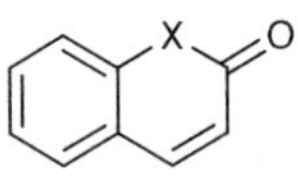

(147) X = N—H
(149) X = O

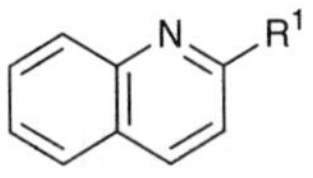

(148)

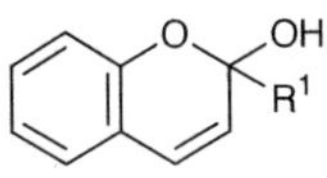

(150)

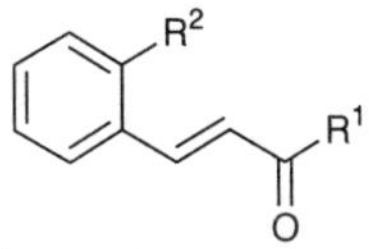

(151) R^1 = OEt, NEt$_2$, Ph, Me, R^2 = NH$_2$
(152) R^1 = as above + OH, R^2 = OH

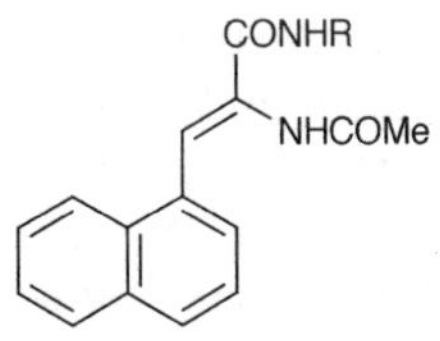

(153) R = H, alkyl, —CH$_2$Ph, —CH$_2$4-OMeC$_6$H$_4$, —CH$_2$4-CF$_3$C$_6$H$_4$

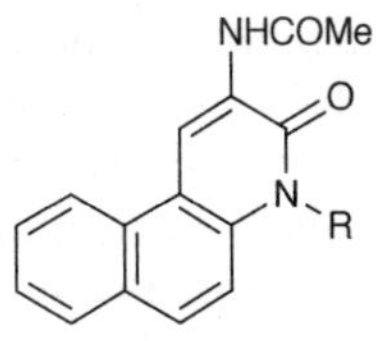

(154)

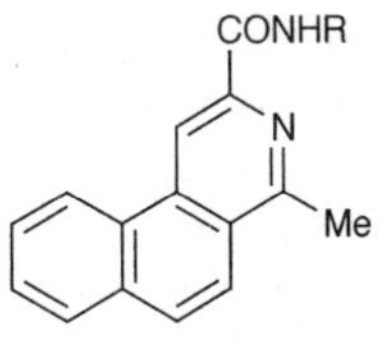

(155)

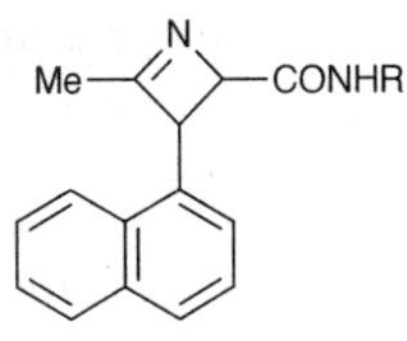

(156)

bulk of the alkyl substituent in (153), and the process is deduced to arise by the electron-transfer mechanism outlined in Scheme 4. The irradiation of the phenol-styrene bichromophore (157) is reported to proceed by way of intramolecular electron transfer and yields a mixture of the dihydrobenzopyran (158) and the dihydrobenzofuran (159).[97] Exciplex emission is observed in acetonitrile for bichromophores having electron donors on the phenol ring, and in these cases the derivatives of (158) greatly predominate with yields of 45-99%. The introduction of an ethyl group on the methylene spacer in (157) provides the alternative route of a di-π-methane rearrangement for (160) in its photochemistry, and indeed the formation of the cyclopropanes (161) as well as the rearranged ethene (162) is also observed in this case.[98]

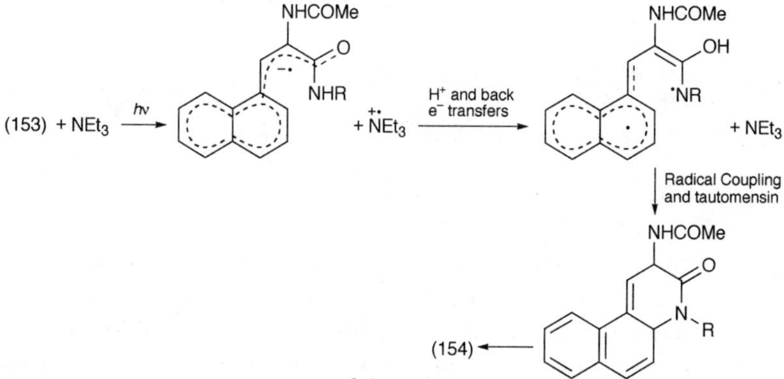

Scheme 4

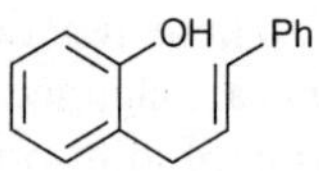

(157)

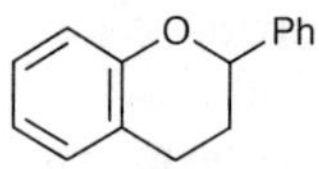

(158)

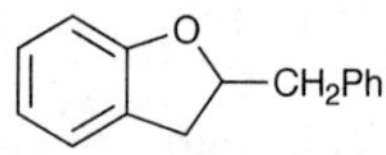

(159)

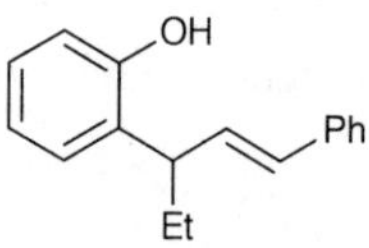

(160)

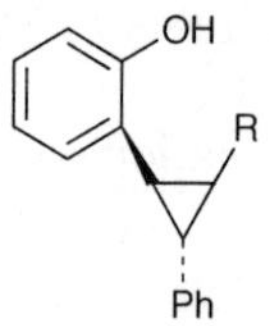

(161) R = ---Et and ◄Et

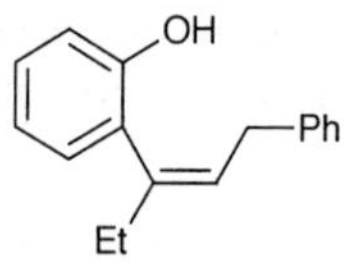

(162)

4-Arylethyl-4-cyanobenzoates undergo Norrish Type II fragmentation on irradiation, but, although 2-(1-naphthyl)ethyl cyanobenzoate exhibits exciplex fluorescence and is unreactive photochemically, the unsubstituted compound (163) undergoes unprecedented photocycloaddition of the ester carbonyl to the naphthalene ring to give isochromenol (164) and (165) products, and shows no exciplex emission.[99] The suggested electron-transfer pathway to (164) and (165) is outlined in Scheme 5. An investigation into the mechanism of the photocyclization of substituted phenylbenzo-1,4-quinones, to give 2-hydroxybenzofuran derivatives (166), has revealed that the reaction proceeds from the $^3n\pi^*$ state to a cyclized intermediate, which in polar or acidified solvents yields (166), but under

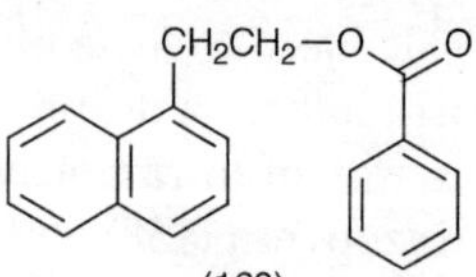

(163)

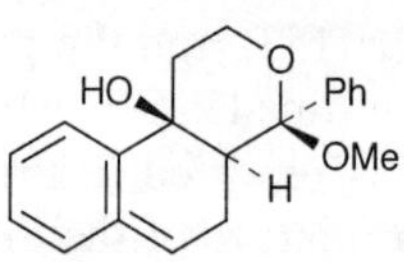

(164)

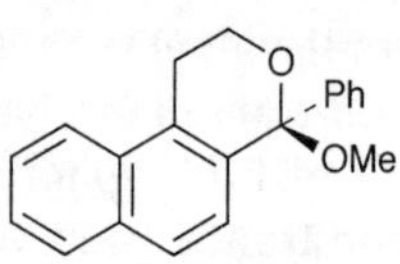

(165)

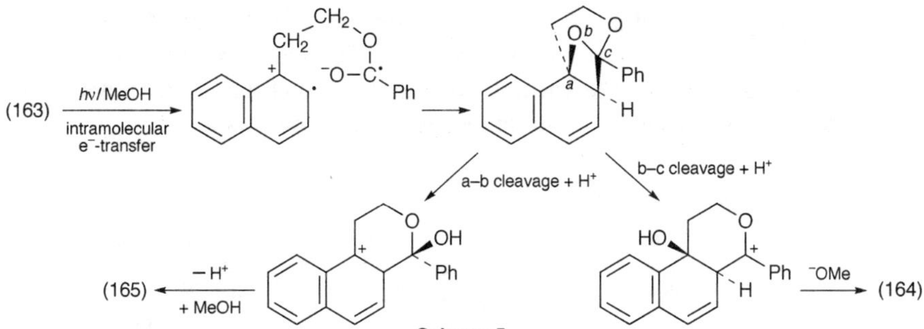

Scheme 5

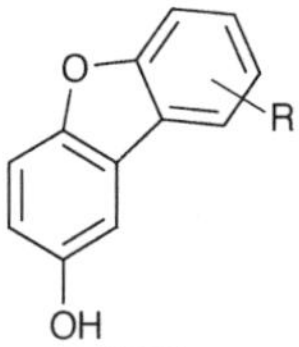

(166)

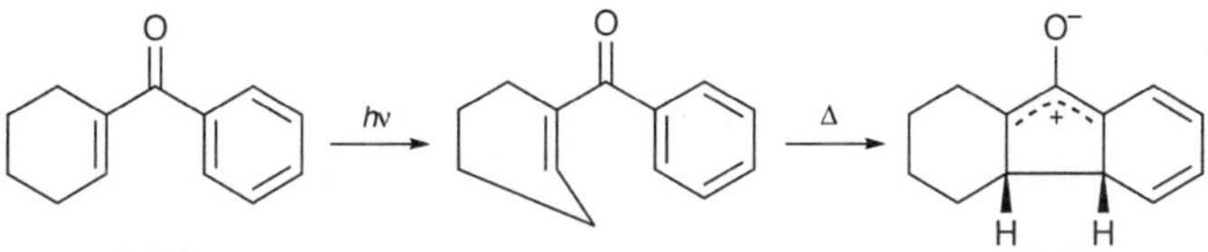

(167) (168)

other conditions cycloreversion to the starting material occurs.[100] The photo-Nazarov cyclization of 1-cyclohexenyl(phenyl)methanone (167) has been re-investigated, and trapping of the 2-oxyallyl intermediate (168), formed on thermal cyclization of the photo-produced *trans* cyclohexene isomer of (167), has been accomplished with various alkenes.[101,102] In the presence of cyclopentadiene, for example, (167) yields (169) and (170), which are photolabile and give the isomers (171) and (172), respectively.

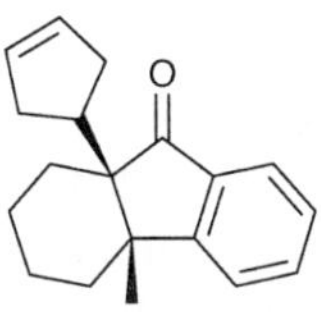

(169)

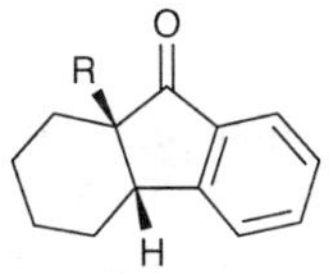

(170) R = 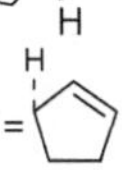 and 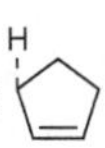

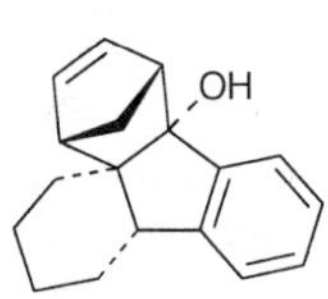

(171)

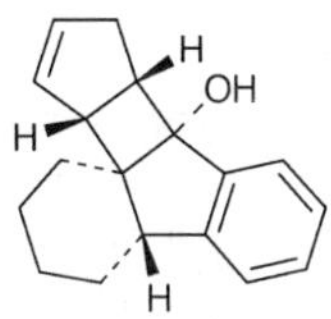

(172)

6 Dimerization Processes

Mechanistic aspects of one of the earliest reported photochemically-induced processes, the photodimerization of anthracene, have been reviewed, and it is evident that this and related systems continue to produce research reports of wide interest and application.[103] The kinetics of the dimerization of anthracene in an inhomogeneous polymer matrix have been studied,[104] and modelling of transient phase halographic gratings of monosubstituted anthracenes in a polymer layer has been carried out.[105] The dimerization of anthracenes (173) substituted with TEMPO radicals has been investigated, and the magnetic properties of the monomers and dimers assessed.[106,107] No photodimerization occurred with (173a and b) but both (173c) and (173d) gave the head-to-tail photodimers

(174). No large magnetic differences were observed in the former monomer-dimer pair, but in the latter case the behaviour varied from the ferromagnetic interactions in (173d) to the variable magnetic interactions in its dimer, which depended on the solvent molecules incorporated in the crystals.[106]

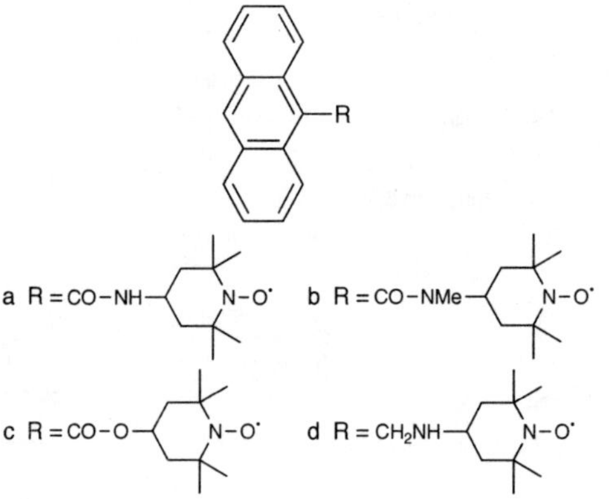

(173)

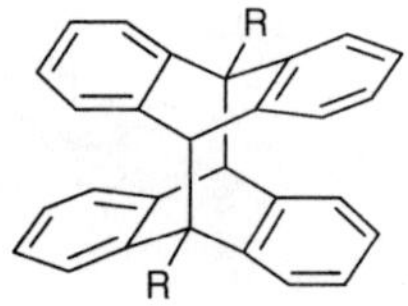

(174)

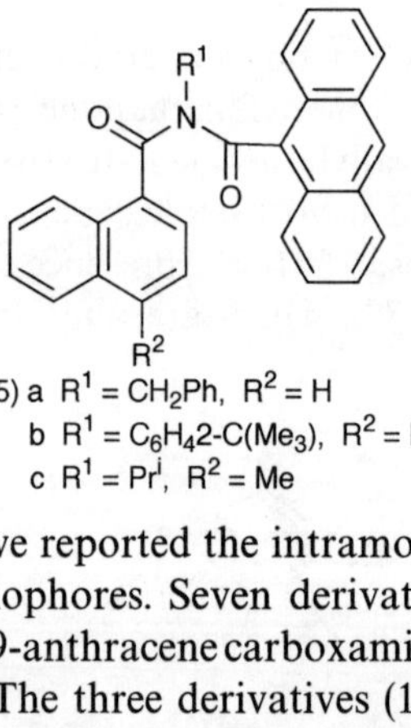

(175) a $R^1 = CH_2Ph$, $R^2 = H$
b $R^1 = C_6H_4$2-$C(Me_3)$, $R^2 = H$
c $R^1 = Pr^i$, $R^2 = Me$

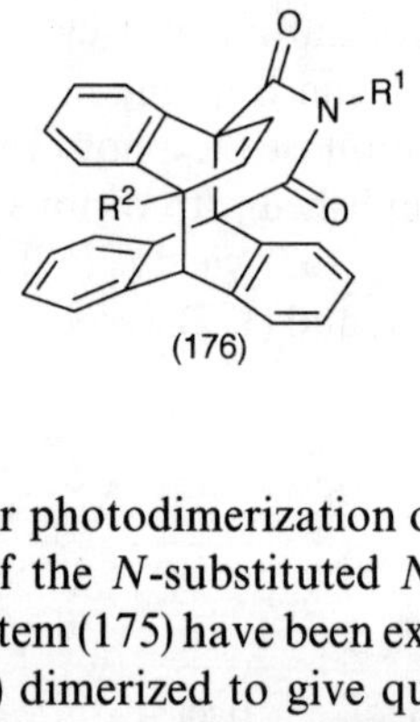

(176)

Two groups have reported the intramolecular photodimerization of anthryl-naphthyl bichromophores. Seven derivatives of the *N*-substituted *N*-(1-naphthalenecarbonyl)-9-anthracene carboxamide system (175) have been examined in the solid state.[108] The three derivatives (175a-c) dimerized to give quantitative yields of (176) and the first two provide the first examples of absolute asymmetric synthesis in ($4\pi + 4\pi$) cycloaddition. The other study was concerned with the photoreactions of the 9-anthryl-1-naphthoyl-poly(ethylene glycol) system (177) in non-polar and polar solvents in the presence of alkali metal ions.[109] Lipophilic interactions in (177) in non-polar solvents, or complexation of the of the polyether chain with the metal ions in polar solvents, bring the two aryl chomophores into close proximity, and intramolecular dimerization occurs. For example, at a concentration of 2×10^{-5} M in cyclohexane, a 100% yield of (178) is obtained, but in tetrahydrofuran no product is observed, although at 1×10^{-3} M in this solvent intermolecular dimerization of the anthracene moieties results.

Pyridines having both electron-donor and -acceptor substituents are photochemically active, and the 2-alkoxy derivatives (179) of nicotinic acid provide the first examples of photodimerization in these systems.[110] The resulting caged dimers (180) are formed in 85-95% yield from benzene solution, and the authors

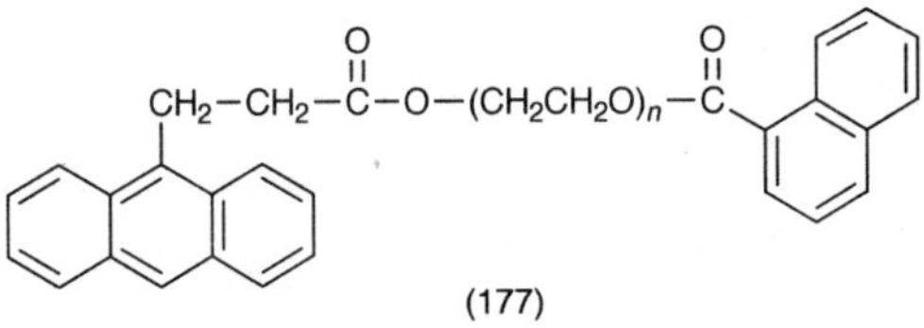

(177)

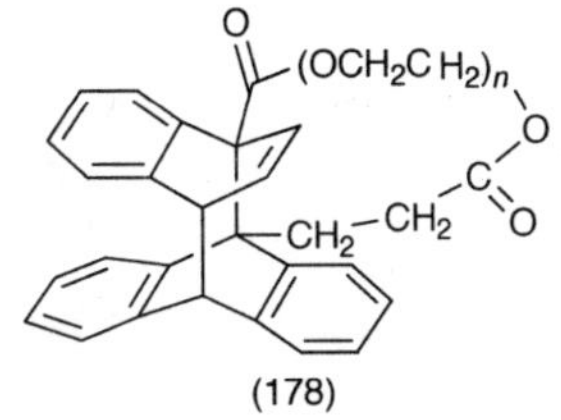

(178)

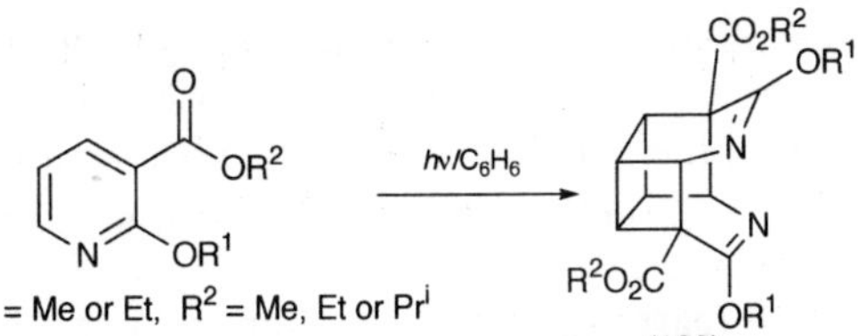

(179) R^1 = Me or Et, R^2 = Me, Et or Pr^i

(180)

suggest that (180) arises from two consecutive $(2\pi + 2\pi)$ photocycloadditions, rather than a $(4\pi + 4\pi)$ process followed by a $(2\pi + 2\pi)$ reaction. Not surprisingly, it is also reported that phenyl-2-methoxynicotinate undergoes a photo-Fries rearrangement.

A novel method for the synthesis of the photopolymer of C_{60} has been described, in which the monomer is recycled by a liquid-phase transport and, for the first time, the $(2\pi + 2\pi)$ photodimer has been extracted from the polymer.[111] Other workers have attempted to characterize spectroscopically the C_{60} polymer produced from solution and report that their results suggest a complex mixture of phases, 'which reduce the molecular symmetry and enrich the Raman and IR spectra'.[112]

7 Lateral Nuclear Shifts

The photo-Fries rearrangement has been used in the synthesis of the diazonamide macrocycle. Thus irradiation of the macrolactam (181) in benzene solution gives good yields of a 2:1 ratio of two atropisomers of the complete core skeleton product (182), with less than 10% of the *para* isomer being formed.[113] It is interesting to note here that preparation of acylated hydroquinones, such as (183) for example, by irradiation of 1,4-quinones in the presence of aldehydes, can be achieved on the 500g scale in 90% yields and in high purity.[114]

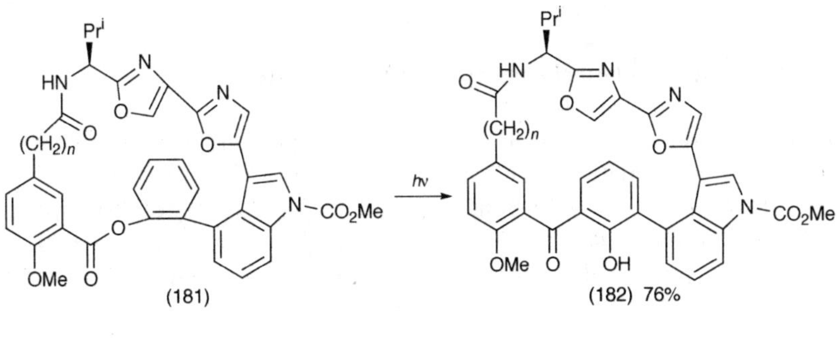

(181)

(182) 76%

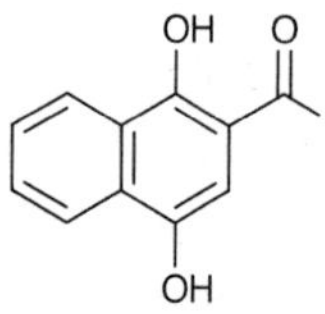

(183)

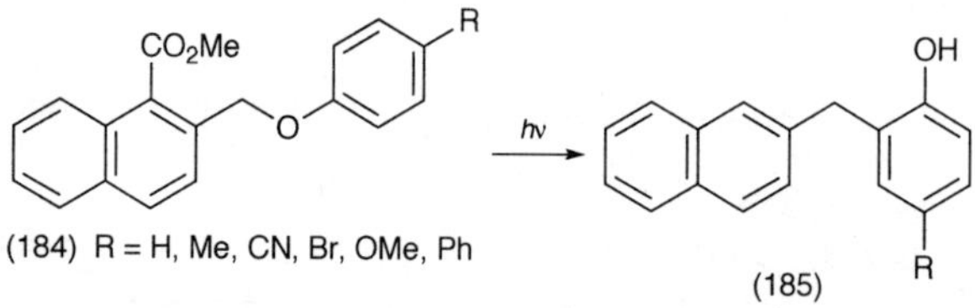

(184) R = H, Me, CN, Br, OMe, Ph

(185)

The intermediate (1-methoxycarbonyl-2-naphthyl)methyl radical has been observed in the photo-Claisen rearrangement of the naphthylmethyl phenyl ethers (184) to (185) by using a two coloured laser flash photolysis technique.[115] Various amounts of the 4-hydroxy substituted isomer are formed, dependent on the nature of the R group, along with phenols and other products from the reactions of phenyl radicals.

The novel photorearrangement of 4-chloromethyl-6,8-dimethylazulene (186) to give 1-chloro-4,6,8-trimethylazulene (187) in deaerated benzene solution is suggested to arise by the mechanism outlined in Scheme 6;[116] and the irradiation of several azoxybenzenes to give the corresponding 2-hydroxyazobenzenes is reported to give higher product yields in the presence of acetic acid, with the relative amounts of, for example, (188) and (189) from 4-methyl azoxybenzene being dependent on the acid concentration.[117]

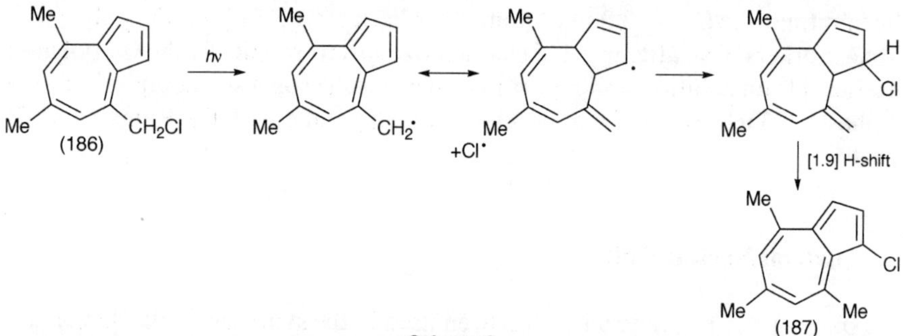

Scheme 6

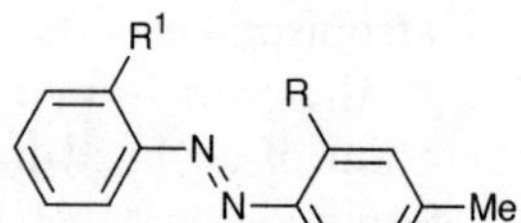

(188) R = OH, R^1 = H
(189) R = H, R^1 = OH

8 Miscellaneous Photochemistry of Aromatic Systems

Quinone methides are commonly cited as intermediates in the photochemistry of arenes which have a methyl or substituted methyl group in the position *ortho* to a functionality (*e.g.* hydroxyl, carbonyl, nitro). A useful summary of these processes with phenols, and of other photochemical methods for forming such intermediates, has been published, with the mechanisms involved and the subsequent chemistry discussed.[118] The authors have also designed biphenyl and terphenyl compounds which can lead to extended methides such as (190). Flash photolysis of the 2-hydroxybenzyl derivatives (191) in aqueous perchloric acid and NaOH solutions, and in acetic acid and phosphate ion buffers, is reported to give the *o*-quinone methide as a short-lived transient, which undergoes hydration to 2-hydroxybenzyl alcohol,[119] and the same group of researchers have described the formation of *o*-quinone-α-phenyl- and -α-(4-anisyl)-methides under similar conditions.[120] The *o*-quinone methides generated on irradiation of aromatic aldehydes and ketones (192) undergo cycloaddition to C_{60}, giving one or both conformers of (193), having pseudoaxial and pseudoequatorial hydroxy groups in ratios which are markedly dependent on the substituents on the benzo and cyclohexene rings.[121]

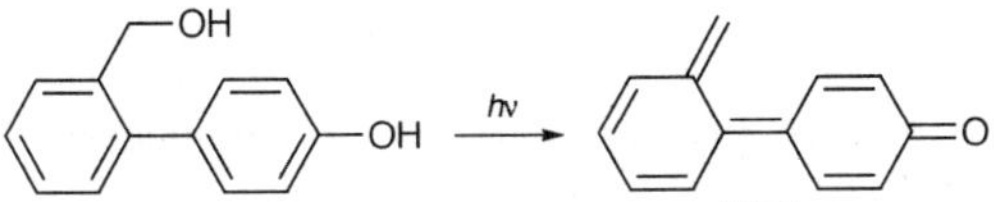

(190)

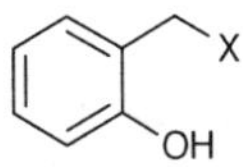

(191) X = OH, $-OC_6H_4$4-CN or $\overset{+}{N}H_3I^-$

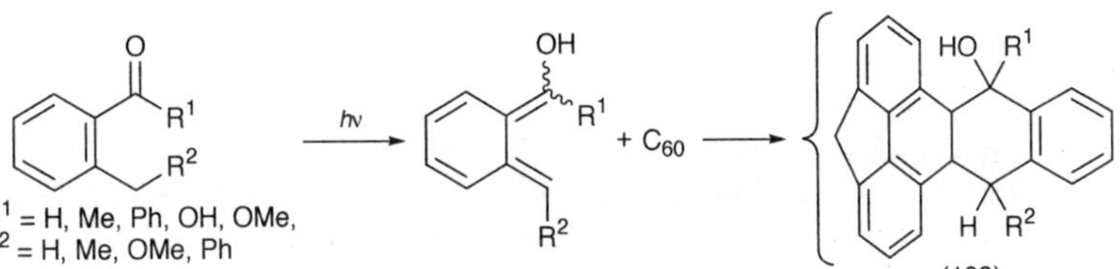

(192) R^1 = H, Me, Ph, OH, OMe, R^2 = H, Me, OMe, Ph

(193)

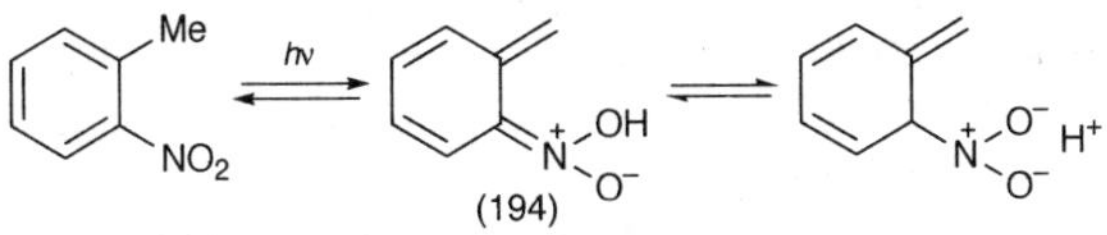

(194)

The thermodynamic and kinetic parameters of the quinonoid *aci*-nitro tautomer (194) of 2-nitrobenzyl compounds have been elucidated from flash photolysis studies, and the pH rate profile for the decay of (194) in aqueous solution has been determined.[122] Other workers have studied this phototautomerism of 2-nitrobenzyl compounds in argon and nitrogen matrices with 254 nm radiation and by DFT calculations.[123] For 2-nitrobenzyl methyl ethers (195) the reaction can be summarized as in Scheme 7. The *o*-quinonoid *aci*-nitro species are the first formed products and undergo secondary photolysis on irradiation in their absorption band (λ_{max} 385-430nm), to give compounds with complex IR spectra which when compared with simulated spectra (DFT calculations) confirms the identity of the reactive species. It is concluded from this study that hydrogen-atom transfer and conformation inversion occur in the excited state.

Biphenylene is renowned for its lack of photoactivity and its weak fluorescence. Irradiation in the gaseous phase of this arene with 193.3 radiation, however, is now reported to result in C-C bond cleavage by absorption of two photons.[124] The diradical (196) arises from a second vibrationally excited 'hot'

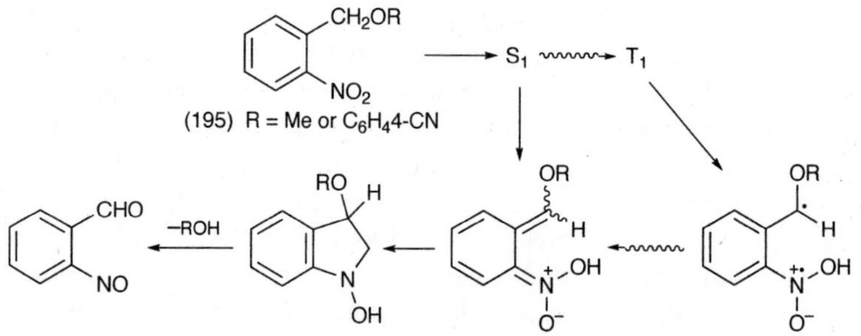

Scheme 7

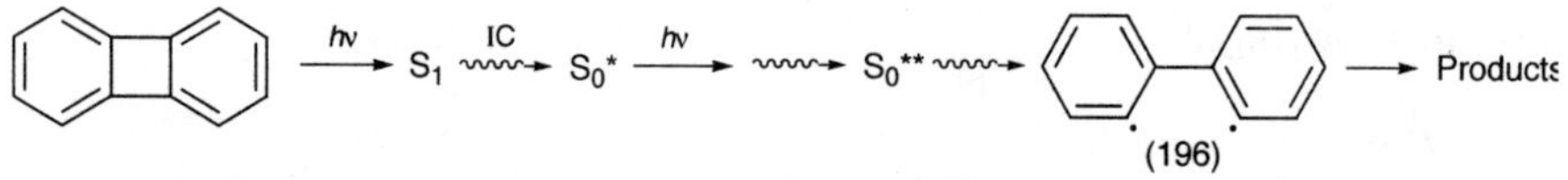

singlet state and undergoes a variety of processes to give phenylacetylene, naphthalene, biphenyl and acenaphthylene as the identified products.

References

1. D. M. Lemal, *Acc. Chem. Res.*, 2001, **34**, 662.
2. S. Tero-Kubota, A. Katsuki and Y. Kobori, *J. Photochem. Photobiol., C*, 2001, **2**, 17.
3. C.-M. Wang, G.-L. Pan, L.-H. Zhao, P. Fan, X.-J. Meng, Y.-F. Ming and M.-G. Fan, *Youji Huaxue*, 2001, **21**, 954.
4. M. D'Auria, *Adv. Heterocycl. Chem.*,2001, **79**, 41.
5. H. E. Zimmerman and E. E. Nesterov, *Acc. Chem. Res.*, 2002, **35**, 77.
6. J. Sivguru, J. R. Scheffer, J. Chandarasekhar and V. Ramamurthy, *J. Chem. Soc., Chem. Commun*, 2002, 830.
7. See for example, J. W. Pavlik, N. Kebede, M. Thompson, A. C. Day and J. A. Barltrop, *J. Amer. Chem. Soc.*, 1999, **121**, 5666.
8. Z. Cao, Q. Zhang, S. D. Peyerimhoff, *Chem.- Eur. J.*, 2001, **7**, 1927.
9. R. Ling, M. Yoshida and P. S. Mariano, *J. Org. Chem.*, 1996, **61**, 4439.
10. H. Lu, Z. Su, L. Song and P. S. Mariano, *J. Org. Chem.*, 2002, **67**, 3525.
11. H. Lu, P. S. Mariano and Y.-F. Lam, *Tetrahedron Lett.*, 2001, **42**, 4755.
12. R. A. King, H. P. Luthi, H. F. Schaefer, F. Glarner and U. Burger, *Chem.- Eur. J.*, 2001, **7**, 1734.
13. M. Boggio-Pasqua, M. J. Bearpark, P. A. Hunt and M. A. Robb, *J. Amer. Chem. Soc.*, 2002, **124**, 1456.
14. J. Ern, M. Petermann, T. Mrozek, J. Daub, K. Kuldova and C. Kryschi, *Chem. Phys.*, 2000, **259**, 331.
15. V. De Waele, U. Schmidhammer, T. Mrozek, J. Daub and E. Riedle, *J. Amer. Chem. Soc.*, 2002, **124**, 2438.
16. L. Gobbi, P. Seiler, F. Diederrich, V. Gramlich, C. Boudon, J.-P. Gisselbrecht and M. Gross, *Helv. Chim. Acta.*, 2001, **84**, 743.
17. J. Cornelisse and R. de Haan, in *Molecular and Supramolecular Photochemistry*, Vol. 8: 'Understanding and Manipulating Excited-State Processes', ed. V. Ramamurthy and K. S. Schanze, Marcel Dekker, New York, 2001, p. 1.
18. J. Cornelisse, *J. Chem. Rev.*, 1993, **93**, 615.

19. S. Bahu, S. A. Fleming, B. Nilsson and T. Turner, *J. Heterocycl. Chem.*, 2001, **38**, 1341.
20. N. Hoffmann and J.-P. Pete, *Synthesis*, 2001, 1236.
21. H. R. Memarian, M. Nasr-Esfahani, H. Gorner and D. Döpp, *Ind. J. Chem. Sect. B: Org. Chem. Incl. Med. Chem.*, 2001, **40B**, 1187.
22. H. R. Memarian, M. Nasr-Esfahani and D. Döpp, *New J. Chem.*, 2001, **25**, 1605.
23. D. Döpp, S. K. Mohamed and A. El-Khawaga, *Helv. Chim. Acta.*, 2001, **84**, 3673.
24. D. Mangion, M. Frizzle, D. R. Arnold and T. S. Cameron, *Synthesis*, 2001, 1215.
25. Y. Yoshimi, S.-I. Konishi, H. Maeda and K. Mizuno, *Tetrahedron Lett.*, 2001, **42**, 3475; Y. Yoshimi, S.-I. Konishi, H. Maeda and K. Mizuno, *Synthesis*, 2001, 1197.
26. S. Kohmoto, Y. Miyaji, M. Tsuruoka, K. Kishikawa, M. Yamamoto and K. Yamada, *J. Chem. Soc., Perkin Trans. 1*, 2001, 2082.
27. K. Chiyonobu, G.-I. Konishi, Y. Inoue and K. Mizuno, *J. Chem. Res., S*, 2001, 135.
28. K. I. Booker-Milburn, P. M. Wood, R. F. Dainty, M. W. Urquhart, A. J. White, H. J. Lyon and J. P. H. Charmant, *Org. Lett.*, 2002, **4**, 1487.
29. S. Jones and J. C. C. Atherton, *Tetrahedron Asymm.*, 2001, **12**, 1117.
30. J. C. C. Atherton and S. Jones, *Tetrahedron Lett.*, 2001, **42**, 8239.
31. T. Noh, H. Yu, Y. Jeong, K. Jeon and S. Kang, *J. Chem. Soc., Perkin Trans. 1*, 2001, 1066.
32. H. Maeda, Y. Miyata and K. Mizuno, *Tetrahedron Lett.*, 2002, **43** 1481.
33. L. Meunier, J.-F. Pilichowski and P. Boule, *Canad. J. Chem.*, 2001, **79**, 1179.
34. J.-P. Aguer, F. Blachere, P. Boule, S. Garaudee and C. Guillard, *Int. J. Photoenergy*, 2000, **2**, 81.
35. P. Coppo, M. Fagnoni and A. Albini, *Tetrahedron Lett.*, 2001, **42** 4271.
36. P. Klan, D. Del Favero, A. Ansorgova, J. Klanova and I. Holoubek, *Environ. Sci. Pollut. Res. Int.*, 2001, **8** 195.
37. S. Rayne, P. Wan, M. G. Ikonomou and A. D. Kostantinov, *Environ. Sci. Tech.*, 2002, **36**, 1995.
38. L. Latterini, F. Elisei, G. G. Aloisi and M. D'Auria, *Phys. Chem. Chem. Phys.*, 2001, **3** 2765.
39. A. I. Shipilov and L. A. Kolpashikova, *Russ. J. Appl. Chem.*, 2001, **74**, 274.
40. D. Vione, V. Maurino, C. Minero, M. Vincenti and E. Pelizzetti, *Chemosphere*, 2001, **44**, 237.
41. M. Tajima, K. Kato, K. Matsunaga and H. Inoue, *J. Photochem. Photobiol., A*, 2001, **140**, 127.
42. F. Kitagawa, M. Murase and N. Kitamura, *Chem. Lett.*, 2001, 786.
43. I. Marchueta, S. Olivella, L. Sola, A. Moyano, M. A. Pericas and A. Riera, *Org. Lett.*, 2001, **3**, 3197.
44. B. Heller, B. Sundermann, H. Buschmann, H.-J. Drexler, J. You, U. Holzgrabe, E. Heller and G. Oehme, *J. Org. Chem.*, 2002, **67**, 4414.
45. F. D. Lewis and R. S. Kalgutkar, *J. Phys. Chem., A*, 2001. **105**, 285.
46. F. D. Lewis, T. L. Kurth and R. S. Kalgutkar, *J. Chem. Soc., Chem. Commun.*, 2001, 1372.
47. J. K. F. Geirsson and A. Kvaran, *J. Photochem. Photobiol., A*, 2001, **144**, 175.
48. M. Takeshita and T. Yamato, *Tetrahedron Lett.*, 2001, **42**, 4345.
49. K. Masutani, R. Irie and T. Katsuki, *Chem. Lett.*, 2002, 36.
50. F. Ortica, D. Levi, P. Brun, R. Guglielmetti, G. Favaro and U. Mazzucato, *Internat. J. Photoenergy*, 2001, **3**, 153.
51. F. B. Mallory, K. E. Butler, A. Berube, E. D. Luzik and C. Mallory, *Tetrahedron*, 2001, **57**, 3715.

52. F. D. Lewis, R. S. Kalgutkar and J. S. Yang, *J. Amer. Chem. Soc.*, 2001, **123**, 3878.
53. D. C. Harrowven, M. I. T. Nunn, N. J. Blumire and D. R. Fenwick, *Tetrahedron*, 2001, **57**, 4447.
54. A. Visnjevac, N. Basaric, B. Kojic-Prodic and M. Sindler-Kulyk, *Acta Crystallogr. Sect. E: Struct. Rep. Online*, 2001, **57**, o1252.
55. N. Basaric, Z. Marinic and M. Sindler-Kulyk, *Tetrahedron Lett.*, 2001, **42**, 3641.
56. I. Skoric, N. Basaric, Z. Marinic and M. Sindler-Kulyk, *Heterocycles*, 2001, **55**, 1889.
57. A. J. Myles and N. R. Branda, *Adv. Func. Materials*, 2002, **12**, 167.
58. M. Irie and S. Kobatake, *Nippon Kessho Gakkaishi*, 2002, **44**, 61.
59. M. M. Krayushkin, L. G. Vorontsova and B. M. Uzhinov. *Int. J. Photoenergy*, 2001, **3**, 25.
60. T. Yamada, K. Muto, S. Kobatake and M. Irie, *I. Org. Chem.*, 2001, **66**, 6164.
61. T. Yamada, S. Kobatake and M. Irie, *Bull. Chem. Soc. Japan*, 2002, **75**, 167.
62. K. Shibata, K. Muto, S. Kobatake and M. Irie, *J. Phys. Chem. A*, 2002, **106**, 209.
63. K. Shibata, S. Kobatake and M. Irie, *Chem. Lett.*, 2001, 618.
64. K. Morimitsu, K. Shibata, S. Kobatake and M. Irie, *J. Org. Chem.*, 2002, **67**, 4574.
65. J. Ern, A. T. Bens, H. D. Martin, K. Kuldova, H. P. Trommsdorff and C. Kryschi, *J. Phys. Chem., A*, 2002, **106**, 1654.
66. L. Giordano, T. M. Jovin, M. Irie and E. A. Jares-Erijman, *J. Amer. Chem. Soc.*, 2002, **124**, 7481.
67. T. B. Norsten, A. Peters, R. McDonald, M. Wang and N. R. Branda, *J. Amer. Chem. Soc.*, 2001, **123**, 7447.
68. T. Yamaguchi, Y. Tanaka, H. Nakazumi and M. Irie, *Shikizai Kyokaishi*, 2001, **74**, 8.
69. K. Matsuda and M. Irie, *Chem.- Eur. J.*, 2001, **7** 3466.
70. K. Matsuda and M. Irie, *Kyushu Daigaku Chuo Bunseki Senta Hokoku*, 2001, **19**, 9.
71. M. M. Krayushkin, F. M. Stoyanovich, O. Y. Zolotarskaya, I. V. Murav'ev, A. Y. Martynkin, L. G. Vorontsova, Z. A. Starikova, V. L. Ivanov and B. M. Uzhinov, *Russ. Chem. Bull.*, 2001, **50**, 110.
72. A. Osuka, D. Fujikane, H. Shinmori, S. Kobatake and M. Irie, *J. Org. Chem.*, 2001, **66**, 3913.
73. T. Kaieda, S. Kobatake, H. Miyasaka, M. Murakami, N. Iwai, Y. Nagata, A, Itaya and M. Irie, *J. Amer. Chem. Soc.*, 2002, **124**, 2015.
74. K. Uchida, A. Takata, S. Nakamura and M. Irie, *Chem. Lett.*, 2002, 476.
75. J. Chauvin, T. Kawai and M. Irie, *Jpn. J. Appl. Phys., Part 1*, 2001, **40**, 2518.
76. H. Guo, F. Zhang, Y. Tang, F. Zhao and G. Wu, *Proc. SPIE-Int. Soc. Opt. Eng.*, 2001, 158.
77. H. Tu and H. Tian, *Proc. SPIE-Int. Soc. Opt. Eng.*, 2001, 171.
78. K. Kasatani, *ITE Lett. Batteries, New Technol. Med.*, 2001, **2**, 220.
79. M. M. Krayushkin, A. Y. Martynkin and N. D. Chuvylkin, *Russ. Chem. Bull.*, 2001, **50**, 381.
80. K. Uchida, D. Guillaumont, E. Tsuchida, G. Mochizuki, M. Irie, A. Murakami and S. Nakamura, *THEOCHEM.*, 2002, **579**, 115.
81. K. Kasatani and T. Fujiwaki, *ITE Lett. Batteries, New Technol. Med.*, 2001, **2**, 215.
82. H. Tian, B. Chen and P.-H. Liu, *Chem. Lett.*, 2001, 990.
83. M. M. Krayushkin, S. N. Ivanov, A. Y. Martynkin, B. V. Lichitsky, A. A. Dudinov and B. M. Uzhinov, *Russ. Chem. Bull.*, 2001, **50**, 116.
84. X. Deng and L. S. Liebeskind, *J. Amer. Chem. Soc.*, 2001, **123**, 7703.
85. G. Bartocci, G. Galiazzo, G. Gennari, E. Marri, Mazzucato and A. Spalletti, *Chem.*

Phys., 2001, **272**, 213.
86. S. Matsumoto, S. Takahashi and K. Ogura, *Heteroat. Chem.*, 2001, **12**, 385.
87. T. Okuyama, Y. Yokoyama and Y. Yokoyama, *Bull. Chem. Soc. Japan*, 2001, **74**, 2181.
88. C. Lafond, O. Pouraghajani, A. Tork, M. Bolte, A.-M. R. Ritcey and R. A. Lessard, *Proc. SPIE-Int. Soc. Opt. Eng.*, 2001, 226.
89. P. J. Campos, M. Caro and M. A. Rodriguez, *Tetrahedron Lett.*, 2001, **42**, 3575.
90. B. R. Kaafarani and D. C. Neckers, *Tetrahedron Lett.*, 2001, **42**, 4099.
91. J.-H. Ho and T.-I. Ho, *J. Chem. Soc., Chem. Commun.*, 2002, 270.
92. J.-K. Luo, M.-P. Cabal, R. F. Federspiel and R. N. Castle, *J. Heterocycl. Chem.*, 2001, **38**, 137.
93. T. A. Grese, M. D. Adrian, D. L. Phillips, P. K. Shetler, L. L. Short, A. L. Glasebrook and H. U. Bryant, *J. Med. Chem.*, 2001, **44**, 2857.
94. P. Formentin, M. J. Sabater, M. N. Chretien, H. Garcia and J. C. Scaiano, *J. Chem. Soc., Perkin Trans. 2*, 2002, 164.
95. T. Horaguchi, N. Hosokawa, K. Tanemura and T. Suzuki, *J. Heterocycl. Chem.*, 2002, **39**, 61.
96. K. Maekawa, T. Igarashi, K. Kubo and T. Sakurai, *Tetrahedron*, 2001, **57**, 5515.
97. C. M. Jimenez, M. A. Miranda and R. Tormos, *Tetrahedron*, 2002, **58**, 115.
98. C. M. Jimenez, M. A. Miranda, R. Tormos and S. Gil, *Eur. J. Org. Chem.*, 2002, 297.
99. K. Morley and J. A. Pincock, *J. Org. Chem.*, 2001, **66**, 2995.
100. C. Belin, S. Bearnais-Barbry and R. Bonneau, *J. Photochem. Photobiol., A*, 2001, **139**, 111.
101. J. Leitich, I. Heise, J. Rust and K. Schaffner, *Eur. J. Org. Chem.*, 2001, 2719.
102. J. Leitich, I. Heise and K. Schaffner, *J. Photochem. Photobiol., A*, 2001, **140**, 207.
103. H. Bouas-Laurent, J.-P. Desvergne, A. Castellan and R. Lapouyade, *Chem. Soc. Rev.*, 2001, **30**, 248.
104. Y. M. Lopatkin, P. A. Kondratenko and I. P. Zharkov, *Theor. Exp. Chem.*, 2001, **37**, 84
105. Y. V. Gritsai and V. V. Mogil'nyi, *Opt. Spectrosc.*, 2001, **90**, 902.
106. S. Nakatsuji, T. Ojima, H. Akutsu and J. Yamada, *J. Org. Chem.*, 2002, **67**, 916.
107. T. Ojima, H. Akutsu, J.-I. Yamada and S. Nakatsuji, *Polyhedron*, 2001, **20**, 1335.
108. S. Kohmoto, Y. Ono, H. Masu, K. Yamaguchi, K. Kishikawa and M. Yamamoto, *Org. Lett.*, 2001, **3**, 4153.
109. M. Xu, L.-Z. Wu, L.-P. Zhang and C.-H. Tung, *Tetrahedron Lett.*, 2001, **42**, 9249.
110. M. Sakamoto, T. Yagi, S. Fujita, T. Mino, T. Karatsu and T. Fujita, *J. Org. Chem.*, 2002, **67**, 1843.
111. S. Pekker, K. Kamaras, E. Kovats, T. Pusztai and G. Oszlanyi, *Synth. Met.*, 2001, **121**, 1109.
112. G. Chambers, K. Henderson, A. B. Dalton, B. McCarthy and H. J. Byrne, *Synth. Met.*, 2001, **121**, 1111.
113. P. Magnus and C. Lescop, *Tetrahedron Lett.*, 2001, **42**, 7193.
114. C. Schiel, M. Oelgemoller and J. Mattay, *Synthesis*, 2001, 1275.
115. Y. Yoshimi, K. Mizuno, H. Maeda, N. Ichinose, T. Tanaka and S. I. Kawanishi, *Chem. Lett.*, 2001, 252.
116. C. A. A. Ruiz, V. G. Toscano, D. de B. Rezende and W. J. Baader, *Tetrahedron Lett.*, 2001, **42**, 2613.
117. J. Yamamoto, R. Hamada and T. Tsuboi, *Nippon Kagaku Kaishi*, 2002, 339.
118. P. Wan, D. W. Brousmiche, C. Z. Chen, J. Cole, M. Lukeman and M. Xu, *Pure Appl. Chem.*, 2001, **73**, 529.

119. Y. Chiang, A. J. Kresge and Y. Zhu, *J. Amer. Chem. Soc.*, 2001, **123**, 8089.
120. Y. Chiang, A. J. Kresge and Y. Zhu, *J. Amer. Chem. Soc.*, 2002, **124**, 717.
121. Y. Nakamura, K. Okawa, S. Minami, T. Ogawa, S. Tobita and J. Nishimura, *J. Org. Chem.*, 2002, **67**, 1247.
122. M. Schworer and J. Wirz, *Helv. Chim. Acta.*, 2001, **84**, 1441.
123. I. R. Dunkin, J. Gębicki, M. Kiszka and D. Sanín-Leira, *J. Chem. Soc., Perkin Trans. 2*, 2001, 1414.
124. T. Yatsuhashi, T. Akiho and N. Nakashima, *J. Amer. Chem. Soc.*, 2001, **123**, 10137.

5
Photo-reduction and -oxidation

BY ANDREW GILBERT

1 Introduction

This review has the format adopted in recent years for photo-reduction and -oxidation processes. Many reactions in these areas, previously described in homogeneous systems, are currently being examined under heterogeneous conditions at surfaces and in cavities, and the publications of these studies are considered here within the relevant reaction-type sections.

2 Reduction of the Carbonyl Group

Benzophenone has been subjected to two-step two-laser excitation which gives access to higher triplet states and the radical cation.[1] Both types of species can be observed under time-resolved spectroscopy, and the former have been shown to undergo intermolecular hydrogen abstraction from propan-2-ols more rapidly than the lowest triplet state. The ketyl radicals formed by sunlight irradiation (10 – 12 days) from the benzophenone derivatives (1) in the presence of acetic acid, in propan-2-ol or diethyl ether solution, yield the corresponding pinacol dimers in a process suggested to form a self-recognition molecular assembly.[2] Similar photodimerization of crystalline 4,4′-dimethylbenzophenone, by way of intermolecular hydrogen abstraction followed by Baeyer-Villiger oxidation of the pinacol, is reported to yield nanostructures.[3] The use of atomic force microscopy in this study has revealed that volcanoes are formed on the crystal face (001) during the photoreduction-dimerization process, and that craters and volcanoes result on the same face during the subsequent oxidation reaction. The reaction kinetics of the radical pairs formed from photoinduced hydrogen abstraction of benzophenone derivatives in micellar solution have been studied by pulsed microwave irradiation.[4] This technique allows the evaluation of all the significant rate constants and, in particular, provides the value for that of the recombination of the ketyl and alkyl radicals (*e.g.* $1.90 \times 10^7 s^{-1}$ for benzophenone) which is difficult to determine accurately by other methods. Aggregation of bile salts such as sodium cholate (2) yields supramolecular systems, the complexity of which has been probed by studying the kinetics of the T_1 states and also of the derived radicals from irradiation of benzophenone and its 4,4′-dimethyl deriva-

Photochemistry, Volume 34

tive in their presence.[5] Production of the ketyl radicals occurs at the primary sites of the aggregates at low concentrations of (2), but at higher concentrations secondary aggregates are present and the hydrogen abstraction by the T_1 ketone is too slow at the binding site to compete with the exit process. In this case, the triplet lifetimes are shortened as a result of an efficient self-quenching process. From such studies, it is deduced that 8 to 13 molecules of (2) are required to define primary and secondary binding sites. The primary reaction from irradiation of naphtho-1,4-quinones included in α-, β- or γ-cyclodextrins is deduced to be hydrogen abstraction from the host framework to give the quinone neutral radicals, which are ejected from the cavity into the aqueous phase and some of which lose a proton to give the radical cation.[6]

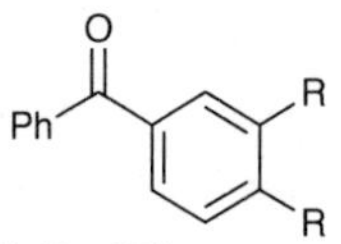

(1) R = OMe,
R–R = –O(CH$_2$)$_2$O–

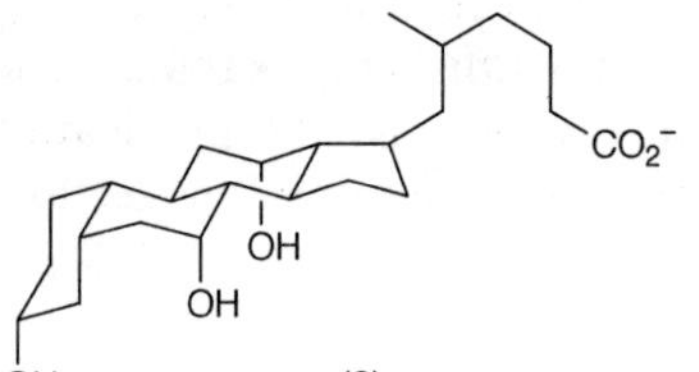

(2)

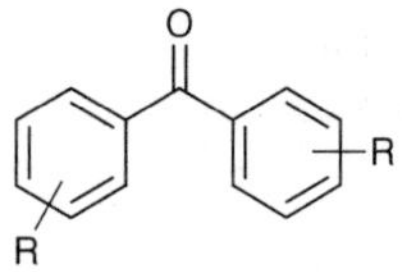

(3) R = H, 4-Cl, 4-OH, 4-OMe, 2-CO$_2$Me, 3-CO$_2$Me

Many studies into the photoreduction of carbonyl compounds are carried out under electron-transfer conditions, and several such systems have been reported during the review period. The photochemistry of acetone has been investigated in the presence of 13 *prim*-, *sec*- and *tert*-aliphatic amines of varying structure.[7] It is deduced from the kinetic data of these systems, supported by product studies, that the T_1 acetone reacts essentially solely by direct hydrogen abstraction, whereas charge-transfer complex and exciplex formation are important routes for the S_1 state. The photoreduction of the 1,2-diketones, camphorquinone and 1-phenylpropa-1,2-dione, catalysed by TiO_2 in alcohol media, is reported to give the corresponding α-hydroxyketones in reasonable yields, which are markedly improved in the presence of water or triethylamine.[8] Studies into the photodegradation of anthraquinone disperse dyes in acetone solution have shown that the process arises by reduction and is accelerated in the presence of triethylamine;[9] and CIDNP and ESR techniques have been used to establish the occurrence of photoinduced hydrogen abstraction in a porphyrin-quinone system in a protic solvent.[10] Acetonitrile solutions of benzo-1,4-quinone and its chlorinated derivatives in the presence of 4,4′dimethoxydiphenylmethane is reported to yield the corresponding radical ions from the triplet state with good quantum efficiency (≥0.72).[11] Back electron transfer dominates in all cases, but both proton transfer to give the radical pair and hydrogen-atom transfer yielding ions also occur; the latter two processes compete significantly more effectively with the back electron transfer when salts are added to the irradiated systems. Two groups report on the influence of room-temperature ionic liquids[12] on the photoreduction of aryl ketones. Dunkin and co-workers have shown that the activation energy for hydrogen abstraction by T_1 benzophenone from imidazolium-based ionic liquids is significantly higher than from conventional solvents and also note that the influence of such liquids on processes involving low polar reactants and transition states may be appreciable and have a subtle origin.[13] Other workers have studied the amine-mediated photoreduction of the benzophenones (3) in ionic liquids and report that, in contrast to the corresponding reactions in conventional solvents, the photoreduction yields mainly the benzhydrol in most cases.[14] It is suggested in these systems, that the reduction process giving both the benzhydrol and benzpinacol derivatives, could be accounted for by an equilibrium between the radical pair (4) and the ion pair (5), in which the benzhydrol results from the latter species.

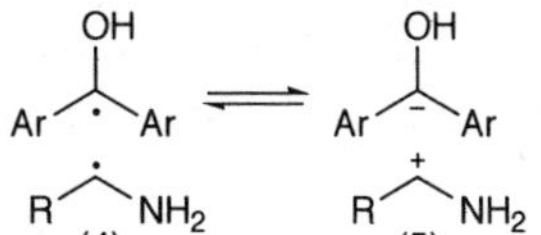

(4) (5)

Coupling of the radicals produced in the photochemical reduction of carbonyl compounds can have synthetic utility. Thus Brule and Hoffmann have shown that ketyl radicals generated in the amine-mediated reduction of aromatic ketones undergo stereoselective addition to (5*R*)-5-(−)-menthyloxy-2[5*H*] furanone (6) to give (7),[15] and Chinese workers have isolated the steroid-quinone coupled compound (8) from irradiation of Δ^5-steroids with benzo-1,4-quinone and its derivatives.[16] In the latter case, 7-hydroxy-derivatives of the Δ^5-steroids are also formed, and in both systems an electron-transfer, proton-transfer, radical-combination sequence is proposed for product formation.

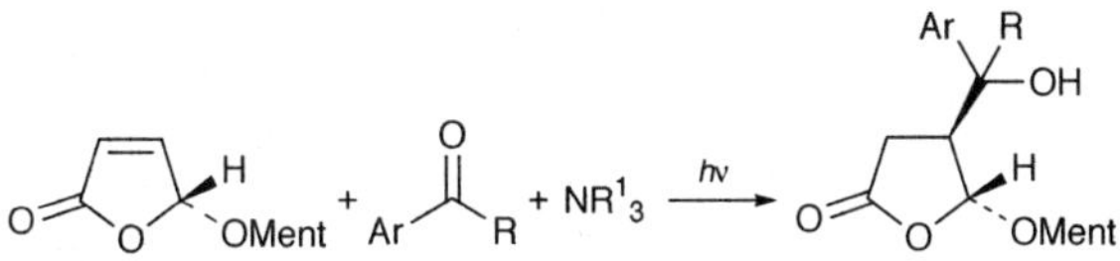

(6) (7)

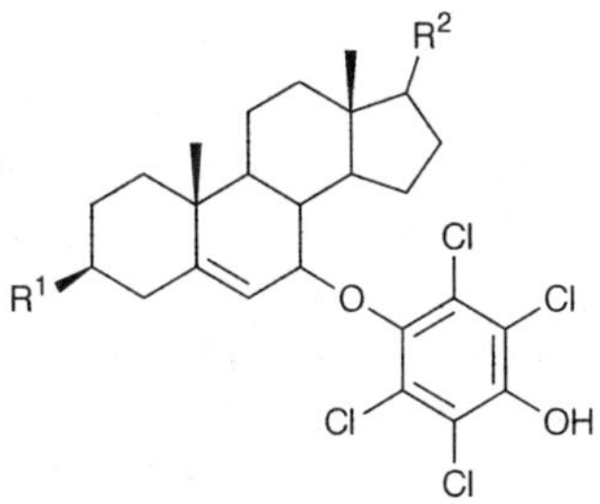

(8) R^1 = OH or OMe, $R^2 = C_8H_{17}$
R^1 = OH, R^2 = O=

Intramolecular hydrogen abstraction is common for carbonyl compounds in a wide range of molecular structures. Such reactions which result in the Norrish Type II process are reviewed in Part II, Chapter 1 of this Report: the present section considers other intramolecular carbonyl reductions. An investigation into the triplet state reactivity of 5-hydroxyflavone (9) has been reported and uses, for the first time, transient absorption and phosphorescence spectroscopies to examine the intramolecular hydrogen-atom transfer yielding the tautomer (10).[17] Kim and co-workers have studied the photochemical reactions of the two structurally analogous aryl ketones 1-(*o*-tolyl)-1-benzoylcyclopropane (11) and 2-(*o*-tolyl)-2-benzoyloxirane (12) and report markedly different behaviours for the two compounds.[18] Thus (11) gave solely the cyclized product (13), and (12) afforded a number of compounds of which only (14) was isolated in sufficient purity to allow its identification. Both products may be considered to arise by initial γ-hydrogen abstraction, but, while the 1,5-diradical from (11) cyclizes efficiently, that from (12) initiates opening of the oxirane to give (15) probably because of the increased acidity and instability compared to the cyclopropane.

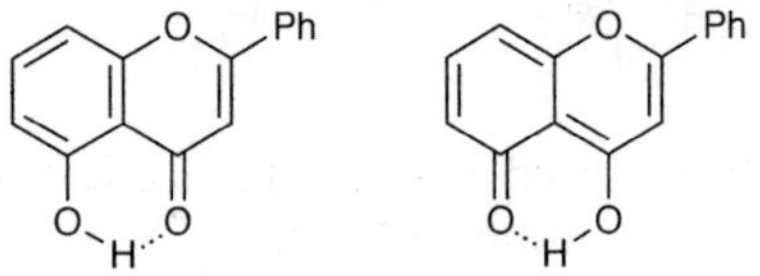

(9) (10)

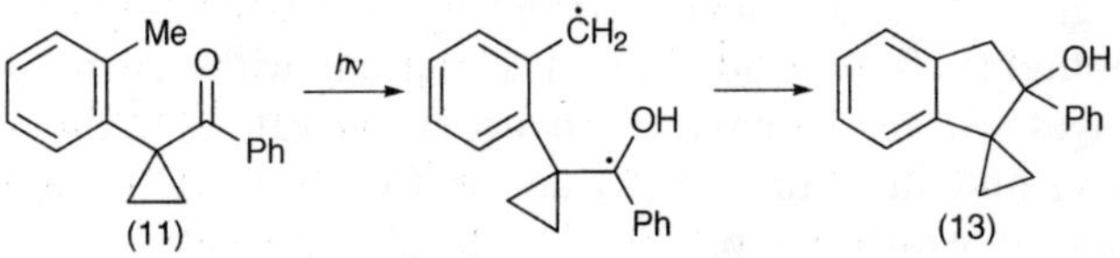

(11) (13)

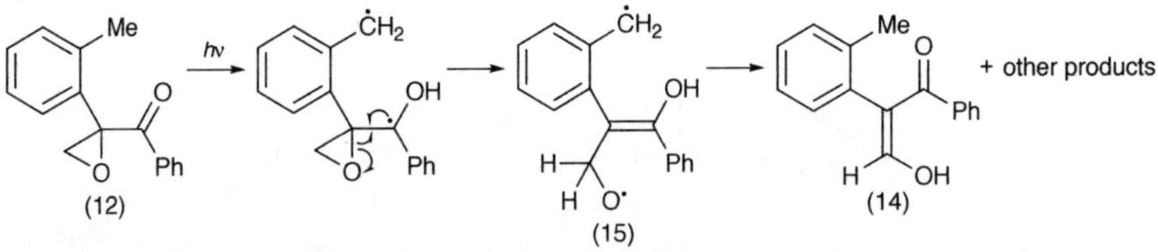

(12) (15) (14)

There have been several examples reported within the year in which intramolecular cyclization of the 1,5-diradical formed following photoinduced δ-hydrogen abstraction has led to structurally complex or synthetically useful compounds. Such reaction of both *o*-alkoxybenzophenone moieties in the macrocycles (16) gives the cyclophanes (17) having tethered benzofuran rings, in yields of 20 to 58%.[19] By a similar route, irradiation of a benzene solution of the macrocycles (18) at 350 nm induces a quadruple photocyclization, which affords the dihydrofuranol ring system in (19).[20] By use of the chiral host (20), Bach and co-workers have succeeded in photocyclizing the prochiral imidazolidinone (21) to give the 1,3-diazabicyclo[3.3.0]octanone with enantiomeric enrichment of some 26% and in good chemical yield.[21] There is a marked preference for formation of the *exo* diastereomers (22) and (23) in toluene solution, but in *tert*-butanol the *endo* isomers (24) and (25) are favoured. From an investigation of the related procedure with α-mesyloxy-β-ketoamides (26), it is reported that not only are the 3,4-dihydro-2*H*-1,3-oxazin-4-ones (27) formed regioselectively,

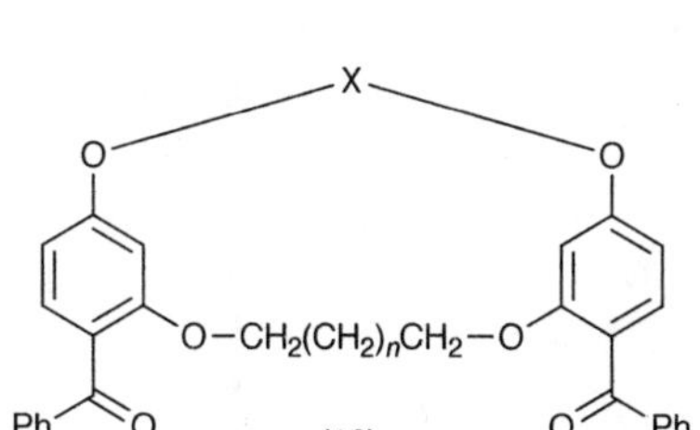

(16)

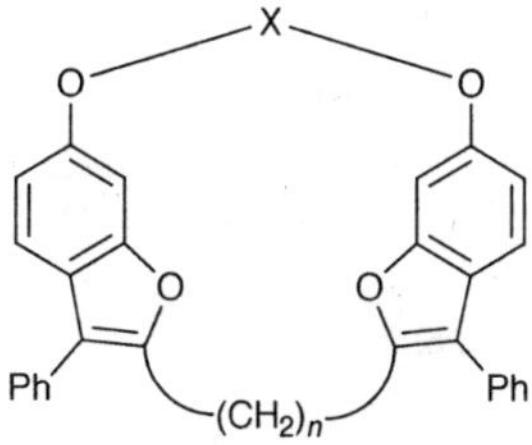

(17)

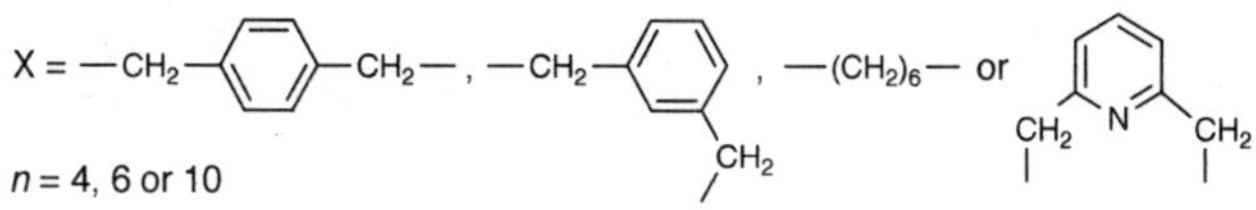

n = 4, 6 or 10

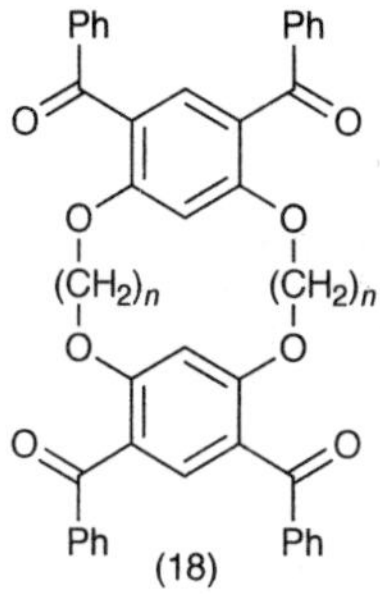

(18)

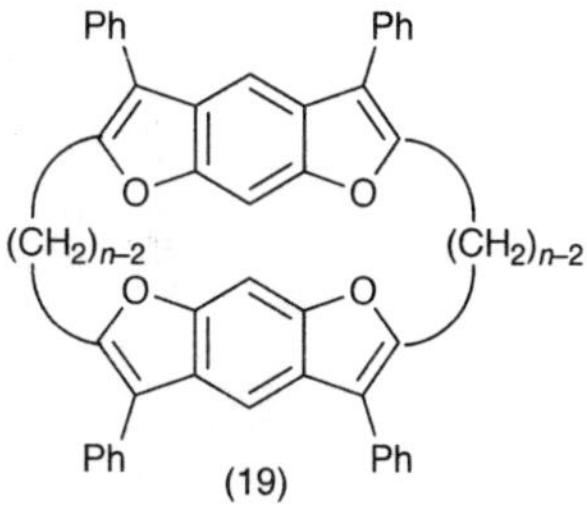

(19)

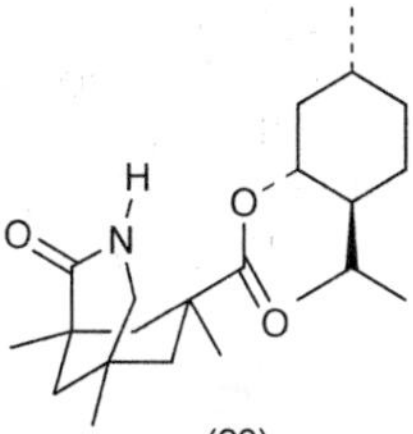

(20)

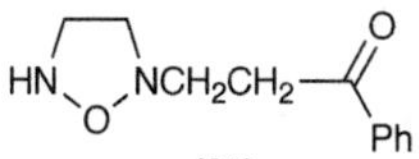

(21)

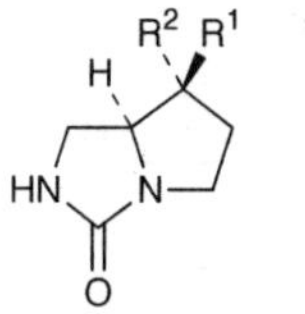

(22) R^1 = OH, R^2 = Ph
(24) R^1 = Ph, R^2 = OH

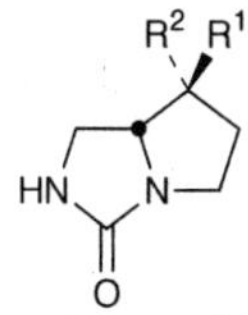

(23) R^1 = Ph, R^2 = OH
(25) R^1 = OH, R^2 = Ph

but also this reaction provides the first example of a C-O bond formation in a Norrish-Yang cyclization; the mechanism is summarized in Scheme 1.[22] It is well documented that 1,4-diradicals produced photochemically from 2-alkylacetophenones and related compounds yield *o*-quinomethanes, which can be trapped by dienophiles. The intramolecular variant of this process has been used in a synthetic strategy leading towards the hamigerans such as (28); in this application, irradiation of (29) gives (30) in 91% yield.[23]

3 Reduction of Nitrogen-containing Compounds

The photoreduction of methyl viologen (MV^{2+}) in Zeolite X exchanged with alkali metal ions has been described in detail.[24] Excitation of this system with 320 nm radiation results in the formation of the cation radical ($MV^{+\cdot}$) in yields

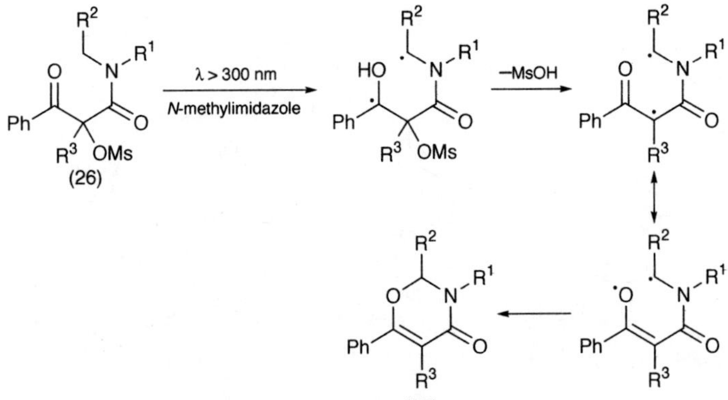

(26) (27)

Scheme 1

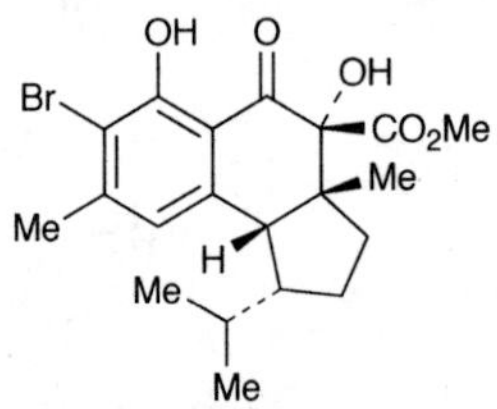

(28)

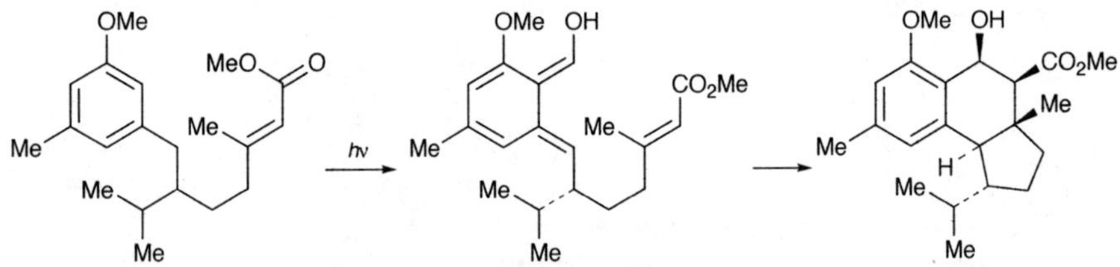

(29) (30)

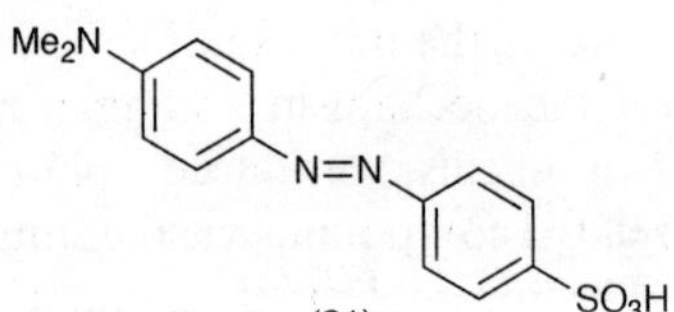

(31)

dependent on the nature of the alkali metal ion exchanged into the zeolite framework, from which it is deduced that the anionic aluminosilicate of the host is the electron donor. Lee and co-workers have investigated the photoreduction of Methyl Orange (31) catalysed by a second dye (Nile Red) adsorbed on TiO_2-Y zeolites.[25] Irradiations were carried out with wavelengths longer than 320 nm, and the photocatalytic effect of the Nile Red was evidenced by the eight-fold increase in reaction compared to the TiO_2-Y zeolite alone. Photoreduction has also been identified as the initial major pathway in the degradation of Disperse

Red (a chlorinated azo dye); this is reported to be followed by cleavage of the azo chromophore.[26]

Broadband irradiation of nitroarenes in the presence of *N,N*-dimethylaminobenzylidene malonic acid derivatives is reported both to reduce the nitro group and to result in demethylation of the amine.[27] The photoreduction of acetonitrile and acetonitrile-water solutions of 1- and 2-nitronaphthalenes and of 1-methoxy-4-nitronaphthalene, in the presence of DABCO, diethylamine or triethylamine using ns laser pulses at 354 nm, has been studied by Goerner and Döpp.[28] The quenching of the T_1 nitronaphthalenes by the amines to give the arene radical anions is close to diffusion controlled in de-oxygenated acetonitrile solution but is slower in the presence of water. At high amine concentrations (>1mM), the α-aminoalkyl radicals react with the ground-state nitroarenes to yield their radicals and thence nitrosonaphthalenes.

4 Miscellaneous Reductions

A new method, using polyethylene matrices at low temperature and high pressure, has been described for studying the mechanism of homogeneously catalysed processes, and has been applied to an investigation of the mechanism of the photocatalytic hydrogenation of dimethyl fumarate (DF) and norbornadiene (NBD), with $Fe(CO)_4(\eta^2\text{-DF})$ and $M(CO)_4(NBD)$ ($M = Cr, Mo$) respectively as the catalytic species.[29] Irradiation of the iron complex in the polyethylene matrix at 150 K gave an intermediate tentatively proposed as $Fe(CO)_3(\eta^4\text{-DF})$, which on warming to 260 K in hydrogen yields $Fe(CO)_3(\eta^2\text{-DF})(\eta^2\text{-}H_2)$; on further warming, this affords dimethylsuccinate. Similar treatment of the NBD complex gives norbornane as the major product. The previously reported photoproducts from the fungicide and insecticide, pentachlorophenol, mostly result from dechlorination, but it has now been reported that photoreduction also occurs to give 2,3,4,5- and 2-hydro-2,3,4,5-pentachlorocyclohexanone.[30]

Two groups have described the photodeoxygenation of sulfoxides. Irradiation of 1,2-benzodiphenylene sulfoxide (32) produces an electrophilic oxidising intermediate, which converts the solvent benzene into phenol by a mechanism suggested to involve combination of a caged hydroxyl radical-aryl radical pair in a stepwise C-H insertion process.[31] The other study involves the 300 nm radiation of aryl sulfoxides in the presence of alkoxides, which gives yields up to 78% of the corresponding sulfide.[32] Other electron donors are effective in this reaction, and *N*-methylcarbazole in methanol solution affords quantitative yields. The authors suggest that the key intermediate is a hydroxysulfuranyl radical and that the S-O cleavage occurs heterolytically. Similar photodeoxygenation of 6-cyanophenanthridine-5-oxide in ethanol solution has been observed in a process, which has been deduced, from proflavine sensitization, to arise from the lowest excited triplet state.[33]

As in previous years, a number of publications, which are relevant to the present section, have appeared on fullerene photochemistry. The photoinduced electron-transfer reactions of C_{60} and C_{70} have been reviewed,[34] as have the photophysics and photochemistries of higher fullerenes and metallofullerenes.[35] In the latter publication it is also reported that C_{76}, C_{82} and C_{84} display transient absorptions in the near-IR region and undergo both photo-reduction and -oxidation processes, while, in contrast to C_{60} and C_{70}, the higher fullerenes have low intersystem-crossing efficiencies. Oxygen is described as having a unique involvement in the electron-transfer process of porphyrin-C_{60} dyads in that it inhibits the back transfer by promoting intersystem crossing of the singlet radical ion pair into the triplet state, but only when the energy of the charged-separated state is lower than that of the individual triplet states.[36] A ferrocene-containing liquid crystal malonate derivative has been grafted onto C_{60}, and this supramolecule undergoes photoinduced electron transfer from the ferrocene to the C_{60} moiety to give a radical pair with a lifetime of several hundred nanoseconds.[37] The photoinduced electron transfer between disubstituted naphthalenes and C_{60} and C_{70} has been studied by laser flash photolysis, and quantum yields and rate constants for the process have been reported.[38] The rates and efficiencies of the transfer are markedly increased for the dihydroxynaphthalenes by the addition of pyridine, but this influence is much lower for the dimethoxy derivative.

Two types of Ti-β-zeolites have been synthesized which display photocatalytic activity for the reduction of CO_2 with water at 323 K, producing methane and methanol.[39] It is deduced that the TiO_2 species are highly dispersed in the framework and have a tetrahedral co-ordination state, and furthermore that the differing water affinity to the zeolite surface has a significant influence on the reactivity and selectivity of the photoreduction process.

5 Oxidation of Aliphatic Compounds

5.1 Singlet Oxygen. – A useful system for the generation of singlet oxygen in aqueous media has been described and comprises C_{60} covalently bound to an insoluble hydrophilic polymer.[40] While there is no triplet-triplet annihilation or oxygen quenching in the dry sample, in aqueous suspension the excited C_{60} sensitizer is quenched very efficiently ($1.9 \pm 0.5 \times 10^8$ $dm^3\ mol^{-1}\ s^{-1}$) giving singlet oxygen.

A detailed theoretical study of the reactions of singlet oxygen with cyclohexa-1,3-diene has been published, and a stepwise diradical pathway to give the endoperoxide with an activation barrier of 26.75 kJ mol^{-1} has been deduced.[41] Of the two ene products, (33) is considered to arise by a concerted pathway while (34) is a product of the same diradical route as the endoperoxide. The addition of singlet oxygen, generated by tetraphenyl porphyrin sensitization, to the cycloheptatriene (35) gives a 90% yield of the endoperoxide (36), which on treatment with triethylamine affords a new isomer (37) of stipatatic acid and (38).[42] Both 1,3-dienes (39) and (40) are reported to give high yields of the corresponding peroxides (41) and (42), respectively, by addition of singlet oxygen, but on treatment of these adducts with cobalt(II) tetraphenylporphyrin, while the former undergoes a 90% conversion into (43), the latter affords solely a mixture of the isomers of (44).[43] The new compounds α- and β-hydroperoxydeoxyar-

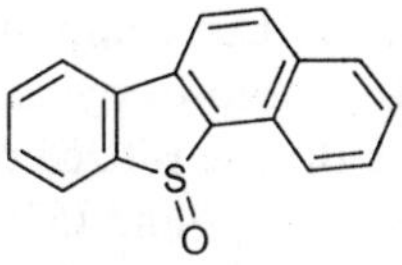

(32)

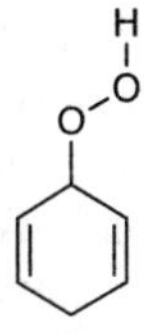

(33)

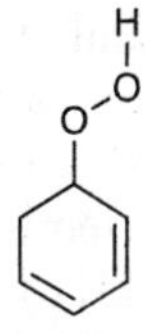

(34)

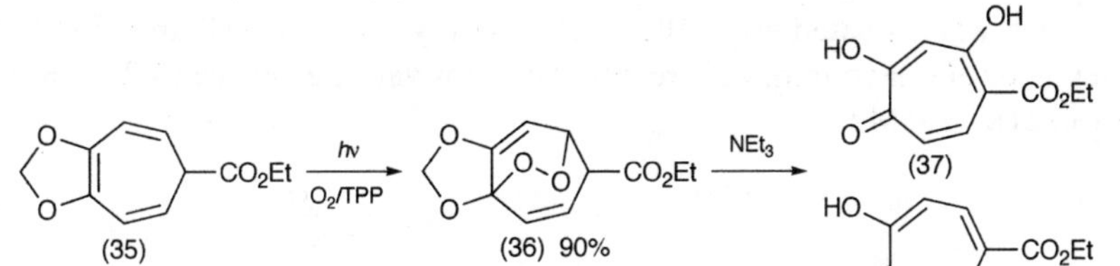

(35) (36) 90% (37) (38)

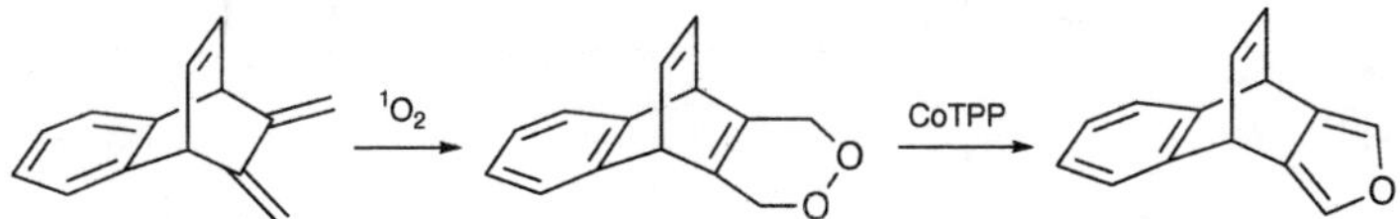

(39) (41) (43) 90%

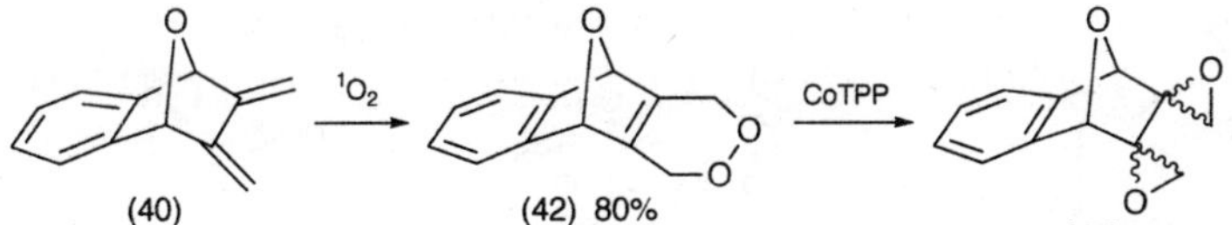

(40) (42) 80% (44)

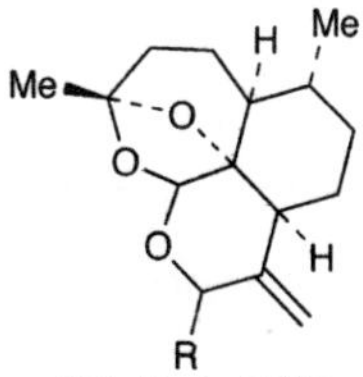

(45) R = α-OOH
(46) R = β-OOH

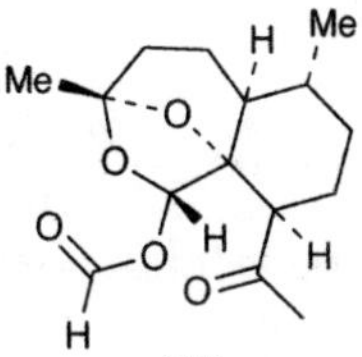

(47)

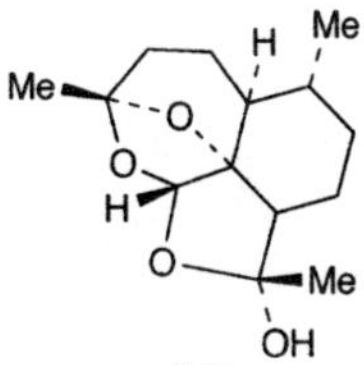

(48)

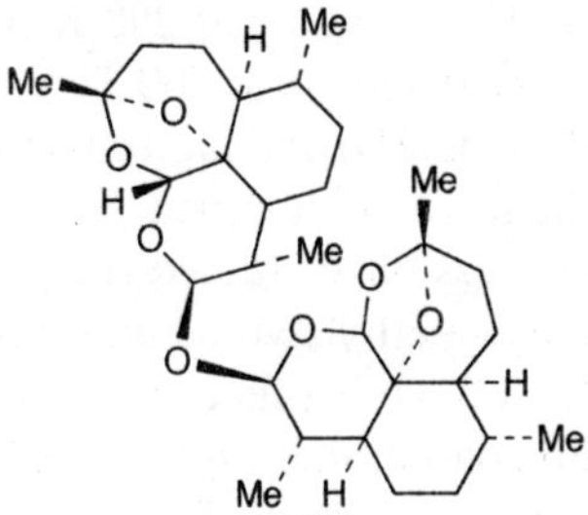

(49)

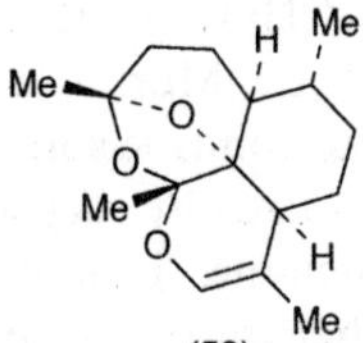

(50)

temisitenes, (45) and (46), respectively, and the formate (47) are reported to be formed, in addition to the two previously identified compounds (48) and (49), from tetraphenyl porphyrin sensitized photo-oxygenation of anhydrodeoxydihydroartemisinin (50).[44] Methylene Blue sensitized oxygenation of the cyclic enol ether (51) yields the novel epoxy compound (52) by the suggested pathway outlined in Scheme 2.[45] The alkenylfullerenes (53) are reported to undergo self-sensitized photo-oxygenation to give (non-isolated) hydroperoxides, which when treated with PPh_3 yield the allyl alcohols (54) and (55).[46] The former alcohol is favoured, with respective ratios varying between 7:3 (R = Me) and 13:12 (R = CMe_3).

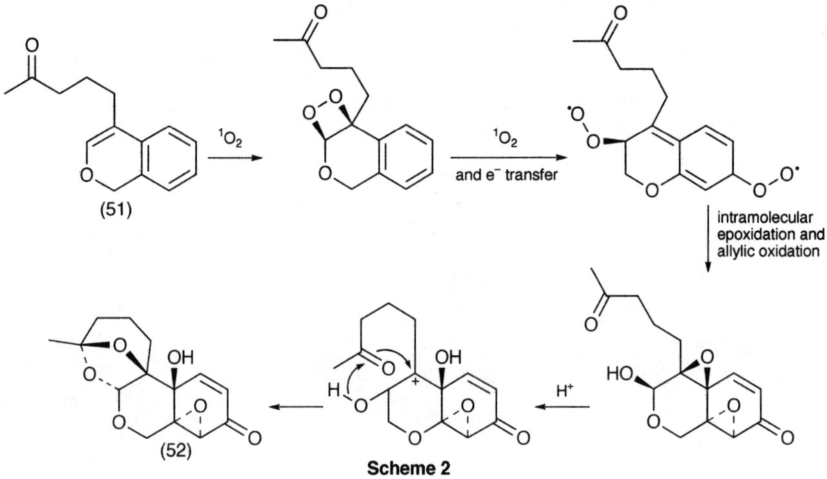

Scheme 2

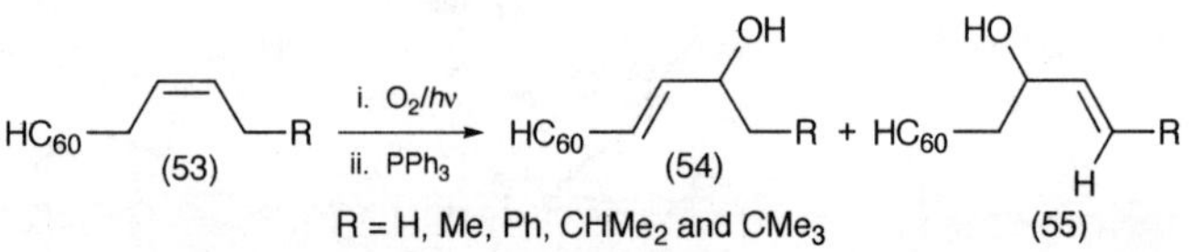

5.2 Other Oxidation Processes. – *n*-Butanal, *n*-pentanal and *n*-hexanal undergo photo-oxidation with 275-380 nm radiation in air at 298 K, with quantum yields in the range 0.4-0.48 at 100 Torr and 0.32-0.38 at 700 Torr, dependent on the substrate.[47,48] Radical (R-CHO → R$^\cdot$ + $^\cdot$CHO) and molecular ($C_xH_yCHO \rightarrow C_{x-1}H_{y-3} + CH_2{=}CHOH$) pathways have been identified yielding the alkene, CO and acetaldehyde. Mori and co-workers have studied the photo-oxygenation reactions of acetonitrile solutions of α-methylstyrene under electron-transfer conditions sensitized by alkylated dimethoxybenzenes.[49,50] The products formed are acetophenone, 1-methyl-1-phenyloxirane, and the ene-product, 3-hydroperoxy-2-phenylprop-1-ene, in a respective ratio of 1:1:0.04, and the mechanism of their formation, which involves reaction of the superoxide ion with the styrene radical cation, is given in Scheme 3. A report in 1988 on the dicyanoanthracene (DCA)-sensitized oxygenation of the vinyloxirane (56) described the

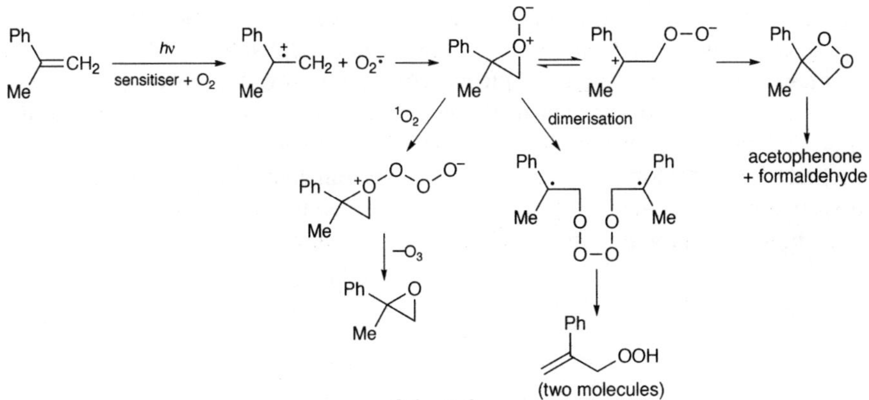

Scheme 3

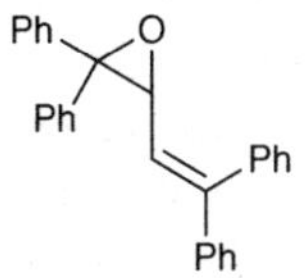
(56)

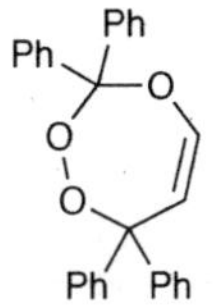
(57)

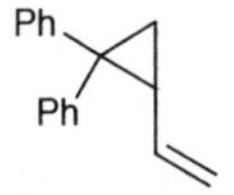
(58)

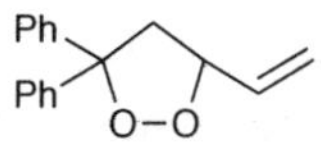
(59)

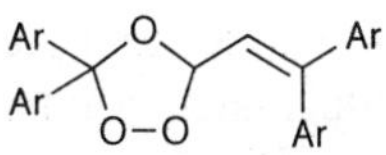
(60)

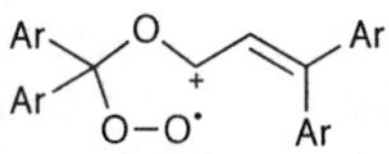
(61)

product as the 1,2,4-trioxepine (57),[51] despite an earlier account that the vinylpropane (58) gave the 1,2-dioxolane (59) under similar conditions.[52] The reactions of tetra-aryl compounds of type (56) have been re-examined, and the product is assigned the expected structure (60), formed by way of electron transfer from (56) to DCA and reaction of the radical cation with triplet oxygen to give (61) and thence (60) by ring closure and back electron transfer.[53] A sensitized electron-transfer pathway involving the superoxide ion $O_2^{-\cdot}$ operates in the efficient route, devised by Içli and co-workers, to the costly perfume, Rose Oxide (62), from citronellol (63) using concentrated sunlight.[54] The sensitizers for this process are the highly stable naphthalene or perylene diimides (*e.g.* (64)) which have been previously developed by these workers as 'solar sensitizers', and the reaction is complete after 60 minutes exposure.

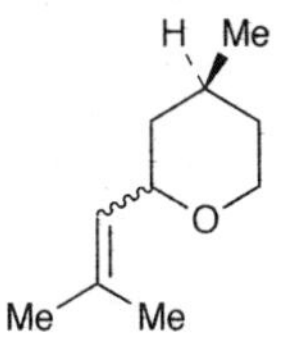
(62)

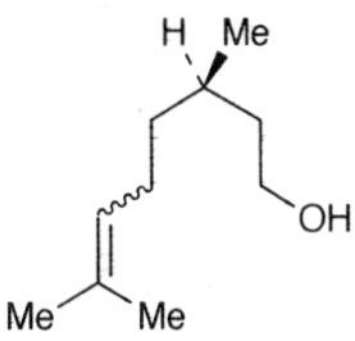
(63)

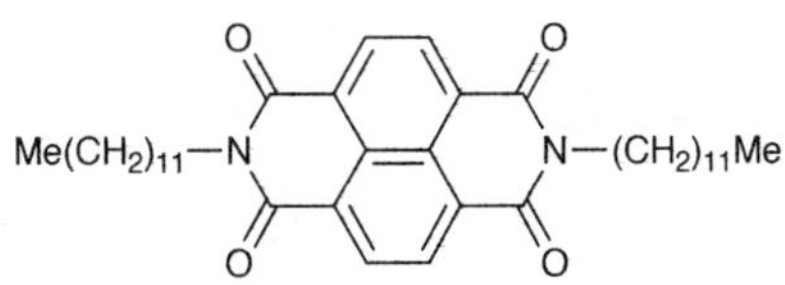
(64)

A new synthetic method for the formation of carbonyl compounds from the corresponding alcohols has been described by Itoh and co-workers.[55] The process involves irradiation of di-*iso*-propyl ether solutions of the alcohol in the presence of iodine to give, for example, cinnamaldehyde in 99% yield from cinnamyl alcohol, as well as (65) from (66), by a pathway involving hydrogen-atom abstraction from the substrate by iodine atoms and reaction of the resulting ketyl radical with molecular oxygen. The kinetics of the photo-oxidative degradation of chlorinated paraffins to give longer chain alkanes have been investigated, and the most efficient conditions involving solutions in 1% acetone-water in the presence of H_2O_2 have been devised.[56]

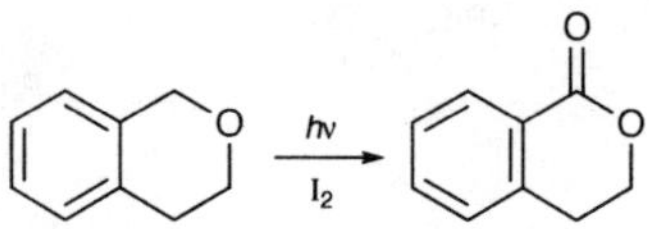

(66) (65)

Photoinduced oxidation reactions of organic compounds over TiO_2 continue to attract considerable attention and the kinetics of such processes have been reviewed.[57] A variety of conditions have been developed for these reactions to accommodate a wide range of substrates, and detailed studies have been carried out on the mechanisms to optimize yields and selectivities. Within the year, reports in this area include the photo-oxidation of short-chain hydrocarbons (C_3 to C_6),[58] cyclohexane,[59] benzhydrol (and reduction of benzil),[60] acetone,[61] polyacrylamide in water,[62] and humic acids in the absence[63] and presence of H_2O_2.[64] The photocatalytic oxidative degradation of chlorinated ethylenes over TiO_2 is also an area of appreciable interest.[65-68] Photo-oxidative processes in the voids of microporous materials have been studied, such as, for example, cyclohexane oxidation in zeolite Y.[69]

6 Oxidation of Aromatic Compounds

6.1 Singlet Oxygen. – Irradiation of the aryl alkoxyfuran (67) under singlet oxygen forming conditions is reported to give unexpectedly the hydroperoxy oxetane (68) accompanied by minor amounts of the enone (69).[70] The usual

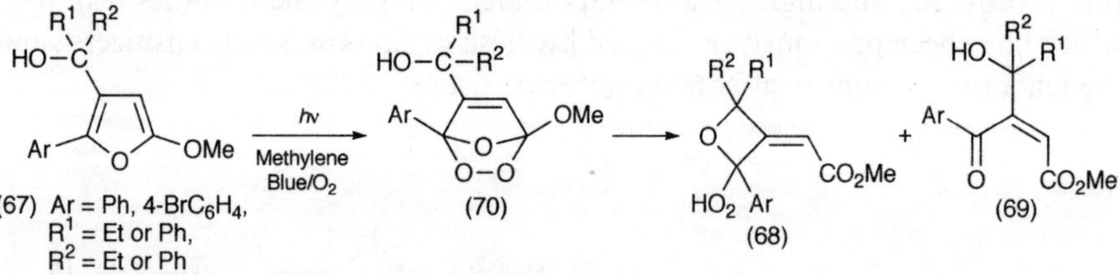

(67) Ar = Ph, 4-BrC$_6$H$_4$, R^1 = Et or Ph, R^2 = Et or Ph (70) (68) (69)

product of such reaction, the endoperoxide (70), however, is considered to be the intermediate which forms (68) by a novel thermal rearrangement. Similarly, the endoperoxides (71) produced from photoinduced oxygenation of 3-acetyl-5-aryl-2-methylfurans (72) are unstable and thermally or photochemically undergo homolytic and heterolytic cleavage to give 2,2-diacetyl-3-aroyloxiranes (73), 3-

acetyl-1-aryl-pent-2-en-1,4-diones (74) and 3-acetyl-1-aryl-2-hydroxypent-2-en-1,4-diones (75).[71] The mechanisms for the formation of these products are summarized in Scheme 4. A further example of the rearrangement of unstable

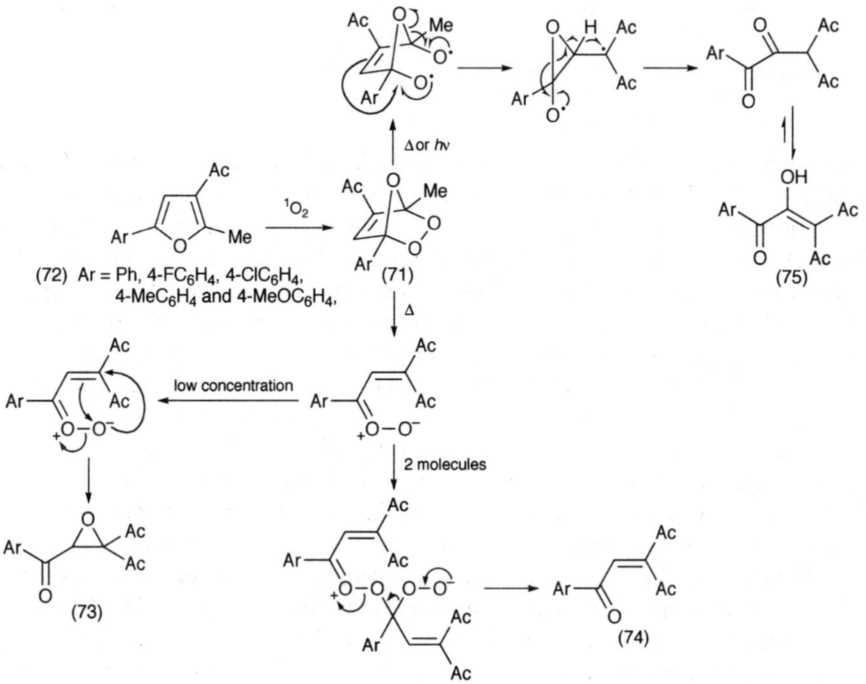

Scheme 4

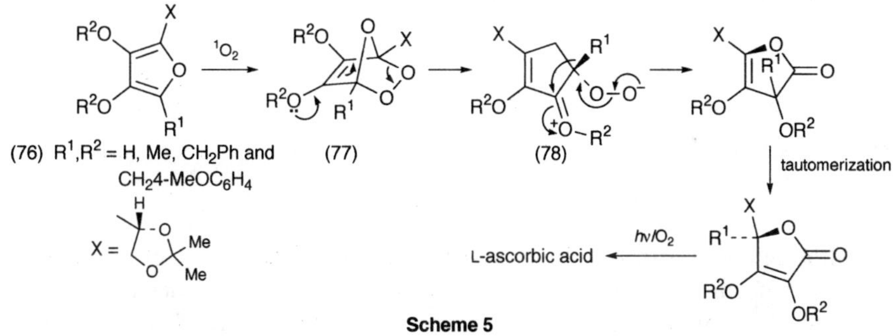

Scheme 5

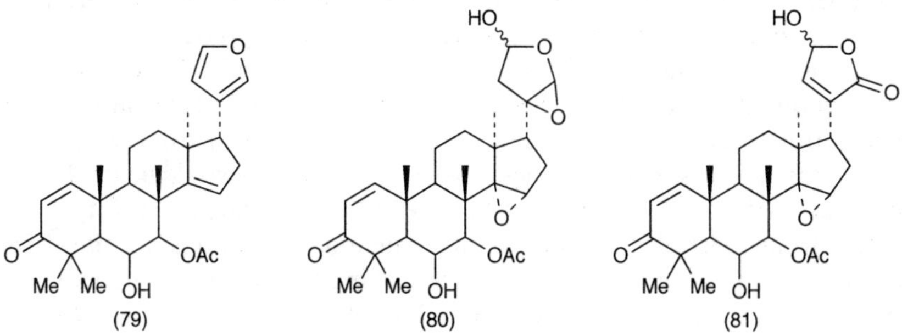

(79) (80) (81)

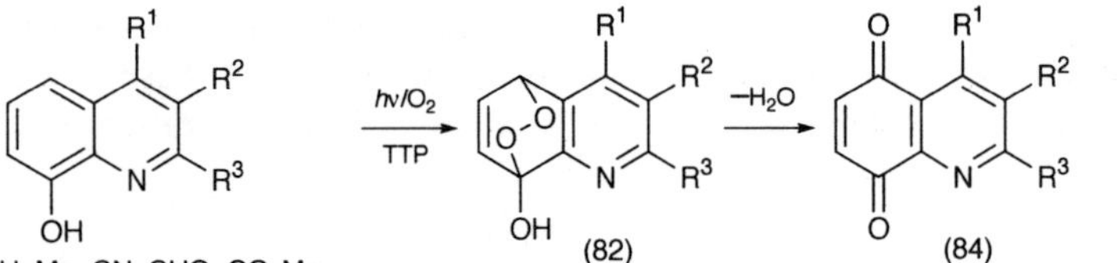

(82) (84)

(83) R^1 = H, Me, CN, CHO, CO_2Me, $CONHPr^i$, $CH_2N(BOC)CH_2Ph$, $CH_2N(BOC)CH_2CO_2Me$

endoperoxides from furans is reported by Hakimelahi and co-workers.[72] In this study, the 3,4-dialkoxyfurans (76) undergo self-sensitized oxygenation to give (77), in which bond cleavage occurs, and following rearrangement of the intermediate (78) and elimination of the photodegradable protecting group, access to Vitamin C and its derivatives is obtained, as shown in Scheme 5. Singlet oxygen is reported to react with the tetranortriterpenoid, nimonol (79) at both the furanyl and cyclopentene moieties, giving (80) and (81) as the final products.[73] It is thus evident that the endoperoxide formed from the furan unit again rearranges, while the latter attack leads to the hitherto unreported formation of an α-epoxide functionality in ring D. Opening of the endoperoxide (82) from photo-oxygenation of substituted 8-hydroxyquinolines (83) leads to an efficient synthesis of the quinoline-5,8-quinones (84) in yields of 50 to 89% dependent on the nature of the substituent.[74] A building block for the Manumycin antibiotics has been devised by Adam and co-workers using a sequence of photo-oxygenation, reduction and Weitz-Scheffer epoxidation.[75] In the first step, irradiation of the readily accessible anilide (85) in acetone-CH_2Cl_2 solution in the presence of tetraphenyl porphyrin and oxygen at 248 K yields the hydroperoxide (86), for subsequent conversion into the required hydroxy-epoxy-enone (87).

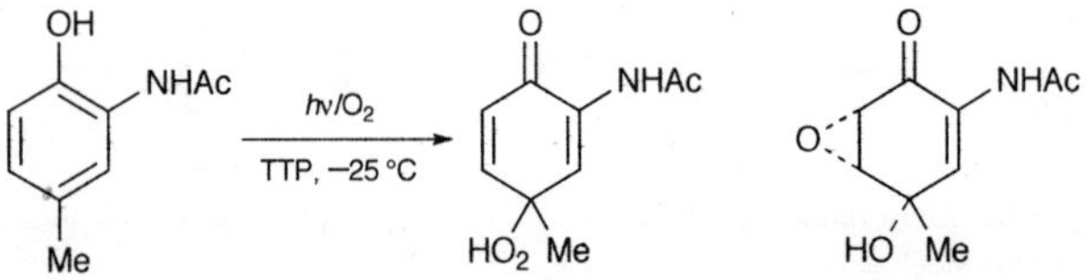

(85) (86) (87)

Although self-sensitized photo-oxygenation of anthracene, for example, is well known, information for the corresponding process with rylenes is sparse. By monitoring fluorescence intensity and spectral changes of terrylene molecules in a *p*-terphenyl host, however, it has been possible to observe the self-sensitized photo-oxygenation of a single fluorophore.[76] A number of products and successive reaction of the same terrylene molecule were observed. The authors also performed quantitative calculations on a number of possible structures for the products, from which it is predicted that endoperoxides arising from reactions at sites A and B in (88) as well as the diepoxide at sites C and D would be stable. The reaction of singlet oxygen with an open-cage fullerene having the moiety (89) is reported, from spectral data and density function calculations, to afford the product moiety (90), which is a 12-membered ring orifice containing enol and carbonyl groups on the fullerene cage.[77] The photodegradation of 4-chloro-

(88)

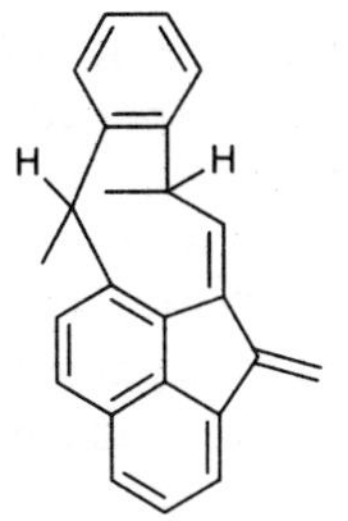

(89)

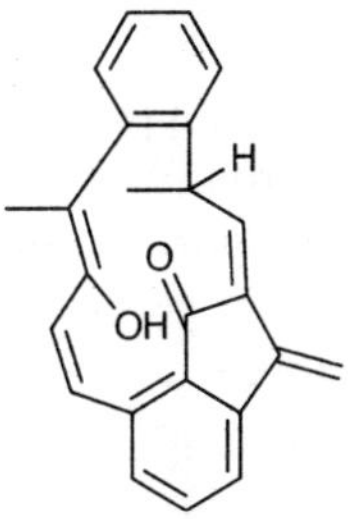

(90)

phenol in aqueous solution sensitized by zinc and aluminium phthalocyanines has been shown to arise predominately by a singlet oxygen mechanism at concentrations of substrate below 10^{-4}mol dm^{-3}, but a Type I mechanism becomes increasingly important at concentrations above 10^{-3}mol dm^{-3}.[78]

6.2 Other Oxidative Processes. – The effects of varying H_2O_2 concentration and the spectral lamp output on the efficiency of photo-oxidative degradation of the common water pollutant, 2,4-dichlorophenoxyacetic acid, have been reported,[79] and it has been demonstrated that sensitized oxidative photolytic degradation of polynuclear aromatic hydrocarbons is less efficient than under conditions involving TiO_2 suspensions.[80] Indeed, many reports have appeared within the year describing the heterogeneous photocatalytic oxidation of aromatic compounds over TiO_2; selected examples of this literature are given below to illustrate the breadth of activity in this area. The efficiency of gas-solid photocatalytic oxidation of benzene and toluene (and also of cyclohexane and cyclohexene) suffers from catalyst deactivation by the formation of carbon deposits on the TiO_2 surface, but it is reported that regeneration can be achieved photochemically in the presence of water as the carbon is converted to CO_2.[81] In aqueous suspensions of TiO_2 at pH 5.6, phenol undergoes 97.8% photo-oxidative degradation with a low power lamp,[82] and this reaction has also been analysed via a kinetic model based on mechanistic considerations which include the influence of dissolved oxygen over a range of experimental conditions.[83] Similar degradation of 4-nitrophenol over TiO_2 shows an almost exponential rate decrease with pH,[84] and has been studied in a membrane process, from which it is revealed that membrane rejection of pollutants and their by-products is an essential consideration.[85] The solar photocatalytic degradation of 4-nitrophenol is recommended as an undergraduate experiment with the praiseworthy objectives of raising student awareness of waste management, a non-contaminating renewable energy source (sunlight), and the use of TiO_2 heterogeneous photocatalysis.[86] Ozer and co-workers have used the oxidation of 1,2-dichlorobenzene in an evaluation of a photosystem involving polyoxometalate- catalysed electron transfer from the conduction band of photoexcited TiO_2 to molecular oxygen, and note from ΔG values that the transfer to the polyoxometalate occurs even when this is endothermic.[87] Similar studies have also been carried out by the same group using polyoxometalate supported on NaY zeolite.[88] The influence of ozone on the TiO_2 catalysed photo-oxidation of pyrrole-2-carboxylic acid, 4-toluenesulfonic acid and naphthalene-1,5-disulfonic acid has been described,[89] and the use of this catalyst system in the treatment of biocontaminated water containing amino acids and proteins has been reported.[90] Leng and co-workers have shown that 4-chloroaniline can be effectively photodegraded using the TiO_2 catalyst immobilized on porous nickel with a binder,[91] and a coating of TiO_2 as a transparent film on silica gel beads is reported to generate a strong oxidative potential on irradiation, which is sufficient to degrade almost all organic compounds to water and CO_2, including dioxins, with an efficiency over 99%.[92]

7 Oxidation of Nitrogen-containing Compounds

1,4-Dihydropyridines (91) are readily aromatized under dye-sensitized photo-oxidation conditions to give pyridines (92), some with the loss of the 4-substituent, in good yields (70-80%).[93] Singlet oxygen is reported to attack the alkaloids papaverine (93), nicotinic acid (94) and ephidrine (95) to give endoperoxides (96) and (97), the indole (98) and the hydrazine (99) in unspecified yields.[94] As for the furan derivatives considered above in Section 6.1, the endoperoxides formed from the pyrroles (100) are thermally labile and give 5-methylenepyrrol-2-(5*H*)-ones (101) in good yields, following acid treatment of the alcohols (102).[95] Ferroud and Rool have described a singlet oxygen mediated new synthetic

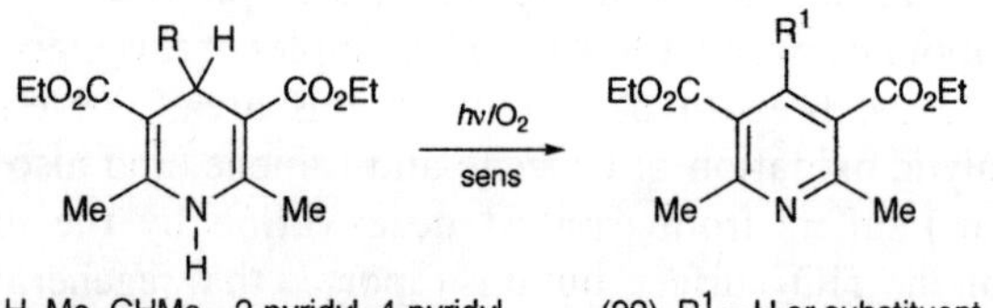

(91) R = H, Me, CHMe$_2$, 2-pyridyl, 4-pyridyl, 2-thienyl, 2-furanyl, 2-pyrrolyl, benzyl, and substituted phenyl

(92) R^1 = H or substituent

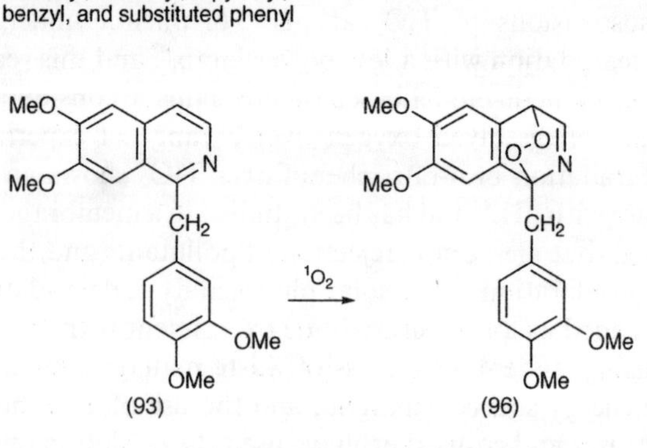

(93) (96)

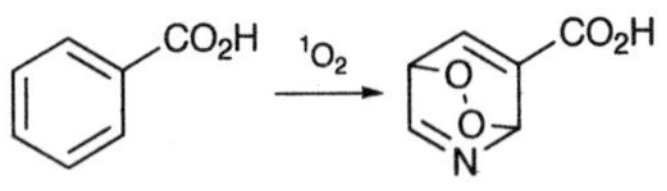

(94) (97)

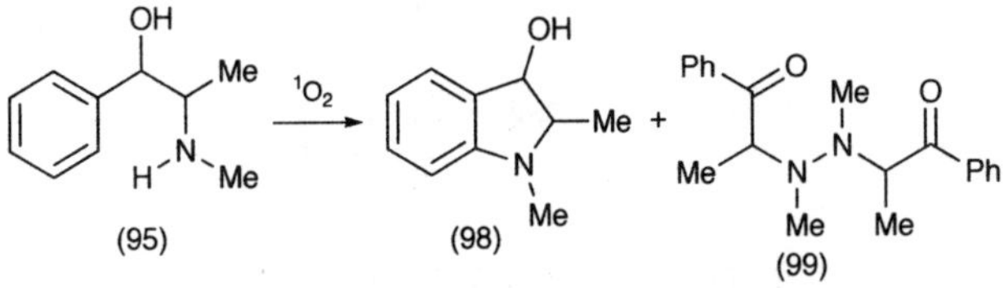

(95) (98) (99)

approach to hydroxyindolenine-catharanthine derivatives.[96] The procedure involves tetraphenylporphyrin as the sensitizer in the presence of oxygen and trimethylsilyl cyanide, with the first step being photocyanation of (103) to (104), which is then attacked by singlet oxygen to give the hydroperoxide intermediate (105), leading to the new alkaloid (106). Oxidation of 2′-deoxyguanosine (107), giving spiroiminodihydantoin (108) as the major product, has been brought about by using either photochemical or thermal procedures.[97] In the former case, the oxidant is triplet excited 2-hydroxyacetophenone in the presence of oxygen, while the thermal reaction involves peroxyl radicals generated from the hitherto unknown dioxetane (109).

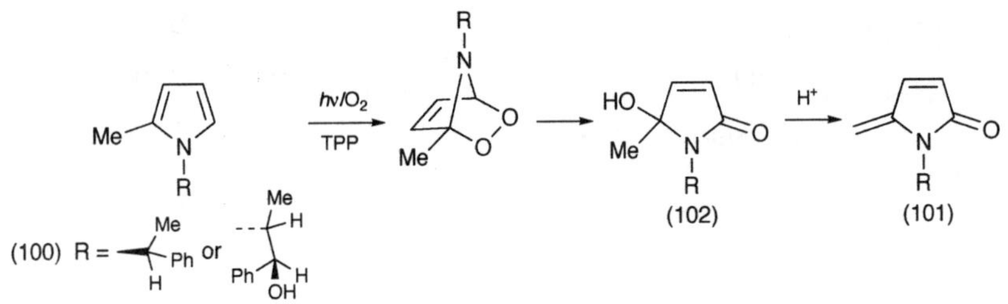

(100) (102) (101)

The oxidation of *N,N*-dibenzylamines in acetonitrile can be sensitized by either the 2,4,6-triphenylpyrylium ion[98] or the 10-methylacridinium ion.[99] The reaction involves the superoxide ion and affords *N*-benzylidenebenzyl amine *N*-oxide (110) and H_2O_2 quantitatively. From an analysis of substituent effects on this process, it is concluded that back electron transfer from the radical of the sensitizer to the radical cation of the hydroxylamine occurs in preference to

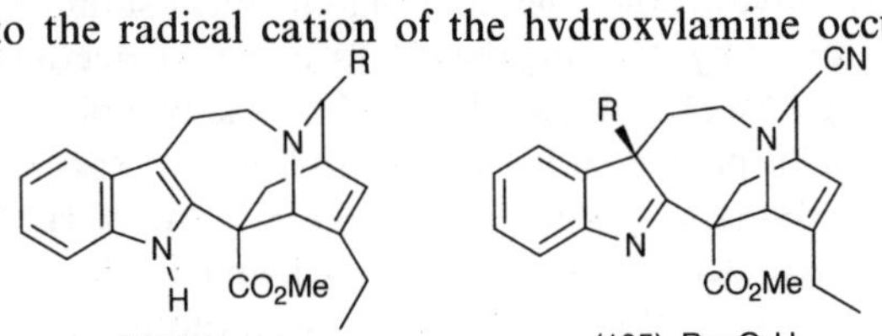

(103) R = H
(104) R = CN

(105) R = O_2H
(106) R = OH

one-electron reduction of oxygen. The role of $O_2^{-\cdot}$ in the riboflavin-sensitized photo-oxidation of 8-oxo-7,8-dihydroguanosine (111) is reported to be pH dependent.[100] The intermediate radical cation of (111) may be trapped by water or

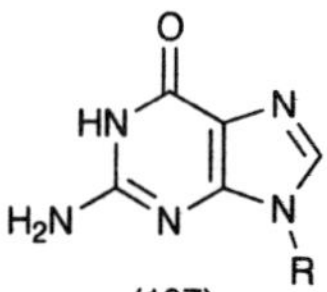
(107)

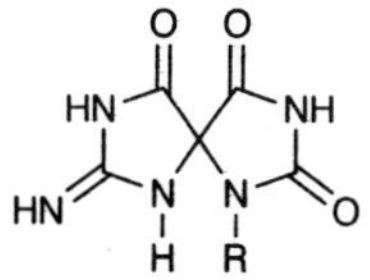
(108)

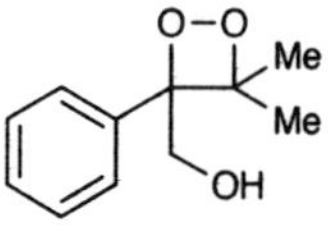
(109)

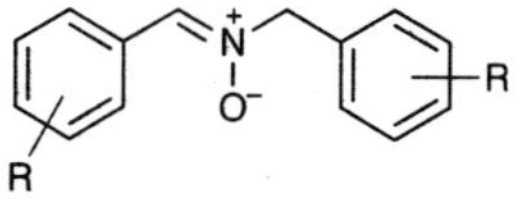
(110) R = H, 4-Me, 4-Cl, 3-Cl, 4-F and 4-CF_3

superoxide. In the former case, the previously described compounds (112) and (113) are formed at pH<7 and pH≥7, respectively, whereas reaction of the radical cation with $O_2^{-\cdot}$ yields the oxaluric acid (114) (pH≤7) and the imidazolone (115) (pH≥8.6). Photo-oxidation of indolizines (116) is reported to proceed by selective excitation of its ground-state charge-transfer complex with molecular oxygen.[101] In benzene solution, (116) forms only tarry material, but in a 1:1 solvent mixture of benzene and methanol, (117) and its geometric isomer and α-phenyl-β-(2-pyridyl)-propenoic acid result in respective yields of 29, 26, and 17%; this change in selectivity is accounted for by reaction of the zwitterionic species (118) with methanol to give the intermediate hydroperoxide (119), as shown in Scheme 6.

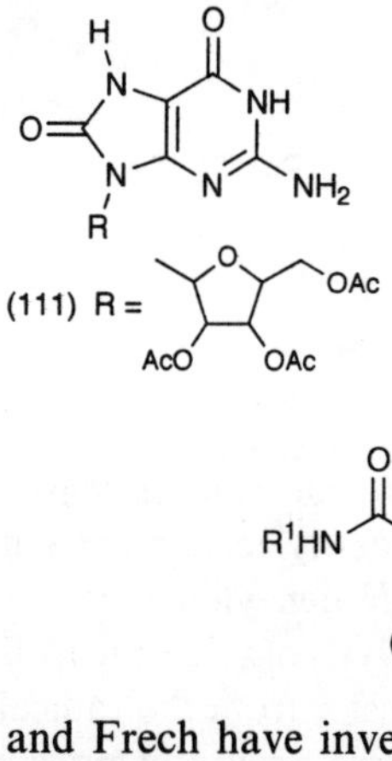
(111) R =

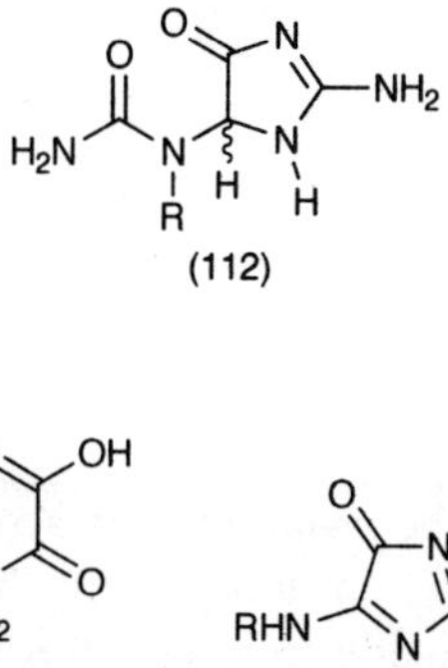
(112)

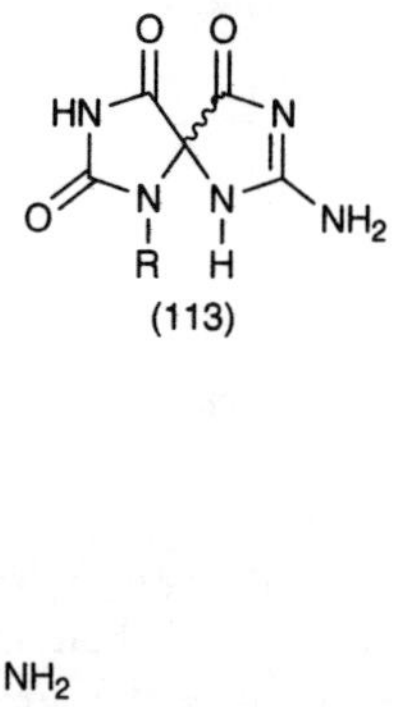
(113)

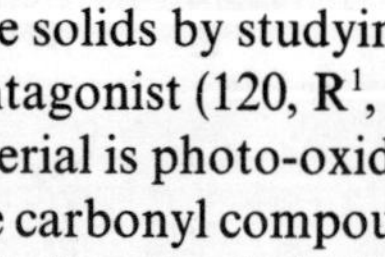
(114)

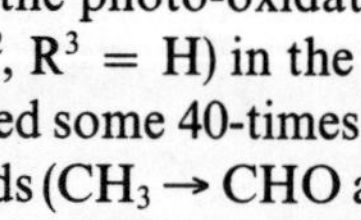
(115)

Qin and Frech have investigated the differing chemical stabilities of amorphous and crystalline solids by studying the photo-oxidative degradation of the α-2 adrenoceptor antagonist (120, R^1, R^2, R^3 = H) in the two physical states.[102] The amorphous material is photo-oxidized some 40-times faster than the crystal form and yields three carbonyl compounds (CH_3 → CHO and –CH_2–NMe–CO– → –CO–NMe–CO–), five alcohols (CH_3→ CH_2OH and R^1, R^2, R^3 = OH in structure (120)), as well as an *N*-oxide. In contrast, although the same three carbonyl compounds result from the crystals, these only yield one alcohol. The authors consider that their results provide some evidence to understand the differing photostabilities of the two states. Photo-oxidation of 9-substituted-10-methyl-9,10-dihydroacridines (121) is reported to occur efficiently with oxygen in

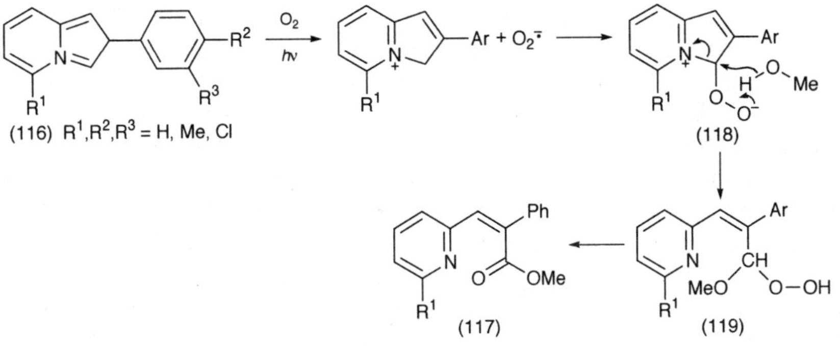

Scheme 6

the presence of dicyanoanthracene and scandium triflate to give either the 9-substituted acridinium ion and H_2O_2 or the unsubstituted acridinium ion and oxygenated products such as ROOH.[103] No reaction is observed in the absence of Sc^{3+}, and the differentiation between the two pathways is found to be in agreement with the influence of the 9-substituent on the cleavage of the radical cation of (121).

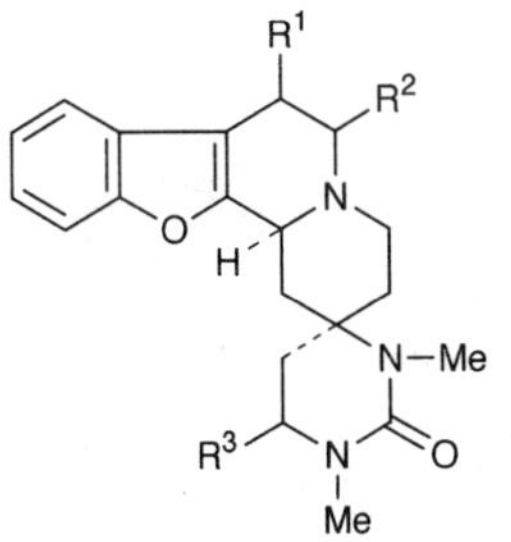

(120)

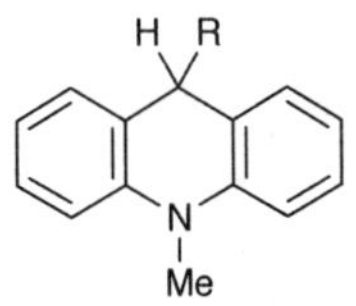

(121) R = H, Me, Et, Ph or CH_2Ph

Photo-oxidative degradation of azo dyes is reported to be brought about efficiently by their irradiation in the presence of H_2O_2,[104,105] and indigo and indigo carmine are totally mineralized to CO_2, NH_4^+, NO_3^-, and SO_4^{2-} on irradiation in the presence of TiO_2.[106] Although Methylene Blue is resistant to biodegradation, in aqueous TiO_2 suspensions in the presence of oxygen it undergoes photocatalytic *N*-demethylation concomitant with oxidative degradation;[107] and mono-, di- and tri-ethanolamines are reported to undergo similar photoinduced mineralization over TiO_2 by way of various intermediates, including a 3-pyrrolidine.[108]

8 Miscellaneous Oxidations

Photo-oxygenation of a number of sulfur-containing compounds has been reported within the year. Using light in the visible region (including solar radiation), Lacombe and co-workers have devised a system for the inexpensive photosensitized production of sulfoxides from sulfides by an electron-transfer mechanism.[109] Singlet oxygen produced by C_{60}-sensitization has been used to oxidize dimethyl sulfide to the sulfoxide,[110] and in a similar process using tetraphenyl porphyrin sensitization, 17 sulfides (122), bearing either anion or radical stabilizing substituents, have been oxidized to the various products outlined in Scheme 7.[111] The observed discrimination between the oxidative cleavage and sulfide oxidation pathways is rationalized in terms of a substituent-directed partitioning in the hydroperoxy sulfonium ylide intermediate, which is considered to exist in both diradical and zwitterionic forms, and furthermore, inter- and intra-molecular pathways for the decomposition of the α-hydroperoxy sulfides are proposed to account for the substituent-dependent production of C-S bond oxidative cleavage compounds. The first example of ethene

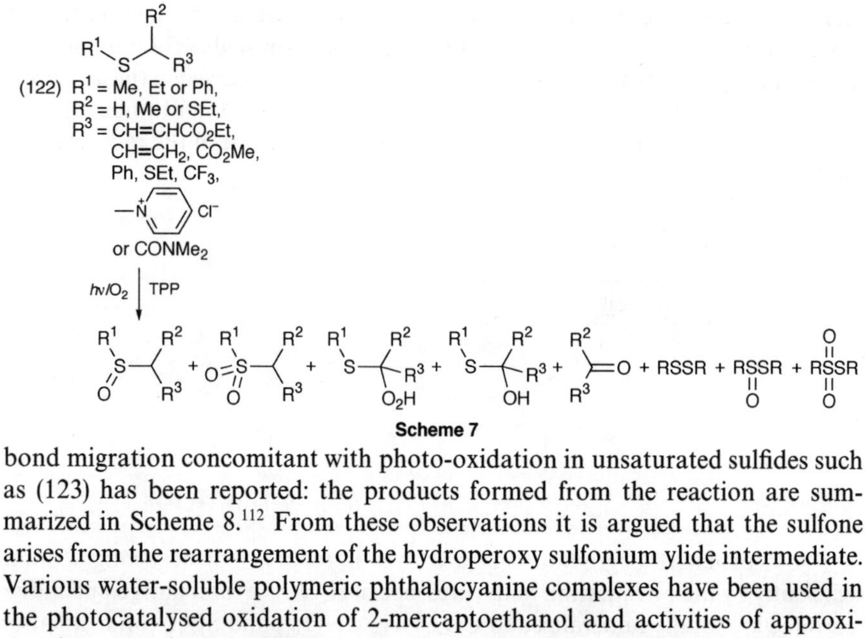

Scheme 7

bond migration concomitant with photo-oxidation in unsaturated sulfides such as (123) has been reported: the products formed from the reaction are summarized in Scheme 8.[112] From these observations it is argued that the sulfone arises from the rearrangement of the hydroperoxy sulfonium ylide intermediate. Various water-soluble polymeric phthalocyanine complexes have been used in the photocatalysed oxidation of 2-mercaptoethanol and activities of approximately two- to four-times greater than those of the mononuclear complexes have been observed.[113] In contrast to previously reported conclusions, the photo-

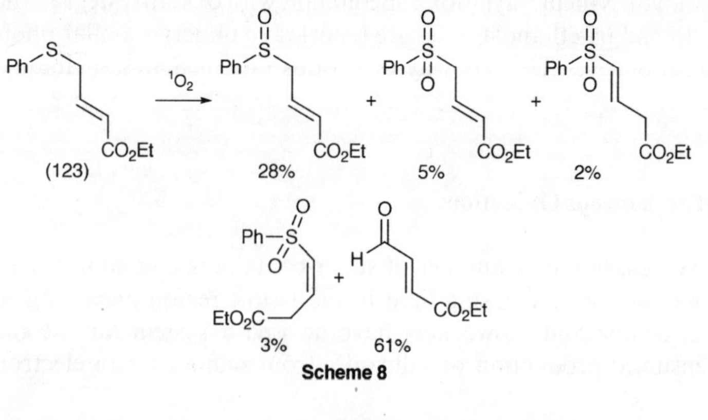

Scheme 8

oxidation ($\lambda = 245$nm) of self-assembled monolayers of mercapto-undecanoic and -propanoic acids has been reported to occur and, in the latter case, irradiation through a mask produced a photopatterned monolayer.[114]

The photosensitized oxidation (*e.g.* by anthraquinone) of dibenzothiophenes is reported to occur only in polar solvents;[115] and the aromatic fraction remaining after biodegradation of crude oil has been observed to be substantially decomposed following exposure to sunlight for several weeks.[116] 4-Methylthiophenol undergoes complete mineralization by photocatalysed oxidation over TiO_2;[117] and irradiation of 2-methylthiophene in aqueous suspensions of TiO_2 affords several intermediates including hydroxythiophenes and dimeric and trimeric compounds.[118] It is relevant to all the studies on TiO_2 photo-catalysed processes to note that the use of ultrasound during these reactions is found to have a pronounced effect on the rate and efficiency of degradation of salicyclic acid.[119]

It has been deduced that, during the photo-oxidation of water by irradiation of aqueous solutions of tolu-1,4-quinone, the observed transient species is a hydroxylating intermediate which can best be considered as a complex between the semiquinone radical and a hydroxyl radical.[120] The reaction conditions of the photodegradation of aminoanthraquinone in cellulose and cellulose-like matrices are reported to dictate whether the process involves oxidation from photogenerated singlet oxygen or reduction associated with a hydrogen abstraction from a host carbinol carbon atom.[121]

Considerable interest continues to be evident in the use of the photo-assisted Fenton reaction, using a H_2O_2-ferrioxalate complex system, for oxidative degradation of organic compounds, particularly pollutants. A review of the extensive literature in this area is considered to be beyond the current treatment of photoinduced oxidative processes, but a brief outline of some such publications is given here to illustrate the scope of investigations in this area. The mechanism of the process continues to be researched, and from studies with highly concentrated waste waters, a 'global' pathway for the degradation of phenol by the photo-Fenton reaction has been proposed.[122] In other work, the use of an iron tetrasulfophthalocyanine-H_2O_2 system has been studied and, according to the suggested mechanism from this investigation, the degradation of organic pollutants can be explained without involving higher valance iron-oxo or iron-peroxo complexes.[123] Reports on the photodegradation of specified compounds under photo-Fenton conditions include those of formic acid in water (modelled and experimentally verified),[124] 4-hydroxybenzoic acid,[125] polyvinyl alcohol,[126] aqueous solutions of *p*-xylene and 2-nitrotoluene,[127] dichloroacetic acid and 2,4-dichlorophenol,[128] seven substituted nitrophenols,[129] and chloroform and tri- and tetra-chloroethylenes.[130] The optimum pH for efficient photodegradation of 4-hydroxyphenylacetic acid is reported to be 3,[131] and intermediates have been identified for the process with 2,4-dichlorophenol.[132] More complex substrates examined in the photo-Fenton reaction include 'dye solutions',[133] Direct Scarlet dyes,[134] humic acids,[135] and Imidacloprid, which is commonly used as an insecticide in Mediterranean agricultural areas.[136] The photocatalytic oxidation of the non-biodegradable 4-nitrotoluene-2-sulfonic acid has been studied in the homogeneous photo-Fenton process and by heterogeneous catalysis using TiO_2 suspensions, in concentrating and non-concentrating solar reactors.[137] The conclusions from this work seem to be that the heterogeneous process followed by biological treatment provides a useful method for total mineralization.

References

1. X. Cai, Z. Han, S. Yao and N. Lin, *Science in China, Series B: Chemistry*, 2001, **44**, 582.
2. H. Mallesha, N. D. Dinesh, K. S. Rangappa, S Shashikanth, N. K. Lokanath, M. A. Sridhar and J. S. Prasad, *Ind. J. Chem. Sect. B: Org. Chem. Incl. Med. Chem.*, 2002, **41B**, 196.
3. Q. D. Zeng, C. Wang, S. d. Xu, C. L. Bai, Y. Li and X. J. Yan, *Appl. Phys. A: Mater. Sci. Process*, 2001, **73**, 247.
4. J. R. Woodward and Y. Sakaguchi, *J. Phys. Chem. A*, 2001, **105**, 4010.
5. O. Rinco, M. H. Kleinman and C. Bohne, *Langmuir*, 2001, **17**, 5781.
6. D. Takamori, T. Aoki, H. Yashiro and H. Murai, *J. Phys. Chem. A*, 2001, **105**, 6001.
7. U. Pischel and W. M. Nau, *J. Amer. Chem. Soc.*, 2001, **123**, 9727.
8. J. W. Park, M. J. Hong and K. K. Park, *Bull. Korean Chem. Soc.*, 2001, **22**, 1213.
9. W. Chu and S. M. Tsui, *J. Environ. Eng. (Reston Va.)*, 2001, **127**, 741.
10. Y. Wang, J. Sun, P. Wang, Q. Tian, B. Yan, G. Xu, Y. Wang and Y. Liu, *Bopuxue Zazhi*, 2001, **18**, 129.
11. T. Del Giacco E. Baciocchi, O. Lanzalunga and F. Elisei, *Chem.- Eur. J.*, 2001, **7**, 3005.
12. See T. Welton, *Chem. Rev.*, 1999, **99**, 2071, for a review of ionic liquids.
13. M. J. Muldoon, A. J. McLean, C. M. Gordon and I. R. Dunkin, *J. Chem. Soc., Chem. Commun.*, 2001, 2364.
14. J. L. Reynolds, K. R. Erdner and P. B. Jones, *Org. Lett.*, 2002, **4**, 917.
15. C. Brule and N. Hoffmann, *Tetrahedron Lett.*, 2002, **43**, 69.
16. C.-G. Yang, Z.-Q. Jiang, D-R. Zhu. S.-P. Wu, L. Yang and L.-M. Wu, *Chinese J. Chem.*, 2001, **19**, 1211.
17. Y. Norikane and T. Arai, *Chem. Lett.*, 2001, 416.
18. H. Kim, T. G. Kim, J. Hahn, D.-J. Jang, D. J. Chang and B. S. Park, *J. Phys. Chem. A*, 2001, **105**, 3555.
19. K. K. Park I. K. Han and J. W. Park, *J. Org. Chem.*, 2001, **66**, 6800.
20. K. K. Park and H. Lim, *Heterocycles*, 2002, **57**, 657.
21. T. Bach, T. Aechtner and B. Neumuller, *J. Chem. Soc., Chem. Commun.*, 2001, 607.
22. P. Wessig, J. Schwarrz, U. Lindemann and M. C. Holthausen, *Synthesis*, 2001, 1258.
23. K. C. Nicolaou, D. Gray and J. Tae, *Angew. Chem., Int. Ed.*, 2001, **40**, 3679.
24. K. T. Ranjit and L. Kevan, *J. Phys. Chem. B*, 2002, **106**, 1104.
25. J. J. Lee, Y. Kim and M. Yoon, *J. Photosci.*, 2001, **8**, 27.
26. C. W. Ma and W. Chu, *Water Res.*, 2001, **35**, 2453.
27. M. Fawi and A. E. Latif, *J. Photochem. Photobiol.*, A, 2001, **141**, 241.
28. H. Goerner and D. Döpp, *J. Chem. Soc., Perkin Trans. 2*, 2002, 120.
29. G. I. Childs, A. I. Cooper, T. F. Nolan, M. J. Carrott, M. W. George and M. Poliakoff, *J. Amer. Chem. Soc.*, 2001, **123**, 6857.
30. R. S. Ray, S. K. Singh, R. K. Hans and P. C. Joshi, *J. Environ. Monitor.*, 2002, **4**, 449.
31. E. Lucien and A. Greer, *J. Org. Chem.*, 2001, **66**, 4576.
32. J. W. Cubbage, T. A. Tetzlaff, H. Groundwater, R. D. McCulla, M. Nag and W. S.

Jenks, *J. Org. Chem.*, 2001, **66**, 8621.
33. K. Tokumura and Y. Matsushita, *J. Photochem. Photobiol., A*, 2001, **140**, 27.
34. M. Fujitsuka and O. Ito, *Kokagaku*, 2000, **31**, 10.
35. M. Fujitsuka, O. Ito, K. Yamamoto and T. Akasaka, *Recent Res. Dev. Phys. Chem.*, 2000, **4**, 135
36. S. Smirnov, I. Vlassiouk, O. Kutzki, M. Wedel and F.-P. Montforts, *J. Amer. Chem. Soc.*, 2002, **124**, 4212.
37. M. Even, B. Heinrich, D. Guillon, D. M. Guldi, M. Prato and R. Deschenaux, *Chem.- Eur. J.*, 2001, **7**, 2595.
38. M. El-Khouly, M. Fujitsuka, O. Ito and M. El-Kemary, *J. Photochem. Photobiol., A*, 2001, **141**, 1.
39. K. Ikeuee, H. Yamashita, M. Anpo and T. Takewaki, *J. Phys. Chem. B*, 2001, **105**, 8350.
40. J. L. Bourdelande, J. Font and R. Gonzalez-Moreno, *Helv. Chim. Acta*, 2001, **84**, 3488.
41. F. Sevin and M. L. McKee, *J. Amer. Chem. Soc.*, 2001, **123**, 4591.
42. A. Dastan, N. Saracoglu and M. Balci, *Eur. J. Org. Chem.*, 2001, **66**, 3519.
43. R. Ozen, F. Kormali, M. Balci and B. Atasoy, *Tetrahedron*, 2001, **57**, 7529.
44. A. M. Galal, S. A. Ross, M. A. El-Sohly, H. N. El-Sohly, F. S. El-Feraly M. S. Ahmed and A. T. McPhail, *J. Nat. Prod.*, 2002, **65**, 184.
45. J.-M. Wu and Y. Li, *Tetrahedron Lett.*, 2001, **42**, 6737.
46. N. Chronakis, G. C. Vougioukalakis and M. Orfanopoulos, *Org. Lett.*, 2002, **4**, 945.
47. J. Tadic, I. Juranic and G. K. Moortgat, *J. Photochem. Photobiol., A*, 2001, **143**, 169.
48. J. Tadic, I. Juranic and G. K. Moortgat, *Molecules*, 2001, **6**, 287.
49. T. Mori, M. Takamoto, T. Wada and Y. Inoue, *Helv. Chim. Acta*, 2001, **84**, 2693.
50. T. Mori, M. Takaamoto, Y. Tate, J. Shinkuma, T. Wada and Y. Inoue, *Tetrahedron Lett.*, 2001, **42**, 2505.
51. S. Futamura and Y. Kamiya, *J. Chem. Soc., Chem. Commun.*, 1988, 1053.
52. S. C. Shim and J. S. Song, *J. Org. Chem.*, 1986, **51**, 2817.
53. M. Kamata, K.-I. Komatsu and R. Akaba, *Tetrahedron Lett.*, 2001, **42**, 9203.
54. H. Dincalp and Içli, *J. Photochem. Photobiol., A*, 2001, **141**, 147.
55. A. Itoh, T. Kodama and Y. Masaki, *Chem. Lett.*, 2001, 686.
56. I.-O. Koh and W. Thiemann, *J. Photochem. Photobiol., A*, 2001, **139**, 205.
57. Y. Ma, M. Zhang and G. Yu, *Huagong Jinzhan*, 2002, **21** 112.
58. C. T. Brigden, S. Poulston, M. V. Twigg, A. P. Walker and A. J. J. Wilkins, *Appl. Catal., B*, 2001, **32**, 63.
59. C. B. Almquist and P. Biswas, *Appl. Catal., A*, 2001, **214**, 259.
60. S. Vijaikumar, N. Somasundaram and C. Srinivasan, *Appl. Catal., A*, 2002, **223**, 129.
61. W. Xu and D. Raftery, *J. Catal.*, 2001, **204**, 110.
62. Y. Chen, J.-M. Cui, B.-H. Wang, S.-Q. Li and J. Liu, *Daqing Shiyou Xueyuan Xuebao*, 2001, **25**, 82.
63. X. Z. Li, C. M. Fan and Y. P. Sun, *Chemosphere*, 2002, **48**, 453.
64. J. H. Tay, D. Chen and D. D. Sun, *Water Sci. Technol.*, 2001, **43**, 319.
65. T. S. Wiltowski, R. D. Howerton, S. B. Lalvani and V. Zamansky, *Energy Sources*, 2001, **23**, 845.
66. D.-K. Lee and I.-C. Cho, *Stud. Surf. Sci. Catal.*, 2001, **133**, 303.
67. D. S. Muggli, M. J. Odland and L. R. Schmidt, J. Catal., 2001, 203, 51.
68. S. Yamazaki, H. Tsukamoto, K. Araki, T. Tanimura, I. Tejedor-Tejedor and M. A. Anderson, *Appl. Catal., B*, 2001, **33**, 109.
69. R. G. Larsen, A. C. Saladino, T. A. Hunt, J. E. Mann, M. Xu, V. H. Grassian and S.

C. Larsen, *J. Catal.*, 2001, **204**, 440.
70. M. R. Iesce, F. Cermola, F. De Lorenzo, I. Orabona and M. L. Graziano, *J. Org. Chem.*, 2001, **66**, 4732.
71. S. Onitsuka, H. Nishino and K. Kurosawa, *Tetrahedron*, 2001, **57**, 6003.
72. G. H. Hakimelahi, M. L. Jain, T. W. Ly, I.-C. Chen, K. S. Ethiraj, J. R. Hwu and A. A. Moshfegh, *J. Org. Chem.*, 2001, **66**, 7067.
73. G. Gopalakrishnan, N. D. P. Singh and V. Kasinath, *Molecules*, 2002, **7**, 112.
74. J. Cossy and D. Belotti, *Tetrahedron Lett.*, 2001, **42**, 4329.
75. W. Adam, H. Kilic and C. R. Saha-Moller, *Synlett*, 2002, 510.
76. T. Christ, F. Kulzer, P. Bordat and T. Basche, *Angew. Chem., Int. Ed.*, 2001, **40**, 4192.
77. Y. Murata and K. Komatsu, *Chem. Lett.*, 2001, 896.
78. K. Ozoemena, N. Kuznetsova and T. Nyokong, *J. Photochem. Photobiol., A*, 2001, **139**, 217.
79. O. M. Alfano, M. J. Martinez, R. J. Brandi and A. E. Cassano, *Water Sci. Technol.: Water Supply*, 2001, **1**, 73.
80. J. Sabate, J. M. Bayona and A. M. Solanas, *Chemosphere*, 2001, **44**, 119.
81. H. Einaga, S. Futamura and T. Ibusuki, *Appl. Catal., B: Environ.*, 2002, **38**, 215.
82. G. Fei, T.-J. Yang and L.-C. Yang, *Xiangtan Daxue Ziran Kexue Xuebao*, 2002, **24**, 60.
83. C. B. Almquist and P. Biswas, *Chem. Eng. Sci.*, 2001, **56**, 3421.
84. J. Lea and A. A. Adesina, *J. Chem. Technol. Biotechnol.*, 2001, **76**, 803.
85. R. Molinari, M. Borgese, E. Drioli, L. Palmisano and M. Schiavello, *Ann. Chim. (Rome, Italy)*, 2001, **91**, 197.
86. J. A. Herrera-Melian, J. M. Dona-Rodriguez, E. T. Rendon, A. S. Vila, M. B. Quetglas, A. A. Azcarate and L. P. Pariente, *J. Chem. Educ.*, 2001, **78**, 775.
87. R. R. Ozer and J. L. Ferry, *Environ. Sci. Technol.*, 2001, **35**, 3242.
88. R. R. Ozer and J. L. Ferry, *J. Phys. Chem., B*, 2002, **106**, 4336
89. E. Gilbert, *Ozone: Sci. Eng.*, 2002, **24**, 75.
90. L. Muszkat, L. Feigelson, L. Bir and K. A. Muszkat, *J. Photochem. Photobiol., B*, 2001, **60**, 32.
91. W. Leng, S. Tong, S. Cheng, J. Zhang and C. Cao, *Chem. J. Internet.*, 2001, **3**, (http://www.chemistrymag.org/cji/2001/032006pe.htm).
92. H. Taoda, *AIST Today*, 2001, **1**, 9.
93. H. R. Memarian, M. M. Sadeghi and A. R. Momeni, *Ind. J. Chem. Sect. B: Org. Chem. Incl. Med. Chem.*, 2001, **40B**, 508.
94. S. Jain, D. Pradhan and A. Upadhyaya, *Ind. J. Chem. Sect. B: Org. Chem. Incl. Med. Chem.*, 2001, **40B**, 1255.
95. A. S. Demir, F. Aydogan and I. M. Akhmedov, *Tetrahedron: Asymmetry*, 2002, **13**, 601.
96. C. Ferroud and P. Rool, *Heterocycles*, 2001, **55**, 545.
97. W. Adam, M. A. Arnold, M. Gruene, W. M. Nau, U. Pischel and C. R. Saha-Moeller, *Org. Lett.*, 2002, **4**, 537.
98. Y. Iijima, S. Obara, Y. Ohba, K. Kubo and T. Sakurai, *J. Photochem. Photobiol., A*, 2001, **143**, 23.
99. Y. Ohba, K. Kubo, T. Igarashi and T. Sakurai, *J. Chem. Soc., Perkin Trans. 2*, 2001, 491.
100. W. Luo, J. G. Muller and C. J. Burrows, *Org. Lett.*, 2001, **3**, 2801.
101. J.-Z. Tian, Z.-G. Zhang, X.-L. Yang, H.-K. Fun and J.-H. Xu, *J. Org. Chem.*, 2001, **66**, 8230.

102. X. Qin and P. Frech, *J. Pharm. Sci.*, 2001, **90**, 833.
103. S. Fukuzumi, S. Fujita, T. Suenobu, H. Imahori, Y. Araki and O. Ito, *J. Phys. Chem., A*, 2002, **106**, 1465.
104. C. R. Lopez, E. A. Gutarra and M. I. Litter, *Chemosphere*, 2002, **48**, 393.
105. D. Georgiou, P. Melidis, A. Aivasidis and K. Gimouhopoulos, *Dyes and Pigments*, 2002, **52**, 69.
106. M. Vautier, C. Guillard and J.-M. Herrmann, *J. Catal.*, 2001, **201**, 46.
107. T. Zhang, T. Oyama, A. Aoshima, H. Hidaka, J. Zhao and N. Serpone, *J. Photochem. Photobiol., A*, 2001, **140**, 163.
108. S. Horikoshi, N. Watanabe, M. Mukae, H. Hidaka and N. Serpone, *New J. Chem.*, 2001, **25**, 999.
109. S. Lacombe, H. Cardy, J. P. Soumillion, N. Soggiu, and M. Simon, *Entropie*, 2000, **36**, 31.
110. R. Lin and I. Chang, *Huaxue*, 2001, **59**, 317.
111. A. Toutchkine, D. Aebisher and E. L. Clennan, *J. Amer. Chem. Soc.*, 2001, **123**, 4966.
112. E. L. Clennan and D. Aebisher, *J. Org. Chem.*, 2002, **67**, 1036.
113. V. Iliev and A. Ileva, *J. Environ. Protect. Ecol.*, 2001, **2**, 603.
114. N. J. Brewer, R. E. Rawsterne, S. Kothari and G. J. Leggett, *J. Amer. Chem. Soc.*, 2001, **123**, 4089.
115. K. Yazu, Y. Yamamoto, K. Miki and K. Ukegawa, *J. Oleo. Sci.*, 2001, 50, 521.
116. H. Maki, T. Sasaki and S. Harayama, *Chemosphere*, 2001, **44**, 1145.
117. M. H. Habibi, S. Tangestaninejad and B. Yadollahi, *J. Ind. Pollut. Control*, 2001, **17**, 67.
118. M. Sokmen, D. W. Allen, A. T. Hewson and M. R. Clench, *J. Photochem. Photobiol., A*, 2001, **141**, 63.
119. L. Davydov, E. P. Reddy, P. France and P. G. Smirniotis, *Appl. Catal., B*, 2001, **32**, 95.
120. A. Pochon, P. P. Vaughan, D. Gan, P. Vath, N. V. Blough and D. E. Falvey, *J. Phys. Chem., A*, 2002, **106**, 2889.
121. R. P. Ponomareva and O. P. Studzinskii, *Russ. J. Gen. Chem.*, 2001, **71**, 759.
122. J. Arana, E. Tello Rendon, J. M. Dona Rodriguez, J. A. Herrera Melian, O, Gonzalez Diaz and J. Perez Pena, *Chemosphere*, 2001, **44**, 1017.
123. X. Tao, W. Ma, T. Zhang and J. Zhao, *Angew. Chem. Int. Ed.*, 2001, **40**, 3014.
124. G. H. Rossetti, E. D. Albizzati and O. M. Alfano, *Indust. Eng. Chem. Res.*, 2002, **41**, 1436.
125. J. Beltran, J. Torregrosa, J. R. Dominguez and J. A. Perez, *J. Chem. Tech. Biotech.* 2001, **76**, 1243.
126. S. H. Bossman, E. Oliveros, S. Gob, M. Kantor, A. Goppert, A. M. Braun, L. Lei and P. L. Yue, *Prog. React. Kinet. Mech.*, 2001, **26**, 113.
127. M. Ghaly, R. Mayer, G. Hartel and R. Haseneder, *Chem. Tech. (Leipzig Ger.)*, 2001, **52**, 228.
128. R. F. Pupo Nogueira, A. G. Trovo and D. F. Mode, *Chemosphere*, 2002, **48**, 385.
129. A. Goi and M. Trapido, *Chemosphere*, 2002, **46**, 913.
130. X. Zhou and Q. Zeng, *Hunan Nongye Daxue Xuebao*, 2001, **27**, 130.
131. J. L. Acero, F. J. Benitez, F. J. Real and A. I. Leal, *Water Sci. Tech.*, 2001, **44**, 31.
132. M. P. Ormad, J. L. Ovelleiro and J. Kiwi, *Appl. Catal., B*, 2001, **32**, 157.
133. N. Deng, X. Liu, Z. Zhang, F. Wu and F. Luo, *Fressen. Environ. Bull.*, 2001, **10**, 638.
134. T. Li and Q. Liu, *Huagong Huanbao*, 2001, **21**, 84.
135. M. Fukushima, K. Tatsumi and S. Nagao, *Environ. Sci. Technol.*, 2001, **35**, 3683.
136. S. Malato, J. Caceres, A. Agueera, M. Mezcua, D. Hernando, J. Vial and A. R.

Fernandez-Alba, *Environ. Sci. Technol.*, 2001, **35**, 4359.
137. S. Parra, S. Malato, J. Blanco, P. Peringer and C. Pulgarin, *Water Sci. Technol.*, 2001, **44**, 219.

6
Photoelimination

BY IAN R. DUNKIN

1 Introduction

This chapter deals with photoinduced fragmentations of organic and selected organometallic compounds, and especially reactions accompanied by loss of small molecules such as nitrogen, carbon monoxide or carbon dioxide. Photodecompositions which produce two or more larger fragments and other miscellaneous photoeliminations are reviewed in the final section. Photofragmentations of carbonyl compounds taking place, for example, by Norrish Type I and II processes, are discussed systematically in Part II, Chapter 1, although some eliminations from carbonyl compounds are also included here, *e.g.* certain decarbonylations and decarboxylations.

Of general interest in this context, a large review of photodissociation of simple molecules in the gas phase has been published,[1] while a theoretical model has been presented, which describes molecular fragmentation into several fragments following absorption of one or several photons from ultrashort laser pulses.[2] Two reviews of photoremovable protecting groups have also appeared.[3,4]

2 Elimination of Nitrogen from Azo Compounds and Analogues

Semiclassical simulations, modified in accord with available experimental data, have been carried out for the events following $n \rightarrow \pi^*$ excitation of the azomethane molecule (MeN=NMe).[5] These showed that, for an isolated molecule, dissociation should take place on the electronic ground-state surface, because radiationless decay is very fast. Moreover, the dissociation is mainly sequential, giving first the methyldiazenyl radical ($MeN_2\cdot$), which fragments rapidly to N_2 and a methyl radical. These studies were extended to simulate the photochemistry of azomethane in aqueous solution, where the fragmentation is suppressed because of vibrational energy loss to the solvent, and the major process is *trans* to *cis* isomerization. The authors claim this as the first simulation of condensed-phase photochemistry with a realistic model of the solvent and a polyatomic solute, and this approach has helped to explain apparent contradictions in experimental results obtained by Lee's and Zewail's groups, which had provided evidence for both slow and fast dissociations.[6,7] The photodissociation

Photochemistry, Volume 34

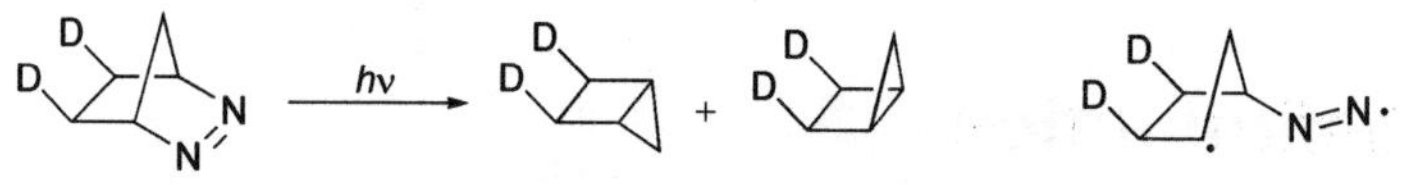

(1) (2*inv*) (2*ret*) (3)

Scheme 1

of *s*-tetrazine into N_2 and two molecules of HCN has also been studied theoretically.[8]

In a continuation of mechanistic investigations on the photo-extrusion of N_2 from diazabicylo[2.2.1]hept-2-ene (DBH), it was found that the stereoselectivity of the formation of (2) from the stereolabelled DBH derivative (1) (Scheme 1) in supercritical media (sc-CO_2, sc-C_2H_6) decreased by a factor of about 2.3 when the pressure was increased up to 200 bar.[9] In all cases the inverted product (2inv) predominated, but in sc-C_2H_6, for example, the rate constant ratio k_{inv}/k_{ret} fell from 10.9 at 40 bar to 4.8 at 200 bar. According to the authors, this observation is best accounted for in terms of the pressure-dependency of the viscosity of the reaction medium, and that the observations imply a stepwise denitrogenation, probably via the singlet diazenyl diradical (3), thus corroborating previous conclusions. The photochemical reactions of the hydroxy benzotriazinone (4) and its tris complex with iron(III) have been examined in solution and in low-temperature (4-77 K) glasses.[10] Photoexcitation at $\lambda > 345$ nm resulted in elimination of N_2. In solution, the first species observed from (4) was the oxime ketene (5), presumably arising by ring opening of the β-lactam (6). In low-temperature glasses, however, EPR revealed the presence of the intermediate diazenyl diradical (7).

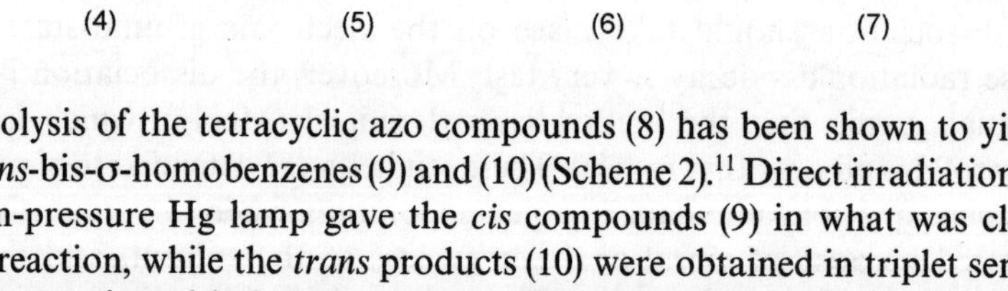

(4) (5) (6) (7)

Photolysis of the tetracyclic azo compounds (8) has been shown to yield *cis*- and *trans*-bis-σ-homobenzenes (9) and (10) (Scheme 2).[11] Direct irradiation with a medium-pressure Hg lamp gave the *cis* compounds (9) in what was clearly a singlet reaction, while the *trans* products (10) were obtained in triplet sensitized reactions starting with the azo compounds (8), the corresponding *cis* compounds (9) or the 1,3,6-cyclooctatrienes (11) which were the thermolysis products of the azo precursors (8).

UV irradiation of the triazolines (12) gave the expected nitrogen-loss products, *i.e.*, the aziridines (13), but only in low (13-17%) yield.[12] The major pathway gave cyclobutane-cleavage products: the triazoles (14), the hitherto unknown pyridazinonorbornadiene (15) and the di-π-methane rearrangement product of the

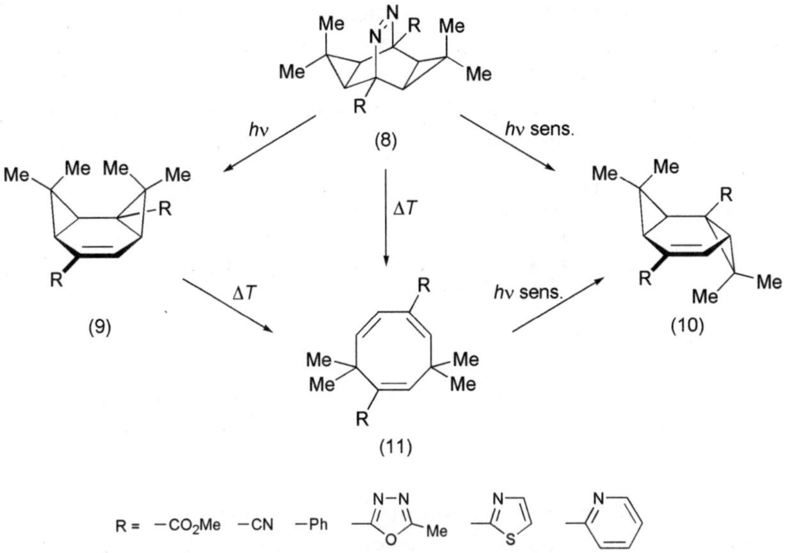

Scheme 2

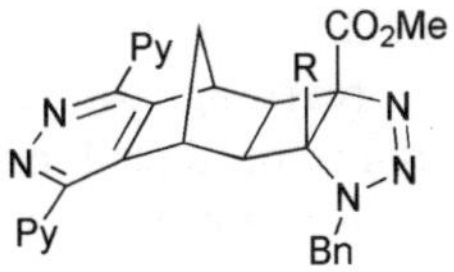

(12)

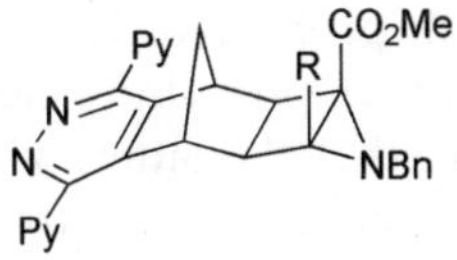

(13)

Py = 2-pyridyl
Bn = benzyl
R = H, CO_2Me

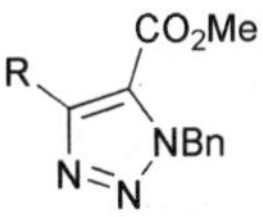

(14)

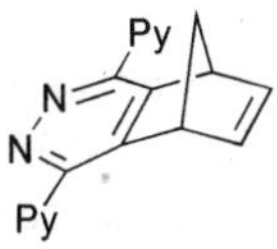

(15)

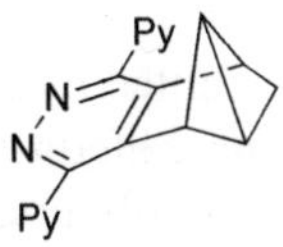

(16)

latter (16). The 1,3-dipolar cycloaddition of diazomethane to (1,1-difluoroallyl)phosphonates gave pyrazolines, which, on photolysis, eliminated N_2 to yield the corresponding cyclopropanes with a diethoxyphosphoryldifluoromethyl substituent.[13] Pyrazole (17), for example, gave the cyclopropane (18).

3 Elimination of Nitrogen from Diazo Compounds and Diazirines

3.1 Generation of Alkyl and Alicyclic Carbenes. – Potential energy surfaces of the ground and some excited states of 3*H*-diazirine and diazomethane have been explored computationally at a high level of theory.[14] An important conclusion was that photoexcitation to the S_1 state of either molecule would be followed by radiationless transition to S_0 via a conical intersection, leaving enough energy for dissociation into N_2 and CH_2 in its lowest singlet state. This should occur on a very short time scale (*ca.* 100 fs). In addition, both precursors can generate excited singlet CH_2 when excited onto the S_2 surface; in this case passing through a conical intersection to the S_1 surface, on which dissociation takes place.

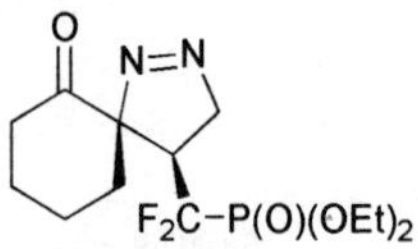

(17)

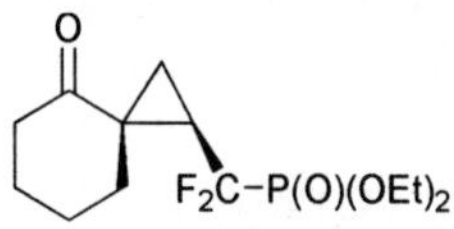

(18)

(19)

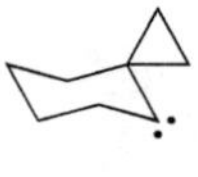

(20)

(21)

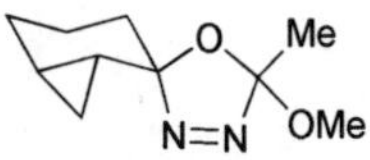

(22)

Conformationally restricted cyclopropylcarbenes (19) and (20), with *syn* or *anti* geometries, repectively, have been generated from their diazo precursors, *e.g.* (21), which were formed *in situ* by photolysis of the corresponding oxadiazolines such as (22).[15] Each carbene had a lifetime of 1-2 ns in cyclohexane at 10°C, and could be trapped with piperidine. The authors claim that these observations support the view that cyclopropylcarbenes do not rearrange to cyclobutenes in solution (0-25°C), and that the latter, when they are produced, are formed in excited state rearrangements of the nitrogen-containing precursors, which by-pass carbene intermediates.

1-Adamantylcarbene (23) and 3-noradamantylcarbene (24) have been generated from the corresponding diazirines, and investigated in laser flash-photolysis (LFP) studies.[16] While (23) can be readily trapped with pyridine to form an ylid, (24) cannot be intercepted by pyridine in LFP experiments. Thus it was deduced that the solution-phase lifetime of (23) is at least 10 times longer than that of (24). Further evidence was obtained which suggested that 3-noradamantylcarbene (24) is formed very inefficiently from its diazirine precursor, which apparently rearranges in the excited state, concurrently with loss of N_2, to form adamantene (25), without the intermediacy of carbene (24).

In nanosecond laser flash-photolysis experiments (193 and 248 nm), Wiberg's silene (27) has been detected directly, as the more prominent transient species (λ_{max} ~265 nm), in the photolysis of the diazo precursor (26; X = N_2) or the ketene (26; X = CO).[17] The second transient species could not be identified fully,

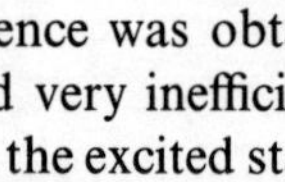

(23)

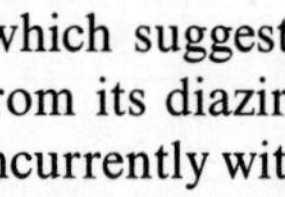

(24)

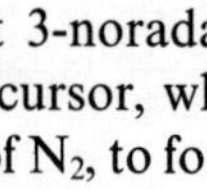

(25)

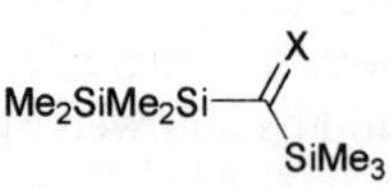

(26)

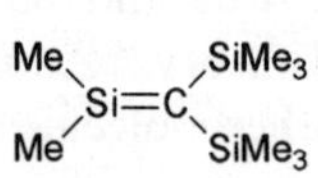

(27)

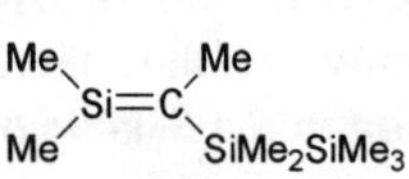

(28)

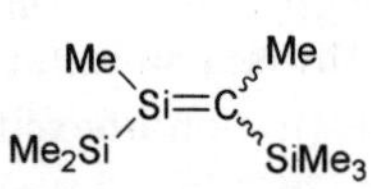

(29)

but was thought to be either (28) or (29, *E* or *Z*). Absolute rate constants were determined for the reactions of (27) and (28) or (29) with MeOH, MeOD, propan-2-ol, cyclohexanol, *tert*-butanol, *tert*-butylamine and acetic acid in hexane solutions at 24°C.

3.2 Generation of Aryl Carbenes. – In LFP experiments, (biphenyl-4-yl)chlorocarbene (30) was detected as a transient (λ_{max} = 360 nm) immediately after the flash, when the corresponding diazirine precursor was photolysed in 2,2,4-trimethylpentane containing 2-vinylpyridine.[18] Carbene (30) then decayed by first-order kinetics to give the pyridinium ylid (31; Ar = biphenyl-4-yl). In preparative-scale experiments, this reaction produced indolizine (32; Ar = biphenyl-4-yl) in 56% yield, by cyclization of (31) and elimination of HCl; cyclopropane products were also obtained (12% total yield).

Naphthoquinone diazides (33; X = H, Me, NO_2) have been photolysed in THF and 1,4-dioxane, yielding a variety of products, depending on the 2-substituent and the solvent.[19] Insoluble copolymers of the resulting carbenes and the solvent predominated in both solvents when X = H or NO_2, but with X = Me polymerization was suppressed and several interesting smaller products resulted: *e.g.*, in dioxane, the novel polyether bridged naphthalene (34) and the spiro compound (35). The naphthoquinone diazide functionality has been incorporated into novel materials for positive resists.[20,21]

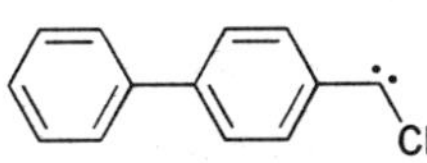

(30)

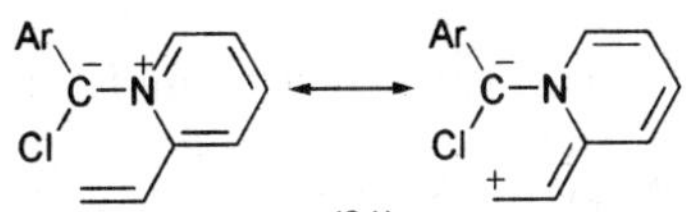

(31)

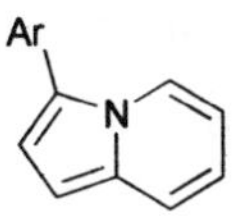

(32)

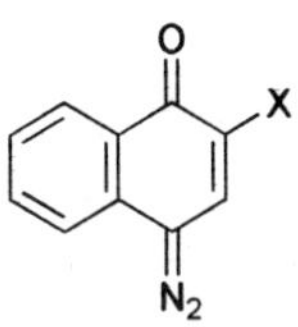

(33)

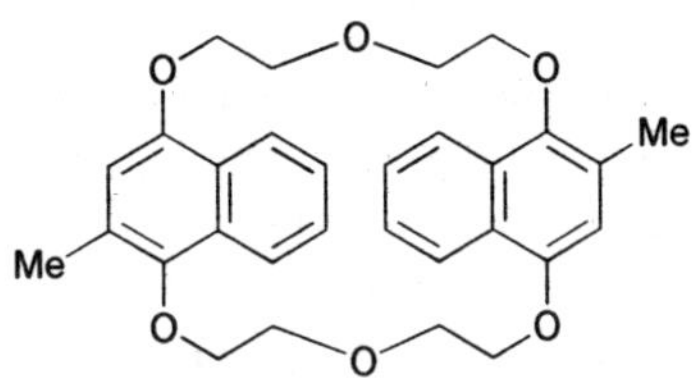

(34)

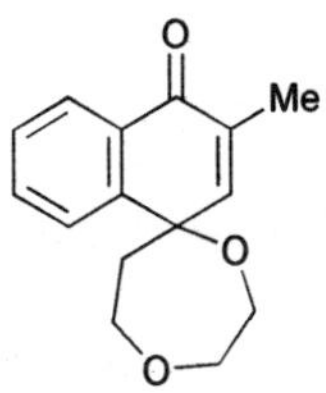

(35)

Photolysis of the silyl phenyldiazomethane compound (36) in methanol gave a 97% yield of the alkene (37; TBS = *t*-$BuMe_2Si$), along with a small amount (2.4%) of the methyl ether (38; TBS = *t*-$BuMe_2Si$).[22] Capture of an intermediate carbene by methanol is proposed to account for the minor product. It is interesting that this observation is at odds with computations at the B3LYP/6-31G* level of theory, which predict that the carbene expected to arise from (36) is not an energy minimum and should rearrange without barrier via silyl migration. Further evidence for the existence of a discrete carbene, however, was provided by the thermolysis of (36) in cyclohexane under O_2, in which substantial amounts of *tert*-butyldimethylsilyl acetophenone were formed, presumably by reaction of the carbene with O_2 to give a carbonyl oxide. In another study, benzophenone

O-oxide (Ph_2COO) was generated by photolysis of Ph_2CN_2 in the presence of O_2 and its reactivity with a range of alcohols determined.[23] Electron donation to the OH group of the alcohol favoured the attack of Ph_2COO, implying that the carbonyl oxide reacts as an electrophile.

Diazirine-base photoreactive fatty acid analogues (39; n = 4, R = H; n = 10, R = H, Me, Et, succinimidyl) have been synthesized, and one of these (n = 10, R = succimidyl) was further elaborated to give the photoreactive galactosylceramide (40).[24] The latter compound decomposed smoothly when irradiated with a black light lamp, and was tested, seemingly successfully, as a photaffinity label in

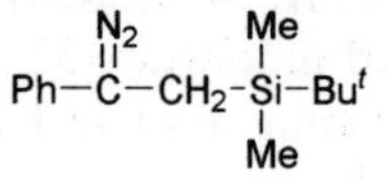

(36)

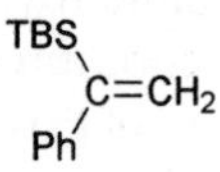

(37)

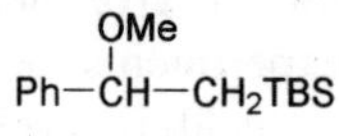

(38)

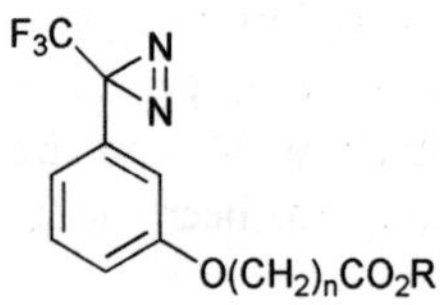

(39)

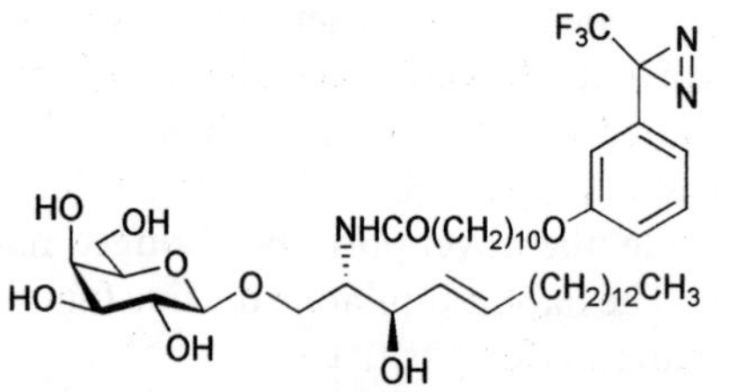

(40)

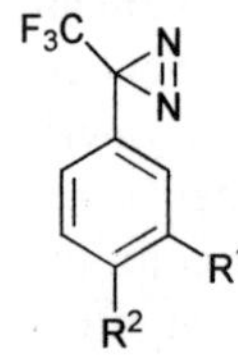

(41)

the detection of rabbit anti-galactosylceramide antibody. Three bifunctional 3-phenyl-3-(trifluoromethyl)diazirines, including tritium-labelled analogues (41; R^1 = H, ^3H; R^2 = CH_2OH, CO_2H, CH_2NH_2) have been synthesized as potential photoaffinity probes.[25]

3.3 Photolysis of α-Diazo Carbonyl and Sulfonyl Compounds. – A photoinduced β-lactam synthesis has been reported, in which diazoketones (42), derived from protected amino acids, underwent photoextrusion of N_2, rearrangement and finally reaction with imines (43) (Scheme 3).[26] The diastereomeric β-lactam products (44) and (45) were formed in yields of 44-84%, and with diastereomeric ratios, (44):(45), ranging from 65:35 to 90:10. Both electron-rich (*e.g.* 43; R^2 = 4-$MeOC_6H_4$) and electron poor imines (*e.g.* R^2 = 4-$O_2NC_6H_4$), as well as heteroaryl imines (R^2 = 2-furyl, 2-thienyl), were successful in the reaction, but none of the aliphatic imines tried led to the formation of β-lactams.

The photochemistry of a series of bis(sulfonyl)diazomethanes (46; Ar = Ph, 4-MeC_6H_4, 4-ClC_6H_4, *c*-C_6H_{11}), which are of interest as photoacid generators in photoresists for deep-UV lithography, has been investigated in solution and in low-temperature matrices.[27] Triplet carbenes (47) could be trapped in solution by reaction with EtOH, cyclohexane or cyclohexene, but in Ar matrices at 10 K no carbenes were detected, only the sulfenes (48) and other rearranged products. Moreover, carbenes could not be trapped by added O_2 in matrices. These observations were in line with theoretical predictions that the singlet carbenes are transition states on the pathway to oxathiirene oxides (49); while the triplet

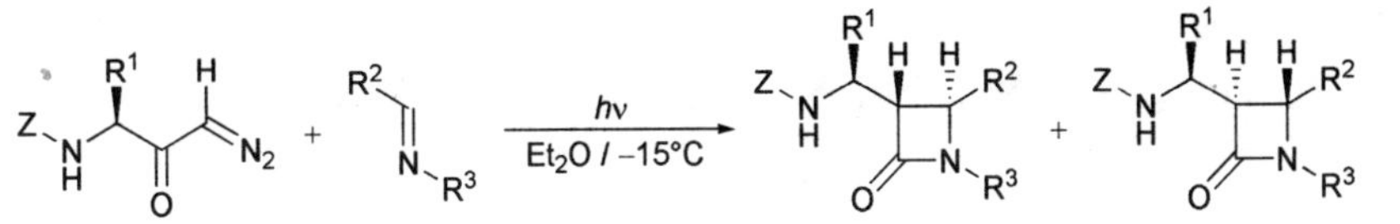

(42) (43) (44) (45)

Z = $PhCH_2OCO$
R^1 = Me, Pr^i, Bu^t

R^2 = aryl, heteroaryl, (*E*)-styryl
R^3 = $PhCH_2$, allyl, Bu^t

Scheme 3

carbenes lie energetically above these singlet transition states, and so are not expected to have a prolonged lifetime, even in low-temperature matrices. Overall, these results seem to demonstrate that the corresponding triplet carbenes are the primary products of the photolysis of the bis(sulfonyl)diazo compounds (46).

4 Elimination of Nitrogen from Azides

On the theoretical front, *ab initio* calculations have been reported for two-dimensional cuts through the full six-dimensional potential energy surfaces of the ground and first excited singlet states of HN_3, with a view to providing insight into the photochemistry of this molecule.[28] Two-dimensional quantum mechanical wavepacket calculations of the photodissociation dynamics were also carried out, and these allowed an examination of the factors influencing the relative branching into the competing H–NNN and HN–NN bond fission channels, which the authors say is in reasonable accord with the available experimental data.

The photolysis of a series of derivatives of *o*-tolyl azide with substituents at the benzylic position (50) has been investigated by matrix-isolation spectroscopy.[29] It was found that introduction of any benzylic substituent possessing a lone pair of electrons (*i.e.*, Y = Br, Cl, MeO, Me_2N) allowed a 1,4-hydrogen shift to occur in the intermediate nitrene, yielding iminoquinone methides (51). On the other hand, additional methyl groups at the benzylic position did not promote this reaction. In cases where the benzylic position was part of a ring system, the size of this ring seemed to be important. Thus, the nitrenotetrahydroisoquinoline (52) rearranged easily, whereas 4-nitrenophthalan (53) did not. The diethylamino derivative (50; X = H, Y = NEt_2) was studied by laser flash photolysis, and the reactivity of the corresponding iminoquinone methide (51; X = H, Y = NEt_2) towards various trapping agents examined.[30] The iminoquinone methide reacted efficiently with electron deficient dienophiles such as maleic anhydride, but not with simple olefins such as cyclohexene. Furthermore, although it showed only low to medium reactivity towards simple primary or secondary amines and simple alcohols, it had high reactivity towards thiols, diols, triols, sugars and cytosine.

1,5-Dinitrenonaphthalene (54) has been generated in argon matrices at 11 K

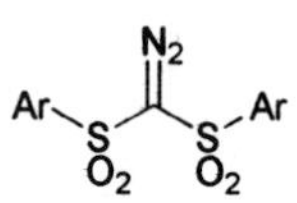

(46)

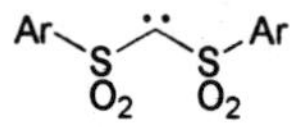

(47)

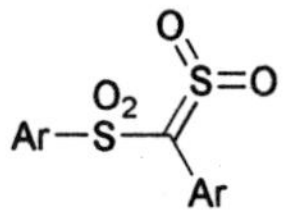

(48)

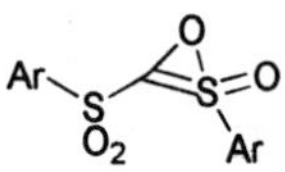

(49)

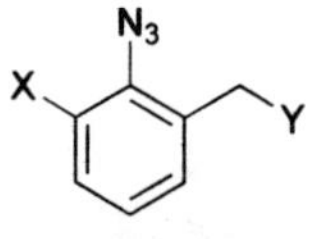

(50)

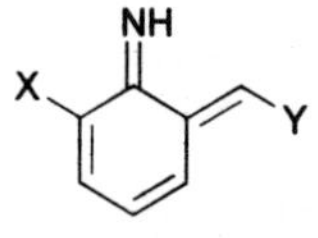

(51)

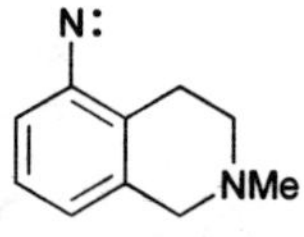

(52)

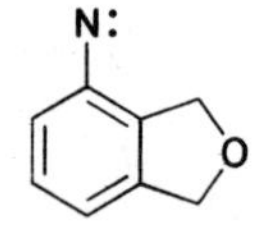

(53)

and in 2-methyltetrahydrofuran glasses at 77 K, by sequential elimination of two N_2 molecules from 1,5-diazidonaphthalene, and it was examined by UV-visible, IR and EPR spectroscopy.[31] The absence of an EPR signal indicated that the ground state of (54) is a singlet diradical. This agreed with theoretical calculations, and the species is probably best represented as the quinoidal diradical (55). A new nitrene to carbene rearrangement has been discovered by EPR spectroscopy in the photolysis of 4-amino-2,6-diazido-3,6-dichloropyridine (56), in which the intermediate quintet dinitrene gave a nitrenodiazacycloheptatrienylidene.[32] A study of the effect of nitrene multiplicity on product yields in the photooxidation of 4,4′-diazidobiphenyl has been reported.[33] The addition of triplet photosensitizers resulted in a substantial increase in the yield of nitroso compounds, with a maximum effect for sensitizers with a triplet energy level of 250 kJ mol^{-1} or higher.

Photolysis of acyl azides with chiral auxiliary groups (57) in the presence of cyclic enol ethers (58) gave the oxazolines (59), which, at least formally, are the result of the cycloaddition of the corresponding aroylnitrenes to the C=C bond of the enol ethers.[34] The cycloaddition occurred *trans* to the OR^2 group in the enol ether, but an analogous cycloaddition of aroylnitrenes to ketones did not take place stereoselectively. Benzoyl azides with a neighbouring amide group underwent an unexpected nitrene insertion into the amide N–H bond.

Azides continue to be exploited in the development of photoaffinity reagents. For example, 3′-azidoabscisic acid (60) has been synthesized as a potential photoaffinity reagent for proteins which bind abscisic acid.[35] Stable in solution in the dark, this azido derivative was decomposed by UV irradiation, and its biological activity was equivalent to that of abscisic acid. An azido, biotinylated affinity reagent for streptavidin has been shown to undergo covalent attachment to a peptide unit in the latter upon photolysis.[36] Preparative routes to the potential photoaffinity reagent, 4-azido-2,3,5,6-tetrafluroaniline, have been described.[37] A novel photocrosslinking resin for the isolation of heparin binding proteins has been developed, which contains a 4-azidophenylamino groups as the photoactive functionality.[38] A 4-azidosalicylamino group has been incorporated into derivatives of gangliosides, which were then radioiodinated to a high specific radioactivity, and the resulting reagents were used for photolabeling of erythrocyte membrane proteins.[39]

Azide photolysis also finds a number of applications in polymer technology.

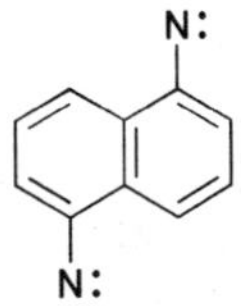
(54)

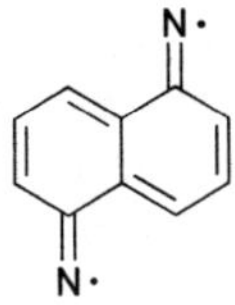
(55)

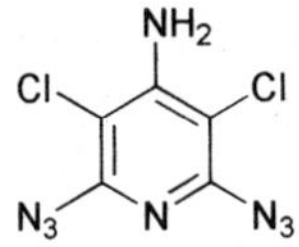
(56)

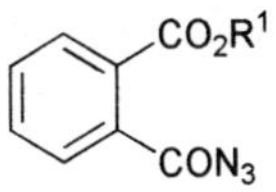
(57)

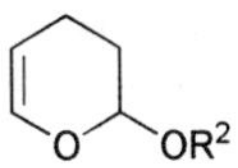
(58)

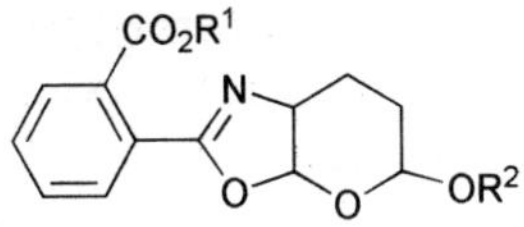
(59)

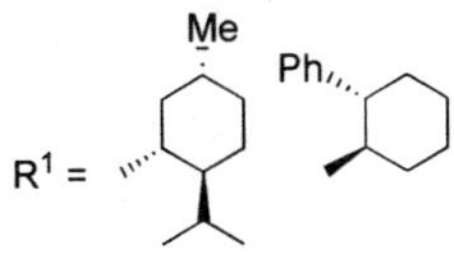

R^2 = Me, Bu^i

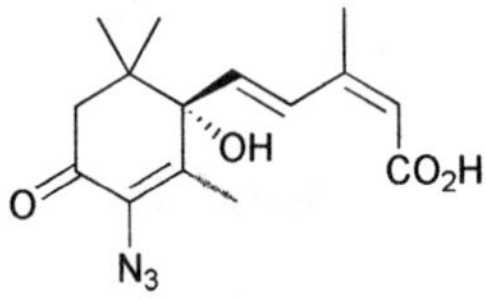
(60)

Aromatic azides with light stabilizer residues or hydrophilic groups, such as glucose or sucrose residues, were synthesized and used for the photolytic surface modification of polyolefins.[40] By photochemical immobilization of the carbohydrates, the hydrophilicity of the polymers was strongly increased. Bulk modifications of polymers were also achieved. Aromatic bisazide compounds have been utilized as photocrosslinkers for a novel silcone ladder polymer.[41] The efficiency of the crosslinking was probed by FT-IR spectroscopy. Gaseous products from the laser photodissociation of a glycidyl azide polymer have been determined by FT-IR, and their reactions with O_2 also examined.[42] Azide-containing photoinitiators for polymerizations have been reported,[43,44] as has an azide-based photosensitive water-soluble polymer.[45]

5 Photoelimination of Carbon Monoxide and Carbon Dioxide

The gas-phase 193 nm photolysis of ketene in the presence of NO has been investigated with the aid of time-resolved mass spectrometry.[46] The main products observed were HCNO and HCN, both of which appear to arise by reaction of CH_2 or HCCO with NO. Relative product yields have been determined at nine different photolysis energies for all three photofragmentation channels of both thermal and vibrationally excited HNCO: $^3NH + CO$, $H + NCO$ and $^1NH + CO$.[47] It was concluded that vibrational excitation impedes the decomposition to $^1NH + CO$ on the S_1 surface and thus increases the number of molecules that cross to S_0 and decay to $H + NCO$.

Further matrix-isolation studies of the generation of benzynes by photolysis of dicarboxylic anhydrides have been published.[48,49] Consecutive decarboxylation and decarbonylation of 1,2-naphthalenedicarboxylic anhydride (61) (Scheme 4) occurred when (61) was photolysed at 355 nm in Ar matrices at 11 K.[48] These processes were quantitatively analysed by UV-visible absorption spectroscopy, and the quantum efficiencies were estimated as 2×10^{-4} and 0.1, respectively. Benzocyclopentadienylideneketene (64) was ultimately formed by further reaction of 1-naphthyne (63) with CO, but the analogous reaction was not observed with the isomeric 2-naphthyne. Two benzediynes (65; X = F, CF_3) have been directly observed in N_2 matrices at 13 K following wavelength-selective photolysis of the benzenetetracarboxylic dianhydrides (66), part of a study of five kinds of

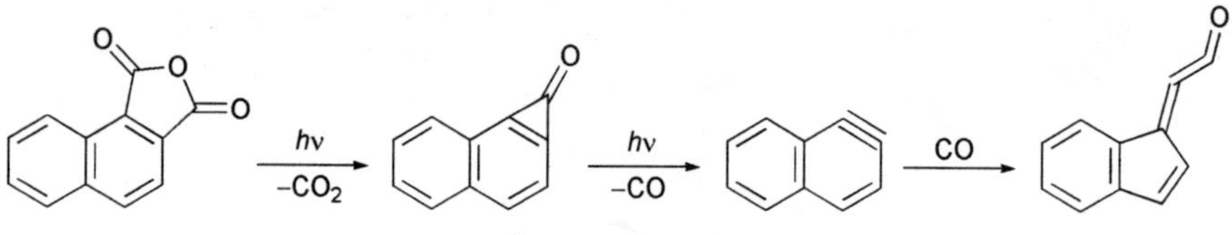

Scheme 4

benzenetetracarboxylic dianhydride derivatives.[49] In the first step, the dianhydrides were converted into benzynedicarboxylic anhydrides, with loss of CO and CO_2, by irradiation at 308 nm; in the second step, 266 nm irradiation resulted in further loss of CO and CO_2, to give the benzdiynes (65). The benzdiynes were ultimately converted into the corresponding hexatriynes (XC≡C–C≡C–C≡CX) by continued irradiation at 266 nm. The other tetracarboxylic dianhydrides studied did not give detectable amounts of benzdiynes, but it was concluded from the observed hexatriyne products that benzdiynes had been generated, but were immediately transformed into the hexatriynes.

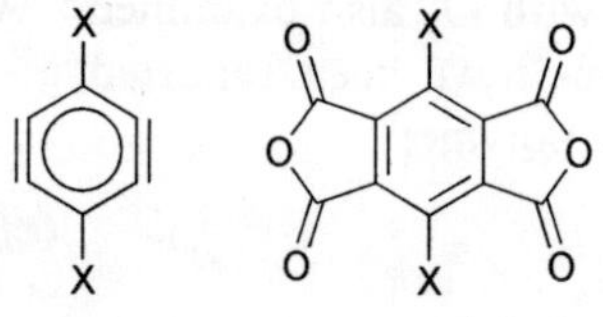

(65) (66)

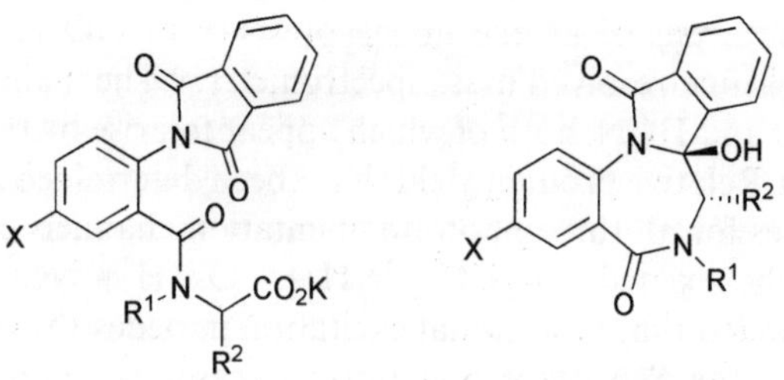

(67) (68)

R^1 = H, Me
R^2 = H, Me, Pr^i, Bu^i, $PhCH_2$
X = H, Cl

Photochemical decarboxylation of amides derived from *N*-phthaloylanthranilic acid and its 4-chlorinated analogue, *e.g.* (67), in aqueous acetone yielded 1,4-benzodiazepines, *e.g.* (68), mostly in good yields, and with excellent diastereoselectivities.[50] Near-UV irradiation of 2-acetyl- and 2-benzoylphenylacetic acids (69; R = Me, Ph; X = H) in benzene solution resulted in their efficient decarboxylation, yielding the corresponding 2-acyltoluenes.[51] The *meta* and *para* isomers did not undergo this process, and analogous esters and amides were essentially photoinert. Interestingly, the *O*-deuteriated acids (69; R = Me, Ph; X = D) gave the monodeuteriated toluene derivatives (70; R = Me, Ph); and the authors suggest the possibility of an intramolecular proton transfer step in the decarboxylation process. Irradiation of 2-phenylpropionic acid in various cation-exchanged faujasites (medium-pressure Hg arc) led to decarboxylation and the predominant formation of racemic (*R,R* + *S,S*) 2,3-diphenylbutane rather than the *meso* isomer (*R,S*).[52] This contrasted markedly with photolysis in isotropic media, and the induced diastereoselectivity was attributed to steric and electronic factors within the zeolite.

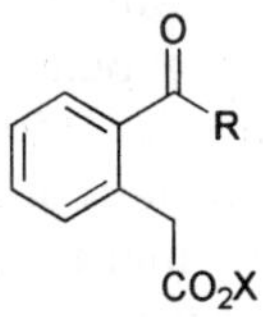

(69)

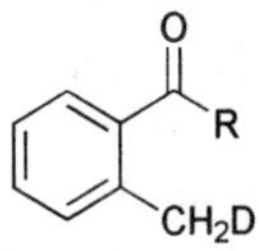

(70)

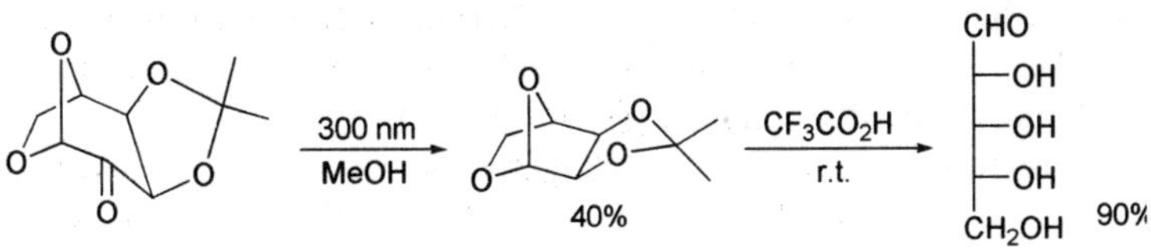

(71) (72) (73)

Scheme 5

A photochemical decarbonylation was the key step in a new route to the aldopentoses, ribose and lyxose, and the aldohexoses, talose and gulose.[53] For example, (71) gave D-ribose (73) via the decarbonylated compound (72), as summarized in Scheme 5. It has been reported that, although styrene glycol carbonate (74; R^1, R^2 = H) and the *trans*-β-methyl analogue (74; R^1 = H; R^2 = Me) extrude CO_2 upon photolysis, to give styrene oxide and phenylacetone ($PhCH_2COMe$), respectively, the β,β-methyl carbonate (74; R^1, R^2 = Me) was apparently inert.[54] A similar surprising contrast in behaviour was also found for the photolysis of styrene oxides and their methylated derivatives. For the latter case, the authors proposed a plausible explanation based on the geometric requirements of the reaction, but it is not clear, at least to this reviewer, how this is to be extended to the carbonates (74).

As mentioned above (Section 3.1), Wiberg's silene (27) has been detected directly as a transient in the flash photolysis of the ketene (26; X = CO), as well as the analogous diazo compound.[17]

5.1 Photoelimination of CO from Organometallic Compounds. – A theoretical model of photofragmentation into several fragments, induced by the absorption of one or several photons from ultrashort pulses, has been applied to the fragmentation of $Fe(CO)_5$, with excellent agreement between the theory and experiment.[2] Photosubstitution of CO in $Fe(CO)_5$ by alcohols and PEt_3 has been studied using femtosecond UV pump IR probe techniques.[55] It was shown, for example, that coordination of the hydroxyl group in alcohols with triplet $Fe(CO)_4$ is generally much faster than observed for comparable singlet species. The molecular structure of transient $Fe(CO)_4$ has been determined using diffraction of ultrashort pulses of electrons.[56] The results indicated that the major product, up to 200 ps, was $Fe(CO)_4$ in the $^1A_1$ state, rather than the ground $^3B_2$ state. Refinement of the structure obtained from the diffraction data gave bond lengths and angles of the transient species to a claimed resolution of about 0.05 Å (5 pm); and the authors' claim that this technique 'should be of significant value in identifying transition configurations and pathways of other reactions' seems well justified. A femtosecond spectroscopic investigation has been made of the photoelimination of CO from $M(CO)_6$ (M = Cr, Mo, W), $Fe(CO)_5$, $Ni(CO)_4$ and $M_2(CO)_{10}$ (M = Mn, Re).[57] The molecules were pumped by one photon at 267 nm, then probed by multiphoton ionization at 800 nm and mass-selective detection of the resulting ions. In contrast to previous beliefs that metal carbonyls of this type are excited to a repulsive potential energy surface, leading to elimination of one or more CO molecules, it was found that only one CO was split off in times typically less than 100 fs, and that this was already a multistep process involving relaxation between excited states. Elimination of a second CO molecule appeared to take place on the S_0 surface of the unsaturated metal carbonyl and required much longer times (> 1 ps).

Photosubstitution of CO in $W(CO)_6$ by alkynes has been investigated by NMR spectroscopy for the alkynes HC≡CH, HC≡CMe and $HC≡CCMe_3$, amongst others.[58] Unstable terminal alkyne complexes of the type $[W(CO)_5(\eta^2\text{-}HC{\equiv}CR)]$, $[W(CO)_4(\eta^2\text{-}HC{\equiv}CR)_2]$ and $[W(CO)_3(\eta^2\text{-}HC{\equiv}CR)_3]$, and vinylidene derivatives $[W(CO)_5(C{=}CHR)]$, were identified and characterized. The complexes $[W(CO)_3(\eta^2\text{-}HC{\equiv}CR)_3]$ were also shown to be involved in the cyclotrimerization ot the alkynes to arenes. Following photolysis of $W(CO)_6$, 2-aminopyrimidine and 4-aminopyrimidine coordinate to $W(CO)_5$ predominantly via the exocyclic amino group (>91%) rather than via N-1, in contrast to what was found for 2-aminopyridine.[59] Secondary photolysis led to low yields of $W(CO)_4$ complexes coordinated to the exocyclic amine and the adjacent endocyclic N atom (N-1 or N-3).

UV photolysis of $Re_2(CO)_{10}$ and benzothiophenes in hexane solution has been shown to produce ring-opened complexes (75; R^1, R^2, R^3 = H; R^1, R^2 = H, R^3 = Me; R^1, R^3 = H, R^2 = Me; R^1, R^2 = Me, R^3 = H).[60] Diamagnetic complexes $[M(CO)_4tmps]_2$ ($(tmps)_2$ = bis[2,2,6,6-tetramethylpiperidyl-(1)]-disulfide (76)) were formed when the metal carbonyls $M_2(CO)_{10}$ (M = Mn, Re) were photolysed in the presence of the disulfide $(tmps)_2$ (76) in *n*-pentane solution.

Photolysis of cyclopentadienylmanganese tricarbonyls such as (77; R^1, R^2 = H, Me) in the presence of a proton source (MeOH or H_2O) was found to lead to

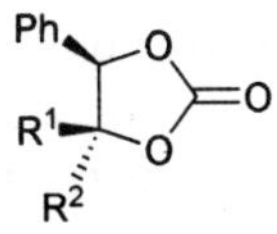

(74)

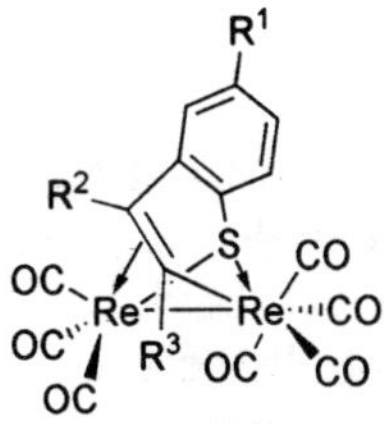

(75)

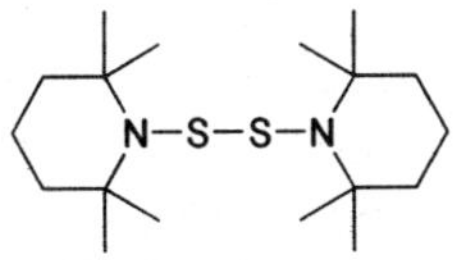

(76)

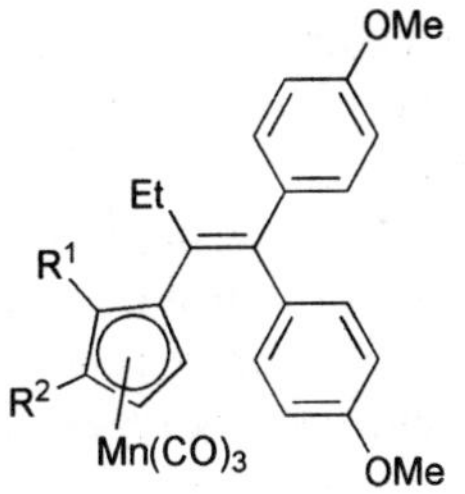

(77)

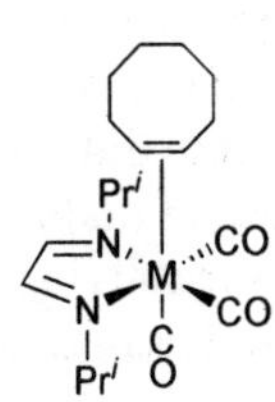

(78)

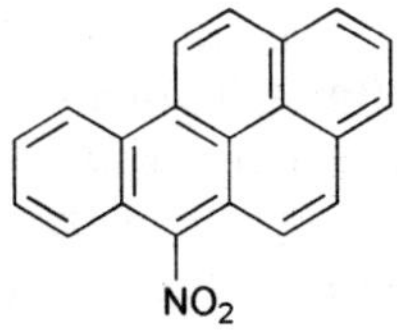

(79)

decomplexation and formation of the free cyclopentadiene in high yields.[61] Attempted decomplexation using Ce(IV) or Fe(III), on the other hand, gave only decomposition products. It was proposed for the photochemical reaction that initial loss of CO and subsequent coordination of MeOH or water facilitated an intramolecular proton transfer to the cyclopentadienyl ring. When $[CpMo(CO)_3(\eta^1\text{-}C_3H_5)]$ (Cp = cyclopentadienyl) was photolysed in low-temperature matrices, one molecule of CO was eliminated, yielding *exo* and *endo* isomers of $[CpMo(CO)_2(\eta^3\text{-}C_3H_5)]$ in approximately equal amounts.[62] Photolysis of $M(CO)_4$(iprop-dab) (M = W, Mo; iprop-dab = 1,4-diisopropyl-1,4-diazabuta-1,3-diene) in the presence of (*E*)-cyclooctene gave high yields of the unprecedented olefin-substituted derivatives (78; M = W, Mo), which were characterized by IR, UV-visible and NMR spectroscopy and by an X-ray diffraction structure determination.[63] Matrix-isolation studies of the Mo precursor, utilizing IR spectroscopy, showed that photolysis resulted in extrusion of one molecule of CO, thereby establishing the generation of a vacant coordination site. Moreover, flash-photolysis experiments indicated the intermediacy of a solvated $Mo(CO)_3$(iprop-dab) fragment. Photolytic CO-substitution by diphosphines ($Ph_2P(CH_2)_nPPh_2$; n = 1-6) has been reported for organoiron derivatives, $CpFe(CO)_2SCOR$ (Cp = cyclopentadienyl; R = alkyl, aryl).[64]

The photodissociation of $[Re(H)(CO)_3(H\text{-dab})]$ (H-dab = 1,4-diaza-1,3-butadiene) has been studied theoretically by wavepacket propagations on ground-state and low-lying excited state potential surfaces calculated by CASSCF/MR-CCI methods.[65] Attention was focused on the Re–H and axial Re–CO bond fissions, and results were correlated with the computed electronic absorption spectrum.

6 Photoelimination of NO

The photodissociation of methyl nitrite on Ag(111) has been simulated using a model with 61 *cis* methyl nitrite molecules adsorbed on a three-layer block of Ag(111), and, from this, attempts were made to reproduce the dynamics of ejected NO molecules.[66] The UV-photodissociation of jet-cooled monomeric nitrosobenzene has been studied by fluorescence excitation spectroscopy of the NO fragments.[67] It was concluded that dissociation occurs after fast internal conversion to the S_0 or S_1 potential energy surfaces, both of which correlate with the electronic ground states of the products, *i.e.*, NO and phenyl radical. 6-Nitrobenzo[*a*]pyrene (79) has been found to release NO upon visible-light irradiation, although other nitrobenzo[*a*]pyrenes did not undergo the same type of photoreaction.[68] MO calculations suggested that a perpendicular conformation of the nitro group with respect to the aromatic ring is important for the release of NO, and the authors believe that these findings could help to develop new types of NO donors.

N-Nitrosodimethylamine (Me_2N–NO) occurs in the environment as a consequence of the activities of a wide range of industries, such as tanning, the manufacture of amines, pesticides, detergents, dyes, rubber, tyres and cosmetics, fish and food processing, and the production of malt beverages. Concerns over health effects associated with exposure to this substance have motivated a study of its photochemistry in aqueous media, with a view to establishing an efficient method for its removal.[69] At pH 3 and 7, direct UV irradiation of Me_2NNO gave dimethylamine and nitrite and nitrate ions as the major degradation products, together with small amounts of formaldehyde, formic acid and methylamine. The photodegradation was much faster at pH 3 than at pH 7, suggesting that protonation occurs prior to N–NO bond cleavage, yielding the dimethylaminium ion ($Me_2NH^{+\cdot}$) and NO in the primary photochemical step.

7 Miscellaneous Photoeliminations and Photofragmentations

7.1 Photoelimination from Hydrocarbons. – Photolysis of acetylene has been carried out with vacuum-UV radiation from a synchrotron, via Rydberg states converging to the first ionization potential at 11.4 eV, and the visible fluorescence of the ethynyl radical (C_2H) was observed.[70] Excitation of several Rydberg states in the range 9-11.4 eV allowed the evolution of this emission to be studied as a function of increasing Rydberg excitation. For most of the excited states, predissociation occurred on a rather low time scale, leaving the C_2H fragment in its $\tilde{A}$ state, and vibrationally hot, with significant excitation in the bending mode. Near-UV photodissociation of allene and propyne, at 203, 209 and 213 nm, has also been investigated.[71] Contrary to previous findings from the photolysis of these molecules at 193 nm, the translational energy spectra associated with the channels forming H atoms were essentially identical for both molecules, and were well reproduced by a statistical model assuming population of all possible vibrational states of the H_2CCCH species. It appears that such behaviour is best accounted for if, for both molecules at the excitation energies employed in this study, internal conversion to, and isomerization on, the ground-state potential energy surface take place before fragmentation. A similar study of the 193 nm photofragmentation of 1,2-butadiene revealed two product channels, giving $CH_3 + C_3H_3$ or $C_4H_5 + H$.[72] The C_3H_3 species in the former process was identified as the propargyl radical, but the C_4H_5 species in the latter could not be identified definitively. As with allene and propyne, the translational energy distributions suggest that both channels result from internal conversion to the electronic ground state followed by dissociation. The photochemistry of cyclohexa-1,4-diene in solution and the gas phase has been studied theoretically using a model postulating the involvement of several conical intersections.[73] One of these was shown to account qualitatively for all the products observed in solution (cyclohexa-1,3-diene, bicyclo[3.1.0]hex-2-ene, hexatriene and benzene + H_2), and also for the elimination of H atoms in the gas phase. The helicopter-type motion of the H_2 product observed previously in the collision-free photolysis of cyclohexa-1,4-diene[74] was not, however, compatible with the properties of this conical intersection, but another one, which appeared to lead naturally to such motion, was proposed and its properties computed.

A review of research on the photodissociation and photoisomerization of a number of small aromatic molecules in molecular beams has been published.[75] The molecules covered were toluene, *m*-xylene, 4-fluorotoluene, α-fluorotoluene and α,α,α-trifluorotoluene. An experimental study of the photodissociation dynamics of benzene at 193 nm has shown that H-atom elimination takes place in a one-photon process, while H_2 and CH_3 elimination channels require two photons.[76] The photodissociation of jet-cooled d_6-benzene has also been examined.[77] In addition to the major channel of D elimination from a one-photon excitation, three channels resulting from two-photon dissociation were found, including two ring-opening channels: elimination of two D atoms, $C_6D_6 \rightarrow CD_3 + C_5D_3$ and $C_6D_6 \rightarrow C_2D_3 + C_4D_3$. Similar studies have been reported for the photodissociation of $PhCD_3$ and $Ph^{13}CH_3$,[78] and for ethylbenzene.[79] In the fragmentation of $PhCD_3$, CD_2H, CDH_2 and CH_3 were observed as fragments, as well as the major products, $PhCD_2 + D$ and $Ph + CD_3$. Moreover, photofragments $^{13}CH_3$ and CH_3 were both observed from $Ph^{13}CH_3$ dissociation. The results suggested that 25% of the excited toluene isomerized to a 7-membered ring (cycloheptratriene) and then rearomatized before dissociation. With ethylbenzene,[79] it appeared that 75% of the molecules dissociated to $PhCH_2 + CH_3$ from an electronic excited state, and 25% from a vibrationally hot electronic ground state. The authors say that this provides the first experimental evidence that alkylbenzenes can dissociate from electronic excited states and not only by a hot-molecule mechanism. Singlet methylene was generated when 1,6-methano[10]annulene was photolysed in a supersonic jet with light of 290-307 nm wavelength.[80] The yield of methylene correlated with the UV absorption spectrum of the $S_0 \rightarrow S_2$ transition, while excitation in the $S_0 \rightarrow S_1$ absorption did not result in photofragmentation. Thus, only the S_2 state generated methylene.

7.2 Photoeliminations from Organohalogen Compounds. – Numerous investigations of the photofragmentations of halomethanes have been reported. Examples include the vibrationally mediated photodissociation of CH_3Cl, which has been studied by detection of the $Cl(^2P_{3/2})$ and $Cl(^2P_{1/2})$ atoms produced;[81] a transient resonance Raman investigation of CH_3I–I, produced in the UV photolysis of iodomethane;[82] and a study of the dynamics of CH_3 fragments from the UV-photodissociation of iodomethane multilayers on GaAs(110).[83] The *iso*-CH_2Br–I species was formed in the A-band excitation of bromoiodomethane in cyclohexane at room temperature, and was detected by picosecond time-resolved resonance Raman spectroscopy.[84] The photodissociation of jet-cooled, vibrationally excited CHF_2Cl molecules has been examined in a time-of-flight mass spectrometer; and the initial vibrational excitation was found to enhance C–Cl and C–H bond cleavage.[85] The effect of symmetry reduction during the 234 nm (A-band) photodissociation of CF_3Br has been investigated, and an estimate of 11% was made for the proportion of molecules which dissociate via a distorted pathway.[86] Dissociation dynamics of CF_2ClBr at 234 nm[87] and of bromoform at 266 nm[88] have also been studied.

In a synthetic application of halomethane photolysis, a novel method has been reported for the trifluoromethylation of aldehydes and ketones by photolysis of CF_3I in the presence of the carbonyl substrate and tetrakis-(dimethylamino)ethene) (TDAE).[89] The reaction ($R^1COR^2 + CF_3I \rightarrow R^1R^2C(OH)CF_3$) was carried out in DMF at low temperatures (–20°C to room temperature) and, in the main, gave good yields (48-90%) of the alcohols. The proposed reaction mechanism involves an initial charge-transfer complex between CF_3I and TDAE that, on irradiation, generates the CF_3^- anion, which is presumably the trifluoromethylating agent. A highly regioselective iodoperfluoroalkylation of allenes, *e.g.* (80), has been reported, which seems to depend on the photoinduced homolysis of the C–F bond in perfluoroalkyl iodides, *e.g.* (81) (Scheme 6).[90] For the examples shown, the products (82) were obtained in yields of 58–88% and with *E*:*Z* ratios ranging from 63:37 to 20:80. Similar reactions were observed with two other perfluoroalkyl iodides: CF_3I and $CF_3(CF_2)_3I$.

The internal energy content of the photofragments HBr and vinylidene

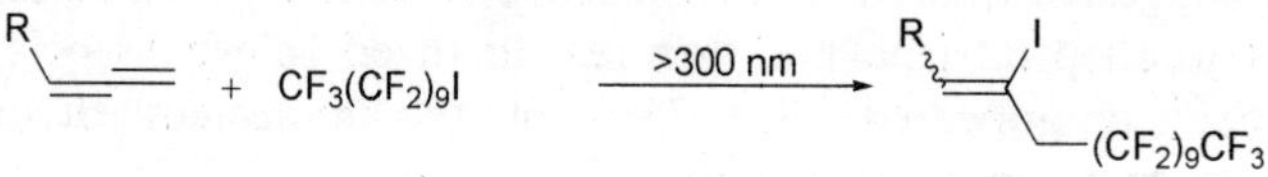

(80) (81) (82)

R = Bu^n, Bu^t, $CH_3(CH_2)_5$, cyclohexyl

Scheme 6

($H_2C{=}C$:), from a minor channel in the 193 nm photolysis of vinyl bromide, has been measured by means of time-resolved FT-IR emission spectroscopy.[91] The results supported a reaction mechanism based on a three-centred HBr elimination process. The vinylidene generated isomerized to acetylene, and acetylene vibrational energy emission bands were detected as a result. The photodissociation of a mixture of *cis*- and *trans*-1,2-dibromoethene at 248 nm has been studied by product translational spectroscopy.[92] Two dissociation channels were found, one giving $Br_2 + C_2H_2$, the other fast and slow Br atoms + C_2H_2, with a branching ratio of about 0.2:0.8. In the major process, the C-Br bond cleavages were not symmetric, producing Br atoms with unequal velocities, but the anisotropy of the products indicated that both reactions occurred within a fraction of a rotational period. The 235 nm photodissociation of CH_3CFCl_2, pre-excited by three, four and five quanta of C–H methyl stretching vibrations, has been studied to investigate the effect of internal excitation on the dynamics of the photofragmentation.[93] The dynamics of the 266 nm photodissociation of CF_3CF_2I has been investigated, using the resonant two-photon ionization technique to determine recoil anisotropy parameters and average translational energies of the eliminated I atoms.[94]

Evidence has been obtained for four significant primary reaction channels in the 193 nm photodissociation of allyl chloride.[95] These were (*i*) a previously unidentified C–Cl bond fission producing low kinetic energy Cl atoms and allyl radicals with high internal energy, (*ii*) an excited state C–Cl bond fission producing Cl atoms with a high kinetic energy, (*iii*) an HCl elimination releasing significant energy to HCl translation, and (*iv*) an HCl elimination producing low kinetic energy HCl. The photodissociations of allyl chloride and allyl bromide at 234 nm have also been studied.[96] Laser induced fluorescence has been utilized to detect CF, CF_2 and CF_3 from the infrared multiphoton dissociation of perfluoropropene.[97]

UV photolysis of dibromoacetonitrile in cyclohexane solution has been investigated by transient resonance Raman spectroscopy with the aid of DFT calculations, and significant amounts of the isomeric Br–NCCHBr species were detected on the nanosecond time scale.[98]

The photodissociation dynamics of iodo-, bromo- and chlorobenzene after excitation with 266 nm light have been studied by femtosecond pump-probe spectroscopy combined with time-of-flight mass spectrometry.[99] The kinetics exhibited single time constants for chlorobenzene (1 ns) and bromobenzene (28 ps), but two time constants were noted for iodobenzene (700 and 350 fs). The 350 fs time constant was due to direct dissociation. The decrease in the other time constants with increasing halogen mass appears to support the assignment of these constants to the decay of initially excited (π,π^*) states to repulsive triplet (n,σ^*) states, promoted by spin-orbit coupling. In the laser ablation of organic solids, an important question is whether molecular fragmentation occurs primarily by direct photodissociation of chemical bonds or by thermal elimination following rapid conversion of the absorbed light energy into heat. The photoinduced fragmentation of chlorobenzene has been modelled theoretically with this question in mind.[100]

Several *o*-(ω-alkenyl)aryl radicals have been generated by photoejection of I atoms from precursor aryl iodides (83), (84; R = H, Ph) and (85), and the rates of ring closure by attack of the radical centres on the tethered double bonds were measured.[101] For the radicals from (83) and (85), the rate constants were 4.2×10^8 and $7.6 \times 10^7\ s^{-1}$, respectively, both ring closures being 6-*exo-trig* processes. For the radical from (84; R = H), the ring closure, a 5-*exo-trig* process, was considerably faster (rate constant $9.6 \times 10^9\,s^{-1}$). Tetrafluoro-*p*-benzyne (86) was generated in neon matrices at 3 K by the 254 nm photolysis of 1,2,3,4-tetrafluoro-3,6-diiodobenzene, in a process involving stepwise loss of the two I atoms.[102] The benzyne was identified by IR spectroscopy with the aid of CASSCF(8,8)/cc-pVDZ computations. At longer wavelengths (260-320 nm), (86) underwent ring opening to FC≡C–CF=CF–C≡CF.

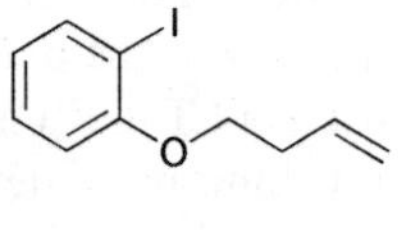

(83)

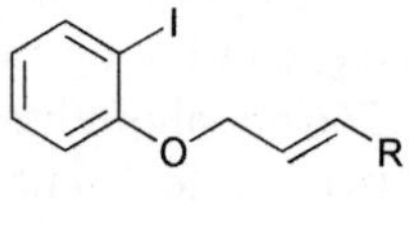

(84)

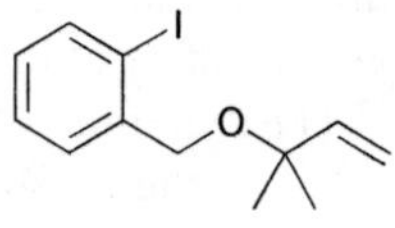

(85)

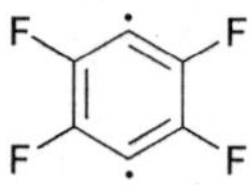

(86)

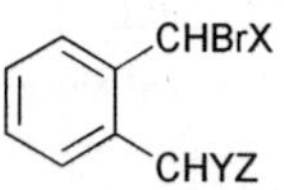

(87)

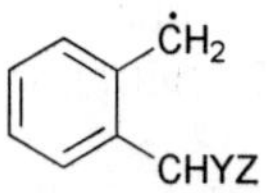

(88)

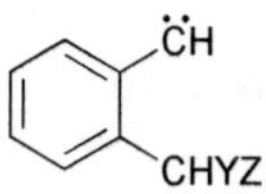

(89)

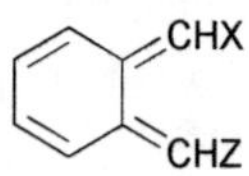

(90)

Flash photolysis of α-brominated *o*-xyxlenes (87; X = Y = Z = H; X = Y = H, Z = Br; X = Br; Y = Z = H; X = Y = Z = Br) at 266 nm has been investigated in non-polar solvents.[103] Depending on the precursor substitution, two transients were observed in each reaction: either a monoradical (88) or a carbene (89), followed by the corresponding *o*-xylylene (quinodimethane) (90). Photoelimination of HCl from the *o*-methylphenacyl chloride (91) in MeOH yields (92) and (93), and has now been shown, in a flash-photolysis study, to proceed via the *Z*-xylylenol (94) (Scheme 7).[104] The final products result from a heterolytic elimination of chloride ion from (94). The homolytic photocleavage of the C-Br bond in 1-bromoacetyl- and 2-bromoacetylnaphthalene has been observed by CIDEP and transient absorption spectroscopy.[105] In the latter case, a transient absorption with λ_{max} ~440 nm was assigned to the triplet state of the parent molecule. After decay of this absorption, a long-lived band with λ_{max} ~380 nm appeared, and this was assigned to the naphthoylmethyl radical. Nevertheless, the yield of this radical was not affected by the concentration of O_2 even though the triplet absorption was quenched by addition of oxygen. It was therefore concluded that the precursor of the radical was an excited singlet state.

7.3 Photofragmentations of Organosilicon and Organogermanium Compounds. – A study has been published of the reactions of di-*tert*-butylsilylene

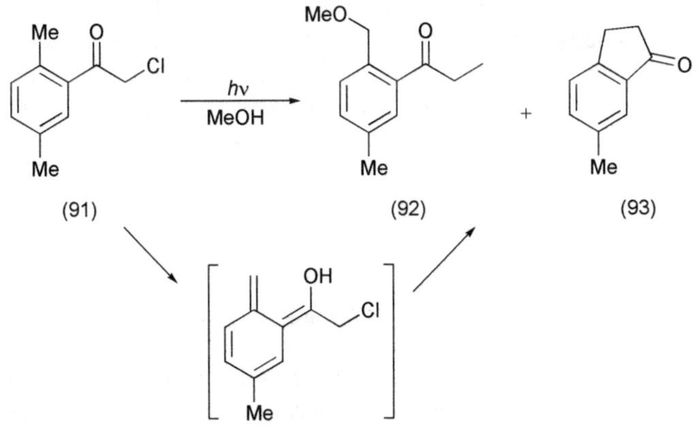

(91) (92) (93)

(94)

Scheme 7

(t-Bu_2Si:), generated by photolysis of hexa-*tert*-butylcyclotrisilane (95), with a number of alkynes and dialkynes linked by C=C bonds.[106] The (*Z*)-hex-3-en-1,5-diyne (96), for example, added two silylenes to give (97), a molecule containing

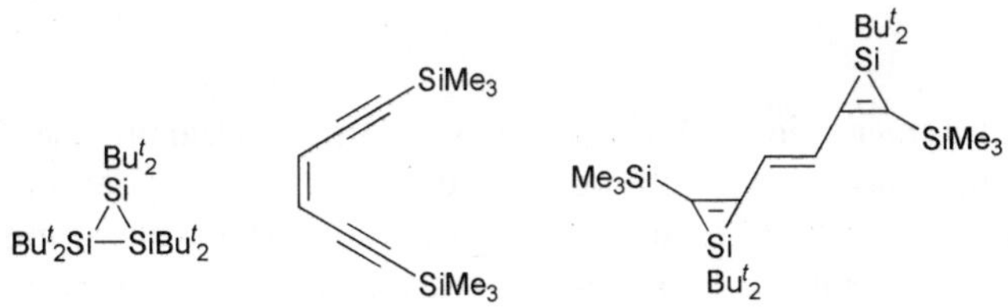

(95) (96) (97)

two silirene rings linked by an *E* carbon-carbon double bond. The stereochemistry of the product was demonstrated by means of an X-ray crystal structure, but the authors do not give an explanation of the observed stereochemical *cis*→*trans* transformation. What is clear, however, is that the silylene intermediate reacted preferentially with the alkyne C≡C bonds rather than the olefinic double bond, and no silirane product was detected. The generation of Wiberg's silene (27), presumably via a silylene, by photoelimination of N_2 or CO from the diazo precursor (26; X = N_2) or the analogous ketene (26; X = CO), respectively, has been mentioned previously in Sections 3.1 and 5. Photoextrusion of silylenes and silyl radicals has been observed in the laser flash photolysis of poylsilenes.[107] The initial yield of silyl radicals seemed to depend on the backbone structure of the polysilene, whereas the yield of silylenes was more or less proportional to the number of linear Si units in the polymer backbone.

Multi-pulse UV laser-induced photolysis of 1,3-disilacyclobutane and oxygen in an excess of a buffer gas (He or N_2) led to the non-explosive chemical vapour deposition of solid methylsilicone films, presumably via reaction of O_2 with silene (H_2Si=CH_2).[108] In contrast, the single-pulse irradiation of the gaseous mixture in the absence of a buffer gas resulted in the *explosive* deposition of solid nanostructured polyoxocarbosilanes poor in hydrogen. 1,1-Diphenylsilene (99; M = Si) and 1,1-diphenylgermene (99; M = Ge) have been generated in hexane solution by laser flash photolysis of the corresponding sila- or germacyclobutanes (98) (Scheme 8).[109] The intermediates (99) were detected directly by

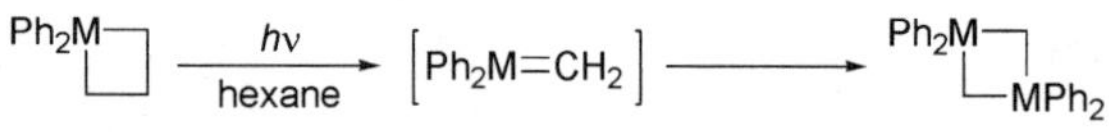

(98) (99) (100)

Scheme 8

absorptions centred on 325 nm, and the kinetics of their dimerization to the head-to-head dimers (100; M = Si, Ge) were determined. The dimerizations of both metallaenes were characterized by near zero Arrhenius activation energies, but the results obtained did not allow a conclusive discrimination to be made between a concerted and a stepwise mechanism for the reaction. A stepwise reaction was, however, considered unlikely. The photolysis of 1,2-digermacyclobutane derivatives with bulky bistrimethylsilylmethyl (disyl) groups (101; Dis = $(Me_3C)_2SiCH$) in toluene solution containing C_{60} gave both the corresponding germylene (Dis_2Ge:) and germacyclopropane (c-$Dis_2GeC_2H_4$) adducts of C_{60}.[110]

7.4 Photofragmentations of Organosulfur, Organoselenium and Organotellurium Compounds. – Multiphoton dissociation and ionization studies of dimethyl disulfide (MeSSMe) have been carried out at 355 nm, using a time-of-flight mass spectrometer.[111] The prominent ion signals observed were due to CH_3^+, S^+, HCS^+ and S_2^+. It was deduced that the dissociation of MeSSMe under these conditions involved both S–S and C–S bond cleavage, while the presence of a signal at m/z 30 was attributed to the elimination of ethane from the excited neutral MeSSMe molecule via a cyclic transition state. Laser-induced photolysis of gaseous diethyl selenium and diethyl tellurium (Et_2M; M = Se, Te) at 248 nm has been shown to be a one-photon process dominated by the elimination of ethene and yielding elemental selenium or tellurium.[112] Both reactions appeared to take place in a stepwise manner via two four-centre transition states (102; R = Et, H; M = Se, Te), and not by homolysis of the M–C bonds, as previously assumed.

7.5 Photolysis of *o*-Nitrobenzyl Derivatives and Related Compounds. – Reviews have appeared of photoremovable protecting groups and linkers, which include sections on *o*-nitrobenzyl derivatives.[3,4] The photolysis of 2 nitrobenzyl methyl ether (103) and of 2-nitrotoluene has been investigated in low temperature matrices by UV-visible absorption and IR spectroscopy, with a view to gaining further insight into the mechanism of the photocleavage of *o*-nitrobenzyl derivatives.[113] In the photolysis of (103), the *o*-quinoid *aci*-nitro intermediate (104) was identified as a very photolabile species with λ_{max} at 430 nm. The IR spectrum of this intermediate clearly indicated, with the aid of DFT

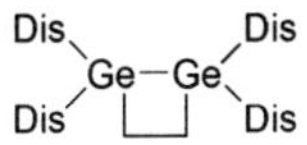

(101)

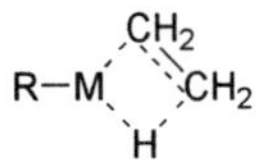

(102)

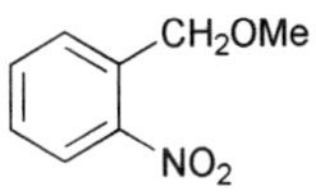

(103)

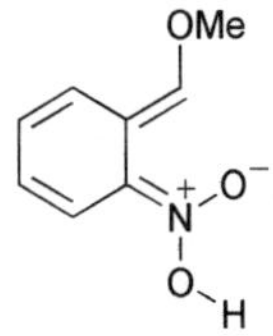

(104)

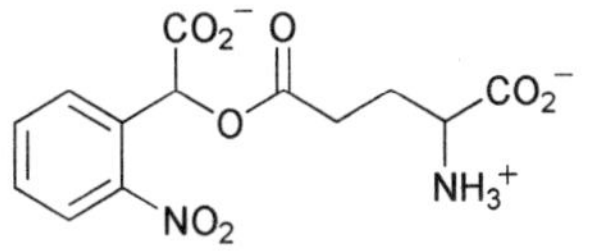

(105)

computations, that it had the *syn-EE* structure shown. It was concluded that H-atom transfer from the benzyl group to the nitro group probably occurred in an electronic excited state. A flash-photolysis study of the *aci*-nitro species generated in the phototautomerization of 2-nitrotoluene has been reported, in which the effects of changing pH were also explored.[114] Time-resolved FT-IR spectroscopy with microsecond resolution was used to investigate the release of glutamate from the caged derivative (105), and in addition kinetic measurements were made on the decay of the 425 nm absorption of the corresponding *aci*-nitro intermediate.[115] The decay of the intermediate was well represented by a single exponential with a time constant of $23 \pm 1\,\mu s$, while the rise in an IR band of the released glutamate was a single exponential with a time constant of $19 \pm 4\,\mu s$. The two rates were clearly in good agreement.

An interesting paper has appeared on the subject of the orthogonality of protecting groups, *i.e.*, the possibility of selectively removing two or more such groups in any chosen chronological order.[116] Orthogonal protection systems can, of course, make use of a combination of photolabile and reagent-sensitive protecting groups, but in the present study, an attempt was made to identify two photocleavable groups, which would respond to different wavelengths. For this approach to succeed, the two photolabile groups must behave as independent chromophores with well separated absorptions, and there must be minimal intramolecular energy transfer when either of these has been excited. The mixed diester (106) (Scheme 9) showed how this might be achieved reasonably successfully: 254 nm irradiation for 5 minutes in MeCN selectively cleaved the dimethoxybenzoin ester to give (107), which was isolated as its methyl ester in 70% yield; while 420 nm irradiation for 24 hours in MeCN selectively cleaved the nitrobenzyl derivative, giving (108), also isolated as its methyl ester in 70% yield.

Several 2-(2-nitrophenyl)ethyl-caged compounds, *e.g.* (109), including caged thymidine nucleosides, have been studied by flash photolysis and under continuous illumination.[117] The effects of solvent basicity and acidity on the cleavage of these compounds were also investigated. Photocleavage occurred via intramolecular H abstraction by the nitro group from the α-position (benzylic) in (109), to give the corresponding *aci*-nitro intermediate. Thereafter, two major reactions competed, depending on the solvent basicity: (*i*) β-elimination of the caged compound from the *anion* of the *aci*-nitro species, giving (110) and R^4OH, or (*ii*) formation of a nitrosobenzene from the undissociated *aci*-nitro species,

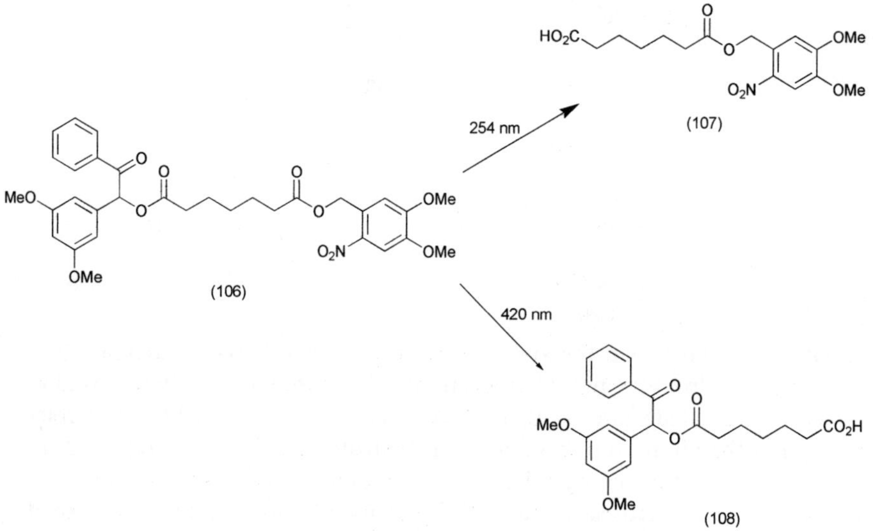

(106) (107) (108)

Scheme 9

without release of the caged compound.

The variety of photocleavable protecting groups based on the *o*-nitrobenzyl system continues to expand. In the period under review, papers have appeared proposing *N*-methyl-*N*-(*o*-nitrophenyl)carbamates (111) for alcohols (ROH),[118] the methylenedioxy propoxycarbonyl functionality (112) for 5′-OH protection of nucleosides,[119] and the (2-nitrofluoren-9-yl)methoxycarbonyl group as a photo-

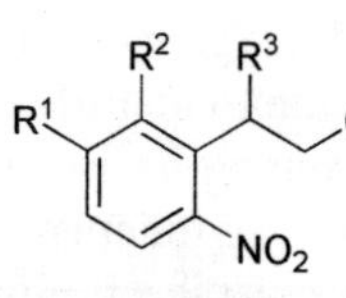

(109)

R^1	R^2	R^3	R^4
H	Cl	H	MeCO
H	H	Me	MeCO
Br	H	Me	CO_2Me
H	Cl	H	CO-5′-*O*-thymidine
H	H	Me	CO-5′-*O*-thymidine

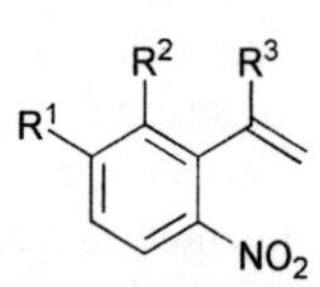

(110)

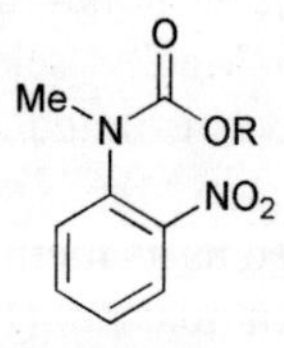

(111)

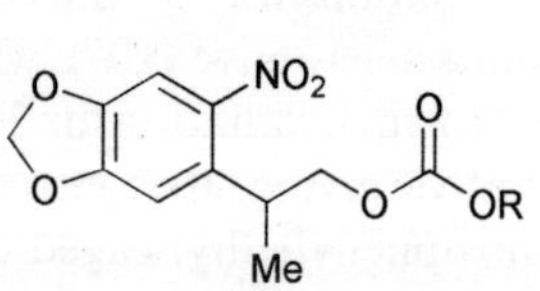

(112)

labile modification of the Fmoc-system.[120] *o*-Nitrobenzyl *p*-nitrophenyl carbonate and an *o*-nitrobenzyl diazo compound have been proposed for protection of amino and carboxylic acid groups, respectively, in steroidal derivatives.[121] A study has been made of *o*-nitroaryldioxolanes as protection systems for pheromones,[122] and *o*-nitromandelyloxycarbonyl (Nmoc) derivatives have been examined as caged amino acids.[123]

Photocleavable 2-nitrobenzylamino-anchored polystyrene-butanediol dimethacrylate supports have been developed for the synthesis of protected peptides, peptide amides and peptide *N*-alkyl amides.[124] A new photolabile linker enabling nucleophilic cleavage of a carboxyl functionality in the presence of primary or secondary amines has been reported (Scheme 10).[125] Thus, release of the bound carboxylate (113) gave the amides (114). An intramolecular version of this reaction led to heterocycles via a cyclorelease mechanism.

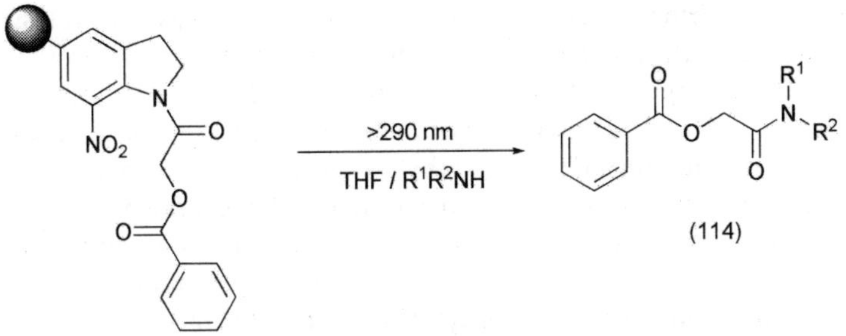

Scheme 10

7.6 Other Photofragmentations. – Photodissociation of methanol has been studied in Ne and Ar matrices at 4 K using VUV synchrotron radiation.[126] Threshold energies of 7.13 ± 0.02 eV (174.0 ± 0.5 nm) and 7.08 ± 0.04 eV (175.0 ± 1.0 nm) were found for the production of OH in Ne and Ar, respectively. The laser photodissociation dynamics of penta-1,4-dien-3-ol at 193 nm have been investigated by laser-induced fluorescence of the rovibrational states of the liberated OH radicals.[127] In the major process, the $^1(\pi,\pi^*)$ state of the alcohol seemed to undergo rapid internal conversion, producing a highly energetic electronic ground state, which was followed by dissociation to pentadienyl ($CH_2{=}CH{-}\dot{C}H{-}CH{=}CH_2$) and OH radicals. A similar study of the photodissociation dynamics of furfuryl alcohol was also reported.[128]

The photodissociation of methyl vinyl ether at 193 nm has been studied in a crossed laser-molecular beam system.[129] Two C–O bond fission channels were observed: a major channel yielding CH_3 + CH_2CHO ($\tilde{A}\,^2A'$), and a minor channel producing CH_3 + CH_2CHO ($\tilde{X}\,^2A''$). Some of the $\tilde{A}$ state vinoxy (CH_2CHO) product underwent secondary dissociation to ketene and H atoms.

A photochemical P–OAr homolytic bond fission, taking place via a resonant two-photon reaction, was observed when aryl diethyl phosphates ($EtO_2P(O)$–OAr; Ar = Ph, 4-$MeOC_6H_4$) were irradiated at 248 nm by a single pulse from a KrF laser.[130] 1,4-Dihydroquinone and phenol products were formed, and this reaction contrasted with the one-photon process, in which O–Ar bond fission occurs.

Energy disposal in the CN ($X\,^2\Sigma^+$) radicals produced in the 157 nm photodissociation of acrylonitrile has been studied.[131] The CN radicals were found in vibrational levels up $v = 6$, and average energies found in vibration and rotation were approximately equal. A similar study of the formation of CN radicals from the 193 nm photodissociation of allyl cyanide has also been reported.[132]

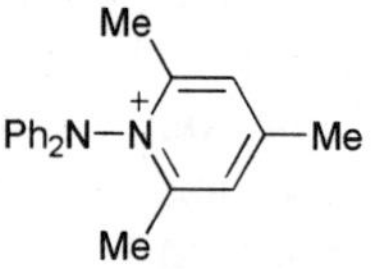

(115)

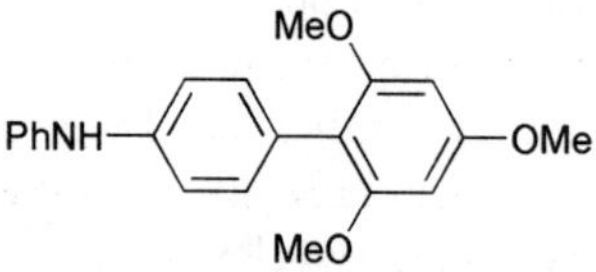

(116)

The *N*,*N*-diphenylnitrenium ion (Ph_2N^+), generated by flash photolysis of the trimethylpyridinium ion (115), has been found to form α-complexes with alkoxybenzenes, leading ultimately to substitution products, such as (116), which was obtained in 44% yield from 1,3,5-trimethoxybenzene.[133] A kinetic study has been made of the reactions of the *N*-methyl-*N*-(4-biphenylyl)nitrenium ion (117), similarly generated by photolysis of a trimethylpyridinium precursor, with various donor-substituted arenes, *e.g.* 1,4-dimethoxy- and 1,3,5-trimethoxybenzene, mesitylene and hexamethylbenzene.[134] In general, three types of products were observed: (*i*) the parent amine, *N*-methyl-*N*-4-biphenylamine, (*ii*) an *ortho* adduct in which the ring position *ortho* to the nitrenium ion centre was bonded to the arene ring, and (*iii*) an *N*-adduct in which the nitrenium ion nitrogen was bonded to the arene. From the measured rate constants, it was concluded that the products derived from an initial π-complex between the nitrenium ion and the arene.

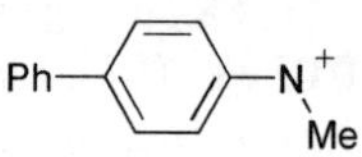

(117)

References

1. H. Sato, *Chem. Rev.*, 2001, **101**, 2687.
2. O. Rubner and V. Engel, *J. Chem. Phys.*, 2001, **115**, 2936.
3. C. G. Bochet, *J. Chem. Soc., Perkin Trans. 1*, 2002, 125.
4. M. Zabadal and P. Klan, *Chemicke Listy*, 2001, **95**, 694; *Chem. Abstr.*, 2002, **136**, 340244w.
5. P. Cattaneo and M. Persico, *J. Am. Chem. Soc.*, 2001, **123**, 7638.
6. S. W. North, C. A. Longfellow and Y. T. Lee, *J. Chem. Phys.*, 1993, **99**, 4423.
7. E. W.-G. Diau, O. K. Abou-Zied, A. A. Scala and A. H. Zewail, *J. Am. Chem. Soc.*, 1998, **120**, 3245.
8. X. Li, S. Anand, J. M. Millam and H. B. Schlegel, *Phys. Chem. Chem. Phys.*, 2002, **4**, 2554.
9. W. Adam, M. Diedering and A. V. Trofimov, *Chem. Phys. Lett.*, 2001, **350**, 453.
10. B. J. Kraft and J. M. Zaleski, *New J. Chem.*, 2001, **25**, 1281.
11. J. Sauer, P. Bäuerlein, W. Ehenbeck, H.-D. Fühlhuber, C. Gousetis and K. Wernthaler, *Eur. J. Org. Chem.*, 2001, 3999.
12. D. Margetic, D. N. Butler and R. N. Warrener, *Aust. J. Chem.*, 2000, **53**, 959.
13. T. Yokomatsu, K. Suemune, T. Murano and S. Shibuya, *Heterocycles*, 2002, **56**, 273.

14. J. F. Arenas, I. López-Tocón, J. C. Otero and J. Soto, *J. Am. Chem. Soc.*, 2002, **124**, 1728.
15. K. Fernamberg, J. R. Snoonian and M. S. Platz, *Tetrahedron Lett.*, 2001, **42**, 8761.
16. E. L. Tae, C. Ventre, Z. Zhu, I. Likhotvorik, F. Ford, E. Tippmann and M. S. Platz, *J. Phys. Chem. A*, 2001, **105**, 10146.
17. T. L. Morkin, W. J. Leigh, T. T. Tidwell and A. D. Allen, *Organometallics*, 2001, **20**, 5707.
18. I. Naito, T. Uryu, H. Sasaki, T. Ishikawa, S. Kobayashi, I. Kamata and A. Oku, *J. Photochem. Photobiol. A*, 2001, **140**, 33.
19. W. Zhang, X. Shao, L. Yang, Z.-L. Liu and Y. L. Chow, *J. Chem. Soc., Perkin Trans. 2*, 2002, 1029.
20. M. Okazaki, Y. Shibasaki and M. Ueda, *Polym, Prepr. (Am. Chem. Soc., Div. Polym. Chem.)*, 2001, **42**, 490; *Chem. Abstr.*, 2001, **135**, 33702q.
21. T. Shimosaki, K. Tabuchi and N. Sano (Konica Chemical Corp., Japan), Japanese Patent 2001 166,466, 22 Jun 2001; *Chem. Abstr.*, 2001, **135**, 45990v.
22. X. Creary and M. A. Butchko, *J. Org. Chem.*, 2002, **67**, 112.
23. A. M. Nazarov, G. A. Yamilova and V. D. Komissarov, *Kinet. Catal. (transl. of Kinetika i Kataliz)*, 2001, **42**, 746; *Chem. Abstr.*, 2002, **136**, 262888t.
24. M. Hashimoto, Y. Hatanaka and K. Nabeta, *Bioorg. Med. Chem. Lett.*, 2002, **12**, 89.
25. Y. Ambroise, F. Pillon, C. Mioskovski, A. Valleix and B. Rousseau, *Eur. J. Org. Chem.*, 2001, 3961.
26. M. R. Linder, W. U. Frey and J. Podlech, *J. Chem. Soc., Perkin Trans. 1*, 2001, 2566.
27. W. Sander, A. Strehl and M. Winkler, *Eur. J. Org. Chem.*, 2001, 3771.
28. P. A. Cook, P. Jimeno, M. N. R. Ashfold, G. G. Balint-Kurti and R. N. Dixon, *Phys. Chem. Chem. Phys.*, 2002, **4**, 1513.
29. G. Bucher, *Eur. J. Org. Chem.*, 2001, 2447.
30. G. Bucher, *Eur. J. Org. Chem.*, 2001, 2463.
31. T. Sato, H. Niino, S. Arulmozhiraja, M. Kaise and A. Yabe, *Chem. Commun.*, 2001, 749.
32. S. V. Chapyshev and P. R. Serwinski, *Mendeleev Commun.*, 2001, **11**, 92.
33. S. V. Zelentsov, A. B. Zhezlov and A. V. Oleinik, *High Energy Chem.*, 2001, **35**, 247.
34. Y. Roeske and W. Abraham, *Synthesis*, 2001, 1125.
35. Y. Todoroki, T. Tanaka, M. Kisamori and N. Hirai, *Bioorg. Med. Chem. Lett.*, 2001, **11**, 2381.
36. T. V. Popova, A. Y. Denisov, M. M. Shakirov, N. I. Komarova, P. V. Alekseyev, M. V. Serebriakova and T. S. Godovikova, *J. Photochem. Photobiol. B*, 2001, **61**, 68.
37. H. P. Spielmann and K. A. H. Chehade (University of Kentucky Research Foundation, USA), US Patent 6,252,096, 26 Jun 2001; *Chem. Abstr.*, 2001, **135**, 61147f.
38. Y. Suda, M. Nakamura, S. Koshida, S. Kusomoto and M. Sobel, *J. Bioact. Compat. Polym.*, 2000, **15**, 468; *Chem. Abstr.*, 2001, **135**, 231655q.
39. T. Pacuszka and M. Panasiewicz, *J. Labelled Compd. Radiopharm.*, 2000, **43**, 1255.
40. S. Knaus, A. Nennadal and B. Froschauer, *Macromol. Symp.*, 2001, **176**, 223.
41. N. Yasuda, S. Yamamoto, Y. Wada and S. Yanagida, *J. Polym. Sci., A*, 2001, **39**, 4196.
42. A. Kumar, Y.-P. Lee and D.-M. Chen, *Combust. Flame*, 2001, **126**, 1736.
43. K. L. Chung, S. C. Lin, W. H. Tseng and H. T. Hung (Industrial Technology Research Institute, Taiwan), Japanese Patent 2001 302,628, 31 Oct 2001; *Chem. Abstr.*, 2001, **35**, 344896n.
44. E. Novikova, A. Kolendo, V. Syromyatnikov, L. Avramenko, T. Prot and K. Golec, *Polimery (Warsaw)*, 2001, **46**, 406; *Chem. Abstr.*, 2002, **136**, 151488t.

45. H. Yoshida (Toppan Printing Co., Japan), Japanese Patent 2002 97,226, 2 Apr 2002; *Chem. Abstr.*, 2002, **136**, 270564z.
46. G. Eshchenko, T. Köcher, C. Kerst and F. Temps, *Chem. Phys. Lett.*, 2002, **356**, 181.
47. H. L. Berghout, S. Hsieh and F. F. Crim, *J. Chem. Phys.*, 2001, **114**, 10835.
48. T. Sato, H. Niino and A. Yabe, *J. Phys. Chem. A*, 2001, **105**, 7790.
49. T. Sato, S. Arulmozhiraja, H. Niino, S. Sasaki, T. Matsuura and A. Yabe, *J. Am. Chem. Soc.*, 2002, **124**, 4512.
50. A. G. Griesbeck, W. Kramer and J. Lex, *Synthesis*, 2001, 1159.
51. M. Sobczak and P. J. Wagner, *Org. Lett.*, 2002, **4**, 379.
52. A. Lalitha, K. Pitchumani and C. Srinivasan, *Tetrahedron*, 2001, **57**, 4455.
53. K. Kadota and K. Ogasawara, *Tetrahedron Lett.*, 2001, **42**, 8661.
54. S. Linder, K. White, M. Palmer, B. Arney and R. White, *Tetrahedron Lett.*, 2002, **43**, 1169.
55. P. T. Snee, C. K. Payne, S. D. Mebane, K. T. Kotz and C. B. Harris, *J. Am. Chem. Soc.*, 2001, **123**, 6909.
56. H. Ihee, J. Cao and A. H. Zewail, *Angew. Chem. Int. Ed.*, 2001, **40**, 1532; *Angew. Chem.*, 2001, **113**, 1580.
57. W. Fuss, S. A. Trushin and W. E. Schmid, *Res. Chem. Intermed.*, 2001, **27**, 447.
58. T. Szymańska-Buzar and K. Kern, *J. Organomet. Chem.*, 2001, **622**, 74.
59. T. W. Stringfield and R. E. Shepherd, *Inorg. Chem. Commun.*, 2001, **4**, 760.
60. M. A. Reynolds, I. A. Guzei and R. J. Angelici, *J. Am. Chem. Soc.*, 2002, **124**, 1689.
61. S. Top, El B. Kaloun, S. Toppi, A. Herrbach, M. J. McGlinchey and G. Jaouen, *Organometallics*, 2001, **20**, 4554.
62. C. Limberg, A. J. Downs, T. M. Greene and T. Wistuba, *Eur. J. Inorg. Chem.*, 2001, 2613.
63. F.-W. Grevels, K. Kerpen, W. E. Klotzbücher, K. Schaffner, R. Goddard, B. Weimann, C. Kayran and S. Özkar, *Organometallics*, 2001, **20**, 4775.
64. I. Jibril, M. El-khateeb, H. Barakat, G. Rheinwald and H. Lang, *Inorg. Chim. Acta*, 2002, **333**, 1.
65. I. Bruand-Cote and C. Daniel, *Chem.- Eur. J.*, 2002, **8**, 1361.
66. S. K. Kim, J. M. White, P. M. Agrawal and D. L. Thompson, *J. Chem. Phys.*, 2001, **115**, 7657.
67. A. Keßler, A. Slenczka, R. Seiler and B. Dick, *Phys. Chem. Chem. Phys.*, 2001, **3**, 2819.
68. K. Fukuhara, M. Kurihara and N. Miyata, *J. Am. Chem. Soc.*, 2001, **123**, 8662.
69. M. I. Stefan and J. R. Bolton, *Helv. Chim. Acta*, 2002, **85**, 1416.
70. S. Boyé, A. Campos, S. Douin, C. Fellows, D. Gauyacq, N. Shafizadeh, P. Halvick and M. Boggio-Pasqua, *J. Chem. Phys.*, 2002, **116**, 8843.
71. R. H. Qadiri, E. J. Feltham, E. H. Cottrill, N. Taniguchi and M. N. R. Ashfold, *J. Chem. Phys.*, 2002, **116**, 906.
72. J. C. Robinson, W. Sun, S. A. Harris, F. Qi and D. M. Neumark, *J. Chem. Phys.*, 2001, **115**, 8359.
73. S. Zilberg and Y. Haas, *Phys. Chem. Chem. Phys.*, 2002, **4**, 34.
74. E. V. Cromwell, D.-J. Liu, M. J. J. Vrakking, A. H. Kung and Y. T. Lee, *J. Chem. Phys.*, 1991, **95**, 297.
75. C.-L. Huang, J.-C. Jiang, S. H. Lin, Y. T. Lee and C.-K. Ni, *Aust. J. Chem.*, 2001, **54**, 561.
76. T. C. Hsu, J. Shu, C. Yang, J. J. Lin, Y. T. Lee and X. Yang, *J. Chin. Chem. Soc. (Taipei)*, 2002, **49**, 1.
77. S.-T. Tsai, C.-L. Huang, Y. T. Lee and C.-K. Ni, *J. Chem. Phys.*, 2001, **115**, 2449.

78. C.-K. Lin, C.-L. Huang, J.-C. Jiang, A. H. H. Chang, Y. T. Lee, S. H. Lin and C.-K. Ni, *J. Am. Chem. Soc.*, 2002, **124**, 4068.
79. C.-L. Huang, J.-C. Jiang, S. H. Lin and Y. T. Lee, *J. Chem. Phys.*, 2002, **116**, 7779.
80. W.-H. Park, S. H. Cho, S. K. Kim and Y. S. Choi, *Bull. Korean Chem. Soc.*, 2001, **22**, 1030; *Chem. Abstr.*, 2002, **136**, 69547c.
81. C. Tao and P. J. Dagdigian, *Chem. Phys. Lett.*, 2001, **350**, 63.
82. Y.-L. Li and D. L. Phillips, *Chem. Phys. Lett.*, 2001, **349**, 291.
83. A. Srivastava and R. M. Osgood, Jr., *Chem. Phys. Lett.*, 2002, **355**, 371.
84. W. M. Kwok, C. Ma, D. Phillips, A. W. Parker, M. Towrie, P. Matousek and D. L. Phillips, *Chem. Phys. Lett.*, 2001, **341**, 292.
85. L. Li, G. Dorfman, A. Melchior and S. Rosenwaks, *J. Chem. Phys.*, 2002, **116**, 1869.
86. T. K. Kim, M. S. Park, K. W. Lee and K.-H. Jung, *J. Chem. Phys.*, 2001, **115**, 10745.
87. S.-H. Lee and K.-H. Jung, *Chem. Phys. Lett.*, 2001, **350**, 306.
88. W.-L. Liu and B.-C. Chang, *J. Chin. Chem. Soc. (Taipei)*, 2001, **48**, 613.
89. S. Aït-Mohand, N. Takechi, M. Médebielle and W. R. Dolbier, Jr.. *Org. Lett.*, 2001, **3**, 4271.
90. A. Ogawa, M. Imura, N. Kamada and T. Hirao, *Tetrahedron Lett.*, 2001, **42**, 2489.
91. D.-K. Liu, L. T. Letendre and H.-L. Dai, *J. Chem. Phys.*, 2001, **115**, 1734.
92. Y. R. Lee, C. C. Chou, Y. J. Lee, L. D. Wang and S. M. Lin, *J. Chem. Phys.*, 2001, **115**, 3195.
93. T. Einfeld, C. Maul, K.-H. Gericke, R. Marom, S. Rosenwaks and I. Bar, *J. Chem. Phys.*, 2001, **115**, 6418.
94. A. V. Baklanov, G. A. Bogdanchikov, M. Aldener, U. Sassenberg and A. Persson, *J. Chem. Phys.*, 2001, **115**, 11157.
95. M. L. Morton, L. J. Butler, T. A. Stephenson and F. Qi, *J. Chem. Phys.*, 2002, **116**, 2763.
96. M. S. Park, K. W. Lee and K.-H. Jung, *J. Chem. Phys.*, 2001, **114**, 10368.
97. L. Rubio and J. A. Torresano, *J. Photochem. Photobiol. A*, 2001, **146**, 1.
98. Y.-L. Li, C. W. Lee, D. Wang, K. H. Leung and D. L. Phillips, *Chem. Phys. Lett.*, 2001, **350**, 78.
99. M. Kadi, J. Davidsson, A. N. Tarnovsky, M. Rasmusson and E. Åkesson, *Chem. Phys. Lett.*, 2001, **350**, 93.
100. Y. G. Yingling, L. V. Zhigilei and B. J. Garrison, *Nucl. Instrum. Methods Phys. Res. B*, 2001, **180**, 171.
101. A. Annunziata, C. Galli, M. Marinelli and P. Pau, *Eur. J. Org. Chem.*, 2001, 1323.
102. H. H. Wenk, A. Balster, W. Sander, D. A. Hrovat and W. T. Borden, *Angew. Chem. Int. Ed.*, 2001, **40**, 2295; *Angew. Chem.*, 2001, **113**, 2356.
103. L. H. Catalani, I. P. de Arrundo Campos and D. de Brito Rezende, *J. Photochem. Photobiol. A*, 2001, **140**, 1.
104. A. P. Pelliccioli, P. Klán, M. Zabadal and J. Wirz, *J. Am. Chem. Soc.*, 2001, **123**, 7931.
105. T. Suzuki, Y. Kaneko, K. Maeda, T. Arai, K. Akiyama and S. Tero-Kubota, *Mol. Phys.*, 2002, **100**, 1469.
106. D. Ostendorf, W. Saak, M. Weidenbruch and H. Marsmann, *Organometallics*, 2002, **21**, 636.
107. S. Seki, S. Tsuji, Y. Matsui and S. Tagawa, *Chem. Lett.*, 2001, 1208.
108. J. Pola, J. Vitek, Z. Bastl and J. Šubrt, *J. Organomet. Chem.*, 2001, **640**, 170.
109. T. L. Morkin and W. J. Leigh, *Organometallics*, 2001, **20**, 4537.
110. Y. Kabe, H. Ohgaki, T. Yamagaki, H. Nakanishi and W. Ando, *J. Organomet. Chem.*, 2001, **636**, 82.

111. R. K. Vatsa, C. Majumder, O. D. Jayakumar, P. Sharma, S. K. Kulshreshtha and J. P. Mittal, *Rapid Commun. Mass Spectrom.*, 2001, **15**, 1941.
112. J. Pola and A. Ouchi, *J. Organomet. Chem.*, 2001, **629**, 93.
113. I. R. Dunkin, J. Gębicki, M. Kiszka and D. Sanín-Leira, *J. Chem. Soc., Perkin Trans. 2*, 2001, 1414.
114. M. Schwörer and J. Wirz, *Helv. Chim. Acta*, 2001, **84**, 1441.
115. Q. Cheng, M. G. Steinmetz and V. Jayaraman, *J. Am. Chem. Soc.*, 2002, **124**, 7676.
116. C. G. Bochet, *Angew. Chem. Int. Ed.*, 2001, **40**, 2071; *Angew. Chem.*, 2001, **113**, 2140.
117. S. Walbert, W. Pfleiderer and U. E. Steiner, *Helv. Chim. Acta*, 2001, **84**, 1601.
118. S. Loudwig and M. Goeldner, *Tetrahedron Lett.*, 2001, **42**, 7957.
119. P. Berroy, M. L. Viriot and M. C. Carré, *Sens. Actuators B*, 2001, **74**, 186.
120. B. Henkel and E. Bayer, *J. Peptide Sci.*, 2001, **2001**, 7 152.
121. S. Watanabe, R. Hiratsuka, Y. Kasai, K. Munakata, Y. Takahashi and M. Iwamura, *Tetrahedron*, 2002, **58**, 1685.
122. L. Ceita, A. K. Maiti, R. Mestres and A. Tortajada, *J. Chem. Res.*, 2001, *(S)* 403; *(M)* 1023.
123. J. P. Kao and F. M. Ross (University of Maryland Biotechnology Unit, USA), US Patent 6,310,083, 30 Oct 2001; *Chem. Abstr.*, 2001, **135**, 331671k.
124. K. S. Kumar, M. Roice and V. N. R. Pillai, *Tetrahedron*, 2001, **57**, 3151.
125. K. C. Nicolaou, B. S. Safina and N. Winssinger, *Synlett*, 2001, 900.
126. B.-M. Cheng, C.-P. Liu, W.-J. Lo and Y.-P. Lee, *Nucl. Instrum. Methods Phys. Res. A*, 2001, **467-468**, 1461.
127. P. K. Chowdhury, *J. Phys. Chem. A*, 2002, **106**, 6223.
128. P. K. Chowdhury, H. P. Upadahyaya and P. D. Naik, *Chem. Phys. Lett.*, 2001, **344**, 292.
129. M. L. Morton, D. E. Szpunar and L. J. Butler, *J. Chem. Phys.*, 2001, **115**, 204.
130. M. Nakamura, A. Ouchi, M. Miki and T. Majima, *Tetrahedron Lett.*, 2001, **42**, 7447.
131. J. Guo, T. Carrington and S. V. Filseth, *J. Chem. Res.*, 2001, **115**, 8411.
132. C. Y. Oh, S. K. Shin, H. L. Kim and C. R. Park, *Chem. Phys. Lett.*, 2001, **342**, 27.
133. S. McIlroy and D. E. Falvey, *J. Am. Chem. Soc.*, 2001, **123**, 11329.
134. D. Chiapperino, S. McIlroy and D. E. Falvey, *J. Am. Chem. Soc.*, 2002, **124**, 3567.

7

Polymer Photochemistry

BY NORMAN S. ALLEN

1 Introduction

This report follows essentially the same format as previous reports. The field of polymer photochemistry continues to be an active area in applied photochemistry, with many topics having strong academic and industrial interest. Photopolymerization and photocuring science and technologies, as before, continue to be developed, particularly with regard toward designing novel and specific initiators and materials for specialist applications. Interest in active ionic initiators and radical-ionic processes continues to expand, while the photocrosslinking of polymers is attractive in terms of enhancing the physical and mechanical properties of electronic materials and the development of liquid crystalline materials. The optical properties of polymers continues to be an active area of strong development, with a continued growth in photochromic and liquid crystalline materials. Last year saw a major literature explosion in LED's (light emitting diodes). In this year's review it represents the largest specialized topic in photochemistry and photophysics, with over 150 articles on PPV's and related materials.

The photooxidation of polymers on the other hand continues to decline at a low profile. Bio- and photodegradable plastics are important for agricultural usage although academic interest here is again minimal. The same applies to polymer stabilization, where commercial applications dominate very much, with emphasis on the practical use of stabilizers. For dyes and pigments stability is a major issue, especially in catalytic chemistry.

2 Photopolymerization

A number of reviews of topical and commercial interest have appeared on the subject of radiation curing science and technology. There continues to be much development in photoinitiators, with articles concentrating on organometallics and anion generators,[1] metallocenes,[2] polymeric systems[3] and laser[4] and general[5] photoinitiation. The influence of heteroatoms in the photocuring of multifunctional monomers and prepolymers has been assessed and reviewed, with the sulfide group having a significant influence over that of the ether functionality.[6,7]

Photochemistry, Volume 34

New lamp technologies are on the scene[8] as are curable paints with improved drying times.[9] New environmentally friendly coatings are also available for UV curing on furniture,[10] while adhesives continue to expand in the UV arena,[11-13] with epoxy[14] and cyanoacrylate[15] resins having specialized interest. Epoxy resins are also highlighted for their non-gas generation during curing,[16] while the modeling of photocuring reactions continues to expand our understanding of reaction rates and gel effects.[17,18] The water resistance of emulsions[19] has important implications, as do photosensitive polyimides for electronic circuitry boards.[20] Sulfur compounds continue to attract interest as photoinitiators, and these have been reviewed in detail.[21]

2.1 Photoinitiated Addition Polymerization. – Photoinitiators, and their performance and mode of action, continue to attract interest. DTCS (*N,N*-diethylaminodithiocarbamoylmethylstyrene) has been found to play an important role as an inimer capable of producing living radicals for vinyl polymerizations.[22] The bifunctional character of *N,N*-dimethyl-4-nitroaniline enables it to operate as an exceptional initiator.[23] The presence of electron donor-acceptor groups allows effective electron transfer to take place as well as offering excitation energy transfer to other tertiary amine molecules such as *N,N*-dimethylaniline. In such cases the latter undergoes rapid consumption in comparison with the former, *i.e.*, at a rate of 165:1. Sulfur compounds such as morpholinyl-2-benzothiazole disulfide are active initiators for polymerizing acrylic monomers.[24-28] This compound undergoes significant degradative initiator transfer as does diphenyl diselenide.[25] Sulfur dioxide, on the other hand, forms charge-transfer complexes with piperazine[26] and *N*-acrylphenothiazine.[27,28] The ratios of the complexes were important, with the piperazine locking into the PMMA chains. Benzophenone with ketosulfone moieties undergo a β-cleavage reaction, while in the presence of an amine it also produces a ketyl radical.[29] The nature of the excited states are important in this reaction, and this was also the conclusion of other work on novel bifunctional benzophenones.[30] In this regard a two-photon absorption process has been considered for the effective polymerization of deep layers of monomers.[31] Here the use of near IR is believed to activate deeper reactions. Although the authors stipulate that no absorption occurs in the near IR region, they nevertheless propose a second absorption process to account for the improved rate. In fact, the process is unlikely to be due to electronic absorption but rather to heat generation through IR absorption. Several coumarins have been found to be effective photoinitiators by undergoing electron transfer with other co-initiators such as oxime esters or titanocene,[32] while coumarin-nitroxide has been found to be an excellent radical trap to give a strongly fluorescent alkoxylamine product.[33] In this case the fluorescence build-up during polymerization is a direct measure of radical generation. Photo-induced radical generation has been confirmed by subphthalocyanines using laser flash photolysis,[34] while a new process involving encapsulation of pyrilium salts has been developed for the controlled release of initiators during polymerization.[35] From the point-of-view of comparisons, the thioxanthone initiator (ITX) has been found to be much less effective than benzophenone in photoinitiation activity,[36] while for azoisobutyronitrile its diffusion rate has been found to control the rate of polymerization.[37] Metal based complexes have been found to be of value in the photopolymerization of aniline.[38-40] The entrapment of ruthenium bipyridyl in a hectorite clay with *N*-phenyl-*p*-phenylenediamine is one such system, which forms a strongly absorbing CT template complex in the interlayers capable of growing polyaniline. Ru has also been used to catalyse the addition of aromatic ketones to vinyl groups for effective polymerization.[41] On the other hand, a complex based on azidopentaaminecobalt (III) undergoes a photoredox mechanism to give free radicals,[42] while monomer complexes of a phosphorus bridged [1]ferrocenophane undergo ring-opening polymerization reactions.[43] A novel cyanine-dye-borate complex has been found to be an excellent IR sensitizer,[44] while eosin-Y has been found to be a powerful photoinitiator in the presence of *N*-methyldiethanolamine.[45] In the latter case, however, dye bleaching was effective only in the presence of a third iodonium salt component. Visible light photosensitizers have also been developed based on azo-dye bound natural polymers such as gelatin.[46] Here the binding of the dye to the polymer played a crucial role in the rate process. In the presence of riboflavin the photopolymerization rate of 2-hydroxyethyl methacrylate has been found to be dependent upon the structure of the co-amine synergist.[47] The type of alkylamino radical was considered to be important. Rose Bengal also forms a 1:2 complex with the ferrocinium salt, which exists mostly as a contact ion-pair.[48] This complex is a highly efficient initiator and has been found to be more effective than the benzoyl phosphine oxides. The phenoxazine dye, resazurin, is also highly effective in the presence of amine co-initiators for the photopolymerization of acrylamide,[49] while methylene blue works well with *N*-methyldiethanolamine and an iodonium salt.[50] The 2-isopropylthioxanthone derivative is also a powerful sensitizer in the presence of ketosulfone derivatives of benzophenone.[51] Diblock copolymers of MMA and VAc (vinyl acetate) have been made using benzophenone as a photointiator,[52] while MMA itself has been unusally photopolymerized with a complex of titanium dioxide and sodium sulfite.[53] Phenolic molecules are shown to reduce not only the rate of photopolymerization of acrylic monomers but also their molecular weight through effective chain termination.[54] Poly(sodium styrenesulfonate-block-*N*-vinylcarbazole) has been successfully prepared through nitroxide mediated living polymerization.[55] Here camphorsulfonic acid and acetic anhydride were used as rate accelerating additives. Similar studies have been undertaken by other workers on polystyrene end-tagged with initator groups.[56] PMMA photopolymerized in the presence of bis(cyclopentadienyl)titanium dichloride and sparteine, as a chelating agent, has been found to be insoluble in most common solvents without any evidence of crosslinking.[57] Other acrylates gave a similar effect, producing polymers of high thermal stability. Other studies of interest include the photopolymerization of methacrylate monomers in a polystyrene matrix,[58] acylphosphine oxide photoinitiators,[59] formation of polyphenols through polymerization of 4-hydroxybenzaldehyde dimethylnitrone,[60] quenching of imidazolyl radicals,[61] poly(methacrylic acid) with camphorquinone moieties[62] and radical transfer using an acetophenone/Fe tris(*N*,*N*-di-Et-dithiocarbamate) system.[63]

Cationic photoinitiators continue to grow in interest particularly with regard to iodonium salts. Diaryliodonium salts are photosensitized by thioxanthone and anthraquinone molecules in the visible spectrum, while triarylsulfonium salts are not.[64-66] Phenothiazine compounds exhibit a similar behaviour,[67] while a number of cationically polymerizable monomers have been made based on enol ether groups,[68] epoxy functional inorganic hybrids,[69] benzyl ether groups[70] and carbazole monomers with vinyl-vinyl ether groups.[71] In this regard, aryl free radicals can also be used to accelerate the rate of cationic polymerization by onium salts.[72] Other cationic initiators include allylic salts,[73-76] 2-benzyl-2-(*N*,*N*-dimethyl-2-ethoxycarbonyl-1-propenyl)ammonium hexafluoroantimonate-1-(4-morpholinophenyl)-butane-1-one[77] and *N*-(2-ethoxycarbonyl-1-propenyl)ammonium hexafluoroantimonate)benzophenone[78] for the photopolymerization of cyclohexene oxide. Oxetanes have been investigated where the rate of initiation of polymerization is high, such as with 2-phenyloxetanes.[79] Rates for monomers with *p*-methoxyphenyl groups were also high. A new photoinitiator, diaryliodonium hexafluorophosphate,[80] has been developed, while in the anthracene sensitized photopolymerization of diphenyliodonium salts 9-phenylanthracene was found to be the main photoproduct.[81] By use of laser flash photolysis, anthracene in its lowest excited singlet state undergoes electron transfer to the diphenyliodonium cation during the initial step of photoacid generation. This process is then followed by fast chemical reactions generating 9-phenylanthracene and protons, and then back electron transfer from the diphenyliodine radical to the anthracene radical cation. The same salts have also been successfully used to photopolymerize 3-methyltetrahydrofuran,[82] while polymers have also been made through the photolysis of susbstituted aryl cyclic sulfonium zwitterionic salts[83] and 2-cyclopropyl-4-methylene-1,3-dioxolanes.[84] Complexes have been formed between phenothiazine and triphenylsulfonium salts,[85] whereas 4-hexyloxysubstituted diazonium salts are highly effective initiators but with high oxygen sensitivity.[86] Other cationic systems include 3-acrylamide-9-ethylcarbazole with diphenyliodonium salts,[87] 2-oxalines,[88] fluoroalkyl propenyl ethers,[89] ((9-carbazolyl)methyl)thiirane with diaryliodonium salts and iron arene complexes[90] and (9-oxo-9*H*-fluorene-2-yl)phenyliodonium hexafluoroantimonate (V) for epoxides.[91]

N-Substituted maleimides have been used to initiate the photopolymerization of mixtures of maleimide and vinyl ether acrylates.[92,93] Maleimides and tertiary amines are also effective initiators for Type II polymerization of acrylates,[94] as are cationic initiators for polyfunctional 1-propenyl ethers[95] and 2,3-dihydrofuran.[96] For maleimide-donor photocopolymerizations, the total electron density, as well as the distribution within the carbon-carbon double bond in the donor monomer, influences the preference and ratio of reactivity.[97] Furthermore, the rate of co-polymerization will also be influenced by the donor structure. Globular latex particles have been made by photopolymerizing MMA and 2,3-diphenylbutadiene in poly(*N*-isopropylacrylamide) (PNIPAM).[98] The latter template undergoes coil-to-globe transitions owing to co-nonsolvency effects, where the polymerization takes advantage of the hydrophobic nanodomains formed by collapsed PNIPAM chains in the solvent system. Apparently, only very low levels of the template polymer were found in the copolymer latex. In a similar fashion, the photopolymerization kinetics of acrylamide have been correlated with monomer organization in various phases of an LLC template.[99] Here the polymerization rate is enhanced in more ordered LLC phases owing to orientation of the monomer molecules. The propagation rate of MMA has also been measured in miniemulsions.[100] In this case the molecular weight exhibits spikes or peaks due to pulses in the initiation process where compartmentalization takes place. Silica functionalized with azo groups has been used to produce bi- and tri-phase poly(styrene)-poly(butadiene),[101] while electrically conductive polymers have been made by photopolymerizing poly(pyrrole) with Group 1B metal particles.[102] Metallic nanostructures were identified in the matrix. In a similar fashion electrically conducting poly(pyrrole)-poly(vinyl alcohol) composities have been made by photopolymerizing with iron-arene salts.[103] Methanol was found to influence the morphology of the matrix and its conductivity. Conductive poly(aniline) has also been prepared by photopolymerizing aniline in the presence of Ru(II) bipyridyl, and methylviologen complexes with *N*-phenyl-*p*-phenylenediamine[104,105] bound into Flemion film. Ladder polymers of *p*-xylylenediammonium sorbate have been prepared by topochemical polymerization,[106] while alkadiyne mono-layers have been formed on graphite surfaces.[107] Other studies of interest include supramolecular structures of bis(methacryloyloxyethylenecarbonyloxyethylene)oxy oligoetheracrylate,[108] synthesis of chloromethyl derivatives with phenylacryloyl groups[109] and photocoplymerization of 9-(2,3-epoxypropyl)carbazole with THF.[110]

Polyaniline is apparently prepared more efficiently under light than thermally,[111] while vinylester resins with hydroxymethyl groups exhibit superior properties after polymerization.[112] Photopolymerization of carbon disulfide has been shown to produce nanoscale gold coloured polythiene films,[113] while super adsorbent polymers based on poly(acrylic acid-sodium acrylate) are more effectively made via photopolymerization.[114,115] Branched polystyrene has been successfully prepared from the photopolymerization of *p*-(phenylselenomethyl)styrene,[116] and good membranes made from the photopolymerization of sodium 4-styrenesulfonate-divinylbenzene.[117] Concentration studies on acrylamide for holographic recordings have shown that an increase in the monomer reduces scattering, but at the disadvantage of a slower reaction rate.[118] The same monomer, when copolymerized with phthalic anhydride, gave different coloured products depending upon the temperature,[119] and also gave nanostructured gels with 2-hydroxyethyl methacrylate.[120] The microhardness of C_{70} crystals can be enhanced through irradiation.[121]

Photoiniferters have received some attention, with a new polymerizable system based on β-methacylyloxyethyl-2-*N*,*N*-diethyldithiocarbamyl acetate being developed for producing living polymers.[122] Branched polystyrenes have also been prepared via the inimer *N*,*N*-diethylaminodithiocarbamoylmethylstyrene,[123] while end functional polystyrenes have been made with phenylseleno groups.[124,125] In this way graft copolymers of PMMA and branched poly(styrene) can be made. The electro-optical behaviour of copolymers of 2-hydroxyethyl acrylate and styrene has been examined,[126] as have deep section polymerizations.[127]

With respect to bleachable photoinitiators, an unsteady one-dimensional model has been developed, accounting for initiator consumption and optical attenuation, and giving rise to relationships for the spatial and temporal variation of the local initiator concentration and its rate.[128] The maximum rates of photoinitiated polymerization of monomers capable of hydrogen bonding have been measured by RT-FTIR. These monomers were found to exhibit higher rates than those which do not form hydrogen bonds.[129] A two-photon induced polymerization has been demonstrated at 710 nm using a thioxanthone and diarylidonium salt in an epoxide resin,[130] while fractal analysis has been applied to STM images of coniferyl alcohol.[131] Polymerization frontals have been monitored via photobleachable initiators,[132] and Raman scattering has been applied to C_{60} single crystals.[133] At high laser intensities some thermal decomposition of the polymer occurs during photopolymerization. Well defined block copolymers of PMMA and polyalkyl acrylates have been synthesized with polynuclear aromatics at the junctions,[134] while steady-state fluorescence measurements have been used to monitor the free radical polymerization of 2-vinylnaphthalene.[135] Here the critical temperature of polymerization was found to be 60°C.

2.2 Photocrosslinking. – Several initiator systems have been examined for photocuring reactions. These include cyanine dye-borate complexes for TMPTA,[136] polyimides with built-in fullerene,[137] safranine dye for styrenesulfonate,[138] *N*,*N*-dimethyl-*N*-(2-hydroxy-3-phenoxypropyl)amine-4-nitrobenzamide for polyethylene glycol diglycidyl ether-pentaerythritol tetrakis(mercaptoacetate) copolymer,[139] quaternary ammonium salts with *N*,*N*-dimethyldithiocarbamate anions for poly(glycidyl methacrylate) films,[140-142] thioxanthones for 1,6-hexanediol diacrylate,[143] oligomeric amines as co-synergists for triacrylates,[144] self-curing mixtures of 1,2,4-trimellitic anhydride, epichlorohydrin and glycidy methacrylate,[145] carbohydrates bound to α-hydroxyalkylphenones,[146] camphorquinone for dimethacrylates,[147] photoacid generating copolymers of *o*-nitrobenzyloxycarbonyl and iminiosulfonate groups,[148] nonaqueous dispersions of particles with photolabile α-heptadecylphenacyl ester stabilizer chains,[149] 4-vinylbenzoyldiphenylphosphine oxide for hydro-gels,[150] quaternary ammonium salts for phenolic resins,[151] and poly(glycidyl methacrylate),[152] iron-arene complexes for epoxides,[153] heptamethine cyanine dyes for coatings,[154] acetophenones for isocyanates[155] and water-based acrylated polyesters.[156] A series of sytrene oxides have been identified as novel curable systems,[157] while a series of poly(cinnam-4′-ylmethyl methacrylate) derivatives undergo linear photoreactions with polarized light, offering useful LC applications.[158] Other systems include the development of photocatalytic coatings with titanium dioxide pigments.[159]

Numerous studies have appeared on cationic photocuring. Trimethylolpropane tripropenyl ethers have been used as reference products for polyfunctional propenyl ethers and display reactivities comparable with those of vinyl ether resins using onium salts as initiators.[160] Temperature has been to shown to have a profound effect on the cationic curing rate of 3,4-epoxycyclohexylmethyl-3,4-epoxycyclohexanecarboxylate,[161] while 3,4-epoxy-1-butene has been found to undergo fast cure owing to the stabilization of the growing cationic chain end by the double bond formed during ring opening.[162] The compound 3-ethyl-3-(2-ethylhexyloxymethyl)oxetane is useful as a UV curable adhesive,[163] while other work has shown that UV curable epoxy acrylate adhesives cure faster under nitrogen than in oxygen, indicating that radical processes possibly play a role in the cure.[164] Oxetanyl-functional silasesquioxanes give pollution free tough coatings,[165] as do resins based on multifunctional acrylates with alkoxysilyl thiols[166] and silicone-epoxies.[167] Several calixarene derivatives with protective trimethylsilyl and other cyclic ether groups have been prepared,[168,169] as have several bifunctional monomers with 2,3-dihydrofuran moieties via the Heck reaction.[170] The latter exhibit very high reactivity during photocationic curing. Alkyl oxetanes possessing excellent compatibility with other monomers have been developed,[171] as have a series of epoxides with functionalities that stabilize radicals, such as benzyl and allyl groups.[172] These stable radicals can apparently induce redox reactions in onium salts, accelerating their decomposition. The addition of an epoxidized soya-bean oil to an aromatic epoxide apparently accelerates the cationic cure rate via a triarylsulfonium salt.[173] Cycloaliphatic diepoxides exhibit the same behaviour, giving a highly insoluble polymer network. The ring-opening of terminal epoxides results in the formation of macrocyclic oligoethers, giving in this case soft segments.[174] Using mixed acrylic and epoxy systems together with free radical and cationic initiators, detailed cure anaylsis showed that the former cure more rapidly.[175] Also, while oxygen was found not to have an affect on the cure rate, cationic cure was influenced by increasing humidity. Diazo resins of poly(phosphoric acid) undergo rapid photoconversion from an ionically soluble into a covalently insoluble system.[176] Other studies of interest include the photocrosslinking of poly(vinylphenol) with epoxy moieties,[177] silane-epoxies,[178] epoxy adhesives,[179] negative working poly(etherimides)[180] and hybrid curing of acrylic-epoxies.[181]

Photocycloaddition reactions have shown that dialkyl (*Z*,*Z*)-muconates undergo three reactions, involving stereoregular polymer formation, isomerism from *Z*,*Z* to *E*,*E* forms, followed by amorphous polymer formation.[182] Liquid crystalline poly(aryl cinnamate) ketocoumarins have been investigated as effective initiators for the 2+2 cycloaddition reaction.[183] Triplet sensitization results in disruption of the nematic mesophase and UV-visible hyperchromism. For dimethyl 1,4-naphthalenediacrylate only polycrystalline forms undergo cycloaddition reactions,[184] while a new polymer of 5-(6-hexanediol cinnamate)-L-glutamate has been synthesized[185]. Fluorine containing distyrylbenzene polymers have been shown to undergo both photoisomerism and 2+2 cycloaddition reactions.[186] The presence of the fluorine groups accelerates the reactions. Studies on *p*-phenylenediacrylic acids have shown that while *p*-mondecyl esters only give oligomers, the monoethyl esters with an amide bond undergo photodimerization.[187]

Mixed monolayers of liquid crystals of hydrophilic cyanobiphenyl with long alkyl chains and a poly(amic acid) alkylamine salt undergo phase separation upon irradiation[188,189] with degradation of the monolayer of poly(amic acid). The morphology of mixtures of a liquid crystal and a poly(ethylene glycol diacrylate) gave an 'islands-sea' structure on polymerization.[190] A mechanism was developed to account for the various types of organizations, depending upon the initiation of the photocuring reaction. LC poly(ester imides) with cinnamoyl groups form Grandjean textures on annealing, with a blue iridescent colour,[191] which is frozen in after curing. Diethyl (*Z*,*Z*)-muconate exhibits a topotactic polymerization on irradiation.[192] At low temperatures below 45°C the reaction is inhibited owing to small changes in packing alignments. Azimuthal anchoring energy coefficients have been measured for poly(vinyl cinnamate),[193] while glass-forming nematic twin molecules have been made.[194] Photoinduced diffusion during photocuring of a chiral-nematic phase yields a cholesteric network in which the helical pitch gradually changes over the cross-section of the film matrix.[195] For bifunctional symmetric and unsymmetric diepoxides with flexible alkylene chains, the phase transition form the crystalline to the liquid crystalline state could be varied with temperature.[196] A polymer network with a liquid crystalline superstructure is formed if the polymerization is carried out in the LC phase. Other studies of interest include ferroelectric copolymers,[197] a hybrid 3-(trimethoxysilyl)propyl methacrylate and a quaternary ammonium surfactant with a polymerizable methacrylate group,[198] a semiflexible polyester with twin-spaced *p*-oxybenzoyl diads[199] and an LC poly(methyl acrylate) containing both a mesogen biphenyl unit and a styryl-2-pyridine unit as a photosensitizer.[200]

Various methodologies in curing processes have been documented. A rotating reaction-surface polymerization reactor has been developed for measuring step-growth and chain-growth polymerizations.[201] High conversions and fast cures were obtained, giving polymers with superior properties. Surfaces for bonding have been modified by treating with acrylated initiators,[202] while photolithographic processes have been modified through the use of composities.[203] Alternative surfaces have been developed with photoreactive groups,[204] and dark coloured coatings with carbon black have been cured effectively.[205] Other pigmented systems have been effectively photocured using mixtures of a xanthene dye and a ferrocinium salt.[206] Effective photocuring systems have also been developed for filament winding,[207] while shrinkage has been measured by TMA.[208] Microlenses have been developed from polysilanes,[209] while other workers have devised processes for developing lenses for telecommunications.[210] A manufacturing process has also been developed for composite carbon-fibre reinforced parts.[211] Epoxy-polyamide formulations were examined in pre-pegs and a model was developed for layered curing. Although cure was affected, the properties of the final material were below the specifications for high performance parts. Optimization of dental resins for cure have been investigated,[212] and the ionic conductivity of methacrylates has been found to increase to a maximum on doping with lithium perchlorate.[213] A novel oligo(poly(ethylene glycol)fumarate) macromer has been developed for functionalizing networks.[214] By locking in to polypeptides, the cured system can be used for developing cell-adhesion specific sequences. Other studies include kinetic evaluations on 3D acrylates,[215] ionic conductance of methacrylates doped with magnesium salts,[216] laser simulated polymerizations,[217] degradable diacrylates,[218] use of high intensity irradiance for reducing initiator concentrations[219] and electrooptical properties of cured films.[220]

Monitoring methodologies for cure continue to be investigated. Isothermal microcalorimetry has been further investigated for real-time measurements of rates of cure for a number of resin systems.[221-223] Structure and functionality are important in this regard. Real-time FTIR and infrared analysis have also been utilized in many ways. These include studies and assessment of surface gradients,[224] conversion rates for dye-borate cationics,[225,226] cycloaddition of bifunctional epoxies with dimethacrylates,[227] curing rates of pure acrylates,[228] stress build-up during cure,[229] comparison with fluorescent probes[230] and rates of thiol and ene consumptions in mixed formulations.[231] In the last study, ene reactions were found to be some 15% faster than thiol reactions, giving rise to a predictive model. Fluorescence probes continue to be widely investigated and developed for monitoring cure rates and microviscosity properties of resins. Blue-shift probes are valuable where diffusion of the non-reacted probe out of the crosslinked regions is of paramount importance for optimum probe response.[232,233] Here a decrease in probe sensitivity was observed when the size of substituents increased, whereas more reactive substituents increased probe sensitivity. Reactive and non-reactive probes have been developed based on nitro-stilbene.[234] Gelation processes are also important where a sudden change in rotational processes occurs. with pyrene being an excellent probe.[235-237] Fluorescent probes in composities can be enhanced through complexation with divalent or trivalent cations close to the particle clusters,[238] while bismaleimides with electron-donating amine moieties exhibit strong self-quenching.[239] The binding of fluorescent probes to glass-fibre composities has been found useful,[240] while, for monofunctional acrylates the cure rate is a single-stage reaction, difunctional systems exhibit a two-stage process.[241] The latter is no doubt due to a second gelation stage. Solvatochromic and rigidchromic fluorescent probes have been developed based on diarylene and butadiene derivatives of 4-dimethylamino-4′-nitrostilbene.[242] Spectral analysis of these probes indicates the existence of an excited state more relaxed than the Frank-Condon state owing to a twisted intramolecular charge transfer, favoured by the co-operative effects of donor-acceptor groups. In other work, probe studies have shown that the primary amino group is several times more reactive than a secondary amine group toward an epoxy system.[243] Both FTIR and fluorescence analysis were found to be consistent in gel-glassy rate determinations.

Factors which influence cure rates have been numerous. *N*-Vinylamides have been found to be more reactive than acyclic amides in the curing of diacrylates,[244] while other studies have investigated the impact of chain transfer on kinetic chain lengths,[245] light intensity and film thickness,[246] shrinkage effects,[247] pressure effects,[248] glycol fragments,[249,250] stress factors[251] and temperature effects.[252] For furfuryl methacrylate, a new stochastic model has been devised called 'CORUB'.[253] A statistical analysis of the model's adequacy shows a good correlation between the experimental data and those modelled stochastically.

Photocrosslinking of polymers continues to attract interest for property enhancements. For polyethylene the use of anthraquinones gives a higher rate of chain scission than that for crosslinking,[254] while the photocrosslinking of polyethylene fibres is higher in nitrogen,[255] and new alkyl-substituted benzophenones have been found as effective photocrosslinking agents.[256] Natural rubber has been photocrosslinked through the use of co-reactive methesacrylate groups,[257] while styrene-butadiene systems can be easily photocrosslinked with a variety of initiators.[258,259] Here low concentrations of photoinitiators can be used for crosslinking thick sections. Poly(vinyl alcohol) has been crosslinked through modification with thymine[260] and uracil[261] derivatives. Photodimerization was important at wavelengths less than 300 nm, with greater crosslinking the longer the spacing units. Poly(vinyl abietate) also undergoes photodimerization through the abietate groups,[262] while PVK has been crosslinked with a Nd:YAG laser[263] and poly(3-hydroxyalkanoates) can be useful for photolithography micropatterning.[264]

Rapid curing of thiol-ene systems occurs in air[265] and, for thiol-acrylates and methacrylates, the latter components cure faster than hydrogen-atom abstraction from the thiol.[266,267] Other applications and materials include the monitoring of butyl acrylate diffusion in polystyrene blocks,[268] development of photocurable membranes responsive to ionic surfactants,[269] preparation and photoisomerism of norbornadiene units on polyesters,[270] photocurable sealants for LCD panels,[271] use of fluorescent dyes for the internal microdiagnosis of polymerized 3D microstructures,[272] development of epoxy and methacrylate monomers with chalcone moieties,[273,274] hybrid sol-gel glasses with Ba titanate and titanium alkoxides for new opto-electronic devices,[275,276] fabrication of 3D photonic crystals with lasers,[277] light curing of acrylic composities,[278] encapsulants for microelectronics,[279] monitoring and optimizing UV curable powder coatings,[280] laquers for DVD's,[281] distortions during prototyping,[282] water-based poly(methacrylate)-modified poly(urethanes),[283] photocurable poly(butadienes) with acid and base stability,[284,285] non-wettable fluorinated acrylics,[286] curing of acrylics under water,[287] yellowing of coatings,[288] curing of multifunctional anhydride resins,[289,290] phase structure of curable blends,[291] crosslinkable poly(caprolactones) for biomaterial applications,[292] use of allyl glycidyl ether to prevent oxygen inhibition in curing,[293] 3D mass transfer in curing of multi-component systems,[294] Langmuir-Blodgett films of L-glutamate[295] and crosslinkable polymers with Rosin groups.[296] Poly(urethanes) with siloxane groups exhibit a high degree of segregation[297] of the flexible units from the hard domains compared with acrylic analogues. Polymers of acetophenone *O*-acryloyloxime undergo photo-initiated thermal crosslinking when sensitized by benzophenone;[298] post-exposure bakes enhanced hardening. The molecular characteristics of siloxane polymers have been shown to influence cure rates,[299] while polymers with phenacyl groups are photocleaved to give carboxyl groups.[300] The latter may then be further reacted with epoxy functionalities. Water-developable acrylic resins have been made by co-reaction with quaternary ammonium salts of phenolic resins,[301,302] and stable optical gratings have been made from amorphous azo-benzene containing polymers.[303] Poly(allylamines) give swellable macroporous gels sensitive to pH,[304] while other developments include the formation of a dianepoxyde[305] and poly(siloxane-acrylics)[306] for litho-applications. Stresses generated during the photocuring of litho systems have been evaluated,[307] while fillers can reduce shrinkage.[308]

Dimethacrylates have been investigated in detail. Evidence has been presented in support of the existence of chain length dependent termination in crosslinking free radical polymerizations.[309,310] Shifting the kinetic chain lengths toward shorter chains had little effect on multiacrylate systems, but for corresponding methacrylates the kinetics changed significantly. Shifting the kinetic chain lengths toward shorter chains led to the delayed onset of a reaction-diffusion-controlled termination behaviour, which could influence network formation in systems above their T_g. In another study, photopolymerization kinetics for dimethacrylates showed the presence of two regions, namely, quick and short, and slow and long.[311] Rigid and flexible dimethacrylates have been mixed and studied, and it was found that the latter exhibited faster rates,[312] while in another study the reverse was found, where Bisphenol A resins cured faster than those based on ethylene glycol.[313] Other work has showed that conversion rates are influenced by the initiator concentration and absorption spectrum, and these in turn are influenced by the light intensity.[314] Coupled with this is the fact that the final properties of the system will depend on the degree of conversion,[315] temeprature and the number of methylene groups in the monomer.[316] Dimethacrylate phosphonates have been made but found to be much less reactive than conventional dimethacrylate systems,[317] whereas for dioxolane derivatives of bisphenol-A no complete ring opening has been found in the presence of a free radical initiator.[318] Hyperbranched dimethacrylate polyesters have been evaluated for use in dental systems,[319] and dendtric polyester acrylates have been made that require no initiator.[320]

Polyimides and polyamides have also attracted some attention. Photosensitive polyimides with benzophenone moieties and hypersensitizing properties have been synthesized and their properties easily controlled for cure.[321] Hyperbranched poly(ether-imides) are useful positive working resists,[322,323] whereas fluorine containing polyimides have excellent near-UV transparency[324,325] and excellent thermal stability. Polyimides with a high content of carboxylic acid groups have enhanced photosensitivity,[326] as do resins with dianhydrides.[327] The incorporation of silica enhanced the physical properties of the cured polyimide materials.[328] Hyperbranched aliphatic acrylated poly(amide esters) cure effectively to give thermally stable systems.[329,330]

2.3 Photografting. – Poly(acrolein) particles have been photografted with tetramethylcyclotetrasiloxane[331] via cleavage of the Si-O bonds. Thin layers of *N,N*-diethylamino)dithiocarbamoylpropyl)(trimethoxy)silane and *N*-isopropylacrylamide have been photografted onto silicone wafers, which exhibit a fast temperature response.[332] Surface modification of poly(imides) has been undertaken with acrylic monomers for potential high tech applications.[333] Maleic anhydride can be grafted to polyethylene films with high adhesion strength at 80°C,[334] while various acidic gases can grafted onto diffent polymer surfaces to give nitrile, amino and sulfonic acid groups.[335] Rare earth metal complexes have been homogenized on polymer surfaces by effectively photografting diketone ligands.[336] Micropatterning on surfaces can be affected provided a photoiniferter is present,[337,338] with acrylic acid giving a better graft than dimethylamino methacrylates. A pore-filling process has been used to produce high-performance composite membranes for selective pervaporation separation of aromatic and non-aromatic solvent vapours.[339] These were based on methacrylate monomers grafted onto poly(acrylonitrile). Vinyl acetate has been photografted onto polyethylene using benzophenone as a photoinitiator.[340] Here preadsorbing the benzophenone to the polymer gave better graft yields. Vinyl monomers have also been photografted onto silica surfaces[341] and poly(propylene).[342] The surface adhesion of polyethylene has also been enhanced by grafting with methacrylamide as well as enhancing its wettability.[343] A method of grafting conjugated polymers onto solid substrates by self-assembled layers has been developed.[344] Here poly(fluorene) was deposited onto electrodeposited carbazole. Thiocarbamate groups have also been photografted to copolymer surfaces,[345] as have thiol-terminated polymers[346] and hydrogels[347] onto gold. Other studies of interest include the production of semi-fluorinated alkane thiol self-assembled monolayers with diacetylene units,[348] photofunctional core shell micro-spheres,[349] acrylated polyurethanes grafted onto acrylic copolymers[350] and methyl acrylate grafted poly(acrylonitrile)[351] for separation of benzene-cyclohexane mixtures.

3 Luminescence and Optical Properties

A number of review articles have appeared covering various aspects of polymer luminescence, optical and conductive properties. These include smart materials based on functionalized polymers,[352] light emitting polymers for visual applications,[353] photophysical characteristics of water soluble polymers,[354] micellar media,[355] diacetylene single crystals,[356] conjugated metal complex polymers,[357] semiconductive polymers,[358] charge-carrier generation[359,360] and intra-chain fluorescence quenching studies.[361]

The chemiluminescence (CL) property of polymers continues to attract a steady interest in the field. Much of the work is directed toward investigating and probing the degradation and oxidation processes in polymer materials. The technique of second time-derivative analysis has been developed for studying chemiluminescence profiles of polypropylene-polybutadiene blends.[362] It is concluded that this technique, when used in conjunction with the integrated CL profile, will enable evaluations of the oxidative induction time (OIT) to be made. For the polymer blends, each phase was found to oxidize independently with separate activation energies. The propagation of CL as fronts through stabilized polymer films could be used as a means of measuring the effectiveness of different stabilizers.[363] The technique was found to be consistent with that undertaken by FTIR analysis. Arrhenius plots have also been obtained for stabilized poly(propylene) films to establish the consistency of measurements and suitable testing conditions.[364] Recycled poly(propylene) films have been shown to exhibit a different CL behaviour to that of conventional virgin material.[365] The former shows threadlike defects as well as a significant change in oxidative stability, which was apparently detectable much earlier than by FTIR and DSC methods. In another study on poly(propylene), the CL intensity has been found to be proportional to the growth in carbonyl species.[366] It was proposed that the CL was due to either energy transfer from the carbonyl species to a more effective emitter or the reaction of peroxides with an oxidation product. Clearly, the nature of the CL species is still elusive. Kinetic evaluations on pre-oxidized poly(propylene) has, however, shown that the rate constants for CL growth match those for fast decomposing hydroperoxides.[367] Similar results have been obtained on the CL analysis of low density polyethylene, where it has been used to analyse trace levels of residual hydroperoxides.[368] Characterization of oxidation rates of polymers and analysis of initiation sites and stress levels have been covered in a number of studies.[369-371] Studies on poly(2,6-dimethyl-1,4-phenylene oxide)[372] and polyesters[373] have shown that the CL behaviour is pre-determined by the oxidation levels and presence of oxygen during measurement. Hydroperoxide sites are found to be responsible for the thermally induced CL in polyesters, with the nature of the aliphatic backbone playing a major role in controlling the CL behaviour. This is further borne out by a study on cellulose,[374] which showed that the CL was dependent not only on the presence of oxygen but also on moisture, no doubt owing to the formation of hydrogen peroxide. In the advanced stages of degradation the CL showed a dependence upon alkali-earth metal ions. Here the metal cations could potentially complex with the glycosidic groups. Concentrations of hydroperoxides have also been evaluated in polyester-melamine resins by CL[375] while CCD devices have been found useful for the detection of CL from PVC formulations.[376] In other applications CL can also detect polyphenols in fabrics,[377] and an electrochemical technique has been developed for the detection of coating integrity based on CL emission.[378] Violet CL emission has been observed from a novel condensation poly(imide)[379] and from fluorophores containing distyrylarylene units.[380] CL has also been observed from light irradiated wood surfaces[381] and found to be correlated with discolouration. The electrical breakdown of epoxy resins gives rise to electroluminescence,[382] while peroxides are associated with their thermally induced oxidative generation of CL.[383] The radiothermoluminescence from different polymers is also dependent upon oxygen pressure.[384]

Polyacetylenes continue to attract some interest. Using Raman scattering, the primary photoexcitation in trans-polyacetylene and polydiacetylene is considered to be a breather exciton, a bound state of an exciton and a localized non-linear lattice excitation associated with stretching vibrations of the polymer backbone.[385] Phenyl substituted polyacetylenes have been evaluated using multi reference singles and doubles CI calculations.[386] Apparently, strong luminescence is due to the fact that the ^{2}Ag two-photon state lies above the optical ^{1}Bu state, in spite of the polyene backbone. Reversed energy ordering in these polymers, as compared to other linear types, is a signature of reduced effective electron correlations. The optical gap of the substituted materials is smaller than that of polyenes of the same length. Also of interest is the observation that the dark conductivity of poly(diacetylene) is enhanced after laser irradiation and that this effect may be due to the growth of surface oxidation products.[387] On the other hand, electrically induced molecular re-orientation can fine-tune the photoluminescence from poly(acetylenes),[388,389] while benzylcarbazole substituted poly(acetylenes) exhibit much weaker emission.[390] Anthryl substituted poly(acetylene) has been made and is black in colour with a visible cut-off at 780 nm,[391] while the photoluminescence of 1-naphthyloxy-1-alkynes gives stronger emission in the monomeric than the polymeric state.[392] The polymerization kinetics of ethylmorpholine pentacosadiynoic amide were measured in floating and supported monolayers at an air-water interface.[393] On irradiation there was an observed blue-to-red polymer colour change with quite distinct levels of fluorescence due to a rearrangement in the packing of the side-chains along the polymer backbone.

Studies on polymer blends have provided valuable information on miscibility. Multiphase behaviour has been observed in copolymer blends of dimethacrylamide and methacrylic acid.[394] Although three main phases were observed, transitions among different phases were discrete with hysteresis loops. Fluorescence analysis indicated the coexistence of hydrophilic and less hydrophilic domains, with their fraction depending upon pH. The monitoring of in-situ blends of methacrylate polymers was found to be inconsistent under shear induced coalescence, whereas, in the absence of shear energy, transfer measurements with donor-acceptor probes were in good agreement.[395] Blend miscibilities of poly(fluorenes) have also been evaluated,[396,397] as have blends of PPV's with poly(ethylene oxide).[398] Excimer formation has been used to determine the miscibility of poly(vinylcarbazole) with PMMA,[399] while complexes of poly(acrylic acid) and poly(vinylpyrrolidone) have been found to be quite stable owing to associated hydrogen bonding.[400] Poly(vinyl alcohol) and poly(vinyl acetate) blends have been prepared on a poly(ethylene) surface and found to give different domains, depending upon the degree of polymer interpenetration.[401] Other studies have looked at the morphology of epoxy-poly(organosiloxane) mixtures where there was some diffusion restrictions[402] and hydrogen bonded poly(acrylics) using the photocontrolled pitch of a cholesteric helix.[403] In the latter, blends with a chiral nematic phase were formed over a wide temperature range and for films with planar orientation a selective light reflection in the visible and IR spectral regions is observed, with a temperature and UV irradiation time dependence.

The properties of dendritic structures of various types have attracted interest from their photophysical and photochemical aspects. Doughnut-like structures have been observed from methyl ethers and phenylenediacetylene derivatives up to 150 nm in height and 8 microns in diameter.[404] Here no correlations were found between the degree of orientation and the formation of doughnut structures. Fluorescent poly(benzyl ether) dendrimers with isobutenyl and naphthoate groups have been made with effective light-harvesting properties,[405] as have similar polymers but with a rhodium (III) porphyrin focal core.[406] Phenylacetylene copolymers with emission quantum yields of up to 0.97 have been made,[407] while dendritic molecules with a 1,1′-binaphthyl core have been found useful for enantioselective recognition of chiral organic molecules.[408] A series of star shaped biocompatible polyesters and polylactides with ruthenium(II) tri-bipyridyl cores have been found to exhibit reversible thermochromic bleaching as well as reactions to external chemical stimuli[409] for sensing applications. Dendritic polyesters have been made with high methacrylate functionality and found to exhibit superior properties after photocuring than linear structures.[410] Cure speeds, however, are significantly slower. For a series of first-generation polyphenylene dendrimers substituted with spatially well-separated peryleneimide chromophores and a terryleneimide trapping chromophore at the rim, over 95% of the total energy absorbed by the molecule is trapped by the outer structures by a FRET process.[411,412] Ethylenediamine with methyl acrylate dendrimers and an anthracene moiety in the core have been developed that undergo a reversible photocyclization process.[413] The photoalignment of 'all-azobenzene' dendrimers has been investigated,[414] whereas polyamidoamine dendrimers with azobenzene groups undergo reversible *trans-cis* isomerism.[415] Dendrimers with a periphery of dibutylamino groups are good donors,[416] while dendrimers with spironaphthoxazine groups undergo not only photoresponsive changes but also reversible complexation with metal ions.[417] Eosin-based dendrimers give less singlet oxygen than the unbound molecule,[418] while pressure affects the luminescence from methoxyphenacetylene dendrimers.[419] Water soluble dendrimers of poly(sulfone-amines) have been found to be highly sensitive to Fe(III) ions with little interference,[420,421] whereas the caging of drug molecules can be undertaken via the use of dendrimers with peripheral carboxy-2-nitrobenzyl groups.[422] New methods have been developed for polyamidoamine,[423] stilbene[424] and polyphenylene[425] dendrimers. Aggregation effects have been found to increase from a series of diblock to star-block methacrylate copolymers,[426] and poly(triazine esters) are photolabile.[427] Hyperbranched poly(ether ketones) with 1,3,5-triazine moieties are amphiphilic, forming self-aggregates at a specific concentration.[428] Similar materials have been made from poly(ethylene oxide) with hydrophobic dendritic carbosilane groups.[429] Block copolymers of third generation could not be dispersed in water, whereas the first and second generations formed micelles with a multi-core lower viscosity centre. Polyamidoamine starburst dendrimers have been shown to form a tight non-polar association, which becomes more densly packed with increasing generation.[430,431] Only weak fluorescence is observed, possibly due to the $n\pi^*$ character of the lowest excited singlet state of molecules. Porphyrin and pyrene molecules have been encapsulated within polymeric shells by ring-opening ε-caprolactone.[432] With increasing chain length the core is isolated further, thus providing a useful methodology for monitoring the effectiveness of encapsulation techniques. Other studies of interest include the development of star shaped polypeptides,[433] polyelectrolytes based on poly(*t*-butyl methacrylate) precursors,[434] fullerenes with iniferter *N*,*N*-diethyldithiocarbamate groups,[435] fullerene star shaped polystyrenes[436] and porphyrin containing polyamides.[437] Finally, by careful design of dendritic molecules the physical properties of the self-assembled structure can be tuned in such a way as to phototrigger micromanipulation, giving rise to the formation of hard spheres.[438] This possibility leads to the bottom-up synthesis of micrometer sized objects through a combination of covalent synthesis and supramolecular organization.

Photochromic systems continue to be highly active. Photoinduced rearrangements readily occur from the 'trans' quinone forms to the 'ana' quinone forms for a styrene derived phenoxyanthraquinone in both solution and film.[439] Polyesters with norbornadiene groups photoisomerize smoothly to the quadricyclane form,[440] as does a norbornene derived polymer with phenoxyquinone groups.[441] Spyropyrans in polymer matrices have significance. Pyran doped polymers have particular value in holographic recording media,[442] with a number of systems having been developed based on spirocyclopropenes,[443] hydroxy-9*H*-xanthen-9-ones,[444] bisindolinobenzospiropyrans linked to thio and carbonyl groups,[445] dicyanomethylene-fulgides,[446] new spiropyrans with thermo and photochromic properties[447] and photochromic acrylates with biphenylene and spiro-oxazine moieties with chiral substituents.[448] In the latter case, the homopolymers did not exhibit mesophases owing to steric hindrance between the side groups. Azo-structures continue to be one of the most studied, with a number simple dopant studies and the grafting of azo dye chromophores to effect optical isomerism.[449-452] Many other studies deal with co-reacted azo-type funtionalities. Hydrophobically modified poly(acrylates) with azobenzene groups have been observed to give unusually high viscosities at certain critical concentrations owing to reversible crosslinks.[453] On irradiation the *trans* to *cis* isomerization gave rise to viscosity changes due to inter-polymer associations. Azo group labelled polyesters also undergo rapid reversible *trans-cis* isomerization,[454] as do poly(malonic esters) with *p*-cyanobenzene groups.[455] In this case unexpected photoinduced birefringence was observed at two laser pumpings of 488 and 365 nm, which were attributed to some other unassigned process and not solely isomerization. PMMA with azo groups also exhibit birefringence with a dependence upon the polarization angle of the laser pump beam and probe beam.[456] Here a model was developed to describe the geometry of the isomerism and to correlate its treatment with the writing-erasing processes of the photoinduce birefringence. Activation energies and hole-buring proceses have also been measured for poly(acrylics) with azo chromophores.[457,458] Photo and thermally reversible isomerization has also been evaluated in amphiphilic copolymers of 4-(4′-dodecylphenylazo)benzoic acid,[459] while in another study relief structures have been imaged by coupling azobenzene groups to urethane-based polymer systems.[460] For blends of LC polymers of alkylhydroxybenzoic acid with 4(4-pyridyloyl)oxyazobenzene, an increase in the concentration of photochromic groups favoured a change in the phase state of the material.[461] In principle it was shown that hydrogen-bonded photochromic blends may be used as optically active media for data recording. For copolymers of *N*-(2-hydroxypropyl)methacrylamide with 6-[4-(4-methoxyphenylazo)phenyl]hexyl methacrylate, the azo functionalities were found to self-organize into dimeric aggregates,[462] while aggregate formation in the PMMA-azobenzene system is highly dependent upon thermal treatment.[463] In this case, the first writing cycle of the aggregated polymer film showed a steep increase in photoinduced birefringence followed by a gradual drop to a certain asymptotic value over longer irradiation periods with polarized light. New write-erase chiral photochromic cholesteric shaped acrylic copolymers have been made with azobenzene groups and chiral menthol or menthone functionalities.[464] The cholesteric helix untwists on irradiation with a corresponding shift in light reflection to longer wavelengths. This process is thermally reversible with the menthyl system. Ionomeric polyurethanes have been made that can undergo an ion exchange with 3(*p*-oxyazobenzene)-sodium-propane-sulfonate groups[465] while a theoretical model has been developed that describes a temperature field which is established within a film of isotropic azobenzene LC side-chain polymer under linear polarized laser beam irradiation.[466] The minimum light intensity required to induce a photo-induced phase transition as well as an offset photo-alignment temperature were calculated. Isothermal luminescence has been observed from PMMA films doped with aromatic amines which had been ionized.[467] On cooling the amine radical cations recombine with trapped electrons emitting light. Similar processes were developed with compounds based on 3-methyl-1-benzothiophen-2-yl moieties,[468] while those based on and *N*-allyl-maleimide did not work.[569] Helical structures have been developed based on poly(fluorenes),[470] while the magnetic properties of photochromic stilbenes with diarylethene units have been investigated.[471] Here the magnetic properties of the molecules are interesting since they can be switched from ferromagnetic to non-magnetic with some ring opening in the isomers also taking place. Copolymers of 3-tetradecylthiophene and 3-*p*-trifluoromethylphenylthiophene exhibit reversible thermal behaviour.[472] Thermal motions of the side chains result in extended conjugation to the phenyl groups along the polymer backbone. Above 70°C the conjugation begins to break down, with the fluoro derivative exhibiting blue shifted emission due to the electron withdrawing effect. Thermally induced phase transitions have been studied for 5,7-dodecadiyne-1,12-diol bis[phenylcarbamate][473] as well as diarylbutadienes.[474] Hinged poly(peptides) have also been made which show uniform *cis* and *trans* isomerism for photochromic moieties that are close to the terminus of the polymer chain.[475] An analytical theory has also been developed for couples undergoing photoisomerization and photoorientation, both for individualized and spectrally overlapping isomers.[476] When diarylethene and spiropyrans are oriented by polarized light, the apparent optical orientation changes sign for the UV versus the visible photochemical transitions. Other studies on chromic materials include polymers with azobenzene side chains,[477] spiropyran doped polystyrene,[478] triazine-stilbene fluorescent brighteners,[479] pyrocarbazoles[480] and LC polymer microspheres.[481]

LC polymers follow on with a similar active interest. A number of LC polyesters have been prepared with 1,3,4-thiadiazole[482,483] and 4-hydroxybenzoic acid and 6-hydroxy-2-naphthoic acid.[484] These polymers form stable enantiotropic smectic and nematic phases. Photoinduced alignment studies have been undertaken on several polyimides,[485] with fluorine and chalcone groups,[486,487] 4-styrylpyridine groups[488] and methyl red dye as dopants.[489] Here pre-tilt angles were found to be related to the degree of imidization, so by adjusting the chemistry of polyimide formation the pre-tilt angle can be varied continuously. Varying the structure of the polyimide produces LC's which will align either parallel or perpendicular to the polarization direction of the UV light. Significant energy transfer (FRET) has been observed between a LC fluorene-based polymer and a non- LC polymer.[490,491] In this case a low threshold photo-pumped amplified spontaneous emission was observed in a waveguide device fabricated with the blend. This low threshold emission originates from a largely reduced self-absorption loss by the efficient energy transfer in the photoluminescent polymer blend with LC properties. PMMA with terphenyl side groups[492] has been made, as have polyethers with stilbene side groups.[493] Both types exhibit LC properties, the latter having bitropic properties, with low and wide phase-transition temperatures as well as good thermal stability. New chloesteric polymer materials possessing double photochromism have been made from copolymers with phenylbenzoate hematogenic and spyropyran side groups.[494] Irradiation and annealing of planarly-oriented films made of these systems leads to the untwisting of a cholesteric helix, inducing an irreversible shift of selective light reflection to the long-wave region owing to the *E-Z* isomerism. The thermally reversible spiropyran gives a separate process but one which could be used in conjunction with the *E-Z* process for photoelectronic recording systems. The LC properties of AB block copolymers of polystyrene and poly(isoprene) with 4-cyanophenylazobenzene side groups is influenced by the microstructure of the materials related to the phase separated coils.[495] Azo grafted ladder-like polysilasesquioxanes normally used for command surface fabrication have an improved photodriven response.[496] The response steepness of an LC cell was found to be two orders of magnitude greater than previous work with an optimum response at 10 mole-% of azo content. The LC alignment in films of polyimide with cinnamoyl and coumarin groups has been found to be induced by the photodimerization processes rather than any *trans-cis* isomerization,[497,498] while the photoinduced alignment of LC polymers with azobenzene groups has been optimized.[499] It was found that the response could be enhanced by chemical tailoring of the polymer structure and physically modifying the alignment procedure. New comb-shaped photochromic and LC ionomers have been made by protonating 4-(10-diethylaminodecyloxy)azobenzene through the carboxylic acid groups of functionalized polymers.[500] Here the growth and decay of induced birefringence for ionomers and photochromic polymers were found to obey similar laws. A series of polyacetylenes containing 4-(*trans-n*-alkylcyclohexanyl-carbonyloxy)phenyl-4-alkynyloxybenzoate side groups have been found to exhibit nematic, smectic A and smectic B phases, while the monomers showed only a nematic phase.[501] The polymers gave green-blue luminescence and had an interdigitated bilayer structure. Homogeneous alignment has been maintained in a series of three ring phenylacetylenes,[502] while a series of acetyl (ethyl) cellulose-acrylic systems reflects one hand of circularly polarized light in a narrow region of wavelengths.[503] LC transitions have been probed via the use of fluorsecent molecules,[504] and nematic diepoxides have been developed for micro-structured devices.[505] Other LC devices include a series of multifunctional chiral photochromic materials[506] and antiferroelectric diacrylate gels.[507] UV irradiation has been used to control and tune the properties of emulsion polymers with azobenzene chromophores,[508] and phase separation has been examined and calculated for LC unsaturated oligoester resins.[509]

The area of LED polymers is the most actively studied in the field. A number of articles have dealt with various aspects of LED's such as their preparation,[510] improvements,[511] ladder structures[512] and engineering aspects.[513] The excited state characteristics of these polymers continue to attract much theoretical and experimental consideration. From configuration-interaction calculations on PPV's, intersystem crossinking from the singlet to the triplet manifold is associated with spin-orbit coupling, giving rise to both singlet and triplet wave functions.[514] Using femtosecond spectroscopy the optical emission form PPV's is described by exponential functions,[515] while transient absoprtion measurements show that charge-carrier generation occurs 100 fs after excitation.[516] The carrier generation is also influenced strongly by the strength of interchain interactions. Structural changes in a single molecule over periods of seconds have been measured by monitoring fluctuations in the wavelength of emission spectra using total internal fluorescence microscopy,[517] while quadratic coupling between ground and excited states has also been measured by fluorescence analysis.[518] The latter results from twisting about the oligomer axis, which is free in the ground state but strongly constrained in the excited state as a planar configuration. A theoretical description of excitions in PPV's has been obtained using the electron-hole Bethe-Salpeter equation,[519] and the Peierls model used to solve the energy and soliton structure.[520] With the Green function method the extent of polarization can be determined,[521] while other work has shown that quenching processes are significant in conjugated polymers,[522] with solid-state materials exhibiting over 50% more intense emission than corresponding solutions.[523,524] In contrast, solutions give stronger emissions owing to interchain quenching process in the solid phase.[525] Energy migration in PPV's is not fully accounted for by the FRET mechanism,[526] with molecular aggregation having a profound effect on the excited state properties of the polymers.[527] The way in which polymer films are cast is also very important.[528] This can influence the nature of any inter-chain interactions and subsequent electroluminescence efficiency. Of particular relevance is the study of inter-chain aggregates in polythiophenes;[529] time-resolved analysis has identified the emissions from intra and inter-chain aggregates. Dynamic quenching of luminescence in films with dense packing is mainly due to excitation energy transfer to aggregates. Here the aggregate contribution to the total luminescence in polythiophene films with dense chain packing is about 50% and is virtually independent of temperature. For PPV's evidence is also presented that photoexcitation of aggregated polymer chains results in the creation of interchain non-emissive polaron pairs in large quantum yield.[530] This effect accounts for the observed lower quantum yields in the solid state and supports the picture of polaron pairs recombining to form the ground state and not the corresponding singlet and triplet states. On the same theme, chain conformation strongly influences the nature of the fluorescence from PPV's.[531] Thus, while extended chains give emission from multiple segments, folded chains give emission from only a few sites. The tightly folded chains enable the polymer to funnel excitons to highly aggregated low energy regions via a 3D FRET. This strong intra-chain coupling causes the polymer to mimic the photophysical behaviour of single chromophores, leading to localized emissions nad fluorescence intermittency. Polymer molecules with specifically designed steric hindrance for backbone contact show higher resistance to solvent induced interactions between segments. Non-solvents force the polymer chains apart causing stacking and aggregation. These buckled chains show red-shifted excimer fluorescence. Such effects can be detrimental to device performance.[532] Inter-chain coupling of PPV derivatives in solution is more predominant for *cis* than for *trans* isomers, with nanotubes offering the best performance for electroluminescent devices. New phenothiazine-quinoline copolymers have smaller band gaps than their corresponding homopolymers, giving blue and red emission bands associated with intramolecular excitons and intermolecular excimers.[533] On a similar theme, interfaces of polyquinolines and polythiophenes exhibit strong quenching due to electron transfer from the quinoline to the thiophene groups.[534] A new weak red emission was observed associated with an unreported exciplex site at the interface. It was suggested that such interaces could have valuable application in photovoltaic devices. Electroactive poly(vinyl diphenylquinoline) has been made and found to be phototunable in acidic media[535] owing to intramolecular excimers. Apparently, in neutral or thin film media, no excimer formation was observed giving rise to electroluminescent sensors. Single layers of carbazole and fluorene containing PPV's have been found to give rise to pure blue emission,[536] and stronger electroluminescence has been observed from D-A bipolar oligomers of PVK with dithienothiophene.[537] Similar results were obtained with alternating thienylene-phenylene copolymers giving red-shifted emission spectra.[538] Varying the thiophenoxy precursor chain length in the PPV is another means of altering the band gap.[539] Another method of enhancing the charge-carrier transport has been to develop a two-step synthesis for introducing electron accepting sulfone groups,[540] while more effective polymers have been developed using polymers with dithiafulvene and bithiazole groups.[541] Thiophene-*p*-phenylene co-oliogomers have been molecularly modified to enhance emission tuning and control their epitaxial crystallization during vapour deposition of thin films.[542] Here films with highly polarized emission were formed, providing useful materials for isotropic applications. Magnetically designed polymers have also been developed to produce dichroic luminescence and electrical anisotropy.[543] These were useful for LC devices. Bipyridyl molecules have been incorporated into water soluble PPV's to examine the emission quenching processes.[544] Quenching increases with increasing charge.

Substitution of solublizing groups, especially those based on the alkoxy functionality, continues to be researched. To avoid the usual problem of phase separation between conjugated and ionic polymer in LED electrochemical cells, flexible poly(ethylene oxide) blocks have been successfully introduced into PPV.[545,546] Emission properties were retained while for a series of alkyl-substituted PPV's the emission quantum yield decreases with increasing chain length.[547] For increasing alkyl chain size there is also a corresponding blue shift in the emission maximum and decrease in electronic delocalization along the chain.[548] On the other hand, enhanced green electroluminescence has been obtained through the development of blends of PVK with alkoxy substituted m-phenylene PPV's.[549] For octyloxy derivatives exciton bands were found to decrease in intensity,[550] while PPV's with 2,3-substituted dibutoxy or dimethyloctylsilyl side-chains gave double layer emitting devices with eletroluminescence efficiency and low turn-on voltages.[551] Blends of methoxy and dioctyloxy PPV's have been found to exhibit amplified spontaneous emission,[552] while copolymers with oligo(ethylene oxide) side chains have been designed to combine high luminescence efficiency with ion-co-ordination ability.[553] PPV's have been found to retain a rod-like structure in poly(styrene),[554] and novel LED electrochemical cells have been made by copolymerization of PPV's with crown-ethers.[555] An LC PPV with two octyloxy-pentylcyclohexyl-phenyloxy side chains hase been made and found to exhibit effective electroluminescence-device characteristics in a sandwich structure test.[556]

Fluorene structures have also attracted some interest. Hyperbranched PPV's with fluorene moieties exhibit short wavelength blue emission with high quantum yield and good thermal stability.[557] Similar copolymers have been found to give much stronger emissions in solution compared with the solid state[558] owing to the latter forming strong aggregates. A colourfast poly-2,8-(indenofluorene-co-anthracene) copolymer has been made which unfortunately undergoes ageing on storage, causing a reduction in emission intensity,[559] with similar problems occurring in other poly(fluorenes) where a small number of dimer sites have been identified.[560] The latter case causes exciton quenching and excimer formation. Other trap free systems based on poly(2,7-(9-fluorenone) have been developed,[561] while in another study dialkylfluorene units give polymers with enhanced electron affinity and ionization potential.[562] To combat the problem of aggregation, copolymers have been developed based on polyfluorene and carbazole units.[563] These disordered polymers gave strong emission with good thermal stability. Polyfluorenes coupled with carbazole units have also been combined with acetylene units[564] and PPV's with phenylphosphne oxide moieties.[565] The latter have well defined emission spectra in the blue and red regions of the spectrum. Highly effective *trans*-poly(9,9-di-*n*-octylfluorene-2,7-vinylene) has also been made by acyclic diene metathesis polymerization using Mo catalysts.[566]

PPV's with oxadiazole groups have also been developed for enhanced LED applications. A biphenyl-oxadiazole derivative has been found to exhibit effective electron-hole transport with good photostability, while a bipyridyl derivative was found to be unstable.[567] Oxadiazole polymers have been found to be excellent energy donors in blends with PPV-carbazole derivatives,[568] giving strong blue-shifted emission. Similar polymers with carbazole and oxadiazole units have been made with green emissions at 530 and 540 nm, respectively.[569] The former had the best device performance, with a low turn-on power. Other oxadiazole polymers with differing content of 2-ethylhexyloxy pendant groups have been found to be stronger emitters than PPV itself, as well as being effective EL devices when layered with In-Sn -oxide coated glass.[570] Other polymers with green emitting oxadiazole[571,572] and 2,4,5-triphenylimidazole[573] chromophores have also been synthesized.

The photophysical characteristics of various structural types of PPV's have been investigated and collated.[574] These include some cyano derivatives[575,576] and block copolymers.[577] Although near unity quantum efficiencies were observed, excimer formation is quite evident in thin coated films. PPV's with dioctyloxy groups behave like a collection of semi-independent excitons with dipoles arranged in the same direction.[578] DC background measurements, however, indicated a few misalignments. Tunable PPV's have been prepared for optoelectronic devices,[579] while a reduction in temperature enhances the EL intensity.[580] Vacuum deposition and doping with iron have also been found to be useful ways of enhancing the luminescence from PPV's.[581] High stereoregularity in poly(phenylacetylenes) gives polymers with specific emissions that are independent of solution concentration when compared to irregular polymers.[582] The integrated fluorescence lifetime of polycrystalline PPV's decreases with increasing temperature, with a corresponding increase in anisotropic decay.[583] In 2-methoxy-5-hexyloxy-substituted PPV's excimer emission is evident and there is no temperature dependence on the fluorescence.[584] Localized chain interactions are responsible for these differences. Similar results were obtained on the same alkoxy derivatives of PPV after spin coating solid films,[585] where aggregate trapping was more important. Other studies include ionic quenching[586] and photoexcitation dynamics in ladder type structures.[587]

Other PPV's have been made with stilbene-*N*-alkylpyridinium chromophores,[588] where the nature of the alkyl group strongly influences the emission quantum yield, with the most efficient being those with long aliphatics. *Z-E* isomerism is also important in PPV's,[589] with azo groups enhancing optical anisotropy owing the side-chain reorientation. PPV's with side-chain vinyl groups have also been prepared and characterized, showing fibrillar and network type morphology.[590] Delayed fluorescence from 1.5% of triplets due to T-T annihilation has been measured for 2-methoxy PPV,[591] while enhanced emission was obtained after copolymerizing PPV with the 1-methoxy derivative.[592] Organometallic modules strongly influence the emission properties of PPV's,[593] while in *m*-phenylene stacking of PPV's is prevented.[594] Alternating rigid block copolymers of PPV's with poly(ethylene glycol) units have been made to enhance solubilization,[595] whereas a new method for PPV synthesis has been devised via a Diels-Alder reaction.[596]

PPV's with amine groups have been investigated. Alternating copolymers of the methoxy-PPV, vinylcarbazole and a triphenylamine-PPV give very high luminescence intensities.[597] Hyperbranched PPV's with triphenylamine hole-transporting groups give good films with high luminescence quantum yields in the red region.[598-600] PPV's with alternating triphenylamine groups give strong green luminescence[601] but high turn-on voltage, while the polymerization of PPV's at low temperatures does not alter their photoluminescence efficiency.[602] The incorporation of alternate *meta*- and *para*-phenylenes in the PPV chain gives a well blue-shifted emission owing to reduced conjugation lengths.[603] Intra- and intermolecular aggregates have been quantified in PPV's,[604] with the former being associated with the luminescence. Monfunctional oligo-PV's with *s*-triazine units aggregate in chiral stacks.[605]

Ionic PPV's and derivatives have been examined. This includes the lyotropic properties of LC poly(pyridinium) salts in protic and aprotic solvents,[606,607] where films were formed easily and luminescence arose in methanol and DMSO. For model systems the presence of potassium nitrate influences the state of aggregation in micellar media,[608] while oligomers of PPV's have been found to behave as more effective sensors to analyte quenchers than their polymer analogues.[609] Carboxy-substituted PPV's have been made with water solubility,[610] giving narrow photoluminescence spectra. Such structures were effective for multilayer self-assembly techniques. In such systems excited state dynamics can be effectively controlled even on the nanometre scale.[611] Energy-transfer processes have been examined in detail in such systems with dopands such as Cu phthalocyanine. This allows the development of controlled photovoltaic devices. A novel soluble derivative of PPV has also been made.[612]

Modified PPV's are also being investigated for more practical applications. This includes doping for example, in polymer-silica hybrids.[613] Results so far indicate the enhanced formation of J-aggregates, a blue-shift in the emission of some 80 nm, a marked increase in efficiency and higher photostability. Similar work has been undertaken on binding into clay.[614] In this case, a 2-D lamellar composite structure efficiently enhanced the electron-hole recombination rate by blocking charge decay. Additionally, the isolation of the polymer chains within a confined geometry by intercalation prevents the excitons from finding low energy trap sites. As mentioned above, PPV's with ethylene glycol groups have improved ionic solvation, thus electrodes with Ca cathodes can achieve high luminance intensity for electrochemical cell behaviour.[615] Polyether-based PPV's with high luminance and good solvent solubility have also been developed,[616] while multicolour devices have been made based on PVK.[617] Fluorescence quenching in diphenyl-substituted PPV's is believed to be due to exciton-exciton annihilation flowed by polaron-pair formation and then further annihiliation.[618] In order to reduce chromophore aggregation in PPV's and enhance emission intensity, planarization has been calculated to give modest shifts of 20-30 nm, which can also preserve the vibronic structure of the monomer.[619] The introduction of diphenylmaleimide structures introduces kinks into the PPV chains[620] and gives twin emissions at 413 and 530 nm. Pendant fullerenes, on the other hand, give two signals due to photoinduced transfer of electrons between the conjugated chains as donors and dangling fullerenes as acceptors.[621] Fully functionalized fullerene PPV's have also been made that are solvent soluble.[622] In poor solvents, cyano-substituted PPV's give red-shifted emission spectra due to increasing inter-chain states,[623] while *cis* and *trans* forms give completely different spectral emissions.[624] Very stable tetra-fluorine PPV derivatives have been made and found to be good emitters with bands at 485 and 515 nm.[625] In contrast to many fluorescence studies, other workers have ascertained the importance of the triplet state through characterization of the phosphorescence emission form PPV derivatives.[626] Metal chelation of PPV's through reaction with 2,2′-bipyridyl groups attenuates electron delocalization.[627] The introduction of silicon-based alternating units also disrupts the electron conjugation in PPV's.[628] Nevertheless, these materials were more easily processed and gave distinct emission spectra dependent upon the turn-on voltage.

Other LED polymers include bifuryl-polyimides with emission yields of up to 15%,[629] 1,3,5-tris-styrylbenzenes,[630] phenyl-capped poly(anilines),[631] poly(anilines),[632] 1,4-bis(3′-methyl-6-octylstyryl)benzene,[633] ionic poly(ethylene oxide),[634] PVK blends,[635] 2,2′-bipyridyl polymers,[636] carbazolylmethacrylates,[637] Pt-doped ethynylenes,[638] poly(pyridine vinylene),[639] Eu-complexed PVK,[640] a poly(naphthyridine-2,6-diyl) complex with $Ru(bpy)_2^{2+}$, [642] and poly(heteroaromatic oxadiazoles),[643]

Polythiophenes have also been widely investigated as LED's. Triplet energy migration has been observed to the longest conjugated segments in 3,4-dibutylthiophenes,[644] while, in poly(3-butylthiophene) layers, mobility has been found to be unaffected by thickness.[645] The morphology of poly(thiophene) derivatives depends upon the nature of the alkyl substituents and and chain length.[646] Here the shorter the chain length the greater the emission intensity owing to a decrease in hopping distance. These materials, however, were highly sensitive to oxygen quenching. Other studies on poly(substituted thiophenes) using near-field fluorescence microscopy have shown that protruding domains are blue-shifted compared with flat areas, with further blue shifts as the irradiation time progressed in the scanning microscope.[647] For a series of poly(3-alkylthiophenes), long chains are considered to enhance the processability of the polymer as well as its emission quantum efficiency.[648] These polymers gave strong emission at 520 nm in chloroform and 570 nm in the solid film. Poly(2-alkyl-benzimidazole)-alt-thiophenes, on the other hand, give twin emission peaks at 450 and 525 nm.[649] For poly(3-(2′-methoxy-5′-octylphenyl)thiophene), highly emissive microcrystalline structures are formed in poor solvents,[650] whereas a series of tetraacetylenic thiophenes give emissions varying from indigo blue through to reddish oranges.[651] Time resolved analysis of the photoluminescence from polythiophenes has indicated a two component decay process,[652] while a series of oligothiophenes with benzyl ether dendrons form well-defined nano-aggregates[653,654] in cooled solutions. The extent of aggregation is reversibly controlled over a small temperature range, with strong bathochromic shifts in the aggregated phase. This suggests that interchain delocalization of the singlet and triplet photoexcitations occurs within the dendritic nanoassemblies. Highly regioregular poly(3-alkylthiophenes) have been made and found to give strong red-orange emission at high conjugation lengths,[655,656] whereas new types of poly(thiophenes) have been developed which emit over the entire visible range of wavelengths.[657] Carboxylic acid substituted poly(thiophenes) exhibit much red-shifted, higher emission intensities in self-assembled layers than in cast films[658] owing to fewer excimer-forming sites. Light-emitting properties have also been controlled by copolymerizing with cyclodiboranes[659] and fluorene units.[660] In the latter, increasing the oligomer chain length or adding carbonyl groups at the chain ends red-shifts the spectra owing to increased electronic delocalization along the molecular frame. Making polyester derivatives had little effect on the spectral properties. Hyperbranched poly(arylenes) with thiophene units give strong blue emission[661] and little evidence for aggregation. Dithienothiophenes have been solubilized in protic solvents while at the same time having good non-polar solvent solubility.[662] They also give rise to strong blue emission. Poly(3-octylthiophenes) give red-shifted emission spectra[663,664] and a lower operating voltage. New silicon-based poly(thiophenes) can be easily spin-cast with a red-shifted green emission at 520 nm.[665] Other polythiophenes are tetracyclohexyl-3″,4″-dihexyl-2,2′:5′,2″:5″,2‴:5‴,2⁗-quinquethiophene-1″,1″-dioxide derivatives,[666] alkoxy-substituted derivatives,[667] terminal-fullerene derivatives[668] and co-fullerene derivatives[669] and poly(3-thiopheneacetic acid).[670]

Dope polymers have been widely investigated as probes of their environment. Pyromethene dyes in acrylic polymers have been studied,[671] as have quinoline derivatives in PVC and HDPE.[672] The latter was supposedly used as a marker for optical identification, but in fact its spectral properties are virtually independent of the environment. Water uptake in polyurethanes has been monitored through fluorescence probes,[673] by and super phosphorescent additives developed for plastics that are better than Zn-sulfide.[674] Conductive patterns have been developed in layers of poly(octamethylene-9,10-anthrylene) through selective irradiation via masks.[675] Subsequent exposure of the films to iodine vapour gave films with conductive and non-conductive areas, which remain quite stable for periods of months. The spectral characteristics of Oxazine-720 in PVK depends upon the nature of the doping process.[676] It is seen that the PVK emission spectra are red-shifted, and there is an increase in intensity with increasing dopant concentration. The distribution of fluorescence lifetimes of embedded dyes in polymer films is shifted to lower values as the film thickness increases.[677] This effect is associated with simple electromagnetic arguments related to the boundary conditions at the interfaces of the polymer film with air and glass. Other probes of interest include 3-coumarin derivatives,[678] polyaniline doped with $H_{3+n}PMo_{12-n}O_{40}$,[679] cellullose esters doped with rhodamine B[680] and fullerene doped polymers.[681] In the latter case continuous laser irradiation results in enhanced blue-shifted emission.

Determination of matrix properties via probes is also useful and has been discussed in some depth.[682] The thermoresponsive behaviour of microgels based on poly(acrylamide) has been monitored through a pyrene probe.[683] It was found that the use of dimethylacrylamide as a hydrophobic modifier raises the lower critical solution temperature of the microgel. Time-resolved analysis on the nanoparticles confirmed the rapid reversibility of the smart thermal response via a two-stage process. The viability of photon-event counting imaging has been applied to fluorescence studies on dyes such as Rhodamine B in sol-gels based on tetraethylorthosilicate.[684] Here cracks or other defects could be easily identified in the gel. The swelling of PMMA gels has also been monitored by a fast transient fluorescence technique.[685] Pyrene lifetimes in the gel were used to measure activation energies for the gel swelling process. The technique has also been extended to studying the helical forms of isotactic polystyrene in *cis*- and *trans*-decalin,[686] as well as polymer temperatures during capilliary rheometry.[687] Pyrene as a probe has also been used to measure the polarity and stability of poly(urethane) micelles.[688] Silica particle growths during sol-gel polymerization have been measured via Rhodamine 6G.[689] Using fluorescence anisotropy, both microviscosity and hydrodynamics can be monitored. The change in the nature of catalyst particles during olefin polymerization on silica can also be monitored via pyrene probes.[690] The same probe has been used to measure molecular relaxation processes in PMMA-silica matrices.[691] There are extensive hydrogen-bonding interactions between the ester carbonyl groups of the PMMA and the silicic acid pore surface. There is also a reduction in the chain entanglements of the polymer and an increase in the free volume due to pore-directed polymerizations. This effect releases several steric hindrances. TICT states of molecular probes also give useful information on chain conformations in polymers.[692] Both free and bound probes will give rise to twin emission peaks, where the close vicinity of the polymer chain is likely to influence the distribution of conformers.

Following on, several studies have appeared using pyrene tagging methods as molecular probes. In pyrene end-labelled poly(dimethylsiloxane) there are at least two types of interactions: constrained and unconstrained.[693] The former is due to peptide interactions and precludes excimer formation, while the second results in classic excimer formation. The extent of pyrene interactions have been measured in supercritical carbon dioxide media, and are found to be closer than when dissolved in good solvents like toluene. In a similar way, the intermolecular associations of an oligoisobutylene with a maleic anhydride function at one of its ends has been investigated in *n*-hexane through the fluorescence analysis of pyrene attached onto the MA moiety via an imide bond.[694] At high oligomer concentrations, more than 95% of the oligomer displaying an amide bond is associated. Without the amide bond, this reduces to 60%. In polar solvents, however, this ratio is further reduced owing to solvent associations. Four generations of poly(aryl ether) monodendrons with pyrene at the focus have been made,[695] where structural changes between the generations of dendrons depend upon the nature of the solvent. From model studies it was suggested that the increased density of larger monodendrons provides a more efficient barrier to the diffusion of oxygen. Dynamic and static photophysical measurements on pyrene attached to a polyethylene:poly(butylmethacrylate)-co-poly(styrene) have been used to measure structure and diffusion dynamics.[696,697] Diffusion entry was measured by adding a fluorescence quencher, ethyl oxalate, which cannot enter the polyethylene matrix. The data indicated the presence of segregated domains with extensive interactions at the domain interfaces. Using pyrene-3-carboxyaldehyde the interaction between sodium dodecyl sulfate and acrylic copolymers has been found to occur at concentrations smaller than that observed for the pure surfactant in water.[698] An increase in polymer concentration lowers the critical aggregation concentration. The cyclization of poly(ethylene oxide) end-tagged with pyrene groups has been monitored, with excimer formation being described by a two state model.[699] At temperatures below 30°C a pyrene dimer is formed, with activation energies close to those for viscous flow activation. The difference between the enthalpy and the binding energy of the excimer in both the globule and coil states reflects the variation of polymer conformations between the cyclized and noncyclized chains. In LC pyrene-labelled cholesterol copolymers, each pyrene molecule is apparently incorporated into the LC domain with restricted access,[700] while in poly(styrene) latexes small hydrophobic pores have been identified through monitoring the extent of pyrene penetration.[701] Via the same probe, it has been shown that thermally reversible phase separation of poly(isopropylacrylamide) is due to coil collapse.[702] Thus, trapping and release of the probe within the polymer coils will give rise to different spectra and emission lifetimes. The folding of pyrene labeled poly(*N*,*N*-dimethylacrylamide) has been described by a blob model.[703] This technique provides valuable details on chain dynamics in solution, including the rate of segmental encounter. Thus, the folding of a polymer chain can be handled by compartmentalizing the polymer coil into subdomains, typical of that used for proteins. Inter-chain interactions have also been measured in an ethylene-propylene random copolymer tagged with 1-pyrenebutanoic acid hydrazide.[704] Using the blob model, 70% of the pyrene groups were found to be closely located, while in an apolar solvent over 87% of the groups are involved in aggregate formation held together by succinic anhydride groups. Poly(*N*-(1-pyrenylmethyl)methacrylate)) behaves in the same way.[705] Hydrophobically modified copolymers of *N*-isopropylacrylamide and *N*-L-valineacrylamide undergo pH dependent phase separation on heating,[706] while in micellar media the quenching of probes by cetyl pyridinium chloride follows the order pyrene>perylene>fluoranthene.[707] The interactions between cationic non-phospholipid liposomes and pyrene-labelled hydrophobically modified sodium poly(2-2-(acrylamido)-2-methylpropane sulfonates) indicates that the polyanions bind to the cationic liposomes.[708] The reversibility of this process was also demonstrated.

Energy transfer and excimer formation of a general interest have been widespread. Sulfonated polystyrene has been shown to give rise to monomer, excimer and dimer emission bands,[709] while the motion of end-capped anthryl groups in polystyrene provides a useful model for predicting the orientational auto-correlation function.[710] Bis(10-phenyl-9-dianthryloxy)methane excimer formation is sensitive to the viscosity of poly(dimethyl)siloxanes,[711] while increasing the styrene content in poly(styrene) by irradiation enhances excimer formation.[712] Polymers with 100% donor-acceptor norbornadiene groups undergo effective isomerism, the effect being enhanced in the presence of a benzophenone triplet sensitizer.[713] Fluorescein and phenolphthalein have been shown to differ markedly in their photoelectric properties,[714] while the donor-acceptor properties of naphthalene and pyrene tagged to poly(ethylene glycol) units on a poly(styrene) chain are affected by external stimuli.[715] In micellar media with increasing salt addition, the chains coil, thus reducing the separation distance between the optical tags. Fullerenes are quenched by electron transfer from polyimides with a strong dependence upon the orientation of the chain.[716] Here *m*-substitution with a high ionization potential exhibits the lowest quenching rate. Microphotoluminescence has been used to measure excitonic luminescence in a single conjugated polymer chain.[717] The data indicates inhomogeneous broadening, very long chains and long-range rapid exciton transfer (<100 ps). Interface distances between films of block acrylic-styrene copolymers have been measured by energy transfer between anthryl and phenanthryl labeled junctions,[718] while the fluorescence quantum yield of 2-*N*-piperidino-5-(2,2′-dicyanovinyl)thiophene is enhanced from 0.0004 in hexane at room temperature to unity in EPA-glass at 77K.[719] The analysis of delayed fluorescence from various conjugated polymers at different temperatures indicates that the emission occurs within a time frame of micro- to milliseconds at 80 K, while the phosphorescence on average occurs over a one second period.[720] This suggests that delayed fluorescence is actually caused primarily by the recombination of geminate electron-hole pairs rather than triplet-triplet annihilation. The effect of an external electrical field in the experiments tends to confirm this deduction. FRET in folded polymer chains has also been under the 'microscope'.[721] Simulations suggest that FRET can act as a good marker of polymer folding only when the Forster radius (R_f) is smaller than the root-mean-square radius (R_0) of the polymer. It is suggested that it may well be important to employ more than one donor-acceptor pair in experiments in order to probe the entire dynamics of chain folding. A new method for the synthesis of polymers with bifluorophores has been developed.[722] In this case it was concluded that nonradiation energy transfer is not feasible if the distance between donor and acceptor exceeds 10 nm. In another analysis of T-T annihilation in conjugated polymers, it was also concluded that this process was effective up to 3% of the total luminescence.[723] Super-quenching has been observed in poly(L-lysine) derivatives tagged with cyanine dyes on each repeat unit.[724] Strong aggregate formation was evident in these polymers, with quenching rates equivalent to or greater than those normally observed with polyelectrolytes. Different types of complexations have been observed in carbazole containing polymers,[725] and FRET dependence upon temperature has been investigated.[726] Activation energies have been measured for main chain segmental motions in anthracene tagged PVK's.[727] Relaxation times were measured through fluorescence anisotropy decay, giving energies greater than those measured for many other polymer systems. It is claimed that PVK has a larger critical molecular weight factor and that the bulky side-chains suppress the conformational transition of the polymer backbone. Excimer formation in PVK copolymers indicates that nearest neighbour interactions are dominant,[728] while acrylic copolymers of PVK give stronger emission than the monomers.[729] The twisting of polymer chains in oligo[2,5-bis(hexadecyloxy-1,4-phenylene]s reduces excimer formation by preventing stacking, whereas hydrophobic interactions due to long aliphatic chains cause excimer formation.[730] Spatial structures have been measured in bifluorophore azomethine polymers,[731] and carrier generation and transport examined in general.[732] Charge-transfer processes in polyimides have been discussed in depth.[733]

Polymer tagging is useful as a macromolecular probe. Naphthalene has been tagged to cyclodextrin for potential light harvesting,[734] while anthracene has been tagged to PMA to determine the efficiency of photodimerization.[735] In anthracene labeled PMMA relaxation times have been determined,[736] whereas spectral broading can be used to determine the size of a polymer molecule.[737] Poly(imidoaryl ethers) have been tagged with bulky *N*-phenyl-naphthalimides and characterized spectroscopically,[738] as have novel 3,4,6-triphenyl-α-pyrone fluorophores[739] where the phenyl group is fixed in a specific orientation for fluorescence. Other characterizable polymers include poly(2-ethynylpyridinium bromide) with propargyl side chains,[740] naphthalene tagged poly(butenoate),[741] dye labeled poly(DL-lactic acid)[742] and proteins,[743] cyanine dyes on nucleic acids,[744] polyphosphazines with carbazolyl side chains,[745] coumarin dendrimers[746] and anthracene labelled epoxy resins.[747] Polystyrene has been shown to give high yields of dimers when formed with dibromomethane[748,749] and, when tagged with fullerene, the emission is blue shifted owing to a loss of conjugation.[750] Poly(acrylics) have also been tagged with naphthalimide dyes and spectrally characterized.[751-753] The fluorescence of poly(imides) has been controlled through the tagging of perylene and quinioxaline units,[754,755] with little aggregation below 2.82 mol-%. The polymers also exhibit complete quenching due to efficient electron-transfer processes. Poly(*N*-isopropylacrylamide) has been tageed with tetraphenylporphyrin groups and found to experience a thermally reversible phase transition upon heating.[756,757] Fluorene has been tagged to polysiloxanes and found to exhibit higher rotational and translational mobilities than for conventional siloxane polymers.[758] Poly(dipropylsilylenes) have been made by vapour deposition[759] and characterized, and dansyl groups on a poly(dimethylsiloxane) have been used to monitor mobility at junction interfaces in liquids and molten polymers.[760] Here the solvent is concluded to play a dramatic role in tuning the local microenvironment that surrounds a polymeric junction point. The biodegradability of poly(amide-anhydrides) can be measured via visibly fluorescent tags,[761] and the spectroscopic properties of vinylcaprolactam copolymers with vinylpyrrolidone have been determined in the presence of cations.[762]

Polymers doped with rare earth metal ions continue to be explored for enhanced optical properties. Poly(acrylic acid)-Eu(III)-dibenzoyl methane exhibits high fluorescence quantum yields,[763] as do ternary Te(III) complexes in styrene-acrylics.[764] Such systems can have useful applications, in particular Ru(II) complexes in polydimethylsiloxanes have been used as 'pressure-sensitive paints'.[765] Water molecules trapped in Eu(II) complexes in poly(methacrylic acid) have been estimated,[766] whereas the stability of Tb(II) ions in *N*-alkylmethacrylamide copolymers depends upon the structural organization of the macromolecules.[767,768] Alkylpyrrole groups influence the emission quantum yields of polypyrrole Ru(III) complexes, with decay rates that are affected by extents of crosslinking.[769] The use of methylviologen for redox processes is still feasible in crosslinked systems. Stern-Volmer quenching measurements by Tl(I) ions have been undertaken on fluorescent dyes doped in sodium dodecyl sulfate,[770] and energy transfer assessed between Ru(II) and Os(II) bipyridyl compexes attached to the ends of a deca(oxoethylene)-33 atom chain.[771] *N*-vinylcarbazole and *N*-vinylpyridine have been complexed with Ir, each showing quenching of the excimer sites.[772] Inorganic phosphors have been doped in polymers to produce light emitting caps for bottles,[773] while lanthanide complexes in 3(trimethoxysilyl)propyl methacrylate give highly stable luminescent matrices.[774] In a similar way, Eu(III) complexes in methacrylate copolymers bound onto silica gel provide a useful medium for intensifying the luminescence output.[775]

Fluorescence probe studies in micellar matrices attract much interest. Micellar size and structural characterization of self-assembled systems have been covered in some depth.[776] Copolymers of poly(isoprene-b-ferrocenylphenylphosphine) form starlike spherical micelles in *n*-hexane with a dense polyferrocene core.[777] The collapse of poly(*N*-isopropyl acrylamide) in water is dependent upon the amount of grafted poly(ethylene oxide).[778] A pyran probe was found to localize in the hydrophobic parts of the copolymer chains and exhibited properties that were highly dependent upon the conformation of the chains during garfting. The effects of poly(ethylene oxide) in triblock copolymers have similarly been studied using a pyrene derivative as a probe,[779] including in one study the amount of water trapped in micelles.[780] Silicone surfactants with hydrophilic amine groups have also been grafted with poly(ethylene oxide) chains and found to exhibit aggregates in aqueous solutions.[781] For related poly(oxyalkylene-substituted siloxane) surfactants in the presence of methyl orange and pyrene probes, micelles are formed with hydrophobic interiors with CMC values that shifted to higher values in mixtures of co-solvents such as ethanol.[782] Interactions between butyl vinyl ether-sodium styrene sulfonate copolymers and oppositely charged ionic surfactants show premicelle formation owing to electrostatic attractions where free probes (pyrene) will migrate from the aqueous to the new hydrophobic sites.[783] A series of novel polyazo electrolytes have been prepared, which exhibited interesting photochromic effects depending upon the functionalization and coating processes.[784] Also, the continuous addition of surfactant destroys the excimer sites. Premicellar formation has also been observed using a pyrene probe for a series of Gemini surfactants,[785] while poly(*N*-acetylethyleneimine) cosurfactants were found to be included in alkyl methacrylate copolymers made by emulsion polymerization.[786] Twin-tailed hydrophobic acrylic monomers have been prepared by micellar methods and, using a dansyl labelled probe, the relative interactive strengths of different surfactants were found to be in the order CTAB>Triton-X>SDS.[787] Binding processes in CTAB micelles have also been investigated using dansyl labelling.[788] Local mobility was confined to the alkyl tails of the surfactant, with polyelectrolyte-surfactant complexes exhibiting an ordered lamella structure. The use of pyrene fluorescence as a probe has shown that KCl decreases the CMC of Triton-X-100 as well as increasing the aggregation number and degree of hydration.[789] The latter is due to water trapped in the micelle, although there were no changes in the micellar micropolarity owing to solubilization of the probe associated with micellar growth. Certainly, electrolyte addition enhanced the microviscosity, which is consistent with micellar hydration. Cyanine dyes have been shown to interact with DNA at low concentrations,[790] while similar dye chrompohores exhibit a much narrower absorption band in poly(vinyl alcohol) than they do in poly(vinyl sulfate).[791] Self-assemblies of fluorescent dyes such as rhodamine B have been demonstrated on zirconium phosphate, where binding reactions give rise to significant enhancement in quantum yields.[792] Energy transfer has been observed not only in these dyes but also in new terpolymers.[793] A sugar cationic surfactant containing a fluorescent coumarin group exhibits a certain chirality, which causes bending forces, resulting in the formation of helixes and tubules,[794] while a series of carbazole amphiphilic copolymers exhibits emission characteristics that are dependent upon the environment,[795] and several binding modes were ascertained for diaminostilbene and polyions.[796] Hydroxyethylcellulose with cationic hydrophobic side-chains form micelle-like aggregates, with hydrogen-bonding playing a major role in the interactions.[797] Amino derivatives of 7-nitrobenz-2-oxa-1,3-diazole have been found to be sensitive to polarity and microviscosity changes for amphiphilic copolymers,[798] whereas poly(methacrylic acid) and poly(vinylpyrrolidone) undergo complexation via a connected cluster model.[799] Fluorescent sensors have been developed which can complex with the ammonium groups of starch,[800] while end-capped poly(sodium-2-acrylamido-2-methylpropanesulfonates) undergo multi-molecular aggregation in salt water solutions.[801,802] In the presence of CTAB and a pyrene salt probe, the latter is excluded from any binding. An ionic derivative of pyrene has been bound to a polyelectrolyte,[803] while other novel probes have been developed.[804] Copolymers of acrylic acid and *N*-dodecylmethacrylamide form micelle-like multipolymer aggregates,[805] while twin-tailed terpolymers reveal viscoeleastic properties consistent with a simple Maxwell model.[806] For a series of poly(allylammonium) chloride salts with differing alkyl side-chains, hydrophobic domains depend upon the length of the alkyl side-chain.[807] Using a pyrene derivative probe, these changes are assigned to aggregation arising from interchain interactions. In fact, polymers with dodecyl side chains give hydrophobic contents where aggregates are independent of the polymer concentration. In the case of *N*-vinylcarbazole labeled methacrylic acid-acenphthylene copolymers, a hypercoiled conformation at low pH makes the fluorescence primarly that of the acenaphthylene excimer.[808] At high pH, however, the copolymers adopted an extended coil conformation, thus decreasing non-radiative energy transfer. Surfactants, micelle formation and aggregation have been critically assessed in some detail.[809]

General polymer luminescence studies are numerous. Crosslinked polyethylene is shown to give emission due to recombination centres,[810] while the luminescence from polypropylene composities depends upon the composite.[811] Novel conjugated poly(1,2-diethynyl-1,1,2,2-tetrmethyldisilanes) are moderately luminescent,[812] while novel luminescent binaphthyl polymers have also been made.[813] Polyanhydrides give fluorescence only when a methane group is in the repeat unit,[814] as do polyamides with a repeat unit of ethidium bromide.[815] Fluorescent dyes have been used to measure polymer melt temperatures during extrusion,[816] while some polyethers give a weak blue fluorescence.[817] Poly(ethyleneterephthalate) gives luminescence due monomeric components,[818] and poly(propylene) doped with 9,10-diphenylanthacene enables the detection of a true infectious profile on thermal oxidation.[819] Polyaniline exhibits fluorescence, which is maintained in a vacuum but removed and red-shifted upon acidification.[820] This also influences the conductivity of the polymer. Electron beam irradiation affects the luminescence from poly(vinyl toluene),[821] while the emission properties of phenyl-substituted siloxanes does not depend upon the phenyl groups.[822] Ultrafast intra-chain vibrational relaxation processes have been measured in poly(fluorene) films.[823]

Whereas composites of the same polymers possess a hierarchy of phases,[824] the composition affects photovoltaic properties. A fibre-optic cure sensor based on evanescent-wave fluorescence spectroscopy has been designed to probe the inter-phase region of GRP composites,[825] and plasma prepared organo-siloxanes give emissions associated with changes from linear to amorphous structures.[826] The same process applied to poly(propylene) gives rise to three spectral components.[827] These are two fast-decay processes, due to radiative de-excitation levels, and a slower carrier-recombination process. The luminescence of the same polymer also depends upon its crystal structure,[828] while that from a series of poly(silanes) depends very much on polymer structure and conformation.[829,830]

Stern-Volmer quenching has been measured in poly(*N*-*p*-phenoxyphenyl)methacrylamide,[831] and aggregation affects the luminescence from polymers from 2,5-dioctyl-*p*-distyrylbenzene,[832] perylenebisimide-polytetrahydrofurans[833] and sulfonated poly(butyl-methacrylate)-polystyrene.[834]

Continuing studies of general optical properties include the luminescence of copoly[arylene-1,2-dioxy-oligodimethylsiloxanylenes],[835] pressure sensitive paints,[836] nanoparticle acrylics,[837] fullerenes,[838] polystyrene particles,[839] polyfluorene,[840] poly(4-vinylbenzyl-thiocyanate),[841] polystyrene films,[842] PMMA,[843,844] stilbenes,[845] complexes of *N*,*N*-dimethylacrylamide and PMA,[846] poly(diallyldiethylammonium chloride),[847] polymers of 9,10-dibromoanthracene and 1,4-diethynylbenzene,[848] poly(Schiff's bases),[849] poly(heteroarylenes)[850] and poly(di-*n*-hexylsilanes).[851] The luminescence of poly(ethylene naphthalate) is associated with the presence of degradation products,[852] as is the case with other materials.[853] Luminescence has been used as a molecular probe for identifying and pin-pointing treeing sites in PMMA.[854] The lasing properties of diamine-xylylene copolymers have been examined,[855] while fluorene-based fluoro-polymers emit blue light except those with a perfluoro-octyl group.[856] Self-quenching of 1-methyl-1,2,3,4,5-pentaphenylsilole in PMMA has been studied,[857] while an overview has been given on photoinduced phenomena in polymer systems.[858]

4 Photodegradation and Photooxidation Processes in Polymers

The photodegradation of polymer materials continues to attract interest, although as in the last few years activity has declined. Thermal degradation and stabilization processes rank much more highly in that particular field. Laser ablation processes continues to grow in interest, with applications in electronics. A useful review on the subject has appeared, giving extensive coverage of techniques and applications.[859]

4.1 Polyolefins. – Polyolefin photooxidation processes is the most actively studied area. Depth profiles have been determined in tropical climates for toughened polypropylene,[860] while other work has covered resistance to discolouration.[861] Iron complexes continue to be investigated as effective pro-oxidants for polyolefin photodegradation.[862,863] Iron is a reknowned catalyst for olefin oxidation, and the problem here has been to optimize processing stability while at the same time producing a formulation which will sensitize photooxidation of the matrix within a given time period. Iron complexes have also been examined in conjunction with starch for applications with poly(ethylene).[864,865] Chemical and physical changes during poly(ethylene) photooxidation continue to be measured,[866] while other work has also dealt with the effects of doping with rubber and manganese stearate.[867] Crosslinking in poly(propylene) during irradiation has been measured,[868] similarly for PP foam material used in snack boxes[869] and for flame retardency.[870] Ethylene-carbon monoxide copolymers have also been investigated for structural and chemical changes during irradiation.[871] Through mass measurements of oxygenated products, it is interesting to note that while peroxy radical termination is evident in photooxidation it is absent in thermal oxidation.[872] Polymer-alkene-fullerene radical products have also been identified during the irradiation of doped polymer materials, using benzophenone as a photosensitizer.[873] Hydroperoxide sites have been enhanced on the surface of polyolefins by grafting with poly(styrene),[874] and poly(ethylene) made via the spherilene technology becomes highly unstable on irradiation, with rapid growth in carbonyl and hydroperoxide groups.[875] Changes in the properties of poly(ethylene) have also been measured under the influence of Sahara winds,[876] while secondary usage of poly(ethylene) gives generally an inferior product with limited usage.[877]

4.2 Polystyrenes and Polyacrylics. – Radical synergy has ben observed in poly(styrene) doped with three initiators.[878] In fluorinated methacrylate formulations, the overall light stability is controlled by the butylmethacrylate component,[879] while irradiation of PMMA can cause chain scissions, giving rise to mechanical failure.[880] With far UV exposure there is also a tendency toward depolymerization.[881] In other work, methacrylates have been shown to be more unstable than acrylates, owing to associated ester chain scissions in the former case.[882]

4.3 Polyesters. – The surface degradation of poly(ethylene terephthalate) has been examined by X-ray methods.[883]

4.4 Polyamides and Polyimides. – Polyimides with *N*-carbonyloxyimide groups undergo chain scission through N-O cleavage as well as breaking of the cyclobutane ring.[884]

4.5 Silicone Polymers. – The fluorescence from poly(methylcyclohexylsilanes) decreases and blue shifts upon irradiation,[885] with incorporation of siloxane groups in the Si-Si bonds. Poly(dimethyldisiloxy) materials with acetophenone groups attached display high instability, showing a 50% reduction in molecular weight over a short period of irradiation.[886] Spectroscopic analysis of poly(silanes) shows that irradiation at longer wavelengths tends to scission the longest segments on polymer chains, while short wavelengths induce degradation of all chain lengths.[887]

4.6 Polyurethanes and Rubbers. – Urethane coatings based on Castor oil develop strong yellowing on UV exposure,[888] while the same is true of aromatic based urethanes rather than aliphatic based systems.[889] The polyether segments are attacked first, forming ester and other volatile products. There is a loss in free-volume and holes during poly(urethane) photodegradation, coupled with a high degree of crosslinking.[890,891] Irradiation of MDI-based urethanes in the presence of nitrogen dioxide causes nitration of the aromatic rings and scission of the methyl substituents followed by crosslinking,[892] while ESR has been used to identify radicals formed during irradiation of poly(urethanes) under different light sources.[893] The melt rheology and post irradiation of poly(octenamer) have been examined,[894] and in styrene-isoprene-styrene blends the isoprene matrix is the most unstable.[895] Initially, chain scission occurs followed later by crosslinking. Spatial heterogeneous degradation processes have been observed in ABS terpolymer,[896] whereas irradiation of natural rubber leads to the formation of liquid rubber as a product[897] via the usual hydroperoxidation reactions.

4.7 Poly(vinyl halides). – PVC used in the construction business has been exposed to various external environments, and it was shown that whilst stress changes slightly there is no change in hardness.[898] Chemical changes were, however, apparent by FTIR, with ozone being the most detrimental external agent.

4.8 Photoablation of Polymers. – Laser ablation is one of the most widely studied areas in the field of polymer degradation. The yield of silyl radicals during KrF laser ablation of poly(silylenes) has been found to be dependent upon the backbone structure of the polymer, whereas the yield of silylenes was found to be proportional to the number of linear Si units in the backbone chains.[899] A fluffy nanostructured powder of poly(hydridomethyldisiloxane) has been prepared by laser CVD of 1,3-dimethyldisiloxane,[900,901] as have other organyltrimethylsilane polymers.[902] Temperature effects on the deposition of laser ablated hexaphenyldisilane have also been investigated, with a view to optimizing film properties.[903] In polyamides and polyimides, the high-fluence implantation of boron ions, mainly via electron stopping, has been found to result in the formation of electron rich π-bonded carbon clusters.[904] In contrast, heavier phosphorus ions, for example, deposit energy predominantly by nuclear collisions to give a lower concentration of π-radicals and a less carbonized surface. Irradiation ablation of poly(methacrylates) with 146 nm light has been found to be dependent upon the structure of the ester groups, while for poly(styrene) no thickness decrease was observed,[905] with some derivatives exhibiting crosslinking. The luminescence from poly(propylene) decreases with laser ablation time,[906] while PVDC has been found to be more sensitive to ablation than PVC.[907] Using poly(ethyleneterephthalate) as a model substrate, it has been shown that new functionalities can be created in microchannels by laser ablation.[908] Nanoscopic aggregation has been studied in laser irradiated amphiphilic random copolymer environments.[909] Structure-property relationships have been established for a series of polymer materials at low irradiation fluences.[910] Polymers with photochemically active groups were found to have the lowest threshold for ablation. The thermal stability of the polymers was not an important parameter. In this way, polymers could be designed for specific ablation processes in order to develop microoptical components. A time-resolved ultramicroscope has been developed in order to monitor the process of pulsed-laser induced melting of 9,10-dicyanoanthracene microcrystals on poly(ethyl methacrylate) films.[911] This instrument is claimed to be able to detect sub-micrometer particles on a time resolution of 400 ps. Crystals with a size of 10 microns were found to be converted into hot liquid droplets, of which some 20% had evaporated. Mixtures of 3,4,9,10-perylenetetracarboxylc dianhydride with Co powder have been found to give fragments upon laser ablation at 308 nm that no longer have carboxylic anhydride groups.[912] IR analysis shows on poly(perinaphthalene) is deposited, giving highly conductive films. Collagen and keratin nave been successfully deposited, with no change in their chemical structure but an increase in their random-coil domans.[913] A polyurethane film has been successfully etched away. with no debris, using 248 nm laser light,[914] while collagen gels have been found to exhibit smaller threshold energies than in their dry state.[915] Here the latent heat of water evaporation absorbs most of the laser energy.

4.9 Natural Polymers. – The discolouration and protection of wood and paper continues to attract some interest. The extensive nature of the subject has been made evident,[916] including modification treatments.[917] Hydroxyl radicals are responsible for the degradation of glycosidic linkages, giving rise to a series of alsonic acids and aldoses.[918] Photochemical bleaching of wood pulp has been found to be much more effective,[919] while acetylation of wood reduces post-photoyellowing.[920] This process inhibits the photolysis of phenoxyl ether groups giving rise to phenoxy radicals. Butyrylation has been reported to be even more effective than acetylation,[921] giving rise to a significant reduction in quinoid products. Modification experiments, utilizing grafting with methacrylic anhydride, also inhibit yellowing.[922] The interesting feature of this study, however, was the observation that degradation still continues and that yellowing is a separate process. Photoyellowing of wood substrates and products has been found to be impaired by 2-hydroxyaromatic stabilizers,[923] while glutathione is a photosensitizer for collagen.[924]

4.10 Miscellaneous Polymers. – Polymers and coatings of various compositions have been investigated and mechanisms reviewed.[925,926] A dual photooxidation process has been proposed for trimethylcyclohexane-polycarbonate,[927] where under short wavelengths a phot-Fries rearrangement is operative as well as chain scission. This polymer was also found to be more photoreactive than normal polycarbonate, owing to the higher incidence of reactive tertiary carbons in the trimethylcyclohexylidene structure. For organic coatings, hardness, T_g and chemical changes by FTIR have all been inter-related during UV ageing.[928] The service lifetime of acrylic-melamine coatings has been measured in a series of ageing apparatus,[929,930] and accelerated weathering tests evaluated for a series of epoxy-acylurethane paints.[931] The photooxidative stability of a series of EB- and UV-cured dialkyl and heterocyclic amine terminated acrylic resins has been examined.[932] UV-cured coatings have been found to be more unstable than those EB-cured, with a strong structural dependence upon the nature of the terminal amine functionality. Thus, dialkylamines were found to be more prone to oxidation than alkanolamines, cycloaliphatic amines and heterocyclic amines. Yellowing is associated with facile oxidation of the amine group by hydrogen atom abstraction, to give unsaturated carbonyl products. Photoirradiated radical yields in acrylic and alkyd paint films have been compared[933] and found to correlate with stability. Fluorinated acrylates have also been studied,[934] as have hot-melt adhesives,[935] UV-cured clear coats[936,937] and epoxy resins.[938]

For a series of poly(benzoxazines), photooxidative stability has been found to depend not only upon the nature of the phenolic substitution but also upon the type of amine functionality present.[939,940] On photooxidation, benzoquinones are formed, giving rise to discolouration. Hydrogen bonding between the phenolic and amine functionalities was also important. Photodegradation depth profiles have been determined in thick section polycarbonates,[941] while the quantum efficiency of luminescence form PPV's after a few minutes irradiation decreases significantly.[942,943] Poly(4′-ethoxyacrylophenone) undergoes random chain scission with a higher quantum yield in solution than in the solid state.[944] Ethyl radicals were formed in high yield due to O-alkyl bond scission, coupled with yellowing due to quinonoid entities. Crosslinking occurs later with a consequent reduction in chain scission due to restricted migration of the free radicals. The fire-retardency of nanocomposites is important,[945] while poly(ethylene glycol) hydrogels are made photoreactive by adding cinnamate groups[946] Super macromolecules are formed in poly(vinyl alcohol) on irradiation in the presence of Fenton's reagent,[947] while high levels of formates are generated during the photooxidation of poly(ethylene oxide),[948] owing to the predominance of peroxy radical termination processes. For polycarbonate-polyurethane structures, photolysis occurs at the C-H bond adjacent to the urethane group, causing rapid chain scission,[949] whereas the microstructure of poly(octenamer) is important in photoooxidation reactions of the polymer.[950] Thus, whilst rapid crosslinking was observed there were no corresponding changes in FTIR functionalities. Hydration effects on the stability of polyacrylics has been determined,[951] and poly(naphthalimides) modified.[952] Other studies of interest include surface anisotropy due to bond breakage,[953] use of DSC methods for hydroperoxide analysis,[954] photohydrolytic compatibility of polyethylene and polycaprolactone[955] and new polymeric sensitizers.[956]

5 Photostabilization of Polymers

The photostabilization of polymer materials for outdoor applications and extended life continues to be an industrial battleground. Consequently, many of the publications are commercial in nature and centre on effects rather than scientific processes. Numerous articles of interest include the surface protection of poly(propylene),[957] new cyanoethylacetophenone derivatives for poly(ethylene) modification,[958] stabilization of rotomolded parts,[959] hindered amine stabilizers (HAS) for glass-reinforced nylon-6,[960] novel additives for poly(propylene) fibres,[961] additives for packaging products,[962] novel cycloaliphatic amines for HAS,[963] absorbers for elastomers,[964] computer-aided design for packages,[965] HAS and absorbers for plastics and coatings in general,[966-976] stabilization of rigid PVC and wood composites,[977] stability of white pigmented ABS,[978] new blends for urethanes,[979] stabilization of natural woods,[980,981] flame retardant ABS,[982] HAS for PVC,[983] novel aryl and heteroaryl methacrylate derivatives,[984] HAS for styrenics,[985] colour interactions of HAS and antioxidants,[986] HAAS for automotive paints,[987] absorbers for fabrics,[988] cycloalkylated products for isoprenes,[989] electronic properties of absorbers,[990] hexaazocyclanes[991] and photo-Fries[992] products for polyesters and enhanced packages for fibres.[993]

In the polyolefin world, the development of metallocene catalysts has attracted much interest, and in this regard stabilization processes have been examined and found to be the same as for conventional polymer grades.[994] During the ageing of polyolefins with a polymeric HAS system, fragments of stabilizers have been identified, with extreme conditions in water and low pH being the most detrimental.[995] HAS grafted to the surface of polyolefins has been claimed to give the best surface protection,[996] while concentration changes in HAS has shown that the original amine decreases slowly, whereas the nitroxyl radicals increase and then after a certain period decrease.[997] It was also shown that, whilst the amine remains uniform throughout the polymer material, the nitroxyl radicals exhibit non-uniformity. Mixtures of HAS structures have been shown to operate synergistically in the light stabilization of polyolefins,[998] as have other stabilizers based on hydroxyphenyltriazine with oxanilide or benzophenone.[999] The incorporation of HAS into urethane coatings has shown that the stabilizer increases the number of urea bridges at the expense of unreacted isocyanate, giving rise to a more photostable material.[1000] The loss of UV absorbers has been measured in clear-coats during UV ageing,[1001,1002] while for UV-cured acrylates HAS are effective radical scavengers.[1003] PVC plates have been made more light-stable by applying a UV curable acrylic coating containing both a HAS and an absorber,[1004] while, in the case of wood pulp, the use of a hydroxyphenylbenzotriazole absorber and a nitroxyl radical has been found to be effective in inhibiting the photolysis of the aryl phenyl ether groups.[1005] Tertiary phosphine complexes of nickel are shown to be highly effective light stabilizers for polystyrene[1006] only when methyl and allyl substituents are present. Finally, UV absorbers give some UV protection to poly(ethylene terephthalate).[1007]

6 Photochemistry of Dyed and Pigmented Polymers

With regard to dyes and pigments, there continues to be moderate interest, with much of the activity centred on the catalytic effects of titanium dioxide. PVC profiles under northern climatic exposures in wet conditions are known in some cases to induce an unacceptable 'pinking' phenomenon.[1008] Here certain grades of titanium dioxide can induce or alter the balance of chemical reactions to form polyene sequences. Photoactivity of titanium dioxide pigments has also been measured via low temperature luminescence, with multiphasic materials showing no evidence for carrier segregation.[1009] This pigment has been used to destroy formaldehyde on carbonized wood surfaces[1010] as well as providing ecological photocatalytic coating surfaces.[1011,1012] Paper containing photocatalytic titanium dioxide pigment decomposes ammonia gas[1013] as well as being useful for the destruction of cyanide ions.[1014] Antibacterial photocatalytic coatings have also been developed,[1015] as have novel processes for photobleaching pulp.[1016] In the photocatalytic decomposition of a reactive red dye, effective decomposition occurred only under visible light exposure,[1017] whereas a zinc oxide pigment doped with Ag was highly effective in decolourizing a blue dye.[1018] Alumina-supported zinc oxide has also been shown to be highly effective in photobleaching a green acid dye with an optimum at a pH of 4.[1019] The weathering behaviour of pigmented polymers has also been discussed.[1020] One interesting observation is that coating of filler particles with a surfactant ehances their ability to stabilize a polymer.[1021]

The natural dye Carthamin has been effectively stabilized using a singlet oxygen nickel complex quencher,[1022] as have a series of triphenylmethane dyes used for carbonless copying papers.[1023] Nickel hydroxyphenylbenzotriazolesulfonate absorbers are also useful in this regard,[1024] as are ester type benzotriazole stabilizers.[1025] Cellophane has been shown to be a good model for the photofading of dyes on paper and cotton,[1026] whilst moisture is important in the photofading of vat dyes in polyesters, with finishing treatments also having a potentially detrimental effect.[1027] Reactive bonded dyes to cotton are more light stable,[1028] while the products of dye fading can also be important in reaction kinetics.[1029] In multilayered coatings of PVC and fluorinated polymers, the photostability of fluorescent dyes has been found to be dependent upon the self-absorption properties of the dyes.[1030] Blends of polymers are also shown to be more effective. The presence of molecular oxygen inhibits the fading of 4-diethylaminoazobenzene dyes in solutions and solid state,[1031] while cyanine dyes with a positive charge are more photostable than corresponding mercocyanines.[1032] Rose Bengal has been used to accelerate the photofading of triphenodioxadine dyes in cellulosic films.[1033]

References

1. C. Kutal, *Coord. Chem. Rev.*, (2001), **211**, 353.
2. K. Morio, *Polym. Prep. (Am. Chem. Soc., Div. Polym. Chem.)*, (2001), **42**, 719.
3. Y. Yagci, *Macromol. Symp.*, (2000), **161**, 19.
4. J. P. Fouassier, *Curr. Trends Polym. Sci.*, (1999), **4**, 163.
5. K. Okawa, *Setchaku no Gijutsu*, (2001), **20**, 35.
6. E. Andrzejewska, *Polimery*, (2001), 46, 88.
7. E. Andrzejewska, *Prog. Polym. Sci.*, (2001), **26**, 605.
8. R.S. Davidson, *Polymer Paint Colour Journal*, (2000), **190**, 30.
9. Y. Suda and S. Dewa, *Kogyo Zairyo*, (2000), **48**, 99.
10. N. Orbey and R. F. Kovar, *Int. SAMPE Tech. Conf.*, (2000), **32**, 932.
11. N. Tanaka, *Kino Zairyo*, (2001), **21**, 42.
12. N. Zhang, Z. Piao, Y. Liang and C. Piao, *Yanbian Daxue Xuebao Ziran Kexueban*, (2000), **26**, 67.
13. M. Mitry and P. Bushendorf, *Rad. Tech. Rep.*, (2000), **14**, 22.
14. Anon, *Kogyo Zairyo*, (2001), **49**, 84.
15. P. J. Courteny, *Adhes. Age*, (2001), **44**, 23.
16. K. Kojima, *Porima Daijesuto*, (2001), **53**, 69.
17. A. Pere and N. Midoux, *Recents prog. Genie Procedes*, (1999), **13**, 237.
18. J. Jakubiak and J. F. Rabek, *Polimery*, (2001), **46**, 10.
19. T. Motoyama, *Setchaku*, (2000), **44**, 337.

20. Y. S. Negi. S. R. Damkale and S. Ansari, *J. Macromol. Sci., Polym. Revs.*, (2001), **41**, 119.
21. A. Wrzyszczynski, F. Scigalski and J. Paczkowski, *Nukleonika*, (2000), **45**, 73.
22. K. Ishizu, A. Mori and T. Shibuya, *Polymer*, (2001), **42**, 7911.
23. A. Costela, I. Garcia-Moreno, O. Garcia and R. Sastre, *Macromol. Chem. Phys.*, (2001), **202**, 2253.
24. P. Ghosh, D. K. Ghosh and G. Pal, *J. Polym. Mat.*, (2000), **17**, 260.
25. T.S. Kwon, K. Suzuki, K. Tagaki, H. Kunisada and Y. Yuki, *J. Macromol. Sci., Pure Appl. Chem.*, (2001), *A38*, 591.
26. H. Zhou, A. Xu, X. Zhang, G. Geng and Q. Gao, *Shangqiu Shizhuan Xuebao*, (2001), **17**, 70.
27. S. Zhang, H. Zhou, G. Yang and Q. Gao, *Huaxue Yanjiu*, (2001), **12**, 29.
28. Q. Gao, R. Li, G. Yang, X. Yu and F. Du, *Ganguang Kexue Yu Guang Huaxue*, (2001), **19**, 116.
29. X. Alonas, C. Grotzinger, J. Lalevee, J. P. Fouassier and M. Visconti, *Eur. Polym. J*, (2001), **37**, 897.
30. S. Y. Yang and M. M. Green, *Polym. Int.*, (2001), **50**, 897.
31. L. R. Denny, J. W. Baur, M. D. Alexander, R. Kannan, S. M. Kirkpatrick and S. J. Clarson, *Int. SAMPE Tech. Conf.*, (2000), **32**, 712.
32. X. Allonas, J. P. Fouassier, M. Kaji, M. Miyasaka and T. Hidaka, *Polymer*, (2001), **42**, 7627.
33. O. G. Ballesteros, L. Maretti, R. Sastre and J. C. Scaiano, *Macromolecules*, (2001), **34**, 6184.
34. T. Urano, E. Ohno-Okumura, K. Sakamoto, T. Hatano, K. Fukui, T. Karatsu and A. Kitamura, *J. Photopolym. Sci. Technol.*, (2000), **13**, 679.
35. E. Y. Komarova, K. Ren and D. C. Neckers, *Polym. Prep. (Am. Chem. Soc., Div. Polym. Chem.)*, (2001), **42**, 737.
36. M. Yin, Y. Sun and W. Yang, *Gaofenzi Cailiao Kexue Yu Gongcheng*, (2000), **16**, 140.
37. M. Terazima, Y. Nogami and T. Tominaga, *Chem. Phys. Lett.*, (2000), **332**, 503.
38. S. Uemura, M. Yoshi, N. Kobayashi and T. Nakahira, *Polym. J.*, (2000), **32**, 987.
39. S. Uemura, T. Shimakawa, N. Kobayashi and T. Nakahira, *J. Mat. Chem.*, (2001), **11**, 267.
40. S. Uemura, S. Tokuda, N. Kobayashi and T. Nakahira, *Synth. Met.*, (2001), **119**, 89.
41. J. M. Mabry, J. Brooks and W. P. Weber, *Polym. Prep. (Am. Chem. Soc., Div. Polym. Chem.)*, (2001), **42**, (2001), **42**, 344.
42. C. Billaud, M. Sarakha and M. Bolte, *J. Polym. Sci., Part A: Polym. Chem. Ed.*, (2000), **38**, 3997.
43. T. Mizuta, M. Onishi and K. Miyoshi, *Organometallics*, (2000), **19**, 5005.
44. S. Zhang, B. Li, L. Tang and Q. Zhou, *Gaofenzi Xuebao*, (2001), **1**, 114.
45. K. S. Padon and A. B. Scranton, *J. Polym. Sci., Part A: Polym. Chem. Ed.*, (2001), **39**, 715.
46. G. A. Eping and I. A. Banerjee, *Polym. Prep. (Am. Chem. Soc., Div. Polym. Chem.)*, (2001), **42**, 492.
47. M. V. Encinas, A. M. Rufs, S. Bertolotti and C. M. Previtali, *Macromolecules*, (2001), **34**, 2845.
48. C. Grotzinger, D. Burget, P. Jacques and J. P. Fouassier, *J. Appl. Polym. Sci.*, (2001), **81**, 2368.
49. L. Villegas, M. V. Encinas, A. M. Rufs, C. Bueno, S. Bertolotti and C. M. Previtali , *J. Polym. Sci., Part A: Polym. Chem.*, (2001), **39**, 4074.
50. K. S. Padon, D. Kim and A. B. Scranton, *Polym. Prep. (Am. Chem. Soc., Div. Polym.*

Chem.), (2001), **42**, 705.
51. J. P. Fouassier, X. Allonas, J. Lalevee and M. Visconti, *J. Polym. Sci., Part A: Polym. Chem. Ed.*, (2000), **38**, 4531.
52. Z. Guo, G. Yang, Z. Huang and J. Huang, *Macromol. Rapid Commun.*, (2001), **22**, 120.
53. J. Qu. S. Zhang, J. Song and W. Huang, *Gaofenzi Xuebao*, (2001), **5**, 656.
54. A. Valdehenito, E. A. Lissi and M. V. Encinas, *Macromol. Chem. Phys.*, (2001), **202**, 2581.
55. M. Nowakowska, S. Zapotocny and A. Karewitz, *Polymer*, (2001), **42**, 1817.
56. Z. Liu, M. Weber and N. J. Turro, *Polym. Prep. (Am. Chem. Soc., Div. Polym. Chem.)*, (2001), **42**, 771.
57. T. Saito, H. Shibata and M. Seno, *J. Appl. Polym. Sci.*, (2001), **80**, 815.
58. J. L. Mateo, M. Calvo and P. Bosch, *J. Polym. Sci. Part A: Polym. Chem. Ed.*, (2001), **39**, 2049.
59. C. Decker, K. Zahouily, D. Decker, T. Nguyen and T. Viet, *Polymer*, (2001), **42**, 7551.
60. M. Heinenberg, M. H. Reihmann and H. Ritter, *Des. Monomers Polym.*, (2000), **3**, 501.
61. T. Urano, M. Ishikawa and H. Itoh, *Imaging Sci. J.*, (1999), **47**, 121.
62. L. Angiolini, D. Caretti and E. Salatelli, *Macromol. Chem. Phys.*, (2000), **18**, 2646.
63. S. H. Qin, D. Q. Qin and K. Y. Qiu, *New J. Chem.*, (2001), **25**, 893.
64. J. V. Crivello and M. Sangermano, *J. Polym. Sci., Part A: Polym. Chem. Ed.*, (2001), **39**, 343.
65. J. V. Crivello and M. Sangermano, *Polym. Prep. (Am. Chem. Soc., Div. Polym. Chem.)*, (2001), **42**, 783.
66. Z. Gomurashvili and J. V. Crivello, *Polym. Prep. (Am. Chem. Soc., Div. Polym. Chem.)*, (2001), **42**, 755.
67. Z. Gomurashvili and J. V. Crivello, *J. Polym. Sci., Part A: Polym. Chem. Ed.*, (2001), **39**, 1187.
68. J. V. Crivello and M. Sangermano, *J. Macromol. Sci., Pure Appl. Chem.*, (2001), *A38*, 487.
69. J. V. Crivello, K. Y. Song and R. Ghoshal, *Chem. Mater.*, (2001), **13**, 1932.
70. J. V. Crivello and R. A. Ortiz, *J. Polym. Sci., Part A: Polym. Chem. Ed.*, (2001), **39**, 3578.
71. Z. Gomurashvili, Y. Hua and J. V. Crivello, *Macromol. Chem. Phys.*, (2001), **202**, 2133.
72. J. V. Crivello, *Polym. Prep. (Am. Chem. Soc., Div. Polym. Chem.)*, (2001), **42**, 773.
73. Y. Yagci, A. Onen and I. Reetz, *Macromol. Symp.*, (2001), 174, 255.
74. Y. Yagci and A. Onen, *Polym. Prep. (Am. Chem. Soc., Div. Polym. Chem.)*, (2001), **42**, 775.
75. A. Onen and Y. Yagci, *Macromol. Chem. Phys.*, (2001), **202**, 1950.
76. M. Onciu, A. Onen and Y. Yagci, *Polym. Int.*, (2001), **50**, 144.
77. Y. Yagci, S. Yildirim and A. Onen, *Macromol. Chem. Ed.*, (2001), **202**, 527.
78. A. Onen and Y. Yagci, *Polymer*, (2001), **42**, 6681.
79. H. Kato and H. Sasaki, *Polym. Prep. (Am. Chem. Soc., Div. Polym. Chem.)*, (2001), **42**, 729.
80. A. Carroy, S. Ilg and T. Bolle, *Farbe Lack*, (2001), **107**, 90.
81. H. Kurata, K. Fujihara, A. Kimura, T. Ohno, M. Matsumura, Y. Hirata and T. Okada, *J. Polym. Sci., Part B: Polym. Phys. Ed.*, (2001), **39**, 2937.
82. S. Mah, J. Choi, H. Lee and S. Choi, *Fibres Polym.*, (2000), **1**, 1.

83. C. E. Hoyle, S. F. Thames, L. R. Mullen and D. L. Schmidt, *J. Polym. Sci., Part A: Polym. Chem. Ed.*, (2001), **39**, 571.
84. A.A. Al-Doaiss, E. Klemm, D. Stadermann and N. Moszner, *Macromol. Chem. Phys.*, (2001), **202**, 270.
85. M. R. Rodrigues and M. G. Neumann, *J. Polym. Sci., Part A: Polym. Chem. Ed.*, (2000), **39**, 46.
86. U. Muller, *Polym. Prep. (Am. Chem. Soc., Div. Polym. Chem.)*, (2001), **42**, 777.
87. H. Pan, J. Zhao, H. Guan, R. Li and Q. Gao, *Huaxue Yanjiu*, (2001), **12**, 49.
88. M. Rusa, G. David, B. C. Simionescu, J. Couve and M. J. M. Abadie, *Macromol. Rapid Commun.*, (2001), **22**, 372.
89. S. Y. Shim, J. H. Kim and D. H. Suh, *Polym. Prep. (Am. Chem. Soc., Div. Polym. Chem.)*, (2001), **42**, 727.
90. V. Andruleviciute, R. Lazauskaite, J. V. Grazulevicius, R. Kavaliunas and A. Stanisauskaite, *Env.Chem. Phys.*, (2000), **22**, 32.
91. A. Hartwig, A. Harder, A. Luhring and H. Schroder, *Eur. Polym. J.*, (2001), **37**, 1449.
92. C. E. Hoyle and S. Jonsson, *Proc. Int. Waterborne, High-Solids, Powder Coat. Symp.*, (2000), *27th*, 16.
93. C. K. Nguyen, R. S. Smith, T. Cavitt, C. E. Hoyle and S. Jonsson, *Polym. Prep. (Am. Chem. Soc., Div. Polym. Chem.)*, (2001), **42**, 707.
94. A. A. Dias, J. F. G. A. Jansen and D. M. Van Dijck, *Surf. Coat. Int.*, (2000), **83**, 502.
95. M. Sangermano, G. Malucelli, R. Bongiovanni, A. Priola, U. Annby and N. Rehnberg, *Polym. Int.*, (2001), **50**, 998.
96. O. Nuyken and H. Braun, *Polym. Prep. (Am. Chem. Soc., Div. Polym. Chem.)*, (2001), **42**, 779.
97. S. Joensson, D. Yang, K. Viswanathan, E. Shier, C. E. Hoyle, K. Belfield and K. Lindgren, *Polym. Prep. (Am. Chem. Soc., Div. Polym. Chem.)*., (2001), **42**, 703.
98. M. R. Pokhrel and S. H. Bossmann, *J. Inf. Rec.*, (2000), **25**, 47.
99. C. L. Lester and C. A. Guymon, *Polym. Prep. (Am. Chem. Soc., Div. Polym. Chem.)*., (2001), **42**, 724.
100. M. Jung, H. Van, M. S. Eelco, T. Julien and A. M. Van Herk, *Macromol. Rapid Commun.*, (2001), **22**, 978.
101. V. Bulacovschi, M. Popa, M. J. M. Abadie and M. Arnautu, *Acta Univ. Cibiniensis, Ser F: Chem.*, (2000), **2**, 19.
102. M. A. Breimer, G. Yevgeny, S. Sy and O. A. Sadik, *Nanolett.*, (2001), ACS ASAP.
103. R. M. Camponanes, E. Bittencourt and J. S. De Carvalho, *Inf. Technol.*, (2001), **12**, 79.
104. S. Uemura, T. Nakahira and N. Kobayashi, *J. Mat. Chem.*, (2001), **11**, 1585.
105. Y. Kim, S. Fukai and N. Kobayashi, *Synth. Met.*, (2001), **119**, 337.
106. A. Matsumoto and S. Nagahama, *Proc. Jpn. Acad., Ser B*, (2001), **77B**, 110.
107. S. Irie, S. Isoda, T. Kobayashi, H. Ozaki and Y. Mazaki, *Probe Microsco.*, (2000), **2**, 1.
108. G. A. Abakumov, S. N. Mensov, A. V. Semenov and S. A. Chesnokov, *Vysokomol. Soedin., Ser A, Ser B.*, (2000), **42**, 1252.
109. E. Rusu, V. Barboiu, A. Airinei, E. Comanita and G. Rusu, *Rev. Roum. Chim.*, (2001), **45**, 451.
110. R. Budreckiene, R. Lazauskaite, R. Kavaliunas and J. Vidas Grazulevicius, *Eur. Polym. J.*, (2001), **37**, 983.
111. A. V. Tadjer, N. P. Gospodinova, G. K. Madjarova, L. K. Terlemezyan and P. T. Mokreva, *Bulg. Chem. Commun.*, (2000), **32**, 66.
112. K. Otsuka, K. Hasagawa, A. Matsumoto, H. Kimura, A. Fukuda, M. Yoshimoto

and H. Fujiwara, *Zairyo*, (2000), **49**, 1282.

113. P. B. Zmolek, H. Sohn, P. K. Gantzel and W. C. Trogler, *J. Am. Chem. Soc.*, (2001), **123**, 1199.
114. J. Lin, Y. Huang and G. Xiang, *Beijing Huagong Daxue Xuebao.*, (2001), **28**, 26.
115. Y. Zhu, R. Ning and Y. Xie, *Gaofenzi Cailiao Kexue Yu Gongcheng*, (2001), **17**, 70.
116. K. Takagi, Y. Nishikawa, T. S. Kwon, H. Kunisade and Y. Yuki, *Polymer J.*, (2000), **32**, 970.
117. S. P. Doguparthy, *J. Membrane Sci*, (2001), **185**, 201.
118. C. Garcia, A. Fimia and I. Pascual, *Appl. Phys. B: Lasers Opt.*, (2001), **72**, 311.
119. A. R. Kan, Z. H. Farooqui and M. Ejaz, *Pak. J. Sci. Ind. Res.*, (2001), **44**, 109.
120. C. L. Lester and C. A. Guymon, *Polym. Mat. Sci.*, (2001), **85**, 200.
121. M. Tachibana, T. Yamaguchi, K. Kojima and H. Sakuma, *J. Appl. Phys.*, (2001), **90**, 470.
122. S. H. Qin and K. Y. Qiu , *Eur. Polym. J.*, (2001), **37**, 711.
123. K. Ishizu and A. Mori, *Polym. Int.*, (2001), **50**, 906.
124. T. S. Kwon, K. Takagi, H. Kunisada and Y. Yuki, *J. Macromol. Sci., Pure Appl. Chem.*, (2000), **37**, 1461.
125. T. S. Kwon, K. Takagi, H. Kunisada and Y. Yuki, *J. Macromol. Sci., Pure Appl. Chem.*, (2000), **38**, 605.
126. X. Fan, X. Xie and Y. Hatate, *Chin. J. Polym. Sci.*, (2001), **19**, 311.
127. G. D. Dixon and N. W. Carlson, *RadTech Rep.*, (1999), **13**, 36.
128. G. Terrones and A. J. Pearstein, *Macromolecules*, (2001), **34**, 3195.
129. J. F. G. A. Jansen, A. A. Dias, M. Dorschu and B. Coussens, *Polym. Prep. (Am. Chem. Soc., Div. Polym. Chem.)*, (2001), **42**, 769.
130. Y. B. Boiko, J. M. Costa, M. M. Wang and S. C. Esener, *Proc. SPIE-Int. Soc. Opt. Eng.*, (2001), **4279**, 212.
131. K. Radotic, M. Tasic, M. Jeremic, Z. Budimlija, J. Simic-Krstic, A. Polzovic and Z. Bozovic, *Gen. Physiol. Biophys.*, (2001), **20**, 33.
132. V. V. Ivanov and C. Decker, *Polym. Int.*, (2001), **50**, 113.
133. M. Sakai, M. Ichida and A. Nakamura, *Fullerene Sci. Technol.*, (2001), **9**, 351.
134. J. Tong, N. Zhou, S. Ni and M. A. Winnik, *Macromolecules*, (2001), **34**, 696.
135. Y. Yilmaz, Y. Yagci and O. Pekcan, *J. Macromol. Sci., Pure Appl. Chem.*, (2001), *A38*, 741.
136. S. Zhang, B. Li, L. Tang, X. Wang, R. Yang and Q. Zhou, *Gaodeng Xuexiao Huaxue Xuebao*, (2001), 22, 1249.
137. Y. Tajima, Y. Shigemitsu, H. Arai, E. Takeuchi and K. Takeuchi, *Synth. Met.*, (2001), **121**, 1167.
138. M. G. Neumann, C. C. Schmitt and H. M. Maciel, *J. Phys. Chem. B.*, (2001), **105**, 2939.
139. S. Katogi and M. Yusa, *J Photopolym. Sci. Technol.*, (2001), **14**, 151.
140. M. Tsnooka, H. Tachi, T. Yamamoto and M. Shirai, *Polym. Prep. (Am. Chem. Soc., Div. Polym. Chem.)*, (2001), **42** 720.
141. M. Tsnooka, H. Tachi, T. Yamamoto, K. Akitomo and M. Shirai, *J. Photopolym. Sci. Technol.*, (2001), **14**, 153.
142. K. Suyama, H. Iriyama, M. Shirai and M. Tsnooka, *J. Photopolym. Sci. Technol.*, (2001), **14**, 155.
143. T. B. Cavitt, B. Phillips, C. K. Nguyen, C. E. Hoyle, S. Joensson and K. Viswanathan, *Polym. Prep. (Am. Chem. Soc., Div. Polym. Chem.)*, (2001), **42**, 709.
144. N. S. Allen, S. V. Ryabov, M. Edge, S. N. Kobylinskyy, A. A. Pokcholenko and Z. V. Voitenko, *Polym. Deg. Stabil.*, (2002), **75**, 229.

145. Z. Feng, Y. Wang and L. Zhao, *Gaofenzi Cailiao Kexue Yu Gongcheng*, (2001), **17**, 66.
146. R. Liska, S. Knaus, H. Gruber and J. Wendrinsky, *Surf. Coat. Int.*, (2000), **83**, 297.
147. Q. Yu, S. Nauman and J. P. Santerre, *J. Appl. Polym. Sci.*, (2001), **82**, 1107.
148. K. Suyama, S. Yamaguchi, M. Shirai and M. Tsnooka, *Polym. Int.*, (2001), **39**, 1708.
149. C. Zhou, M. A. Winnik and T. Jao, *J. Polym. Sci., Part A: Polym. Chem. Ed.*, (2001), **39**, 2642.
150. J. H. De Groot, K. A. Dillingham, H. Deuring, H. J. haitjema, K. Hodd and S. Norrby, *Polym. Mater. Sci. Eng.*, (2001), **85**, 53.
151. X. Tan, J. Guo and H. Xie, *Shihua Jishu Yu Yingyong*, (2000), **18**, 195.
152. H. Tachi, T. Yamamoto, M. Shirai and M. Tsnooka, *J. Polym. Sci., Part A: Polym. Chem. Ed.*, (2001), **39**, 1329.
153. X. Chen and Y. Chen, *Zhongshan Daxue Xuebao, Ziran Kexueban*, (2000), **39**, 122.
154. T. Hatano, K. Fukui, T. Karatsu, A. Kitamura and T. Urano, *J. Photopolym. Sci. Technol.*, (2000), **13**, 697.
155. H. Tokuda and B. F. Rainer, *DIC Tech. Rev.*, (2000), **6**, 49.
156. M. Visconti and M. Cattaneo, *Prog. Org. Coatings*, (2000), **40**, 243.
157. N. A. Nikolic and R. A. V. Schultz, *Polym. Prepr. (Am. Chem. Soc., Div. Polym. Chem.)*., (2001), **42**, 743.
158. J. Lee, H. Kim and H. Kim, *Bull. Korean Chem. Soc.*, (2001), **22**, 179.
159. T. Nakamura, *Technocosmos*, (2001), **14**, 34.
160. G. Malucelli, M. Sangermano, A. Priola, U. Annby and N. Rehnberg, *FATIPEC Congress*, (2000), *25th*, 161
161. R. T. Olsson, H. E. Bair, V. Kuck and A. Hale, *Polym. Prepr. (Am. Chem. Soc., Div.Polym. Sci.)*, (2001), **42**, 797.
162. M. Sangermano, S. N. falling and J. V. Crivello, *J. Macromol. Sci., Pure Appl. Chem.*, (2001), *A38*, 919.
163. H. Sasaki, *Polym. Prepr. (Am. Chem. Soc., Div. Polym. Chem.)*., (2001), **42**, 731.
164. Q. Wang, P. Zhu and Z. Li, *Polym. Prepr. (Am. Chem. Soc., Div. Polym. Chem.)*., (2001), **42**, 216.
165. H. Suzuki and H. Sasaki, *Polym. Prepr. (Am. Chem. Soc., Div. Polym. Chem.)*., (2001), **42**, 733.
166. T. Maruyama, N. Kusumoto and K. Seko, *J. Photopolym. Sci. & Technol.*, (2001), **14**, 165.
167. K. Song, J. V. Crivello and R. Ghoshal, *Polym. Prepr. (Am. Chem. Soc., Div. Polym. Chem.)*., (2001), **42**, 785.
168. T. Nishikuboto, A. Kameyama, K. Tsutsui and S. Kishimoto, *J. Polym. Sci., Part A: Polym. Chem. Ed.*, (2001), **39**, 1481.
169. T. Nishikuboto, A. Kameyama, K. Tsutsui and S. Kishimoto, *J. Polym. Sci., Part A: Polym. Chem. Ed.*, (2001), **39**, 1169.
170. O. Nuyken and H. Braun, *Des. Monomers Polymers*, (2001), **4**, 19.
171. K. Yoneda and H. Sasaki, *J. Photopolym. Sci. Technol.*, (2001), **14**, 171.
172. J.V. Crivello and R. Acosta, *J. Polym. Sci., Part A: Polym. Chem. Ed.*, (2001), **39**, 2385.
173. C. Decker, T. N. Viet and H. P. Thi, *Polym. Int.*, (2001), **50**, 986.
174. S. Warwel, E. Fehling and M. Kunz, *Eur. J. Lipid Sci., Technol.*, (2001), **103**, 133.
175. C. Decker, T. V. Nguyen, D. Decker and E. Weber-Koehl, *Polymer*, (20011), **42**, 5531.
176. Q. Li, J. Chen, Z. Li and W. Cao, *Gaofenzi Xuebao*, (2001), **1**, 134.
177. H. Okamura, C. Harada, S. Morishita, M. Shirai, M. Tsnooka, T. Fujiki, S.

Kawasaki and M. Yamada, *J. Photopolym. Sci. Technol.*, (2001), **14**, 159.
178. F. Wang and D. C. Neckers, *Macromolecules*, (2001), **34**, 6202.
179. M. Stumbeck, *Adhaes.-Kleben Dichten*, (2000), **44**, 28.
180. M. Okazaki, Y. Shibasaki and M. Ueda, *Chem. Lett.*, (2001), **8**, 762.
181. M. Chen, Q. Chen, S. Xiao and X. Hong, *Ganguang Kexue Yu Guang Huaxue*, (2001), **19**, 208.
182. K. Tashiro, S. Nakamoto, S. Saragai, A. Matsumoto and T. Tsubouchi, *Polymer*, (2001), **42**, 6747.
183. D. Creed, C. E. Hoyle, J. Lin, A. M. Peeler, P. Subramanian and V. Krishnan, *J. Polym. Sci., Part A: Polym. Chem. Ed.*, (2000), **39**, 134.
184. G. N. Gerasimov, T. V. Kiryanova, V. K. Belskii, A. I. Stash, S. A. Lebdev, N. V. Kozlova and L. I. Trakhtenberg, *Zh. Fiz. Khim.*, (2001), **75**, 538.
185. X. Zhang, H. Menzel and L. Mannfred, *Hecheng Huaxue*, (2000), **8**, 519.
186. V. Stehmel, B. Stiller, B. Strehmel, A. Ananda and D. C. Neckers, *Polym. Prep. (Am. Chem. Soc., Div. Polym. Chem.)*, (2001), **42**, 749.
187. F. Nakanishi and J. Nagasawa, *Busshitsu Kogaku Kogyo Gijutsu Kenkyusho Hokoku*, (2000), **8**, 89.
188. M. H. Kim and J. D. Kim, *Thin Solid Films*, (2001), **385**, 142.
189. M. H. Kim and J. D. Kim, *Synth. Met.*, (2001), **117**, 177.
190. T. Anazawa and M. Yamada, *Kawamura Rikagaku Kenkyusho Hokoku*, (2000), 117.
191. B. Sapich, J. Stumpe, H. R. Kricheldorf, A. Fritz and A. Schoenhals, *Macromolecules*, (2001), **34**, 5694.
192. S. Saragai, K. Tashiro, S. Nakamoto, T. Kamae, A. Matsumoto and T. Tsubouchi, *Polym. J.*, (2001), **33**, 199.
193. M. Vifan, I. D. Olenik, A. Mertelj and M. Copic, *Phys. Rev. E: Stat. Nonlinear, Soft Matter Phys*, (2001), **63**, 061709.
194. K. Kurschner, P. Strohriegl, P. Van de Witte and J. Lub, *Mol. Cryst. Liqu. Cryst. Sci. Technol Sect A.*, (2000), **352**, 301.
195. D. J. Broer, J.A. M. M. Van Haaren and J. Lub, *Proc. SPIE-Int. Soc. Opt. Eng.*, (2000), **4107**, 59.
196. G. Schnurpfeil, A. Harder and H. Schroder, *Macromol. Chem. Phys.*, (2001), **202**, 180.
197. G. H. Hsiue, Y. A. Sha and R. J. Jeng, *Macromol. Chem. Phys.*, (2001), **202**, 287.
198. J. H. Norton and C. A. Guymon, *Polym. Prep. (Am. Chem. Soc., Div. Polym. Chem.).*, (2001), **42**, 753.
199. A. Arcioni, C. Bacchiocchi, M. D'Elia, R. tarroni and C. Zannoni, *Mol. Cryst. Liq. Cryst. Sci. Technol., Sect A.*, (2001), **362**, 279.
200. W. S. Kim, J. W. Lee, Y. W. Kwak, J. K. Lee, Y. T. Park and S. D. Yoh, *Polym. J.*, (2001), **33**, 643.
201. K.V. K. Boodhoo, W. A. Dunk and R. J. Jachuk, *Polym. Prepr. (Am. Chem. Soc., Div. Polym. Chem.).*, (2001), **42**, 813.
202. M. Kunz and M. Bauer, *Vak. Forsch. Prax.*, (2001), **13**, 115.
203. M. Haberer, G. Zak and C. B. Park, *Solid Freeform Fabr. Symp. Proc.*, (2000), 333.
204. O. Prucker, K. Muller and J. Ruhe, *Mater. Res. Soc. Symp. Proc.*, (2001), **629**, FF9.8.1.
205. Z. Jovanovic, J. Lahaye, H. Laver, S. Meqert and C. te Walvaart, *Eur. Coat. J.*, (2000), **10**, 16.
206. L. Catilaz-Simonin and J. P. Fouassier, *J. Appl. Polym. Sci.*, (2001), **79**, 191.
207. E. Lackey, R. Charware, K. Inamdar, V. Patil, S. Rahatekar, B. Skelton, J. Smith and L. Worrel, *Int. SAMPE Tech. Conf.*, (2000), **32**, 959.

208. H. E. Bair, M. L. Schilling, V. L. Colvin, A. Hale and N. J. Levinos, *Proc. NATAS Annu. Conf. Therm. Anal. Appl.*, (2000), *28th*, 554.
209. Y. Sakurai, S. Okuda, N. Nagayama and M. Yokoyama, *J. Mat. Chem.*, (2001), **11**, 1077.
210. C. Croutxe, O. Soppera and D. J. Lougnot, *Appl. Surf. Sci.*, (2000), **168**, 89.
211. J. Kinard and J. Colton, *Polym. Compos.*, (2000), **21**, 918.
212. H. Lu, L. G. Lovell and C. N. Bowman, *Polym. Prep. (Am. Chem. Soc., Div. Polym. Chem.)* (2001), **42**, 763.
213. K. Yamada, T. Haraguchi, H. Yamane, S. Ide, K. Isomura, K. Takehara, N. Iwamoto and T. Kajiyama, *Kitakyushu Kogyo Koto Senmon Gakko Kenkyu Hokoku*, (2001), **34**, 115.
214. S. Jo, H. Shin and A. G. Mikos, *Biomacromolecules*, (2000), **88**, 5071.
215. A. Bratus, A. Turovski, Yu. Medvedevskikh, I. Semenyuk and G. Zaikov, *Int. J. Polym. Mater.*, (2001), **48**, 371.
216. N. Yoshimoto, Y. Tomonaga, M. Ishikawa and M. Morita, *Electrochim. Acta*, (2001), **46**, 1195.
217. R. M. Garipov, A. A. Efremova, S. S. Yu, R. Deberdeev and G. E. Zaikov, *Russ. Polym. News*, (2000), **5**, 1.
218. D. Wang and D. C. C. Lam, *J. Mater. Sci., Lett.*, (2000), **19**, 2109.
219. D. Skinner, *Surf. Coat. Sci.*, (2000), **83**, 508.
220. C. Artal, M. B. Ros, J. L. Serrano, N. Pereda, J. Etxebarria, C. L. Folcia and J. Ortega, *Macromolecules*, (2001), **34**, 4244.
221. P. A. Bondarenko, K. A. Dotsenko, V. N. Vatulyov and V. V. Magdinets, *Kompoz Polim. Mater.*, (2000), **22**, 60.
222. O. P. Batog, V. N. Vatulyov and V. V. Magdinets, *Kompoz Polim. Mater.*, (2000), **22**, 17.
223. E. Andrzejewska, *Macromol. Symp.*, (2001), **171**, 243.
224. D. Bielinski, P. Glab and L. Slusarski, *Polimery*, (2001), **46**, 494.
225. S. Zhang, B. Li, L. tang, R. yang and Q. Zhou, *Setachu no Gijutsu*, (2000), **13**, 407.
226. B. Lin, Z. Zhang, L. Tang and Q. Zhou, *Polimer J.*, (2001), **33**, 263.
227. D. H. Choi, S. J. Oh, S. Y. Ban and K. Y. Oh, *Bull. Korean Chem. Soc.*, (2001), **22**, 1207.
228. I. V. Khudyakov, M. B. Purvis and N. J. Turro, *Polym. Prep. (Am. Chem. Soc., Div. Polym. Chem.)*, (2001), **42** 767.
229. A. A. Stolov, T. Xie, J. Penelle, L. Shaw and H. D. Stidham, *Polymer Eng. Sci.*, (2001), **41**, 314.
230. C. Peinado, E. F. Salvador, J. Baselga and F. Catalina, *Macromol. Chem. Phys.*, (2001), **202**, 1924.
231. N. B. Cramer and C. N. Bowman, *J. Polym. Sci., Part A: Polym. Chem. Ed.*, (2001), **39**, 3311.
232. W. F. Jager and O. van den Berg, *Polym. Prep. (Am. Chem. Soc., Div. Polym. Chem.)*, (2001), **42**, 807.
233. R. Vatanparast, S. Li and H. Lemmetyinen, *J. Appl. Polym. Sci.*, (2001), **82**, 2607.
234. W. F. Jager, M. Wallin and M. V. Fernandez, *Polym. Prep. (Am. Chem. Soc., Div.Polym. Chem.)*, (2001), **42**, 751.
235. M. Wang, J. Qiang, Y. Fang and D. Hu, *Yingyong Huaxue*, (2000), **17**, 603.
236. O. Peckan, D. Kaya and M. Erdogan, *J. Appl. Polym. Sci.*, (2001), **80**, 1907; O. Peckan and D. Kaya, *J. Appl. Polym. Sci.*, (2001), **81**, 3161.
237. D. Cho, D. S. Kim and J. K. Lee, *Polymer (Korea)*, (2001), **25**, 199.
238. V. Bekiari, E. Stathatos, P. Lianos, U. L. Stangar, B. orel and P. Judeinstein,

Monatsh. Chem., (2001), **132**, 97.
239. X. Zhang, F. S. Du, Z. C. Li and F. M. Li, *Macromol. Rapid Commun.*, (2001), **22**, 983.
240. J. Gonzalez-Benito, F. Mikes, J. Bravo, A. J. Aznar and J. Baselga, *J. Macromol. Sci., Phys.*, (2001), *B40*, 429.
241. P. Bosch, A. Fernandez-Arizipe and J. L. Mateo, *Macromol. Chem. Phys.*, (2001), **20**, 1961.
242. C. Peinado, E. F. Slavador, F. Catalina and A. E. Alonzo, *Polymer*, (2001), **42**, 2815.
243. F. Mikes, J. Gonzalez-Benito and J. Baselga, *J. Macromol. Sci., Phys.*, (2001), *B40*, 405.
244. C. W. Miller, K. Viswanathan, S. Jonsson, C. Nason, W. F. Kuang, D. Yang, B. Kess and C. E. Hoyle, *Polym. Prep. (Am. Chem. Soc., Div. Polym. Chem.)*, (2001), **42**, 811.
245. K.A. Berchtold, T. M. Lovestaed and C. N. Bowman, *Polym. Prep. (Am. Chem. Soc., Div. Polym. Chem.)*, (2001), **42**, 759.
246. H. Arai, Y. Tajima and K. Takeuchi, *J. Photopolym. Sci. Technol.*, (2001), **14**, 143.
247. H. E. Bar, M. L. Schilling, V. L. Colvin, A. Hale and N. J. Levinos, *Annu. Conf. Therm. Anal. Appl.*, (1999), *27th*, 99.
248. M. Sakai and A. Nakamura, *Phys. Rev. B: Condens. Matter Mater. Phys.*, (2001), **64**, 033413/1.
249. P. I. Shapoval, Ya. P. Kovals'kii, O. V. Suberlyak, G. O. Marshalok and Yu. I. Yachishin, *Dopov. Nats. Akad. Nauk Ukr.*, (2000), **190**, 162.
250. A. Bratus, A. Turovki, Yu. Medvedeskikh, E. Zagladko and G. Zaikov, *Polym.-Plast. Technol. Eng.*, (2001), **40**, 181.
251. A.A. Stolov, T. Xie, J. Penelle and S. L. Hsu, *Macromolecules*, (2001), **34**, 2865.
252. A. R. Kitsya, A. N. Bratus, Yu. G. Medvedevskikh, A. A. Turovskii, G. V. Gafiichuk and G. E. Zaikov, *Gongneng Gaofenzi Xuebao*, (2001), **7**, 18.
253. J. A. E. Corzo, N. Davidenko and R. Alvarez, *Comput. Theoret. Polym. Sci.*, (2001), **11**, 89.
254. A. Shyichuk and G. Tokaryk, *Polimery*, (2000), **45**, 548.
255. G. Tan, Z. Chen, J. Yu and Z. Liu, *Zhongguo Fangzhi Daxue Xuebao*, (2000), **26**, 91.
256. Q. Wu and B. Qu, *Polym. Eng. Sci.*, (2001), **41**, 1220.
257. V. D. Ramos, D. Derouet and L. Y. Leila, *Nat. Polym. Compos., [Proc. 3rd Int. Symp., Workshop Prog. Prod. Process. Cellul. Fibres Nat. Polym.]*, (2000), 51.
258. Q. Gao, Q. Wang, C. Qiao and R. Li, *Gongneng Gaofenzi Xuebao*, (2001), **14**, 177.
259. C. Decker and T. V. Nguyen, *Euromat 99, Biannu. Meet. Eur. Mater. Soc, (FEMS).*, (1999), **13**, 286.
260. Y. Inaki and H. Hiratsuka, *J. PhotoPolym. Sci. Technol.*, (2000), **13**, 739.
261. Y. Ohtani, Y. Inaki and M. Miyata, *J. Photopolym. Sci. Technol.*, (2001), **14**, 295.
262. W. Kim, H. Jang, H. Hong and K. Seo, *Macromol. Rapid Commun.*, (2001), **22**, 825.
263. G. Zhang and W. Liang, *Huabei Dianli Daxue Xuebao*, (2000), **27**, 75.
264. S. N. Kim, S. C. Shim. D. Y. Kim, Y. H. Rhee and Y. B. Kim, *Macromol. Rapid Commun.*, (2001), **22**, 1066.
265. C. E. Hoyle, M. Cole, M. Bachemin, W. Kuang, B. Yoder, C. Nguyen and S. Jonsson, *Polym. Prep. (Am. Chem. Soc., Div. Polym. Chem.)*, (2001), **42**, 697.
266. N. B. Cramer and C. N. Bowman, *Polym. Prep. (Am. Chem. Soc., Div. Polym. Chem.)*, (2001), **42**, 699.
267. L. Lecamp, F. Houllier, B. Youssef and C. Bunel, *Polymer*, (2001), **42**, 2727.
268. C. F. Jasso, J. Valdez, J. H. Perez and O. Laguna, *Macromol. Symp.*, (2001), **80**, 1343.
269. J. Sanchez and M. Del Valle, *Electroanalysis*, (2001), **13**, 471.

270. Y. Konno, A. Kameyama, T. Nishikubo and Y. Nagai, *J. Polym. Sci., Part A: Polym. Chem. Ed.*, (2001), **39**, 2683.
271. K. Teramoto and Y. Mori, *Koen Yoshishu-Nippon Setachaku Gakkai Nenji Taikai*, (1999), *37th*, 143.
272. H. Sun, T. Tanaka, K. Takada and S. Kawata, *Appl. Phys. Lett.*, (2001), **79**, 1411.
273. D. H. Choi, S. J. Oh, B. H. Cha and J. Y. Lee, *Eur. Polym. J.*, (2001), **37**, 1951.
274. R. Balaji and S. Nanjundan, *React. Funct. Polym.*, (2001), **49**, 77.
275. R. Popielarz, C. K. Chaing, R. Nozaki and J. Obrzut, *Polymer*, (2001), **628**, CC11.5.1.
276. O. Soppera, C. Croutxe-Barghorn and D. J. Lougnot, *New J. Chem.*, (2001), **25**, 1006.
277. S. Shoji and S. Kawata, *Proc. SPIE-Int. Soc. Opt. Eng.*, (2000), **3937**, 12
278. E. V. Sitzmann, D. A. Bramer and W. Al-Akhdar, *RadTech. Rep.*, (2000), **14**, 28.
279. K. K. Baikerikar and A. B. Scranton, *J. Appl. Polym. Sci.*, (2001), **81**, 3449.
280. S. Padaki and R. W. Buehner, *Proc. NATAS Annu. Conf. Therm. Anal. Appl.*, (1999), 27th, 775.
281. V. V. Krongauz and C. P. Chawla, *RadTech. Rep.*, (2001), **15**, 34.
282. Y. Duan, S. Wang, H. Chen and B. Lu, *Huaxue Gongcheng (Xi'an, China)*, (2000), **28**, 53.
283. J. Yang, Z. Zeng, Z. Wang and Y. Chen, *Yingyong Huaxue*, (2001), **18**, 631.
284. B. Yang and B. Schaeffer, *Polym. Prep. (Am. Chem. Soc., Div. Polym. Chem.)*, (2001), **42**, 795.
285. J. L. Mateo, M. Calvo and P. Bosch, *J. Polym. Sci., Part A: Polym. Chem. Ed.*, (2001), **39**, 2444.
286. B. S. Shemper and L. J. Mathias, *Polym. Prep. (Am. Chem. Soc., Div. Polym. Chem.)*, (2001), **42**, 461.
287. P. Dolez, M. Marek and B. J. Love, *J. Appl. Polym. Sci.*, (2001), **82**, 546.
288. K. Studer and R. Koniger, *Eur. Coat. J.*, (2001), **1**, 26.
289. K. S. Anseth and D. J. Quick, *Macromol. Rapid Commun.*, (2001), **22**, 564.
290. A. K. Burkoth and K. S. Anseth, *Polym. Drugs Drug Delivery Syst.*, (2001), 181.
291. K. Murata and T. Anazawa, *Kawamura Rikagaku Kenkyusho Hokoku*, (2001), 11.
292. G. John and M. Morita, *Mat. Sci. Eng.*, (2000), *C13*, 91.
293. Y. Zhang, F. Huang, D. Ruan and R. Xia, *Zhongguo Kexue Jishu Daxue Xuebao*, (2000), **30**, 201.
294. R. Caputo, A. V. Sukhov, C. Umeton and R. F. Ushakov, *J. Exp. Theor. Phys.*, (2001), **92**, 28.
295. X. Zhang, M. Henning and M. L. Mannfred, *Hecheng Huxaue*, (2001), **9**, 44.
296. W. Kim, S. Ryu, K. Hong, D. Lee, K. Min, L. Park, K. Seo and I. Kang, *Polymer (Korea)*, (2000), **74**, 757.
297. V. I. Shtompel, S. V. Ryabov, A. F. Maslyuk, G. K. Bereznitskii and Yu. Yu. Kercha, *Vysokomol. Soedin., Ser. A, Ser. B*, (2000), **42**, 1221.
298. K. Suyama, Y. Miyamoto, t. Matsuoka, S. Wada and M. Tsunooka, *Polymer Adv. Technol.*, (2000), 11, 589.
299. C. Iojoiu, M. Ropot, M. J. M. Abadie, V. Harabagiu, M. Pinteala and B. C. Simionescu, *Isr.–Technol. Advantages.*, (200), **2**, 6.
300. K. Inomata, S. Kawasaki, A. Kameyama and T. Nishikubo, *J. Polym. Sci., Part A: Polym. Chem. Ed.*, (2001), **39**, 530.
301. Z. Gan, X. Tan and H. Xie, *Huazhong Ligong Daxue Xuebao*, (2000), **28**, 109.
302. H. Xie, M. Xiao and J. Guo, *Macromol. Mater. Eng.*, (2000), **283**, 48.
303. H. Takase, A. Natansohn and P. Rochon, *J. Photopolym. Sci. Technol.*, (2001), **14**,

189.
304. R. R. Dsa, S. Samal, S. Choi and K. E. Geckeler, *Macromol. Rapid Commun.*, (2001), **22**, 850.
305. O. P. Batog, *Kompoz Polim. Mater.*, (1999), **21**, 41.
306. Y. Matsuura, T. Tamai, K. Matsukawa, H. Inoue, T. Hamamoto, H. Toyota and K. Sato, *J. Photopolym. Sci. Technol.*, (2001), **14**, 175.
307. D. Karalekas, D. Rapti, C. Kontomitros, D. Zacharopoulos and E. E. Gdoutos, *Int. SAMPE Tech. Conf.*, (2000), **32**, 68.
308. Y. Duang, S. Wang, D. Li, X. Fan and B. Lu, *Cailiao Kexue Yu Gongcheng*, (2001), **19**, 39.
309. K.A. Berchtold, B. Hacioglu, L. Lovell, J. nei and C. N. Bowman, *Macromolecules*, (2001), **34**, 5103.
310. K. A. Berchtold, B. Hacioglu, L. Lovell, J. Nei and C. N. Bowman, *Polymer*, (2001), **42**, 4925.
311. Yu. G. Medvedevskikh, E. A. Zaglad'ko, A. N. Bratus and A. A. Turovskii , *Dopov. Nats. Akad. Nauk Ukr.*, (2000), **10**, 148.
312. J. W. Stansbury and S. H. Dickens, *Polymer*, (2001), **42**, 6363.
313. O. Yu, S. Nauman, J. P. Santerre and S. Zhu, *J. Mater. Sci.*, (2001), **36**, 3599.
314. L. Lecamp, P. Lebaudy, B. Youssef and C. Bunel, *Polymer*, (2001), **42**, 8541.
315. N. Havard, E. Dargent, P. Lebaudy, L. Lecamp and J. Grenet, *J. Therm. Anal. Calorim.*, (2000), **61**, 701.
316. E. Andrzejewska, M. B. Bagacki and M. Andrzejewski, *Polimery*, (2001), **46**, 549.
317. U. Quittmann, L. Lecamp, W. El Khatib, B. Youssef and C. Bunel, *Macromol. Chem. Phys.*, (2001), **202**, 628.
318. M. H. Sherif, E. Klemm, W. Gunther and D. Stadermann, *Polym. Polym. Compos.*, (2001), **9**, 327.
319. Q. Wan, S. R. Schricker and B. M. Culbertson, *J. Macromol. Sci., Pure Appl. Chem.*, (2000), *A37*, 1317.
320. B. Pettersson, *Farbe Lacke*, (2001), **107** , 123.
321. L. Yang, Z. Li, G. Cheng and K. Yao, *Gongneng Cailiao*, (2000), **31**, 196.
322. H. Wei, H. Kou, W. Shi, H. Yuan and Y. Chen, *Polymer*, (2001), **42**, 6741.
323. M. Okazaki, Y. Shibasaki and M. Ueda, *Polym. Prep. (Am. Chem. Soc., Div. Polym. Chem.)*, (2001), **42**, 490.
324. S. Lin, B. Ho, L. Lai, S. Hong and K. Cheng, *Annu. Techn. Conf.-Soc. Plast. Eng.*, (2000), *58th*, 3778.
325. A. Ya. Vainer, K. M. Dyumaev, Kh. B. Kaminarskaya and E. Yau. Kornienko, *Dokl. Akad. Nauk*, (2000), **375**, 60.
326. T. Nakano, H. Iwasa, N. Miyagawa, s. Takahara and T. Yamaoka, *J. Photopolym. Sci. Technol.*, (2000), **13**, 715.
327. L. Liu, Z. Zhu, Z. Wang and L. Sun, *Gaofenzi Cailiao Kexue Yu Gongcheng*, (2000), **16**, 54.
328. Z. Zhu, J. Yin, F. Cao, X. Shang and Q. Lu, *Adv. Mater. (Weinheim, Germany)*, (2000), **12**, 1055.
329. D. Lin, W. Shi, K. Nie and Yu. Zhang, *J. Appl. Polym. Sci.*, (2001), **82**, 1630.
330. D. Lin, H. Kou, W. Shi, H. Yuan and Y. Chen, *J. Appl. Polym. Sci.*, (2001), **82**, 1637.
331. H. Morita and H. Tonooka, *J. Photopolym. Sci. Technol.*, (2001), **14**, 203.
332. L. Liang, P. C. Rieke, G. E. Fryxell, J. Liu, M. H. Engehard and K. L. Alford, *J. Phys. Chem. B.*, (2000), **104**, 11667.
333. C. Zhu, Q. Lu and J. Yin, *Chem. J. Internet*, (2001), **3**, No page given.
334. J. Gao, J. Lei, Z. Su, B. Zhang and J. Wang, *Polymer J*, (2001), **3**, 147.

335. P. Feiertag, T. Kavc, U. Meyer, I. Gsoels, W. Kern, I. Rom and F. Hofer, *Synth. Met.*, (2001), **121**, 1371.
336. J. Li, H. Zeng and Q. Yu, *Chem. J. Internet*, (2001), **3**, No page given.
337. N. Luo, A. Metters, K. Anseth and N. C. Bowman, *Polym. Prep. (Am. Chem. Soc., Div. Polym. Chem.)*, 42, 711.
338. H. Ma, R. H. Davis and N. C. Bowman, *Polymer*, (2001), **42**, 8333.
339. J. Frahn, G. Malsch and H. H. Schwartz, *Macromol. Symp.*, (2001), **164**, 269.
340. J. Deng, W. Yang and B. Ranby, *J. Appl. Polym. Sci.*, (2001), **80**, 1426.
341. Y. Shirai, K. Shirai and N. Tsubokawa, *J. Polym. Sci., Part A: Polym. Chem. Ed.*, (2001), **39**, 2157.
342. S. Y. Kim, Y. J. Kim, O. H. Kwon, Y. C. Nho and C. N. Choi, *J. Korean Fibre Soc.*, (2000), **37**, 582.
343. K. Yamada and M. Hirata, *Polym. Prep. (Am. Chem. Soc., Div. Polym. Chem.).*, (2001), **42**, 757.
344. C. Xia and R. C. Advincula, *Chem. Mater.*, (2001), **13**, 1682.
345. S. Takahashi and T. Kasemura, *Koen Yoshishu-Nippon Setchaku Gakkai Nenji Taikai*, (2000), *38th*, 85.
346. D. J. Dyer, T. Zhao and J. B. Green, *Polym. Prep. (Am. Chem. Soc., Div. Polym. Chem.).*, (2001), **42**, 178.
347. A. Laschewsky, O. Ouari, C. Mangeney and L. Jullien, *Macromol. Symp.*, (2001), **164**, 323.
348. E. M. Cheadle, D. N. Batchelder, S. D. Evans, H. L. Zhang, H. Fukushima, S. Miyashita, M. Graupe, A. Puck, O. E. Shmakova, R. Colorado and T. R. Lee, *Langmuir*, (2001), **17**, 6616.
349. K. Ishizu and S. Shiratori, *J. Mater. Sci. Lett.*, (2000), **19**, 2105.
350. Matsuda, N. Koura, H. Kamusewitz and D. Paul, *J. Membrane Sci.*, (2000), **179**, 69.
352. M. Leclerc, *Can. Chem. News*, (2000), **52**, 22.
353. J. Cornil and J. Luc-Bredas, *Chim. Nouv.*, (2000), **18**, 3183.
354. D. Liaw, C. C. Huang and E. T. Kang, *Curr. Trends Polym. Sci.*, (1999), **4**, 117.
355. R. Zana, *Amphiphilic Block Copolymers*, (2000), 221.
356. M. Schwoerer, E. Dormann, T. Vogtmann and A. Feldner, *Macromol. Syst: Microsc. Interact. Macroscop. Prop., Final Rep. Collab. Res. Centre 213, 'Topospezifische Chem. Toposel. Spektrosk. Macromolekulsyst: Mikrosk. Wechselwirkung Makrosk. Funkt.'*, (2000), 122.
357. W. E. Jones, L. Hermans and B. Jiang, *Mol. Supramolecular Photochem.*, (1999), 1-29.
358. A. J. Epstein, *Springer Ser. Mater. Sci.*, (2001), **41**, 3.
359. E. Frankevich, H. Ishii, Y. Hamanaka, T. Yokoyama, A. Fujii, S. Li, K. Yoshino, A. Nakamura and K. Seki, *Synth. Met.*, (2001), **119**, 495.
360. M. Kaneko, *Prog. Polym. Sci.*, (2001), **26**, 1101.
361. J. Sung and S. Lee, *J. Chem. Phys.*, (2001), **115**, 9050.
362. P. K. Fearon, D. J. Whiteman, N. C. Billingham and S. W. Bigger, *J. Appl. Polym. Sci.*, (2001), **79**, 1986; S. W. Bigger, P. K. Fearon, D. J. Whiteman, T. L. Phease and N. C. Billingham, *Polym. Prep. (Am. Chem. Soc., Div. Polym. Chem.)*, (2001), **42**, 375.
363. P. Eriksson, T. Reitberger, G. Ahlblad and B. Stenberg, *Polym. Deg. Stab.*, (2001), **73**, 177.
364. L. Matisova-Rychla and J. Rychly, *Polym. Deg. Stab.*, (2001), **73**, 393.
365. G. Ahlblad, P. Gijsman, B. Terselius, A. Jansson and K. Moller, *Polym. Deg. Stabil.*, (2001), **73**, 15.
366. I. Blackely and G. A. George, *Macromolecules*, (2001), ACS, ASAP.

367. L. Rychla and J. Rychly, *Polym. Anal. Deg.*, (Eds. A. Jimenez and G. E. Zaikov), Nova Science Publishers, Inc., Y., (2000), 123-134.
368. P. K. Fearon, S. W. Bigger and N. C. Billingham, *Polym. Prep. (Am. Chem. Soc., Div. Polym. Chem.)*, (2001), **42**, 352.
369. D. Kockott, *Polym. Test.*, (2001), **20**, 725.
370. L. Matisova-Rychla and J. Rychly, *Polym. Prep. (Am. Chem. Soc., Div. Polym. Chem.)*, (2001), **42**, 377.
371. J. Wang and Q. Lei, *Harbin Ligong Daxue Xuebao*, (2001), **6**, 1.
372. L. Matisova-Rychla, J. Rychly and G. A. George, *Polym. Deg. Stabil.*, (2002), **75**, 385.
373. N. S. Allen, G. Rivalle, M. Edge, T. Corrales and F. Catalina, *Polym. Deg. Stabil.*, (2001), **75**, 237.
374. M. Strlic, J. Kolar, B. Pihlar, J. Rychly and L. Matisova-Rychla, *Polym. Deg. Stabil.*, (2001), **72**, 157.
375. C. A. Lukey, *Prog. Org. Coat.*, (2001), **41**, 129.
376. L. Woo, S. Y. Ding and C. L. Sanford, *Polym. Prep. (Am. Chem. Soc., Div. Polym. Chem.)*, (2001), **42**, 866.
377. N. Honma, C. Sakai and Y. Maeshima, *Shizuoka-ken Hamamatsu Kogyo Gijitsu Senta Kenkyu Hokoku*, (2000), **10**, 41.
378. F. E. Buttrey, M. Mallace and H. N. McMurray, *Eur. Fed. Corros. Publ.*, (2000), **28**, 139.
379. H. W. Rhee, C. Kim and M. S. Gong, *Bull. Korean Chem. Soc.*, (2001), **22**, 727.
380. C. W. lee, S. W. Joo, O. Kim, J. Ko and M. S. Gong, *Dyes Pigments*, (2002), **52**, 37.
381. M. Ohkoshi and H. Saijo , *Makuzai Gakkaishi*, (2001), **47**, 152.
382. V. Griseri, L. A. Fothergill, C. Laurent and G. Teyssedre, *J. Phys. D: Appl. Phys.*, (2001), **34**, 2534.
383. O. L. Kramer, E. M. Battalova, A. I. Voloshin and Yu. A. Prochukhan, *Vysokomol. Soedin., Ser. A, Ser. B*, (2001), **43**, 530.
384. A. S. Smolyanskii, I. P. Shelukhov, A. P. Podsoblyaev and V. I. Tupikov, *Nucl. Instrum. Methods Phys. Res., Sect B.*, (2001), **173**, 326.
385. G. S. Kanner, Z. V. Vardeny, G. Lanzani and L. X. Zheng, *Synth. Met.*, (2001), **116**, 71.
386. A. Shukla, H. Ghosh and S. Mazumdar, *Synth. Met.*, (2001), **116**, 87.
387. A. B. Pavlinov and I. P. Raevskii, *Tech. Phys. Lett.*, (2000), **26**, 1096.
388. W. Ge, J. W. Y. Lam and B. Z. Tang, *Appl. Phys. Lett.*, (2001), **78**, 1652.
389. B. Z. Tang, J. W. Y. Lam, X. Kong, F. Salhi, K. K.. Cheuk, H. Kwok, Y. Huang and W. Ge, *Proc. SPIE-Int.Soc. Opt. Eng.*, (2000), **4107**, 24.
390. P. P. S. Lee, Y. P. Dong, K. K. L. Cheuk, F. S. W. Chau and B. Z. Tang, *Polym. Prep. (Am. Chem. Soc., Div. Polym. Chem.)*, (2001), **42**, 502.
391. S. M. A. Karim, R. Nomura, H. Kajii, R. Hidayat, K. Yoshino and T. Masuda, *J. Polym. Sci., Part A: Polym. Chem. Ed.*, (2000), **38**, 4717.
392. P. P. S. Lee, K. K. L. Cheuk, Y. P. Dong, F. S. W. Chau and B. Z. Tang, *Polym. Prep. (Am. Chem. Soc., Div. Polym. Chem.)*, (2001), **42**, 600.
393. U. G. Ulrich and J. Peltonen, *Langmuir*, (2001), **17**, 1518.
394. M. Anaka, H. Noda, R. Motokawa and T. Nakahira, *Polymer*, (2001), **42**, 09887.
395. O. Vorobyova and M.A. Winnik, *J. Polym. Sci., Part A: Polym. Phys. Ed.*, (2001), **39**, 2302.
396. J. Morgado, E. Moons, R. Friend and F. Cacialli, *Adv. Mater.*, (2001), **13**, 810.
397. A. R. Buckley, M. D. Rahn, J. Hill, J. Cabanillas-Gonzales, A. M. Fox and D. D. C. Bradley, *Chem. Phys. Lett.*, (2001), **339**, 331.

398. L. M. Carvalho, L. F. Santos, F. E. G. Guimaraes, D. Goncalves, A. S. Gomes and R. M. Faria, *Synth. Met.*, (2001), **119**, 361.
399. F. Ishizaki, s. Machida and K. Horie, *J. Photopolym. Sci. Technol.*, (2001), **14**, 215.
400. H. Kaczmarek, A. Szalla and A. Kaminska, *Polymer*, (2001), **42**, 6057.
401. E. G. Granados, J. Gonzalez-Benitos, J. Baselga, D. Dibbern-Brunelli, T. D. Z. Atvars, I. Esteban and I. F. Pierola, *J. Appl. Polym. Sci.*, (2001), **80**, 949.
402. J. C. Cabanelas, J. Gonzales-Benito, J. Bravo and J. Baselga, *Makromol. Rapid Commun.*, (2001), **22**, 694.
403. A. Yu. Bobrovski, N. I. Boiko and V. P. Shibaev, *Vysokomol. Soedin., Ser. A, Ser. B.*, (2000), **42**, 1495.
404. S. Masuo, H. Yoshikawa, T. Asahi, H. Masuhara, T. Sato, D. L. Jiang and T. Aida, *J. Phys. Chem. B*, (2001), **105**, 2885.
405. H. Tokuhisa, E. Koyama, Y. Nagawa and K. Hiratani, *Chem. Commun.*, (2001), **7**, 595.
406. S. Kimata and T. Aida, *Tetrahedron Lett.*, (2001), **2**, 4187.
407. D. L. Jiang, T. Sato and T. Aida, *Chin. J. Polym. Sci.*, (2001), **19**, 161.
408. L. Z. Gong, Q. Hu and L. Pu , *J. Org. Chem.*, (2001), **66**, 2358.
409. P. S. Corbin, M. P. Webb, J. E. McAlvin and C. L. Fraser, *Biomacromolecules*, (2001), **2**, 223.
410. H. Huang, J. Z. Zhang and W. F. Shi, *J. Appl. Polym. Sci.*, (2001), **80**, 499.
411. M. Maus, R. De, M. Lor, T. Weil, S. Mitra, U. M. Wiesler, A. Herrmann, J. Hofkens, T. Vosch, K. Muellen and F. C. DeSchryver , *J. Am. Chem. Soc.*, (2001), **123**, 7668.
412. G. De Belder, G. Schweitzer, S. Jordens, M. Lor, S. Mitra, J. Hofkens, S. De Feyter, M. van de Auweraer, A. Herrmann, T. Weil, . Mullen and F. C. DeSchryver, *ChemPhysChem.*, (2001), **2**, 49.
413. Y. Takaguichi, T. Tajima, K. Ohta, J. Motoyoshiya and H. Aoyama, *Chem. Lett.*, (2000), **12**, 1388.
414. S. Wang and R. Advincula, *Polym. Mater. Sci. Eng.*, (2001), **85**, 611.
415. S. Ghosh and A. K. Banthia, *Tetrahedron Lett.*, (2001), **42**, 501.
416. J. L. Segura, R. Gomez, N. Martin and D. M. Nazario, *Org. Lett.*, (2001), **3**, 2645.
417. L. Zhang, F. Huo, L. Wu, Z. Wang and X. Zhang, *Macromol. Chem. Phys.*, (2001), **202**, 1618.
418. V. Balzani, P. Ceroni, S. Gestermann, M. Gorka, C. Kauffmann, M. Maestri and F. Vogtle, *ChemPhysChem.*, (2000), **1**, 224.
419. A. Zhu, P. Bharathi, H. G. Dricakmer and J. S. Moore, *J. Polym. Sci., Part A: Polym. Chem. Ed.*, (2001), **39**, 2859.
420. C. Gao and D. Yan, *Chin. Sci. Bull.*, (2000) **45**, 1760.
421. C. Gao, D. Yan, B. Zhang and W. Tang, *Polym. Prep. (Am. Chem. Soc., Div. Polym. Chem.)*., (2001), **42**, 538.
422. S. Watanabe, M. Sato, S. Sakamoto, K. Yamaguichi and M. Iwamura, *J. Am. Chem. Soc.*, (2000), **122**, 12588.
423. Z. Lei, Q. Han, Z. Yang, R. Wang and Y. Wang, *Xibei Shifan Daxue Xuebao, Ziran Kexueban*, (2001), **37**, 44.
424. T. Mizutani, M. Ikegami, R. Nagahata and T. Arai, *Chem. Lett.*, (2001), **10**, 1014.
425. T. Weil, U. M. Wiesler, A. Herrmann, R. Bauer, J. Hofkens, F. C. DeSchryver and K. Muellen, *J. Am. Chem. Soc.*, (2001), **123**, 8101.
426. S. Pascual, A. Narrainen and D. M. Haddleton, *Polym. Prep. (Am. Chem. Soc., Div. Polym. Chem.)*., (2001), **42**, 434.
427. M. Eigner, H. Komber and B. Voit, *Macromol. Chem. Phys.*, (2001), **202**, 245.
428. S. Y. Cho, Y. Chang, J. Kim, S. C. Lee and C. Kim, *Macromol. Chem. Phys.*, (2001),

202, 263.
429. Y. Chang and C. Kim, *J. Polym. Sci., Part A: Polym. Chem.*, (2001) **39**, 402.
430. C. L. Larson, *Appl. Spectrosc.*, (2001), **55**, 679.
431. D. L. Richter-Egger, J. C. Landry, A. Tesfai and S. A. Tucker, *J. Phys. Chem. A.*, (2001), **105**, 6826.
432. S. Hecht, N. Vladimirov and J. M. J. Frechet, *J. Am. Chem. Soc.*, (2001), **123**, 18.
433. H. A. Klock, J. R. Hernandez, S. Becker and K. Mullen, *J. Polym. Sci., Part A: Polym. Chem. Ed.*, (2001), **39**, 1572.
434. C. Mengel, W. H. Meyer and G. Wegner, *Macromol. Chem. Phys.*, (2001), **202**, 1138.
435. J. D. He, J. Wang, S. D. Li and M. K. Chen, *J. Appl. Polym. Sci.*, (2001), **81**, 1286.
436. D. A. Sykmanov, Yu. F. Biryulin, L. V. Vinogradova and V. N. Zgonnik, *Semiconductors*, (2001), **35**, 643.
437. G. J. Capitosti, S. J. Cramer, C. S. Rajesh and D. A. Modarelli, *Org. Lett.*, (2001), **3**, 1645.
438. G. C. Dol, K. Tsuda, J. Weener, M. J. Bartels, T. Asavei, T. Gensch, J. Hofkens, L. Latterini, A. P. H. J. Schenning, B. W. Meijer and F. C. DeSchryver, *Angew. Chem., Int. Ed.*, (2001), **40**, 1710.
439. S. Y. Ju, K. D. Ahn, D. K. Han, D. H. Suh and J. M. Kim, *J. Photosci.*, (2000), **7**, 131.
440. A. Ikeda, A. Kameyama, T. Nishikubo and T. Nagai, *Macromolecules*, (2001), **34**, 2728.
441. J. M. Kim, H. Y. Shin, K. H. Park, T. H. Kim, S. Y. Ju, D. K. Han and K. D. Ahn, *Macromolecules*, (2001), **34**, 133.
442. A. Tork, F. Boudreault, M. Roberge, A. M. Ritcey, R. A. Lessard and T. V. Galstian, *Appl. Opt.*, (2001), **40**, 1180.
443. Y. Tan, S. A. Ahmed, H. Durr, V. Huch and A. Abdel-Wahab, *Chem. Commun.*, (2001), **14**, 1246.
444. P. J. Coelho, L. M. Carvalho, J. C. Silva, A. M. F. Oliveira-Campos, A. Samat and R. Guglielmetti, *Helv. Chimica Acta*, (2001), **84**, 117.
445. S. R. Keum, Y. K. Choi, M. J. Lee and S. H. Kim, *Dyes Pigments*, (2001), *50*, 171.
446. Z. Sun, R. S. Hosmane, M. Tadros, *J. Heterocycl. Chem. Ed.*, (2000), **37**, 1439.
447. K. C. Feng and J. Griffiths, *Adv. Col. Sci. Technol.*, (2001), **4**, 12.
448. H. Hattori and T. Uryu, *Liq. Cryst.*, (2001), **28**, 25.
449. F. Poncin-Epaillard, J. Beunet, A. Bulou and J. F. Bardeau, *J. Polym. Sci., Part A: Polymer Chem. Ed.*, (2001), **39**, 3052.
450. N. Sekar, *Colourage*, (2001), **48**, 49.
451. H. Yang, Y. He, X. Wang and Q. Zhou, *Gaofenzi Xuebao*, (2000), **6**, 808.
452. G. F. Wang, F. Gan, J. Wang, L. Yang, Z. Xu and Y. Heqing, *Gangxue Xuebao*, (2001), **21**, 495.
453. I. Porcar, P. Sergot and C. Tribet, *ACS Symp. Ser.*, (2001), **780**, 82.
454. J. Kronek, J. Luston, F. Bohme and H. Komber, *Macromol. Symp.*, (2001), **170**, 301.
455. W. J. Joo, C. H. Oh, S. H. Song, P. S. Kim and Y. K. Han, *J. Phys. Chem. B.*, (2001), **105**, 8322.
456. L. Z. Zhang, Z. G. Cai, V. Ninulescu, K. Jin and Z. X. Liang, *Chin. J. Polym. Sci.*, (2001), **19**, 255.
457. V. D. Athawale and S. M. Deshpande, *J. Polymer Mater.*, (2001), **18**, 69.
458. V. M. Churikov, *JETP Lett.*, (2000), **72**, 415.
459. K. Matczyszyn, K. Palewska and J. Sworakowski, *Mol. Phys. Rep.*, (1999), **25**, 86.
460. Y. He, X. Wang and Q. Zhou, *Polym. Prep. (Am. Chem. Soc., Div. Polym. Chem.)*, (2001), **42**, 274.
461. E. B. Barmatov, A. V. Medvedev, S. A. Ivanov, M. V. Barmatova and V. P. Shibaev,

Vysokomol. Soedin., Ser. A, Ser. B., (2001), **43**, 468.
462. K. Sugiyama and K. Sono, *J. Appl. Polym. Sci.*, (2001), **81**, 3056.
463. B. J. Kim, S. Y. Park and D. H. Choi, *Bull. Korean Chem. Soc.*, (2001), **22**, 271.
464. A. Yu. Bobrovsky, N. I. Boiko, V. P. Shibaev and J. Springer, *Liq. Cryst.*, (2001), **28**, 919.
465. T. Buruiana, E. C. Buruiana and I. Diaconu, *Bulg. J. Phys.*, (2000), **27**, 49.
466. X. Wang, W. Shen, Q. Zhang, T. Li and X. Cheng, *Mol. Cryst. Liq. Cryst. Sci. Technol., Sect A*, (2000), **350**, 225.
467. M. Yamamoto, H. Ohkita and S. Ito, *Chin. J. Polym. Sci.*, (2001), **19**, 129.
468. G. Pan, M. G. Fan, P. Fan, H. Z. Wang and Z. C. Wei, *Chem. Commun.*, (2001), **18**, 1744.
469. H. Tian and H. Y. Tu, *Adv. Mater.*, (2000), **12**, 1597.
470. H. Tang, M. Fujiki, M. Motonaga and K. Torimitsu, *Polym. Prep. (Am. Chem. Soc., Div. Polym. Chem.)*, (2001), **42**, 440.
471. F. Dietz and N. Tyutyulkov, *Chem. Phys.*, (2001), **265**, 165.
472. S. J. Lee and H. Lee, *Synth. Met.*, (2001), **121**, 1691.
473. T. Kondoh, C. Itoh and K. Tanimura, *J. Phys. Soc. Jpn.*, (2001), **70**, 1817.
474. A. K. Singh, *J. Indian Chem. Soc.*, (2001), **78**, 53.
475. A. J. Williams and V. K. Gupta, *J. Polym. Sci., Part B: Polym. Phys. Ed.*, (2001), **39**, 2759.
476. Z. Sekkat, H. Ishitobi, D. Yasumatsu and S. Kawata, *Proc. SPIE-Int. Soc. Opt. Eng.*, (2000), **4106**, 133.
477. J. Liu and Q. Zhang, *Gaofenzi Cailiao Kexue Yu Gongcheng*, (2001), **17**, 15.
478. P. Uznanski, *IEEE Trans. Dielectr. Electr. Insul.*, (2001), **8**, 507.
479. I. Grabchev and V. Bojnov, *Z. Naturforsch., A: Phys. Sci.*, (2000), **55**, 833.
480. M. M. Oliveira, L. M. Carvalho, C. Moustrou, A. Samat, R. Guglielmetti, A. M. F. Oiliveira-Campos, *Mag. Res. Chem.*, (2001), **39**, 637.
481. D. R. Cairns, M. Sibulkin and G. P. Crawford, *Appl. Phys. Lett.*, (2001), **78**, 2643.
482. M. Sato, M. Notsu, S. Nakashima and Y. Uemoto, *Macromol. Rapid Commun.*, (2001), **22**, 681.
483. M. Sato and Y. Uemoto, *Macromol. Rapid Commun.*, (2000), **21**, 1220.
484. S. J. Lukacs, *J. Phys. Chem.*, (2001), **105**, 3372.
485. K. Usami, K. Sakamoto and S. Ushioda, *J. Appl. Phys.*, (2001), **89**, 5339.
486. S. W. Lee, T. Chang and M. Ree, *Macromol. Rapid Commun.*, (2001), **22**, 941.
487. S. J. Sung, H. T. Kim, J. W. Lee and J. K. Park, *Synth. Met.*, (2001), **117**, 277.
488. J. Y. Hwang, D. S. Seo, J. H. Son and D. H. Suh, *Jpn. J. Appl. Phys.*, Part 2 (2001), **40**, 761.
489. J. L. West, L. Su, M. Nishikawa, W. B. Michino0ri and Y. A. Reznikov, *Proc. SPIE-Int. Soc. Opt. Eng.*, (2000), **414**, 250.
490. Y. C. Kim, T. W. Lee, O. Ok. Park, C. Y. Kim and H. N. Cho, *Adv. Mater.*, (2001) **13**, 646.
491. Y. C. Kim, T. W. lee, H. N. Cho, J. M. Hong and C. Y. Kim, *Synth. Met.*, (2001), **121**, 1753.
492. L. Oriol, M. Pinol, J. L. Serrano, C. Martinez, R. Alcala, R. Cases and C. Sanchez, *Polymer*, (2001), **42**, 2737.,
493. X. Liang, L. Zhang, J. Wang, Z. Cai and Z. Liang, *Gaofenzi Xuebao*, (2001), **1**, 109.
494. A. Yu. Bobrovskii, N. I. Boiko and V. P. Shibaev, *Vysokomolekulyarnye Soedineniya Ser. A, Ser. B.*, (2001), **43**, 1805.
495. K. Moriya, M. Nakagawa, G. mao and C. K. Ober, *Macromol. Rapid. Commun.*, (2000), **21**, 1309.

496. B. Kong, L. Cui, P. Xie, R. Zhang, C. He and N. T. Chung, *Liqu. Cryst.*, (2000), **27**, 1683.
497. M. Ree, S. I. Kim and S. W. Lee, *Synth. Met.*, (2001), **117**, 273.
498. Y. Tian, E. Akiyama, Y. Nagase, A. Kanazawa, O. Tsutsumi and T. Ikeda, *Trans. Mater. Res. Soc. Jpn.*, (2000), **25**, 1091
499. O. Tsutsumi, Y. Wu, J. Mamiya and T. Ikeda, *Proc. SPIE-Int. Soc. Opt. Eng.*, (2000), **4107**, 1.
500. E. B. Barmatov, A. V. Medvedev, S. A. Ivanov, M. V. Barmatova and V. P. Shibaev, *Vysokomol. Soedinya., Ser. A, Ser. B.*, (2001), **43**, 261.
501. C. H. Ming and C. S. Hsu, *J. Polym. Res.*, (2001), **8**, 159.
502. C. Sekine, K. Iwakura, M. Minai and K. Fujisawa, *Liquid Crystals*, (2001), **28**, 1505.
503. S. Shimamaoto, Y. Uraki and Y. Yasumitsu, *Cellulose*, (2000), **7**, 347.
504. T. Hattori, N. Hanai, H. Inouye and H. Nakatsuka, *Jpn. J. Appl. Phys.*, (2001), **40**, 4928.
505. D. Wohrle, A. Hartwig, G. Schnurpfeil, A. Harder and H. Schroeder, *Polym. Adv. Technol.*, (2000), **11**, 739.
506. V. P. Shibaev, Yu. A. Bobrovsky and N. I. Boiko, *Macromol. Symp.*, (2001), **174**, 319.
507. M. C. Artal, M. B. Ros, L. J. Serrano, M. R. del la Fuente and M. A. Perez-Jubindo, *Chem. Mater.*, (2001), **13**, 2056.
508. I. Porcar, P. Perrin and C. Tribet, *Langmuir*, (2001), **17**, 6905.
509. E. Grosicka and M. Mucha, *Proc. SPIE-Int. Soc. Opt. Eng.*, (2000), **414**, 4147.
510. R. Peetz, O. Narwark, O. Herzog, S. Brocke and E. ThronCsanyi, *Synth. Met.*, (2001), **119**, 539.
511. Z. Peng, *Polym. News*, (2000), **25**, 185.
512. G. Lanzani, S. De Silvestri, G. Cerullo, S. Stagira, M. Nisoli, W. Graupner, G. Leising, U. Scherf and K. Mullen, *Semicond. Polym.*, (2000), 235.
513. T. Mark, M. Inbasekaran, *Adv. Mater.*, (2000), **12**, 1737.
514. D. Beljonne, Z. Shuai, G. Pourtois and J. L. Bredas, *J. Phys. Chem.*, (2001), **105**, 3899.
515. M. W. McCutcheon, J. F. Young, A. G. Pattantyus-Abraham and M. O. Wolf, *J. Appl. Phys.*, (2001), **89**, 4376.
516. D. Moses, A. Dogariu and A. J. Heeger, *Synth. Met.*, (2001), **116**, 19.
517. K. Shinohara, S. Yamaguichi and T. Wazawa, *Polymer*, (2001), **42**, 7915.
518. M. I. Sluch, A. Godt, U. H. F. Bunz and M. A. Berg, *J. Am. Chem. Soc.*, (2001), **123**, 6447.
519. J. W. van der Horst, P. A. Bobbert, P. H. L. de Jong, M. A. J. Michels, L. D. A. Siebbeles, J. M. Warman, G. H. Gelinck and G. Brocks, *Chem. Phys. Lett.*, (2001), **334**, 303.
520. W. Barford and R. J. Bursill, *Synth. Met.*, (2001), **119**, 251.
521. K. Chen, X. Sun and T. F. George, *Synth. Met.*, (2001), **119**, 275.
522. E. J. W. List, C. H. Kim, W. Graupner, G. Leising and J. Shinar, *Synth. Met.*, (2001), **119**, 511.
523. C. Belton, D. F. O'Brien, W. J. Blau, A. J. Cadby, P. A. Lane, D. D. C. Bradley, H. J. Byrne, R. Stckmann and H. H. Horhold, *Appl. Phys. Lett.*, (2001), **78**, 1059.
524. C. Belton, D. O'Brien, A. Cadby, P. lane, H. J. Byrne, R. Stockmann, H. H. Horhold, D. D. C. Bradley and J. W. Blau, *Synth. Met.*, (2001), **119**, 567.
525. R. Jakubiak, M. Yan, W. C. Wan, B. R. Hsieh and L. J. Rothberg, *Isr. J. Chem.*, (2000), **40**, 153.
526. E. J. W. List, C. Creely, G. Leising, N. Schulte, A. D. Schluter, U. Sherf, K. Mullen and W. Graupner, *Synth. Met.*, (2001), **119**, 659.

527. G. Lanzani, G. Cerullo, S. Stagira and S. De Silvestri, *J. Photochem. Photobiol: Part A: Chem. Ed.*, (2001), **144**, 13.
528. T. Q. Nguyen, R. Y. Yee and B. J. Schwartz, *J. Photochem. Phot biol: Part A: Chem. Ed.*, (2001), **144**, 21.
529. A. Ruseckas, E. B. Namdas, M. Theander, M. Svensson, A. Yartsev, D. Zigmantas, M. R. Andersson, O. Inganas and V. Sundstrom, *J. Photochem. Photobiol: Part A: Chem. Ed.*, (2001), **144**, 3.
530. P. Wang, C. J. Collinson and L. J. Rotherberg, *J. Polym. Sci., Part A: Polym. Chem. Ed.*, (2001), 144, 63.
531. T. Huser and M. Yan, *J. Photochem. Photobiol: Part A: Chem. Ed.*, (2001), **144**, 43.
532. A. B. Dalton, J. N. Coleman, M. in het Panhuis, B. McCarthy, A. Drury, W. J. Blau, B. Paci, J. M. Nunzi and H. J. Byrne, *J. Photochem. Photobiol: Part A: Chem. Ed.*, (2001), **144**, 31.
533. S.A. Jenekhe, L. Lu and M. M. Alam, *Macromolecules*, (2001), **34**, 7315.
534. M. M. Alam and S. A. Jenekhe, *J. Phys. Chem.*, (2001), **105**, 2479.
535. L. Lu and S. A. Jenhke, *Macromolecules*, (2001), **34**, 6249.
536. T. Ahn and H. K. Shim, *Synth. Met.*, (2001), **121**, 1663.
537. O. K. Kim, H. Y. Woo, K. S. Lee, J. K. Kim, D. Y. Kim, H. K. Shim and C. Y. Kim, *Synth. Met.*, (2001), **121**, 1607.
538. J. M. Xu, S. C. Ng and O. H. S. Chan, *J. Mat. Res.*, (2001), **16**, 1235.
539. D. H. Hwang, J. H. Chang, H. K. Shim and T. Zyung, *Synth. Met.*, (2001), **119**, 393.
540. H. Detert and E. Sugioni, *Synth. Met.*, (2001), **122**, 15.
541. J. W. Park, J. H. Lee, S. I. Ko, T. W. Kim, T. Uemura and Y. Chujo, *Synth. Met.*, (2001), **121**, 1689.
542. H. Yanagi, T. Morikawa, S. Hotta and K. Yase, *Adv. Mater.*, (2001), **13**, 313.
543. I. Osaka, H. Goto, K. Itoh and K. Akagi, *Synth. Met.*, (2001) **119**, 541.
544. D. Wang, J. Wang, D. Moses, G. C. Bazan, A. J. Heeger, J. H. Park and Y. W. Park, *Synth. Met.*, (2001), **119**, 587.
545. H. Q. Wang, Y. Q. Zhang, F. D. Zeng, G. X. Wang and D. S. Shan, *Gaofenzi Cailiao Kexue Yu Gongcheng*, (2000), **16**, 130.
546. H. Q. Wang, Y. Q. Zhang, F. D. Zeng, G. X. Wang and D. S. Shan, *Gongneng Gaofenzi Xuebao*, (2000), **13**, 393.
547. J. Zheng, G. He, C. Yang, L. Huang and Y. Li, *J. Appl. Polym. Sci.*, (2001), **80**, 1299.
548. W. Y. Huang, W. Gao, T. K. Kwei and Y. Okamoto, *Macromolecules*, (2001), **34**, 1570.
549. E. E. Guerel, Y. Pang and F. E. Karasz, *Polym. Prep. (Am. Chem. Soc., Div. Polym. Chem.)*, (2001), **42**, 185.
550. C. Dridi, N. Blel, A. Chaieb, M. Majdoub, M. S. Roudesli, J. Davenas, H. B. Ouada and H. Maaref, *Eur. Polym. J.*, (2001), **37**, 683.
551. R. E. Martin, F. Geneste, R. Riehn, B. S. Chuah, F. Cacialli, A. B. Holmes and R. H. Friend, *Synth. Met.*, (2001), **119**, 43.
552. H. H. Horhold, H. Tillmann, C. Bader, R. Stockmann, J. Nowotny, E. Klemm, W. Holzer and A. Penzkofer, *Synth. Met.*, (2001), **119**, 199.
553. J. Morgado, F. Cacialli, R. H. Friend, B. S. Chuah, H. Rost and A. B. Holmes, *Macromolecules*, (2001), **34**, 3094.
554. J. D. White, J. H. Hsu, W. S. Fann, S. C. Yang, G. Y. Pern and S. A. Chen, *Chem. Phys. Lett.*, (2001), **338**, 263.
555. H. Hang, D. Liu and X. Wang, *Gaofenzi Xuebao*, (2001), **4**, 433.
556. J. H. Park, C. h. Lee, K. Akagi, H. Shirakawa and Y. W. Park, *Synth. Met.*, (2001), **119**, 633.

557. H. Peng, Z. Xie, J. Luo, L. Cheng, K. Xu, D. Jia, H. Kwok and B. Z. Tang, *Polym. Prep. (Am. Chem. Soc., Div. Polym. Chem.)*, (2001), **42**, 230.
558. L. O. Palsson, C. Wang, A. P. Monkman, M. R. Bryce, G. Rumbles and I. D. W. Samuels, *Synth. Met.*, (2001), **119**, 627.
559. D. Marsitsky, J. C. Scott, J. P. Cheng, V. Y. Lee, R. D. Miller, S. Setayesh and K. Mullen, *Adv. Mater.*, (2001), **13**, 1096.
560. M. Theander, D. M. Johansson, A. Ruseckas, D. Zigmantas, M. R. Andersson, V. Sundstrom and O. Inganas, *Synth. Met.*, (2001), **119**, 615.
561. F. Uckert, Y. H. Tak, K. Mullen and H. Bassler, *Adv. Mater.*, (2000), **12**, 905.
562. A. Charas, J. Morgado, J. M. G. Martinho, L. Alcacer and F. Cacialli, *Chem. Commun.*, (2001), **13**, 1216.
563. C. Xia and R. C. Advincula, *Macromolecules*, (2001), **34**, 5854.
564. J. B. Kim, Y. K. Han, J. M. Hong, Y. C. Kim, S. J. Kim and H. N. Cho, *Synth. Met.*, (2001), **119**, 105.
565. V. Deimede, J. K. Kallitsis and T. Pakula, *J. Polym. Sci., Part A: Polym. Chem.*, (2001), **39**, 3168.
566. K. Nomura, H. Morimoto, Y. Imanishi, Z. Ramhani and Y. Geerts, *J. Polym. Sci., Part A: Polym. Chem. Ed.*, (2001), **39**, 2463.
567. C. Wang, M. Kilitziraki, L. O. Palsson, M. R. Bryce, A. P. Monkman and I. D. W. Samuels, *Adv. Funct. Mater.*, (2001), **11**, 47.
568. S. H. Jin, J. E. Jung, D. K. Park, B. C. Jeon, S. K. Kwon, Y. H. Kim, D. K. Moon, S. H. Kim and Y. S. Gal, *Eur. Polym. J.*, (2001), **37**, 921.
569. Y. R. Hong, D. W. Lee, K. Kim, J. I. Jin, C. E. Lee, H. M. Lee, Y. Park, B. H. Shon and J. K. Park, *Macromol. Symp.*, (2001), **175**, 169.
570. D. W. Lee, K. Y. Kwon, J. L. Jin, Y. Park, Y. R. Kim and I. W. Hwang, *Chem. Mater.*, (2001), **13**, 565.
571. M. Zheng, L. Ding, E. E. Guerel, P. M. Lahti and F. E. Karasz, *Polym. Prep. (Am. Chem. Soc., Div. Polym. Chem.)*, (2001), **42**, 280.
572. J. Lu, K. Miyatake, A. R. Hlil and A. S. Hay, *Macromolecules*, (2001), **34**, 5860.
573. K. Yagi, C. F. Soong and M. Irie, *J. Org. Chem.*, (2001), **66**, 5419.
574. W. Holzer, A. Penzkofer, R. Stockmann, H. Meysel, H. Liebegott and H. H. Horhold, *Polymer*, (2001), **42**, 3183.
575. M. M. de Souza, G. Rumbles, D. L. Russell, I. D. W. Samuel, S. C. Moratti, A. B. Holmes and P. L. Burn, *Synth. Met.*, (2001), **119**, 635.
576. R. Mullner, L. Noirez, E. Zojer, F. Stelzer and G. Leising, *Mater. Res. Soc. Symp. Proc.*, (2000) **598**, BB3.16/1.
577. E. M. Kyllo, T. L. Gustafson, D. K. Wang, R. G. Sun and A. J. Epstein, *Synth. Met.*, (2001), **116**, 189.
578. J. D. White, J. H. Hsu, S. C. Yang, W. S. Fann, G. Y. Pern and S. A. Chen, *J. Chem. Phys.*, (2001), **114**, 3848.
579. K. Alimi, P. Molinie, M. Majdoub, J. C. Bernede, J. L. Fave, H. Bouchriha and M. Ghedira, *Eur. Polym. J.*, (2001), **37**, 781.
580. D. M. Johansson, M. Theander, G. Srdanov, G. Yu, O. Inganaes and M. R. Andersson, *Macromolecules*, (2001), **34**, 3716.
581. A. Talaie, Y. K. Lee, D. J. Choo, K. M. Kim, S. H. Park, G. Huh, J. Y. Lee, T. Taguchi, E. Maeder and N. Esmaili, *Iran. Polym. J.*, (2000), **9**, 169.
582. H. Xu, S. Guang, S. Zhang, B. Tong and B. Tang, *Gaofenzi Xuebao*, (2001), **2**, 186.
583. T. G. Bjorklund, S. H. Lim and C. J. Bardeen, *J. Phys. Chem. B*, (2001), **105**, 11970.
584. Y. Nabetani, M. Yamasaki, A. Miura and N. Tamai, *Thin Solid Films*, (2001), **393**, 329.

585. J. Wang, E. K. Miller, D. Moses and A. J. Heeger, *Synth. Met.*, (2001), **119**, 591.
586. S. Stagira, C. Gadermaier, G. Lanzani, G. Cerullo, M. Zavelani-Rossi, U. Scherf, G.Leising and S. de Silvestri, *Synth. Met.*, (2001), **119**, 609.
587. J. A. Mikroyannidis, *Macromol. Chem. Phys.*, (2001), **202**, 2367.
588. F. J. Lange, M. Leuze and M. Hanack, *J. Phys. Org. Chem.*, (2001), **14**, 474.
589. T. Matsui, T. Nagata, A. Fujii, M. Onoda, M. Teraguichi, T. Masuda and K. Yoshino, *Synth. Met.*, (2001), **119**, 599.
590. G. Brizius, N. G. Pschirer, W. Steffen, K. Stitzer, H. C. zur Loye and U. H. F. Bunz, *J. Am. Chem. Soc.*, (2000), **122**, 12435.
591. A. P. Monkman, H. D. Burrows, I. Hamblett and S. Navaratnam, *Chem. Phys. Lett.*, (2001), **340**, 467.
592. A. Talaie, Y. K. Lee, G. Huh, K. M. Kim, H. Y. Jeong, D. J. Choo, J. Y. Lee and J. Jang *Mater. Sci. Eng.*, (2001), **B85**, 199.
593. W. Stefen and U. H. F. Bunz, *Macromolecules*, (2000), **33**, 9518.
594. A. A. Sarker, E. Guerel, M. Zheng, P. M. Lahti and F. E. Karasz, *Macromolecules*, (2001), **34**, 5897.
595. Z. R. Wagner, T. K. Roenigk and F. E. Goodson, *Macromolecules*, (2001), **34**, 5740.
596. M. L. Keshtov, A. L. Rusanov, V. V. Kireev, A. A. Kirillov, S. V. Keshtova and B. M. Uzhinov, *Vysokomol. Soedin., Ser. A, Ser. B*, (2001), **43**, 957.
597. Y. Qiu, L. Duan, X. Hu, D. Zhang, M. Zheng and F. Bai, *Synth. Met.*, (2001), **123**, 39.
598. F. Bai, M. Zheng, T. Lin, J. Yang, Q. He, Y. Li and D. Zhu, *Synth. Met.*, (2001), **119**, 179.
599. H. Peng, K. Xu, J. Luo and B. Z. Tang, *Polym. Prep. (Am. Chem. Soc., Div. Polym. Chem.)*, (2001), **42**, 560.
600. S. Tanaka, K. Takeuchi, M. Asai, T. Iso and M. Ueda, *Synth. Met.*, (2001), **119**, 139.
601. H. C. Li, Y. H. Geng, S. W. Tong, Z. Y. Xie, R. Hua, G. P. Su, L. X. Wang, X. B. Xing and F. X. Wang , *Synth. Met.*, (2001), **119**, 149.
602. M. R. Andersson, D. m. Johansson, M. Theander and O. Inganas, *Synth. Met.*, (2001), **119**, 63.
603. P. A. Lane, S. M. Lipson, A. J. Cadby, D. F. O'Brien, S. J. Martin, W. J. Blau and D. D. C. Bradley , *Synth. Met.*, (2001), **119**, 661.
604. S. Yokojima, D. H. Zhou and G. H. Chen, *Chem. Phys. Lett.*, (2001), **333**, 397.
605. A. P. H. J. Schenning, P. Jonkheijm, E. Peeters and E. W. Meijer, *J. Am. Chem. Soc.*, (2001), **123**, 409.
606. P. K. Bhowmik, R. A. Burchett, H. Han and J. J. Cebe, *Macromolecules*, (2001), **34**, 7579.
607. P. K. Bhowmik, R. A. Burchett, H. Han and J. J. Cebe, *J. Polym. Sci., Part A: Polym. Chem. Ed.*, (2001), **39**, 2710.
608. J. L. Yang, H. Z. Lin, M. Zheng and F. L. Bai, *Chin. Chem. Lett.*, (2001) **12**, 369.
609. B. S. Gaylord, S. Wang, A. J. Heeger and G. C. Bazan, *J. Am. Chem. Soc.*, (2001), **123**, 6417.
610. A. Fujii, T. Sonoda, T. Fujisawa, R. Ootake and K. Yoshino, *Synth. Met.*, (2001), **119**, 189.
611. R. Schroeder, J. R. Heflin, H. Wang, H. W. Gibson and W. Graupner, *Synth. Met.*, (2001), **119**, 1521.
612. S. Valiyavettil and C. Baskar, *Polym. Mater. Sci. Eng.*, (2001), **84**, 1079.
613. J. Tang, Y. Wang, D. Yu, Z. Zhou, C. Wang and B. Yang, *Mater. Lett.*, (2001), **50**, 371.
614. T. W. Lee, O. O. Park, J. J. Kim, J. M. Hong and Y. C. Kim, *Chem. Mater.*, (2001), ACS, ASAP.

615. J. Morgado, F. Cacialli, R. H. Friend, B. S. Chuah, H. Rost, S. C. Moratti and A. B. Holmes, *Synth. Met.*, (2001), **119**, 595.
616. I. K. Spiliopoulos and J. A. Mikroyannidis, *Macromolecules*, (2001), **34**, 5711.
617. S. Shirai and J. Kido, *J. Photopolym. Technol.*, (2001), **14**, 317.
618. W. Holzer, A. Penzkofer, R. Stockmann, H. Meysel, H. Liebegott and H. H. Horhold, *Synth. Met.*, (2002), **125**, 343.
619. M. Levitus, K. Schmieder, H. Ricks, K. D. Shimizu, U. H. F. Bunz and M. A. Garcia-Garibay, *J. Am. Chem. Soc.*, (2001), **123**, 4259.
620. N. C. Yang and D. H. Suh, *Macromol. Rapid Commun.*, (2001), **22**, 335.
621. R. A. Marcos, M. T. Rispens, J. C. Hummelen and R. A. J. Janssen, *Synth. Met.*, (2001), **119**, 171.
622. S. Wang, Y. Li and D. Zhu, *Proc. Electrochem. Soc.*, (2000), 208.
623. M. Onciu, I. Constantin-Chiriac and E. Rusu, *Roum. Chem. Q. Rev.*, (2000), **7**, 13.
624. J. H. Hsu, W. Fann, H. F. Meng, E. S. Chen, E. C. Chang, S. A. Chen and K. W. To, *Chem. Phys.*, (2001), **269**, 367.
625. G. Kwak and T. Masuda, *Macromol. Rapid Commun.*, (2001), **22**, 846.
626. L. H. Gan, Y. M. Wang, Y. Xu, N. K. Goh and Y. Y. Gan, *Macromolecules*, (2001), **34**, 6117.
627. D. Hertel, S. Setayesh, H. Nothofer, U. Scherf, K. Muller and H. Bassler, *Adv. Mater.*, (2001), **13**, 65.
628. E. S. Manas and L. X. Chen, *Chem. Phys. Lett.*, (2000), **331**, 299.
629. S. H. Jung, H. K. Kim, S. H. Kim, Y. H. Kim, S. C. Jeoung and D. Kim, *Macromolecules*, (2001), **33**, 9277.
630. H. K. Park and M. Ree, *Synth. Met.*, (2001), **117**, 197.
631. B. Behnisch and M. Hanack, *Synth. Met.*, (2001), **121**, 1661.
632. J. Y. Shimano and A. G. MacDiarmid, *Synth. Met.*, (2001), **119**, 365.
633. E. O. Kim, S. B. Park, S. Heo and J. S. Lee, *J. Korean Chem. Soc.*, (2001), **45**, 156.
634. L. H. Gan, Y. Y. Gan and C. Y. Liau, *Eur. Polym. J.*, (2001) **37**, 1479.
635. V. Ananiashvili, R. Akhvlediani and Kh. Gaprindashvili, *Izv. Akad. Nauk Gruz., Ser. Khim.*, (2000), **26**, 192.
636. J. L. Yan, Y. Zhao and C. Zhu, *Bandaoti Xuebao*, (2001), **22**, 423.
637. D. A. M. Egbe, M. Ayuk, A. M. Amer and E. Klemm, *Des Monomers Polym.*, (2001), **4**, 169.
638. D. Bogdal, A. Bororn and J. Pielichowski, *Zesz. Nauk. Politech. Slask., Chem.*, (1999), **140**, 45.
639. J. S. Wilson, A. Kohler, R. H. Friend, M. K. Al-Suti, M. R. A. Al-Mandhary, M. S. Khan and P. R. Raithby, *J. Chem. Phys.*, (2000), **113**, 7627.
640. S. Gillissen, M. Jonforsen, E. Kesters, T. Johansson, M. Theander, M. R. Andersson, O. Inganaes, L. Lutsen and D. Vanderzande, *Macromolecules*, (2001), **34**, 7294.
641. Q. Ling, M. Yang, Z. Wu, X. Zhang, L. Wang and W. Zhang, *Polymer*, (2001), **42**, 4605.
642. T. Yamamoto, H. Nakajima, N. Hayashida, K. Shiraishi and H. Kokubo, *Polymers Adv. Technol.*, (2000), **11**, 658.
643. S. C. Ng, M. Ding, H. S. O. Chan and W. L. Yu, *Macromol. Chem. Phys.*, (2001), **202**, 8.
644. E. Colabella, A. Borghesi, R. Tubino, M. Catellani and S. Luzzati, *Synth. Met.*, (2001), **119**, 613.
645. S. S. Pandey, W. Takashima, T. Endo, M. Rikukawa and K. Kaneto, *Synth. Met.*, (2001), **121**, 1561.
646. G. Casalbore, N. Camaioni, V. Fattori, A. M. Fichera, M. C. Gallazi, A. Geri, E.

Girott and G. Giro, *Synth. Met.*, (2001), **121**, 1575.
647. N. Kurokawa, H. Yoshikawa, H. Masuhara, N. Hirota and K. Hyodo, *J. Microsc.*, (2001), **202**, 420.
648. Y. Liu, Y. Xu and D. Zhu, *Macromol. Chem. Phys.*, (2001), **202**, 1010.
649. I. Nurulla, T. Morikita, H. Fukimoto and T. Yamamoto, *Macromolecules*, (2001), **202**, 2335.
650. M. Theander, M. Svensson, A. Ruseckas, D. Zigmantas, V. Sundstrom, M. R. Andersson and O. Inganas, *Chem. Phys. Lett.*, (2001), **337**, 277.
651. G. T. Hwang, H. S. Son, J. K. Ku and B. H. Kim, *Org. Lett.*, (2001), **3**, 2469.
652. T. Kobayashi, J. Hamazaki, M. Arakawa, H. Kunugita, T. Endo, M. Rikukawa, K. Sanui and K. Ema, *J. Phys. Soc. Jpn.*, (2001), **70**, 2517.
653. J. J. Apperloo, P. R. L. Malenfant, J. M. J. Frechet and R. A. J. Janssen, *Synth. Met.*, (2001), **121**, 1259.
654. J. J. Apperloo, R. A. J. Janssen, R. L. malenfant and J. M. J. Frechet, *J. Am. Chem. Soc.*, (2001), **123**, 6916.
655. A. Bolognesi, C. Botta and W. Porzio, *Monatsh. Chem.*, (2001), **132**, 121.
656. A. Bolognesi, W. Porzio, A. Provasoli, C. Botta, A. Comotti, P. Sozzani and R. Simonutti, *Macromol. Chem. Phys.*, (2001), **202**, 2586.
657. G. Barbarella, L. Favaretto, G. Sotgiu, M. Zambianchi, C. Arbizzani, A. Bongini, M. Mastragostino,. Gigli and R. Cingolani, *Proc. SPIE-INT. Soc. Opt. Eng.*, (2000), **4134**, 37.
658. S. Lee, Y. Kang and C. Lee, *Synth. Met.*, (2001), **117**, 257.
659. M. Miyata, N. Matsumi and Y. Chujo, *Macromolecules*, (2001), **34**, 7331.
660. M. Belletete, J. F. Morin, S. Beupre, M. Ranger, M. Leclerc and G. Durocher, *Macromolecules*, (2001), **34**, 2288.
661. J. W. Y. Lam, J. D. Luo, H. Peng, Z. L. Xie, K. T. Xue, Y. P. Dong, L. Cheng, C. F. Qiu, S. Hoi and B. Z. Tang, *Chin. J. Polym. Sci.*, (2001), **19**, 585.
662. L. Peters, T. Sano, J. J. Morrison, N. Feeder, A. B. Holmes and A. Kraft, *Synth. Met.*, (2001), **119**, 175.
663. S. H. Shin, A. Y. Jeong, J. H. Kim, H. Lee and D. P. Kim, *Polymer (Korea)*, (2001), **25**, 137.
664. S. H. Shin, A. Y. Jeong, H. Lee and D. P. Kim, *Kongop Hwahak*, (2001), **123**, 348.
665. N. S. Baek, S. H. Jung, D. J. Oh, G. T. Hwang and B. H. Kim, *Synth. Met.*, (2001), **121**, 1743.
666. G. Gigli, M. Anni, M. Theander, R. Cingolani, G. Barbarella, L. Favaretto and O. Inganas, *Synth. Met.*, (2001), **119**, 581.
667. M. Svensson, M. Theander, O. Inganas and M. R. Andersson, *Synth. Met.*, (2001), **119**, 113.
668. A. Cravino, G. Zerza, H. Neugebauer, N. S. Sacrifici, M. Magini, S. Bucella, M. Svensson and M. R. Andersson, *Chem. Commun.*, (2000), **24**, 2487.
669. S. Luzzati, M. Panigoni and M. Catellani, *Macromolecules*, (2001), **116**, 171.
670. J. M. De Souza and E. C. Pereira, *Synth. Met.*, (2001), **118**, 167.
671. A. Bergmann, W. Holzer, R. Stark, H. Gratz, A. Penkofer, F. Amat-Guerri, A. Costela, I. Garcia-Moreno and R. Sastre, *Chem. Phys.*, (2001), **271**, 201.
672. R. Rogge and S. R. Ahmad, *J. Opt. Electron Adv. Mater.*, (2000), **2**, 247.
673. K. Hakala, r. Vatanparast, E. Vuorimaa and H. Lemmetyinen, *J. Appl. Polym. Sci.*, (2001), **82**, 1593.
674. J. A. Bodi and A. Burgess, *Annu. Techn. Conf-Soc. Plast. Eng.*, (2000), *58th*, 2611.
675. V. Sinigersky, K. Mullen, M. Klapper and I. Schopov, *Adv. Mater.*, (2000), **12**, 1058.
676. H. Li, J. Peng, X. Chen, X. Liu and A. Hu, *Gaofenzi Cailiao Kexue Yu Gongcheng*,

(2001), **17**, 164.

677. R. Vallee, N. Tomczak, H. Gersen, E. M. H. P. Van Dijk, M. F. Garcia-Parajo, G. J. Vancso and N. F. Van Hulst, *Chem. Phys. Lett.*, (2001), **348**, 161.
678. P. Hrdlovic and M. Kaholek, *Chem. Listy*, (2001), **95**, 102.
679. S. Wang, H. He, X. Cui, J. Gong and L. Yu. Qu, *Huaxue Xuebao*, (2001), **59**, 1163.
680. C. Charcosset and J. C. Bernengo, *Eur. Phys. J: Appl. Phys.*, (2000), **12**, 195.
681. G. Z. Li, N. Minami and Y. Ichio, *Polym. Eng. Sci.*, (2001), **41**, 1580.
682. K. Iwai, *Kakagaku*, (2001), **32**, 26.
683. N. J. Flint, S. Gardebrecht, I. Soutar and L. Swanson, *ACS Symp. Ser.*, (2001), **780**, 208.
684. K. Suhling, G. Hungerford, R. W. Airey and B. L. Morgan, *Measur. Sci. Technol.*, (2001), **12**, 131.
685. M. Erdogan and O. Peckan, *Polymer*, (2001), **42**, 4973.
686. H. Itagaki, *Macromol. Symp.*, (2001), **166**, 13.
687. A. J. Bur, S. C. Roth and H. Lobo, *Annu. Techn. Conf.-Soc. Plast. Eng.*, (2001), *59th*, 3071.
688. J. Liu, D. Li, X. Ren and D. Tian, *Gaofenzi Xuebao*, (2001), **1**, 127.
689. D. J. S. Birch, C. D. Geddes, J. Karolin and K. Wynne, *Proc. SPIE-Int. Soc. Opt. Eng.*, (2001), **4252**, 97.
690. M. Stork, A. Herrmann, T. Nemnich, M. Klapper and K. Mullen, *Angew. Chem., Int. Ed.*, (2000), **39**, 4367.
691. I. M. Kalogeras and A. Vassilikou-Dova, *J. Phys. Chem.*, (2001), **105**, 7651.
692. A. Bajorek and J. Paczkowski, *Polimery*, (2000), **45**, 792.
693. M. A. Kane, S. Pandey, G. A. Baker, S. A. Perez, E. J. Bukowski, D. C. Hoth and F. V. Bright , *Macromolecules*, (2001), **34**, 6831.
694. A. K. Mathew, J. Duhamel and J. Gao, *Macromolecules*, (2001), **34**, 1454.
695. J. M. Riley, S. Alkan, A. Chen, M. Shapiro, W. A. Khan, W. R. Murphy and J. E. Hanson, *Macromolecules*, (2001), **34**, 1797.
696. C. Kosa, M. Danko, A. Fiedlerova, P. Hrdlovic, E. Borsig and R. G. Weiss, *Macromolecules*, (2001), **34**, 2673.
697. M. Danko, P. Hrdlovic and E. Borsig, *J. Macromol. Sci., Pure Appl. Chem.*, (2001), *A38*, 467.
698. V. A. O. Tiera, J. M. Tiera and M. G. Neumann, *J. Dispersion Sci. Technol.*, (2001), **22**, 177.
699. J. P. S. Farinha, S. Picarra, K. Miesel and J. M. G. Martinho, *J. Phys. Chem.*, ACS, ASAP.
700. S. I. Yusa, K. Kakimoto, T. Yamamoto and Y. Morishima, *Macromol. Rapid Commun.*, (2001), **22**, 253.
701. Y. K. Gong and K. Nakashima, *Chem. Commun.*, (2001), **18**, 1772.
702. C. K. Chee, S. Rimmer, I. Soutar and L. Swanson, *Polymer*, (2001), **42**, 5079.
703. S. Kanagalingam, C. F. Ngan and J. Duhamel, *Polym. Prep. (Am. Chem. Soc., Div. Polym. Chem.)*, (2001), **42**, 606.
704. V. Vangani, J. Drage, J. Mehta, A. K. Mathew and J. Duhamel, *J. Phys. Chem.*, (2001), **B105**, 4827.
705. Y. Yan, T. Feng and Z. Tong, *Haxue Shijie*, (2000), **41**, 597.
706. C. Poncet-Legrand and F. M. Winnik, *Polymer J. (Tokyo).*, (2001), **33**, 277.
707. D. Patra and A. K. Mishra, *Polycyclic Aromatic Compd.*, (2001), **18**, 367.
708. M. Mizusaki, Y. Morishima and F. M. Winnik, *Polymer*, (2001), **42**, 5615.
709. Z. Liao, X. Chen and J. Shen, *Guilin Gongxueyuan*, (2000), **20**, 260.
710. Y. Katsumoto, F. Tsunomori, H. Ushiki, L. Letamendia and J. Rouch , *Eur. Polym.*

J., (2001), **37**, 475.
711. J. L. Pozzo, L. Fuchtjohann, A. Ferraro, O. Provot, H. Bouas-Laurent and J. P. Desvergne, *Polycyclic Aromat. Compd.*, (2001), **18**, 293.
712. C. E. Hoyle, J. Zhao, B. Orler, W. Kuang and C. L. McDonald, *J. Macromol. Sci., Pure Appl. Chem.*, (2001), *A38*, 527.
713. N. Kawashima, A. Kameyama, T. Nishikubo and T. Nagai, *J. Polym. Sci., Part A: Polym. Chem. Ed.*, (2001), **39**, 1773.
714. A. Boguta and D. Wrobel, *J. Fluorescence*, (2001), **11**, 129.
715. K. N. Jayachandran, S. Maiti and P. R. Chatterji, *Polymer*, (2001), **42**, 6113.
716. Y. Tajima, Y. Shigemitsu, M. Hoshino and K. Takeuchi, *J. Photopolym. Sci. Technol.*, (2001), **14**, 147.
717. T. Guillet, J. Berrehar, R. Grousson, J. Kovensky, C. Lapersonne-Meyer, M. Schott and V. Voliotis, *Phys. Rev. Lett.*, (2001), **87**, 087401.
718. Y. Rharbi and M. A. Winnik, *Macromolecules*, (2001), **34**, 5238.
719. S. L. Bondareva, V. N. Knyuksho, N. V. Masalov and K. Dzilinski, *Opt. Spektrosc.*, (2001), **91**, 59.
720. D. Hertel, Yu. V. Romanovskii, B. Schweitzer, U. Scherf and H. Bassler, *Macromol. Symp.*, (2001), **175**, 141.
721. G. Rinivas, A. Yethiraj and B. Bagchi, *J. Phys. Chem. B.*, (2001), **105**, 2475.
722. S. A. Siling, S. V. Shamahin, I. A. Ronova, A. Yu. Kovaleski, A. B. Grabchev, I. Yu. Tsiganova and V. I. Yuzhakov, *J. Appl. Polym. Sci.*, (2001), **80**, 398.
723. J. Shinar, J. Partee, E. J. W. List, B. L. Uhlhorn, C. H. Kim, W. Graupner and G. Leising, *Mol. Crsyt. Liq. Cryst. Sci. Technol. Sect A.*, (2001), **361**, 1.
724. R. M. Jones, T. S. Bergstedt, C.Buscher, D. McBranch and D. Whitten, *Langmuir*, (2001), **17**, 2568.
725. G. Tamulaitis, V. Gulbinas, A. Undzenas and L. Valkunas, *Environ. Chem. Phys.*, (1999), **21**, 89.
726. G. Srinivas and B. Bagchi, *Los Alamos Natl. Lab., Prep. Arch. Condens. Matter*, (2001), **1-19**, 0105138.
727. H. Aoki, H. Katayama, M. Kamigaito and M. Sawamoto, *Polymer J.*, (2001), **33**, 464.
728. J. Gallego, D. Perez-Foullerat, F. Medicuti and W. L. Mattice, *J. Polym. Sci., Part A: Polym. Phys.*, (2001), **39**, 1272.
729. Q. Gao, Y. Cui, X. Yu and F. Du, *Gaofenzi Xuebao*, (2001), **3**, 392.
730. D. Vyprachticky, V. Cimrova, L. Machova and V. Pokorna, *Collect. Czech. Chem. Commun.*, (2001), **66**, 1473.
731. S. A. Siling, S. V. Shamahin, I. A. Ronova, A. Yu. Kovaleski, A. B. Grachev, O. Yu. Tsiganova and V. I. Yuzhakov, *Int. J. Polym. Mater.*, (2001), **49**, 115.
732. H. Li, J. Peng, X. Xu and H. Xia, *Wuli Huaxue Xuebao*, (2001), **17**, 329.
733. M. Hasegawa and K. Horie, *Prog. Polym. Sci.*, (2001), **26**, 259.
734. M. Tamura, D. Gao and A. Ueno, *Chem.-Eur. J.*, (2001), **7**, 1390.
735. C. Bratschkov, P. Karpuzova, K. Mullen, M. Klapper and I. Schopov, *Polym. Bull.*, (2001), **46**, 345.
736. A. B. Kirpach, E. V. Anufrieva, I. Soutar and L. Swanson, *M Vysokomol. Soedin., Ser. A, Ser. B.*, (2000), **42**, 1657.
737. M. N. Berberan-Santos, E. N. Bodunov and J. M. G. Martinho, *Opt. Spectrosc.*, (2000), **89**, 876.
738. Y. Meng, A. R. Hill and A. S. Hay, *Polym. Adv. Technol.*, (2001), **12**, 206.
739. K. Hirano, S. Minakata and M. Komatsu, *Chem. Lett.*, (2001), **1**, 8.
740. Y. S. Gal, W. C. Lee, S. Y. Kim, J. W. Park, S. H. Jin, K. N. Koh and S. H. Kim, *J.*

Polym. Sci., Part A: Polym. Chem. Ed., (2001), **39**, 712.

741. Y. Dong, J. W. Y. Lam and B. Z. Tang, *Polym. Prep. (Am. Chem. Soc., Div. Polym. Chem.)*, (2001), **42**, 572.
742. S. Ponsart, J. Coudane, J. L. Morgat and M. Vert, *J. Labelled Compd. Radiopharm.*, (2001), **44**, 677.
743. D. V. Nicoloau, T. Taguchi and H. Suzuki, *Proc. SPIE-Int. Soc. Opt. Eng.*, (2000), **4200**, 49.
744. S. M. Yarmoluk, V. B. Kovalska, D. V. Kryvorotenko, A. O. Balanda and T. Y. Ogul'chansky, *Spectrochim. Acta Part A.*, (2001), **57A**, 1533.
745. Z. Li, J. Qin, X. Deng and Y. Cao, *J. Polym. Sci., Part A: Polym. Chem. Ed.*, (2001), **39**, 3428.
746. A. Adronov, D. . Robello and J. M. J. Frechet, *J. Polym. Sci., Part A: Polym. Chem. Ed.*, (2001), **39**, 1366.
747. S. B. Brichkin, V. F. Razumov, L. M. Bogdanova and B. A. Rozenberg, *Vysokomol. Soedin., Ser. A, Ser. B.*, (2000), **42**, 791.
748. E. S. Tilman and T. E. Hogen-Esch, *Rolym. Prep. (Am. Chem. Soc., Div. Polym. Chem.)*, (2001), **42**, 576.
749. K. Alberty, E. Tillman, S. Carlotti, S. E. Bradforth and T. E. Hogen-Esch, *Rolym. Prep. (Am. Chem. Soc., Div. Polym. Chem.)*, (2001), **42**, 584.
750. L. V. Vinogradova, V. N. Zgonnik, D. A. Sykmanov and Yu. F. Biryulin, *Vysokomol. Soedin., Ser. A, Ser. B.*, (2001), **43**, 1002.
751. M. Chen, H. Yan, Y. Hu and Y. Shen, *Huadong Ligong Daxue Xuebao*, (2001), **27**, 195.
752. I. Grabchev, V. Bojinov and R. Betcheva, *J. Appl. Polym. Sci.*, (2001), **81**, 2643.
753. I. Grabchev and V. Bojinov, *Polym. Deg. Stabil.*, (2000), **70**, 147.
754. M. Thelakkat, P. Poesch and H. W. Schmidt, *Macromolecules*, (2001), **34**, 7441.
755. F. Sussmeier and H. Langhals, *Eur. J. Org. Chem.*, (2001), **3**, 607.
756. J. Li, X. Shi and S. Wu, *Gaofenzi Xuebao*, (2000), **6**, 763.
757. Yu. S. Avlasevich, V. N. Knyukshto, O. G. Kulinkovich and K. N. Solov'ev, *J. Appl. Specrosc.*, (2000), **67**, 663.
758. H. J. Egelhaaf, E. Holder, P. Herman, D. Oelkrug and E. Lindner, *J. Mat. Chem.*, (2001), **11**, 2445.
759. K. Obata, K. Sakamoto and M. Kira, *Macromolecules*, (2001), **34**, 2739.
760. C. A. Munson, M. A. Kane, G. A. Baker, S. Pandey, S. A. Perez and F. V. Bright, *Macromolecules*, (2001), **34**, 4624.
761. H. L. Jiang, K. J. Zhu and L. J. Dai, *Polym. Int.*, (2001), **50**, 722.
762. E. V. Anufrieva, M. G. Krakovyak, T. V. Sheveleva, V. B. Luschik, R. A. Gromova, V. D. Pautov, Yu. E. Kirsch, N. A. Yanul and T. N. Nekrasova, *Vysokomol. Soedin., Ser. A, Ser. B.*, (2000), **42**, 1837.
763. J. Tang and W. Zhang, *Gaofenzi Xuebao*, (2001), **4**, 480.
764. D. Wang, Q. Lin, L. S. Fu, H. R. Li, Z. X. Chu and B. Yang, *Gaodeng Xuexiao Huaxue Xuebao*, (2001), **22**, 695.
765. X. Lu, I. Manners and M. A. Winnik, *'New Trends in Fluorescence Spectroscopy'*, ed. B. Valeur and J. C. Brochon, Springer-Verlag, Berlin, (2001), 229.
766. Y. Takahashi and T. Kimura, *Bull. Chem. Soc. Jpn.*, (2001), **74**, 1899.
767. E. V. Anufrieva, T. N. Nekrasova, M. G. Krakovyak, T. D. Ananeva and V. B. Lushchik, *Vysokomol. Soedin., Ser. A, Ser. B.*, (2001), **43**, 875.
768. E. V. Anufrieva, T. N. Nekrasova, T. D. Ananeva, R. A. Gromova, V. B. Lushchik and M. G. Krakovyak, *Vysokomol. Soedin., Ser. A, Ser. B.*, (2000), **42**, 994.
769. H. Laguitton-Pasquier, A. Martre and A. Deronzier, *J. Phys. Chem. B.*, (2001), **105**,

4801.
770. G. V. mel'nokov, S. N. Shtykov, I. Yu. Goryacheva and E. V. Fedorenko, *Russ. Chem. Bull.*, (2001), **50**, 986.
771. S. J. A. Pope, C. R. Rice, M. D. Ward, A. F. Morales, G. Accorsi, N. Armaroli and F. Barigelletti, *J. Chem. Soc., Dalton Trans.*, (2001), **15**, 2228.
772. J. S. Lee, N. G. Kang, Y. S. Cho, J. H. Ahn, C. Lee and J. J. Kim, *Polym. Prep. (Am. Chem. Soc., Div. Polym. Chem.)*, (2001), **42**, 448.
773. T. Odaki, K. Takagi, K. Hashimoto and Y. Toda, *Shikizai Kyokaishi*, (2001), **74**, 495.
774. Y. H. Li, H. J. Zhang, S. B. wang, Q. G. Meng, H. R. Li and X. H. Chuai, *Thin Solid Films*, (2001), **385**, 205.
775. B. Yan and H. Zhang, *J. Adv. Mater.*, (2001), **33**, 39.
776. M. Vasilescu, A. Caragheorgheopol and H. Caldararu, *Adv. Colloid Inter. Sci.*, (2001), **89**, 169.
777. L. Cao, I. Manners and M. A. Winnik, *Macromolecules*, (2001), **34**, 3353.
778. J. Virtanen, H. Lemmetyinen and H. Tenhu, *Polymer*, (2001), **42**, 9487.
779. R. Wang, H. Knoll, F. Rittig and J. Kaerger, *Langmuir*, (2001), **17**, 7464.
780. K. Nakashima and K. Takeuchi, *Appl. Spectrosc.*, (2001), **55**, 1237.
781. P. L. Kuo, S. S. Hou, C. K. Teng and L. W. Liang, *Colloid Polym. Sci.*, (2001), **279**, 286.
782. Y. Lin and P. Alexandridis, *Polym. Prep. (Am. Chem. Soc., Div. Polym. Chem.)*, (2001), **42**, 229.
783. E. T. Imazaki, C. C. Schmitt and M. G. Neumann, *Langmuir*, (2001), **17**, 3486.
784. L. Wu, X. Tuo, H. Cheng, Z. Chen and X. Wang, *Macromolecules*, (2001), **34**, 8005.
785. J. H. Mathias, M. Rosen and L. Davenport, *Langmuir*, (2001), **17**, 6148.
786. G. David, F. Ozer, B. C. Simionescu, H. Zareie and E. Piskin, *Eur. Polym. J.*, (2001), **38**, 73.
787. G. L. Smith and C. L. McCormick, *Langmuir*, (2001), **17**, 1719.
788. B. Ren, Z. Tong, F. Gao, X. Liu and F. Zeng, *Polymer*, (2001), **42**, 7291.
789. J. A. Molinar-Bolivar, J. Aguiar and C. C. Ruiz, *Mol. Phys.*, (2001), **99**, 1729.
790. T. Y. Ogul'chansky, M. Y. Losytskyy, V. B. Kovalska, V. M.Yashchuk and S. M. Yarmoluk, *Spectrochim. Acta, Part A.*, (2001), **57**, 1525.
791. A. Miura, Y. Yanagawa and N. Tamai, *J. Microsc.*, (2001), **202**, 425.
792. C. V. Kumar and A. Chaudhari, *Microporous Mesoporous Materials*, (2000), **41**, 307.
793. G. L. Smith and C. L. McCormick, *ACS Symp. Ser.*, (2001), **780**, 38.
794. M. Blanzat, S. Massip, V. Speziale, E. Perez and L. I. Rico-Lattes , *Langmuir*, (2001), **17**, 3512.
795. R. Rutkaite, G. Buika and J. V. Grazulevicius, *J. Photochem. Photobiol., Part A. Chem. Ed.*, (2001), **138**, 245.
796. D. Pines, J. Cohen, Y. Zhao, V. Khutorsky, Z. Priel and E. Pines, *Springer Ser. Chem. Phys.*, (2001), **66**, 588.
797. S. E. Burke and R. Palepu, *Carbohydrate Polymers*, (2001), **45**, 233.
798. A. Alonsa, F. Catalina, E. F. Salvador and C. Peinado, *Macromol. Chem. Phys.*, (2001), **202**, 2293.
799. S. Liu, Y. Fang, D. Hu, G. Gao and J. Ma, *J. Appl. Polym. Sci.*, (2001), **82**, 620.
800. A. Castellan, A. Nourmamode, V. Sartor, S. Grelier, P. Looten and M. Norell, *J. Photochem. Photobiol., Part A: Chem. Ed.*, (2001), **141**, 183.
801. M. Mizusaki, Y. Morishima, B. B. Raju and F. M. Winnik, *Eur. Phys. J. E.*, (2001), **5**, 105.

802. C. Wang, Q. Sun, Z. Tong, X. Liu, F. Zeng and S. Wu, *Colloid Polym. Sci.*, (2001), **279**, 664.
803. C. Y. Wang, Q. L. Sun, L. Qi, Z. Tong, X. X. Liu, F. Zeng and F. Gao, *Gaodeng Xuexiao Huaxue Xuebao*, (2001), **22**, 1265.
804. K. Hayashi, *TechnoCosmos.*, (2001), **14**, 27.
805. Y. Sato, A. Hashidzume and Y. Morishima, *Macromolecules*, (2001), **34**, 6121.
806. G. L. Smith and C. L. McCormick, *Macromolecules*, (2001), **34**, 5579.
807. M. J. Tiera, T. V. aP. De Oliveira, E. C. De Toldeo and G. L. De Sena, *Colloid Polym. Sci.*, (2001), **278**, 1052.
808. Y. Yin, Y. Fang, D. D. Hu and G. L. Gao, *Wuli Huaxue Xuebao*, (2001), **17**, 776.
809. M. da G. Miguel, H. D. Burrows, S. J. Formishino and B. Lindman, *J. Mol. Struct.*, (2001), **563**, 89.
810. G. Teyssedre, C. Laurent, A. Aslanides, N. Quirke, L. A. Dissado, G. C. Montanari, A. Campus and L. Martinotto, *IEEE Trans. Dielectr. Electr. Insul.*, (2001), **8**, 744.
811. R. Qi, W. Pang, K. Hu, Q. Zhu and G. Zhou, *Gaofenzi Cailiao Kexue Yu Gongcheng*, (2001), **17**, 102.
812. N. Matsumi, and Y. Chujo, *Polym. J.*, (2001), **33**, 383.
813. X. Wu, Y. Liu and D. Zhu, *Synth. Met.*, (2001), **121**, 1699
814. H. L. Jiang, K. J. Zhu and L. J. Dai, *Makromol. Rapid Commun.*, (2001), **22**, 414.
815. Y. Fukishima, *Polym. Bull.*, (2001), **45**, 465.
816. A. J. Bur, M. G. Vangel and S. C. Roth, *Polym. Eng. Sci.*, (2001), **41**, 1380.
817. Z. Mazurak, B. Morejko-Buz, D. Neugebauer, A. Stolarzewicz and V. Getautis, *Proc. SPIE-Int. Soc. Opt. Eng*, (2000), **4087**, 650.
818. G. Teyssedre, J. Menegotto and C. Laurent, *Polymer*, (2001), **42**, 8207.
819. G. Goerge, I. Blakey, B. Goss and M. Grigg, *Polym. Prep. (Am. Chem. Soc., Div. Polym. Chem.)*, (2001), **42**, 402.
820. J. Y. Shimano and A. G. MacDiarmid, *Synth. Metals*, (2001), **123**, 251.
821. L. Torrisi, *Radiat. Eff. Defects Solids.*, (2001), **154**, 89.
822. M. W. Backer and U. C. Pernisz, *Polym. Prep.(Am. Chem. Soc., Div. Polym. Chem.)*, (2001), **42**, 122.
823. S. C. J. Meskers, J. Hubner, M. Oestreich and H. Bassler, *Chem. Phys. Lett.*, (2001), **339**, 223.
824. R. Stevenson, R. G. Milner, D. Richards, A. C. Arias, J. D. Mackenzie, J. J. M. Friend, D. J. Kang and M. Blamire, *J. Microsc.*, (2001), **202**, 433.
825. D. L. Woerdeman and R. S. Parnas, *Appl. Spectrosc.*, (2001), **55**, 331.
826. P. Horvath, F. Schauer, I. Kuritka, O. Salyk, M. Weiter, N. Dokoupil, S. Nespurek and V. Fidler, *Monatsh. Chem.*, (2001), **132**, 177.
827. M. Duran, F. Massines, G. Tessedre and C. Laurent, *Surf. Coat. Technol.*, (2001), **142**, 743.
828. R. Qi, C. Zhou, Q. Zhu and Q. Zhou, *Gaofenzi Cailiao Kexue Yu Gongcheng*, (2001), **17**, 159.
829. S. Nespurek, F. Schauer and A. Kadaschuk, *Monatsh. Chem.*, (2001), **132**, 159.
830. S. Toyoda and M. Fujiki, *Macromolecules*, (2001), **34**, 2630.
831. H. Guan, Q. Gao, J. Shi, R. Li, G. Yang and X. Yu, *Huaxue Yanjiu*, (2000), **11**, 13.
832. K. Henderson, A. B. Dalton, g. Chambers, A. Drury. S. Maier, A. G. Ryder, W. Blau and H. J. Byrne, *Synth. Met.*, (2001), **119**, 555.
833. E. E. Neuteboom, R. A. J. Janssen and E. W. Meijer, *Synth. Met.*, (2001), **121**, 1283.
834. Z. Liao, Y. Chen, J. Zhao and J. Shen, *Hecheng Xiangjiao Gongye* (2001), **24**, 300.
835. J. M. Mabry, M. K. Runyon and W. P. Weber, *Macromolecules*, (2001), **34**, 7264.
836. B. F. carroll, J. P. Hubner, K. S. Schanze and J. M. Bedlek-Anslow, *J. Visualisation*,

(2001), **4**, 121.
837. C. J. Ellison, D. B. Hall, K. E. Miller, J. S. Royal and J. M. Torkelson, *Polym. Prep. (Am. Chem. Soc., Div. Polym. Chem.)*, (2001), **42**, 226.
838. F. Cataldo, *Fullerene Sci. Technol.*, (2000), **8**, 633.
839. S. I. Matsushita, T. Miwa and A. Fujishima, *Stud. Surf. Sci. Catal.*, (2001), **132**, 845.
840. R. Shroeder, B. Ullrich, W. Graupner and U. Scherf, *J. Phys. Condens. Matter*, (2001), **13**, 313.
841. T. Kavc, W. Kern, C. Zenz, G. Leising, G. Kranzelbinder and E. Toussaere, *Monatsh. Chem.*, (2001), **132**, 531.
842. C. C. White, K. B. Migler and W. L. Wu, *Polym. Eng. Sci.*, (2001), **41**, 1497.
843. J. Cheng, R. Fang and Z. Lin, *Gaofenzi Cailiao Kexue Yu Gongcheng.*, (2001), **17**, 21.
844. X. Liu, Q. Shen and Z. Cao, *Guanxue Xuebao*, (2000), **20**, 991.
845. X. M. Wang, Y. G. Zhou, W. T. Yu and M. H. Jiang, *Chin. Chem. Lett.*, (2001), **12**, 1001.
846. R. Rutkaite, G. Buika, V. Grazuleviciene and J. V. Grazulevicius, *Environ. Chem. Phys.*, (1999), **21**, 94.
847. D. W. Kim, A. Blumstein, J. Kumar and S. K. Tripathy, *Chem. Mater.*, (2001), **13**, 243.
848. K. Shinohara and H. Higuchi, *Trans. Mater. Res. Soc. Jpn.*, (2000), **25**, 825.
849. Z. Mazurak, K. Gornik and D. Sek, *Proc. SPIE-Int. Soc. Opt. Eng.*, (2000), **4087**, 645.
850. S.A. Siling, I. A. Ronova, V. A. Madii, E. I. Lozinskja. J. E. Borisevich, S. V. Vinogradova, V. A. Kizel, A. V. Reznichenko and V. N. Kokin, *Synthetic Properties of Heterocyclic Compounds*, ed. A. Svetlana Siling, G. E. Zaikov, Nova Science, N. Y., (2001), 115-119.
851. M. Shimizu, S. Suto, T. Goto, A. Watanabe and M. Matsuda, *Phys. Rev. B: Condens. Matter Phys.*, (2001), **63**, 073403.
852. G. Teyssedre, C. Laurent, P. Jonnard and C. Bonnelle, *Annu. Rep.-Conf. Electr. Insul. Dielectric Phenom.*, (2000), **2**, 543.
853. Y. Ehara, H. Kishida and T. Ito, *Annu. Rep.-Conf. Insul. Dielectric Phenom.*, (2000), **1**, 321.
854. D. Yamazaki, Y. Ehara, H. Kishida and T. Ito, *Annual Rep.-Conf. Insul. Dielectric Phenom.*, (2000), **1**, 317.
855. W. Holzer, A. Penzkofer, H. H. Horhold, D. Raabe and M. Helbig, *Opt. Mater.*, (2000), **15**, 225.
856. H. Ameshima, N. Nemoto and T. Endo, *J. Polym. Sci. : Part A: Polym. Chem. Ed.*, (2001), **39**, 3143.
857. J. Luo, Z. Xie, J. W. Y. Lam, C. Qiu, H. S. Kwok, X. Zhan, Y. Liu, D. Zhu and B. Z. Zhan, *Polym. Prep. (Am. Chem. Soc., Div. Polym. Chem.)*, (2001), **42**, 228.
858. X. Sun, R. L. Fu, K. Yonemitsu and K. Nasu, *Phys. Rev. A: At. Mol. Opt. Phys.*, (2001), **64**, 032504/1.
859. H. Sato and S. Nishio, *J. Photochem. Photobiol., Part C: Revs Ed.*, (2001), **2**, 139.
860. J. R. White and T. J. Turton, *Polym. Prep. (Am. Chem. Soc., Div. Polym. Chem.)*, (2001), **42**, 390.
861. H. Wang, *Huaxue Yangjiu*, (2000), **11**, 19.
862. S. Islam, *J. Bangladesh Acad. Sci.*, (2001), **25**, 25.
863. S. Islam, *J. Bangladesh Acad. Sci.*, (2000), **13**, 135.
864. L. Dai, S. Zhou, W. Li, J. Lin, Q. Yao, Y. Zou, Z. Wnag, H. Hao, H. Zhang, Y. Ning and J. Guo, *Xiamen Daxue Xuebao Ziran Kexueban*, (2000), **39**, 358.
865. X. Liang, D. Jian and H. Zhang, *Guangxi Daxue Xuebao, Ziran Kexueban*, (1999),

24, 284.
866. F. Carrasco, P. Pages, S. Pascual and X. Colom, *Eur. Polym. J.*, (2001), **37**, 1457.
867. F. Khabbaz and A. C. Albertsson, *J. Appl. Polym. Sci.*, (2001), **79**, 2309.
868. D. Y. Stavichnaya, A. V. Shiichuk and D. P. White, *Ukr. Khim. Zh.*, (2001), **67**, 123.
869. C. Chen, Z. Chen and H. Yin, *Huxaue Fenxi Jiiang*, (2000), **9**, 14.
870. H. Inata, K. Hashizume and K. Takeda, *Kobunshi Ronbunshu*, (2001), **58**, 233.
871. F. Severini, R. Gallo, L. Landro, M. Pegoraro, L. Brambilla, M. Tommasini, C. Castiglioni and G. Zerbi, *Polymer*, (2001), **42**, 3609.
872. J. L. Philippart and J. L. Gardette, *Polym. Deg. Stabil.*, (2001), **73**, 185.
873. B. Qu, S. M. Chen and L. M. Dai, *Appl. Mag. Reson.*, (2000), **19**, 59.
874. H. Liu, I. Keen, L. Rintoul and G. A. George, *Polym. Deg. Stabil.*, (2001), **72**, 543.
875. L. Guadagno, C. Naddeo, V. Vittoria, G. Camino and C. Cagnani, *Polym. Deg. Stabil.*, (2001), **72**, 175.
876. N. Hassini, K. Guenachi, A. Homou, J. M. Saiter, S. Marais and E. Beucher, *Polym. Deg. Stabil.*, (2001), **75**, 247.
877. F. P. La Mantia and J. L. Gardette, *Polym. Deg. Stabil.*, (2002), **75**, 1.
878. V. B. Ivanov and E. U. Khavina, *Plast. Massy*, (2000), **12**, 23.
879. M. lazzari, O. Chiantore and V. Castelvetro, *Polym. Int.*, 92001), **50**, 863.
880. X. Colom, F. Nogues, J. Valldeparas, F. Carrillo and A. Gordillo, *Revista de Metalurgia*, (2001), **37**, 190.
881. M. A. Bruk, M. V. Kondrateva and A. M. Sergeev, *Vysokomol. Soedin., Ser. A, Ser. B.*, (2000), **42**, 1751.
882. Y. Itoh, T. Goshima and H. Shirai, *J. Appl. Polym. Sci.*, (2001), **82**, 2227.
883. Anon, *Hyogo-kenritsu Kogyo Gijusu Senta Kenkyu Hokokusho*, (2000), **9**, 77.
884. C. Chung, S. Cho and S. Y. Oh, *Mol. Cryst. Liq. Cryst. Sci. Technol, Sect. A: Mol. Cryst. Liq. Cryst.*, (2001), **370**, 177.
885. H. P. Hu, Q. Y. Chen, K. L. Huang and D. B. Chen, *Hunan Shifan Daxue Ziran Kexue Xuebao*, (2000), **23**, 46.
886. J. R. Sargent, S. K. Gupta and W. P. Weber, *Polym. Prep. (Am. Chem. Soc., Div. Polym. Chem.)*, (2001), **42**, 92.
887. Y. Skryshevski, N. Ostapenko, A. Kadashchuk, A. Vakhin and S. Suto, *Mol. Cryst. Liq. Cryst. Sci. Technol., Sect A.*, 92001), **361**, 37.
888. R. P. Singh, N. S. Tomer and S. V. Bhadraiah, *Polym. Deg. Stabil.*, (2001), **73**, 443.
889. M. P. Luda, R. Tauriello and G. Camino, *Eur. Coat. J.*, (2000), **74**, 76.
890. R. Zhang, P. E. Mallon, H. Chen, C. M. Huang, J. Zhang, Y. Li, Y. Wu, T. C. Sandreczki and Y. C. Jean, *Prog. Org. Coat.*, (2001), **42**, 244.
891. R. Zhang, P. E. Mallon, H. Chen, C. M. Huang, J. Zhang, Y. Li, Y. Wu, T. C. Sandreczki and Y. C. Jean, *Polym. Mater. Sci. Eng.*, (2001), **85**, 109.
892. E. B. Orler, D. A. Wrobleski, D. W. ayne and B. L. Bennett, *Polym. Prep. (Am. Chem. Soc., Div. Polym. Chem.)*., (2001), **42**, 348.
893. Y. He, J. P. Yuan, H. Cao, R. Zhang, Y. C. Jean and T. C. Sandreczki, *Prog. Org. Coat.*, (2001), **42**, 75.
894. S. Commereuc, L. Gonon and V. Verney, *Proc. Int. Congr. Rheol.*, 13th (2000), **1**, 142.
895. S. Commereuc, L. Gonon and V. Verney, *Int. J. Polym. Anal. Charac.*, (2000), **6**, 75.
896. S. Schlinck and J. G. Bokria, *Polym. Prep. (Am. Chem. Soc., Div. Polym. Chem.)*., (2001), **42**, 244.
897. S. S. Solansky, R. P. Singh, D. Reyx and I. Campistron, *J. Rubber Res.*, (2000), **3**, 76.
898. N. Belhaneche-Bensemra and N. Ouazene, *Polym. Prep. (Am. Chem. Soc., Div. Polym. Chem.)*., (2001), **42**, 862.

899. S. Seki, S. Tsuji, Y. Matsui and S. Tagawa, *Chem. Lett.*, (2001), **12**, 1208.
900. J. Polar, A. Ouchi and Z. Bastl, *Chem. Vap. Dep.*, (2001), **7**, 19.
901. J. Polar, M. Urbanova, A. Galikova, A. Galik, Z. Bastl, J. Subrt, A. Ouchi, P. Papagiannakopoulos and H. Beckers, *Mem. Inst. Sci. Ind. Res., Osaka Univ.*, (2000), **57**, 205.
902. J. Pola, M. Urbanova, Z. Bastl, J. Subrt, M. Sakuragi, A. Ouchi and H. Morita, *Polymer*, (2001), **42**, 1311.
903. M. Susuki, M. yamaguchi, L. Ramonat and X. Zeng, *J. Photochem. Photobiol., Part A: Chem. Ed.*, (2001), **145**, 223.
904. V. N. Popok, I. I. Azarko, V. B. Odzhaev, A. Toth and R. I. Khaibullin. *Nucl. Instrum. Methods Phys. Res. Sect B*, (2001), **178**, 305.
905. T. Shinozuka, M. Shirai and M. Tsunooka, *J. Photopolym. Sci. Technol.*, (2000), **13**, 751.
906. Y. Ohki, T. Toyoda, Y. Hama, W. Wei and F. Massines, *Proc. Int. Conf. Prop. Appl. Dielectric Mater.*, (2000), **1**, 439.
907. C. S. C. de Castro, G. B. Gerardo and E. E. Monteiro, *Quim. Nova*, (2001), **24**, 303.
908. F. Bianchi, Y. Chevolot, H. J. Mathieu and H. H. Girault, *Anal. Chem.*, (2001), **73**, 3845.
909. T. Kino, S. Machida, K. Horie, S. Yusa and Y. Morishima, *Macromol. Chem. Phys.*, (2001), **202**, 1331.
910. T. Lippert, C. David, J. T. Dickinson, M. hauer, U. Kogelschatz, S. C. Langford, O. Nuyken, C. Phipps, J. Robert and A. Wokaun, *J. Photochem. Photobiol., Part A: Polym. Chem. Ed.*, (2001), **145**, 145.
911. K. Saitow, H. Banjo, N. Ichinose, S. Kawanishi, H. Masuhara and H. Fukumura, *J. Photochem. Photobiol., Part A: Polym. Chem. Ed.*, (2001), **145**, 159.
912. S. Nisho, K. Tamura, Y. Tsujine, T. Fukao, M. Nakano, A. Matsuzaki, H. Sato and T. Yamabe, *J. Photochem. Photobiol., Part A: Polym. Chem. Ed.*, (2001), **145**, 165.
913. Y. Tsuboi, N. Kimoto, M. Kabeshita and A. Itaya, *J. Photochem. Photobiol., Part A: Polym. Chem. Ed.*, (2001), **145**, 209.
914. T. Masubuchi, H. Fukumura, H. Masuhara, K. Suzuki and N. Hayashi, *J. Photochem. Photobiol., Part A: Polym. Chem. Ed.*, (2001), **145**, 215.
915. K. Tsunoda, M. Sugiura, M. Sonoyama, H. Yajima, T. Ishii, J. Taniyama and H. Itoh, *J. Photochem. Photobiol., Part A: Polym. Chem. Ed.*, (2001), **145**, 195.
916. D. N. S. Hon, *Wood Cellulose Chem., 2nd* Ed, Marcel Dekker, N. Y., (Ed: Hon), (2001), 513-546.
917. A. T. Mamadalimov, P. K. Khabibullaev and M. Shermatov, *Uzb. J. Phys.*, (1999), **1**, 465.
918. D. F. Guay, B. J. W. Cole, R. C. Fort, M. C. Hausman, J. M. Genco, T. J. Elder and K. R. Overly, *J. Wood. Chem. Technol.*, (2001), **21**, 67.
919. E. Afonso do Nascimento, *Proc. Braz. Symp. Chem. Lignins Other Wood Compos.*, (1995), *4th*, 204.
920. C. Heitner, R. St. John Manley, B. Ahvazi and J. Wang, *J. Pulp Paper Assoc.*, (2001), **27**, 325.
921. S. T. Chang and H. T. Chang, *Holzforschung*, (2001), **55**, 255.
922. C. A. S. Hill, N. S. Cetin, R. F. Quinney, H. Derbeyshire and R. J. Ewen, *Polym. Deg. Stabil.*, (2001), **72**, 133.
923. C. Li and A. J. Ragauskas, *J. Pulp. Pap. Sci.*, (2001), **27**, 202.
924. A. Sionowska, *Polym. Deg. Stabil.*, (2001), **73**, 107.
925. J. L. Gardette, *Environmen. Sci. Poll. Contr. Ser.*, (2000), **21**, 671.
926. E. Rusa, E. Comanita and M. Onciu, *Roum. Chem. Q. Rev.*, (2000), **7**, 241.

927. A. Rivaton, B. Mailhot, J. Soulestin, H. Varghese and J. L. Gardette, *Polym. Deg. Stabil.*, (2002), **75**, 17.
928. M. Irigoyen, P. Bartolomeo, F. X. Perrin, E. Aragon and J. L. Vernet, *Polym. Deg. Stabil.*, (2001), **74**, 59.
929. T. Nguyen, J. Martin, E. Byrd and E. Embree, *Polym. Prep. (Am. Chem. Soc., Div. Polym. Chem.)*, (2001), **42**, 420.
930. J. W. Martin, T. Nguyen, E. Byrd, B. Dickens and N. Embree, *Polym. Deg. Stabil.*, (2002), **75**, 193.
931. F. X. Perrin, M. Irigoyen, E. Aragon and J. L. Vernet, *Polym. Deg. Stabil.*, (2001), **72**, 115.
932. N.S. Allen, M. C. Marin, M. Edge, D. W. Davies, J. Garrett and F. Jones, *Polym. Deg. Stabil.*, (2001), **73**, 119.
933. M. Kawaraya, M. Yabe and K. Tsubouchi, *Shikizai Kyokaishi*, (2000), **73**, 594.
934. V. Castelvetro, M. Aglietto, F. Ciardelli, O. Chiantore and M. Lazzari, *ACS Symp. Ser.*, (2001), **787**, 129.
935. N. J. Earhart, *Hot Melt Symp., Proc.*, (2000), 155.
936. A. Valet and C. Decker, *RadTech. Rep.*, (2000), **14**, 40.
937. A. Valet and C. Decker, *Mod. Paint. Coat.*, (2001), **91**, 29.
938. E. Takeuchi, Y. Tajima and K. Takeuchi, *J. Photopolym. Sci. Technol.*, (2001), **14**, 139.
939. J. A. Macko and H. Ishida, *Polymer*, (2001), **42**, 6317.
940. J. A. Macko and H. Ishida, *Macromol. Chem. Phys.*, (2001), **202**, 2351.
941. T. J. Turton and J. R. White, *Plast. Rubb. Compos.*, (2001), **30**, 175.
942. L. ma, Y. Gao, D. Wu, B. Wang and B. Zhang, *Huaxue Tongbao*, (2001), **5**, 285.
943. K. tada and M. Onoda, *Thin Solid Films*, (2001), **393**, 358.
944. N.A. Weir and M. Delaney-Luu, *Eur. Polym. J.*, (2001), **37**, 1339.
945. A. Tidjani and C. A. Wilkie, *Polym. Deg. Stabil.*, (2001), **74**, 33.
946. Y. Zheng, F. M. Andreopolous, M. Micic, Q. Huo, S. M. Pham and R. M. Leblanc, *Adv. Funct. Mater.*, (2001), **11**, 37.
947. S. H. Bassmann, E. Oliveros, S. Gob, M. Kantor, A. Goppert, A. M. Braun, L. Lei and P. L. Yue, *Prog. React. Kinetics Mech.*, (2001), **26**, 113.
948. S. Morlat and J. L. Gardette, *Polymer*, (2001), **42**, 6071.
949. B. Claude, L. Gonon and J. L. Gardette, *Polymer Testing*, (2001), **20**, 771.
950. S. Commereuc, L. Gonon and V. Verney, *Int. J. Polym. Anal. Charact.*, (2000), **6**, 59.
951. M. Thouvenin, I. Linossier, O. Sire, J. J. Peron and K. Vallee-Rehel, *Macromolecules*, (2002), **35**, 489.
952. D. Sek, E. Schab-Balcerack, e. Grabiec and M. J. M. Abadie, *Mater. Res. Innovations*, (2001), **4**, 93.
953. M. Oh-e, D. Kim and Y. R. Shen, *J. Chem. Phys.*, (2001), **115**, 5582.
954. Y. Ouldmetidji, L. Gonon, S. Commereuc and V. Verney, *Polym. Test.*, (2001), **20**, 765.
955. Y. Ke, J. Bei, W. Ze and S. Wang, *Gaofenzi Xuebao*, (2001), **1**, 73.
956. M. Nowakowska, M. Kepczyriski and K. Szczubialka, *Pure Appl. Chem.*, (2001), **73**, 491.
957. V. Chisari, A. Pollicino, A. Recca and v. Siracusa, *J. Polym. Eng.*, (2000), **20**, 191.
958. A. S. Dzhafarov, *Azerb. Khom. Zh.*, (1998), **2, 51.**
959. T. Steele and L. Davis, *Annu. Tech. Conf.-Soc. Plast. Eng.*, (2000), *58th*, 1409.
960. H. H. Chin and J. H. Botkin, *Annu. Tech. Conf.-Soc. Plast. Eng.*, (2000), *58th*, 2673.
961. O. Kuvshinnikova, R. E. Lee and N. A. Favstritsky, *Annu. Tech. Conf.-Soc. Plast. Eng.*, (2000), 58th, 2836.

962. N. J. Earhart, *Adhes. Age*, (2000), **43**, 40.
963. P. Gijsman, *Polym. Prep. (Am. Chem. Soc., Div. Polym. Chem.)*, (2001), **42**, 386.
964. L. Davis, *Annu. Tech. Conf.-Soc. Plast. Eng.*, (2000), *58th*, 3281.
965. E. V. Germashev, V. E. Derbisher, Y. L. Zotov, M. N. Tsapleva, E. V. Konnova and P. M. Vasil'ev, *Plasticheskie Massy*, (2001), **7**, 36.
966. J. E. Pickett, *Polym. Prep. (Am. Chem. Soc., Div. Polym. Chem.)*, (2001), **42**, 344.
967. Anon, *Plast. Addit. Comp.*, (2000), **2**, 18.
968. J. E. Pickett and S. Nespurek, *Environmen. Sci. Poll. Contr. Serv.*, (2000), **21**, 163.
969. I. Vulic, J. Stretanski and B. Sanders, *Polym. Polym. Compos.*, (2000), **8**, 529.
970. D. Megert, *Polym. Paint Col. J.*, (2000), **190**, 9.
971. L. Liu, *Tanxingti*, (2001), **11**, 13.
972. A. Wegmann, *Res. Discl.*, (2001), **445**, P732.
973. J. M. Eng, L. H. Davis and I. Vulic, *Polyp. Text. World Cong.*, (2000), **295.**
974. J. E. Pckett, *Plast. Coat.*, (2001), 73.
975. R. Ravichandran and R. Iyengar, *Plast. Coat.*, (2001), 57.
976. J. Pospisil and S. Nespurek, *Prog. Polym. Sci.*, (2000), **25**, 1261.
977. L. M. Matuana, D. P. Kamdem and J. Zhang, *J. Appl. Polym. Sci.*, (2001), **80**, 1943.
978. L. Rosik, *Plasty Kauc.*, (2000), **37**, 356.
979. D. Rogez, *Verfkroniek*, (2000), **73**, 23.
980. V. Godard, M. C. Triboulot, A. Merlin and X. Deglise, *Pint. Acabados Ind.*, (2000), **42**, 36.
981. H. Turkulin, V. Jirous-Rajkovic and A. Bogner, *Surf. Prop. Durability Exter. Wood Buil. Compos. Int. Conf.*, (1999), **3**, 3/1.
982. X. Wang and L. Ning, *Beijing Huagong Daxue Xuebao*, (2000), **27**, 17.
983. Y. Ye, F. Liu, F. Bai, Q. Wang and J. Kong, *Zhongguo Suliao*, (1999), **13**, 74.
984. V. G. Siromyatnikov, O. Yu. Kolendo, I. O. Savchenko, L. O. Vretik, L. P. Paskal, O. V. Demchenko, E. G. Bezdushna and K. Golets, *Ukr. Khim. Zh.*, (2000), **66**, 41.
985. R. P. Singh, A. V. Prasad and J. K. Pandey, *Macromol. Chem. Phys.*, (2001), **202**, 672.
986. H. L. Wang, C. Wang and W. X. Chen, *Yingyong Huaxue*, (2001), **18**, 548.
987. J. L. Gerlock, A. V. Kucherov and C. A. Smith, *Polym. Deg. Stabil.*, (2001), **73**, 201.
988. H. P. haerri, D. Haenzi and J. J. Donze, *Melliand Textilber*, (2001), **82**, E16.
989. Ch. K. Rasulov, F. A. Nabiev, L. B. Zeinalov and R. K. Babaeva, *Aerb. Khim. Zh.*, (2000), **2**, 52.
990. P. Pipiraite, *Leit. Fiz. Z.*, (2000), **40**, 291.
991. I. V. Savenkova, *Izv. Vyssh. Uchebn. Zaved., Khim. Khim. Teknol.*, (2000), **43**, 59.
992. M. Lin and C. Le, *Polym. Prep. (Am. Chem. Soc., Div. Polym. Chem.)*, (2001), **42**, 3.
993. J. R. Pauquet and A. G. Oertli, *Chem. Fibres Int.*, (2000), **50**, 377.
994. F. Gugumus and N. Lelli, *Polym. Deg. Stabil.*, (2001), **72**, 407.
995. N. Haider and S. Karlsson, *Polym. Deg. Stabil.*, (2001), **74**, 103.
996. S. M. Desai, J. K. Pandey and R. P. Singh, *Macromol. Symp.*, (2001), **169**, 121.
997. P. Franchi, M. Lucarini, G. F. Pedulli, M. Bonora and M. Vitale, *Macromol. Chem.*, (2001), **202**, 1246.
998. F. Gugumus, *Polymer Deg. Stabil.*, (2002), **75**, 295.
999. F. Gugumus, *Polymer Deg. Stabil.*, (2002), **75**, 309.
1000. M. P. Luda, G. camino, E. Laurenti, S. Novelli, T. Temtchenko and S. Turri, *Polym. Deg. Stabil.*, (2001), **73**, 387.
1001. C. A. Smith, J. L. Gerlock and R. O. Carter, *Polym. Deg. Stabil.*, (2001), **72**, 89.
1002. J. L. Gerlock, A. V. Kucherov and M. E. Nicholas, *J. Coat. Technol.*, (2001), **73**, 45.
1003. C. Decker, K. Zahouily and A. Valet, *Macromol. Mater. Eng.*, (2001), **286**, 5.

1004. C. Decker, *Polym. Prep. (Am. Chem. Soc., Div. Polym. Chem.)*, (2001), **42**, 858.
1005. C. R. Beaton and D. S. Argyropoulos, *Photochem. Photobiol.*, (2001), **73**, 605.
1006. M. Edge, P. Faulds, A. McMahon, G. C. Ranger and D. Turner, *Eur. Polym. J.*, (2001), **37**, 349.
1007. G. J. M. Fechine, M. S. RAbello and R. M. Souto-Maior, *Polym. Deg. Stabil.*, (2002), **75**, 153.
1008. J. Lemaire, N. Siampiringue, R. Chaigneau, P. Delprat and G. Parmeland, *Polym. Prep. (Am. Chem. Soc., Div. Polym. Chem.), (2001)*, 42, *856.*
1009. R. janes, M. Edge, J. Rigby, D. Mourelatou and N. S. Allen, *Dyes Pigments*, (2001), **48**, 29.
1010. H. Tokoro and S. Sake, *Mater. Sci. Res. Int.*, (2001), **7**, 132.
1011. S. Sawano, *Kogyo Toso*, (2000), **164**, 10.
1012. K. Takahama, *Kino Zairyo*, (2001), **21**, 23.
1013. M. Morikawa and M. Takahashi, *Ehime-ken Kogyokei Shiken Kenkyu Kikan Kenkyu Hokoku*, (2000), **38**, 91.
1014. S. P. Hudson, M. Shirkhan and C. A. Pickles, *Trans. Inst. Min. Metal., Sect. C.*, (2000), **109**, C137.
1015. T. Yasunaga and E. Iwamura, *KOBELCO Technology Review*, (2001), **24**, 18.
1016. M. G. H. Terrones, A. E. H. Machado, R. Ruggiero and A. Castellan, *Braz. Symp. Chem. Lignins Other Wood Compos., Proc., 6th, meeting date 1999*, 352.
1017. Y. Yu and C. H. Langford, *Langmuir*, **17**, 897.
1018. P. Peralta-Zamora, G. C. A. K. Gouvea, F. Wypych, S. G. Moraes and N. Duran, *Braz. Symp. Chem. Lignins Other Wood Compos., Proc., 6th, meeting date 1999*, 136.
1019. S. Sakthivel, B. Neppolian, M. Palanlchamy, B. Arabindo and V. Murugesan, *Water Sci. Technol.*, (2001), **44**, 211.
1020. J. Lemaire, N. Siampiringue and D. Fromageot, *Caoutch. Plast.*, (2001), **78**, 32.
1021. P. Anna, Gy. Marosi, I. Ravadits and M. A. Maatoug, *Polym. Deg. Stabil.*, (2001), **73**, 463.
1022. H. Oda, *Color Technol.*, (2110), **117**, 204.
1023. M. A. Caine, R. W. McCabe, L. Wang, R. G. Brown and J. D. Hepworth, *Dyes Pigments*, (2002) **52**, 55.
1024. H. Oda, *Dyes Pigments*, (2001), **48**, 233.
1025. H. Oda, *Dyes Pigments*, (2001), **48**, 151.
1026. L. C. Abbott, P. MacFaul, L. Jansen, J. Oakes, J. R. L. Smith and J. N. Moore, *Dyes Pigments*, (2001), **48**, 49.
1027. M. D. Teli and U. Sayed, *Book Pap.-Int.Conf. Exh.*, (1999), 340.
1028. D. rastogi, K. Sen and M. Gulrajani, *Colour. Technol.*, (2001), **117**, 193.
1029. T. Takagishi and N. Katsuda, *Book Pap.-Int. Conf. Exh.*, (1999), 358.
1030. I. Baumberg, O. Berezin, A. Dradkin, B. Gorelik, L. Kogan, M. Voskobojnik and M. Zaidman, *Polym. Deg. Stabil.*, (20001), **73**, 403.
1031. T. Caronna, F. Fontana, B. Marcandalli and E. Selli, *Dyes Pigments*, (2001), **49**, 127.
1032. S. Yang, H. Tan, H. Xiao, X. Shang, X. Gong, S. Yao and K. Chen, *Dyes Pigments*, (2001), **49**, 93.
1033. T. Hihara, Y. Okada and Z. Morita, *Dyes Pigments*, (2001), **50**, 185.